AF529466

WASTEWATER
Treatment Technologies and
ENVIRONMENT

WASTEWATER
Treatment Technologies and ENVIRONMENT

— Vol. 1 —

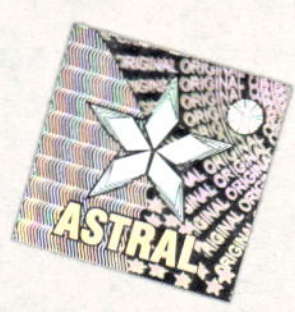

— Editors —

Dr. S. N. Kaul

Dr. Lidia Szpyrkowicz

Professor (Dr.) Arvind Kumar

— Assocaite Editors —

N. N. Rao

D. M. Dharmadhikari

National Environmental Engineering Research Institute

Nagpur - 440 020

2004

DAYA PUBLISHING HOUSE

Delhi - 110 035

ISBN 81-7035-342-4 (Set)
ISBN 81-7035-343-2 (Vol. 1)
ISBN 81-7035-344-0 (Vol. 2)

Published by : **Daya Publishing House**
1123/74, Deva Ram Park
Tri Nagar, Delhi - 110 035
Phone : 27383999
Fax : (011) 27382329
e-mail : dayabooks@vsnl.com
website : www.dayabooks.com

Showroom : 4762-63/23, Ansari Road, Darya Ganj
New Delhi - 110 002
Phone: 23245578, 23244987

Laser Typesetting : **Classic Computer Services**
Delhi - 110 035

Printed at : **Chawla Offset Printers**
Delhi - 110 052

PRINTED IN INDIA

Preface

The past few years have seen the emergence of a growing, widespread desire in this country, and indeed everywhere, that positive actions be taken to restore the quality of our environment, and to protect it from the degrading effects of all forms of pollution - air, noise, solid waste, and water. Since pollution is a direct or indirect consequence of waste, if there is no waste, there can be no pollution, and the seemingly idealistic demand for "zero discharge" can be construed as a demand for zero waste. However, as long as there is waste, we can only attempt to abate the consequent pollution by converting it to a less noxious form. In those instances, in which a particular type of pollution has been recognized, three major questions usually arise:

1. How serious is the pollution?
2. Is the technology to abate it available? and
3. Do the costs of abatement justify the degree of abatement achieved ?

The principal intention of this book on *Wastewater Treatment Technologies and Environment* is to help in formulating useful answers to the second and third of these questions, and to examine their costs in the light of the real level of benefits afforded.

The traditional approach of applying tried-and-true solutions to specific pollution problems has been a major factor contributing to the success of environmental engineering, and in large measure has accounted for the establishment of a methodology of pollution control. However, realizing the already great complexity of current environmental problems, and understanding that, as time goes on, these issues will become even more complex and interrelated; render it imperative that intelligent planning of pollution abatement systems must be undertaken. Prerequisite to such planning is an understanding of the performance, potential, and limitations of the various methods of pollution abatement available for environmental engineering. We believe that the unification of the concepts and engineering methodology found in this book is a logical step in the evolution of environmental engineering, particularly for wastewater treatment.

The treatment of various engineering systems presented will show how an engineering formulation of the subject flows naturally from the fundamental principles and theory of chemistry, physics and mathematics. This emphasis on fundamental science is based on the recognition that engineering practice has of necessity in recent years become more firmly based of scientific principles, rather than depending largely on our empirical accumulation of facts, as was earlier the case. It was not intended, though, to neglect empiricism when such data lead quickly to the most economic design; certain engineering systems are not readily amenable to fundamental scientific analysis, and in these instances we have resorted to less science in favour of more art and empiricism.

An engineer must understand science within a context of applications. It is, therefore, necessary to equip oneself with promising potential technologies. The book on *Wastewater Treatment Technologies and Environment* is the outcome of scientific collaboration between National Environmental Engineering Research Institute (NEERI, India); and Department of Environmental Sciences, University of Venice, Venice, Italy; under 'Bilateral Programme of CSIR, India and CNR, Italy Collaborative Project'.

The editors are pleased to acknowledge the encouragement and support received from their colleagues at NEERI and Department of Environmental Sciences, University of Venice, Venice, Italy, during the conceptual stages. The editors wish to thank the contributing authors for their time and effort, and for having borne patiently our numerous queries and comments. Last but not the least, the editors appreciate the meticulous efforts rendered by Mr. G. B. R. Sharma, Mr. C. R. Bakre and Mr. P. A. Ingle in the preparation and production of this book.

One of the authors, Professor (Dr) Arvind Kumar wishes to express his sincere gratitude to Professor Indu Dhan, Hon'ble Vice Chancellor and Professor Binodanand Jha, Hon'ble Pro-Vice Chancellor of S. K. University, Dumka who have been the source of constant inspiration. He is also thankful to his wife, Professor Kumari Bimla and sons, Kumar Pallav Ramachandran and Kumar Prasun Ramakrishnan whose natural smiles extended him relief from this tiresome endeavour.

It is my sincere duty to express my heartiest thankfulness to Shri Anil Mittal, Proprietor, Daya Publishing House, Delhi for taking keen interest in publishing of this book.

Dr. S. N. Kaul

Dr. Lidia Szpyrkowicz

Professor (Dr.) Arvind Kumar

Contents

— VOLUME 1 —

1

Kinetic and Hydrodynamic Aspects of the Removal of Pollutants from Wastewater in Electrochemical Reactors

LIDIA SZPYRKOWICZ

Department of Environmental Sciences, University of Venice, Dorsoduro 2137, 30123 Venice, Italy

Introduction

Research on the use of electrochemical methods for the treatment of industrial wastewater has demonstrated that many pollutants, organic and inorganic, can be effectively removed by anodic oxidation or cathodic reduction. Electrochemical reactors have been successfully used since the early '60s for the recovery of pollutants from metal plating factories, for silver recovery from photographic baths and in other processes of metal recovery (Rajeshwar *et al.*, 1994). Among various applications of electrochemical processes, those in which oxidation of pollutants can be achieved is particularly important, as it can offer an alternative to traditional chemical oxidation. Thus electrochemical processes have been considered recently for other environmental applications, ideally requiring total combustion of organic pollutants to CO_2 and H_2O or modification of their structure, resulting in improved biodegradability (Comninellis, 1992; Simond *et al.*, 1997; Foti *et al.*, 1999). Application of electrochemical reactors proved feasible for the purification of wastewater from tanning industry (Naumczyk *et al.*, 1996), of wastewater containing cyanides (Ho *et al.*, 1989), phenols (Comninellis and Negrini, 1995), chlorophenols (Polcaro and Palmas, 1997), wastewater containing dyes (Szpyrkowicz *et al.*, 2001a) and deriving from dye production (Abdo and Al-Ameeri, 1987) and others (Murphy *et al.*, 1982). It results that the composition of the wastewater, and in particular the type of pre-treatment it had undergone before electro-oxidation, can influence strongly the performance of electrochemical oxidation. For example studies conducted at the University of Venice (Szpyrkowicz *et al.*, 1994, Szpyrkowicz *et al.*, 1995) with graphite, Ti/Pt, Ti/Pt-Ir, Ti/RuO_x, Ti/PbO_2 and other anode materials proved that purifying tannery wastewater by an electrochemical process can be efficient, particularly if the process is applied for the pre-treated effluents. The rate of removal of a single polluting species proved to be a function of the electrocatalytic properties of the anode material.

In fact, the type of the oxidative process involved in the removal of a pollutant is a function of the material of the anode and application of different anodic materials with different electrocatalytic properties can affect reactor treatment efficiencies. For electrochemical processes occurring via a direct reaction of the pollutant at the electrode surface a term *direct electrolysis* is generally used. The term *indirect electrolysis* refers to those processes in which a redox couple, whose oxidised form is being continuously renewed by an anodic heterogeneous reaction, is involved in a homogeneous phase reaction with the pollutant. The reactors which operate by applying direct or indirect electrochemical oxidation (or reduction) are influenced by stirring in a completely different way.

Electrochemical Reactors and Direct Electro-oxidation

Fig. 1 explains the concept of a direct electrochemical reaction applied for the pollutant abatement. The graph shows the anodic process as an example.

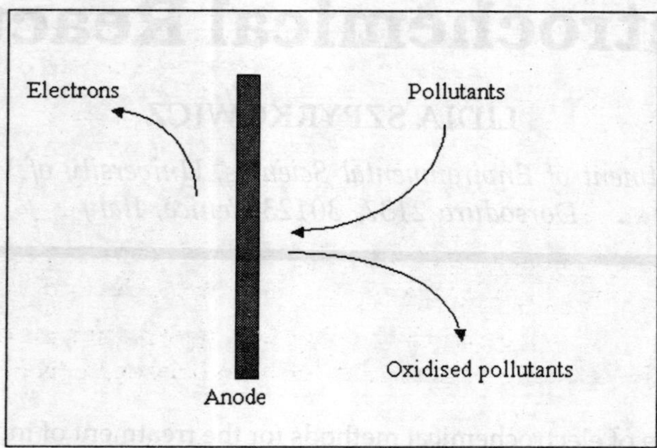

Fig. 1: The Scheme of the Removal of Pollutants by Direct Electro-oxidation

As can be seen oxidation occurs directly on an electrode by the application of appropriate potential, without the involvement of other substances. The interplay of electron transfer and mass transfer is fundamental: mass transport and the electron transfer are linked in series. At low overpotentials transport rates are normally fast compared with the electrochemical reaction kinetics and electron transfer is the rate determining step [see Fig. 2 (Pletcher and Walsh, 1993)]. Current density (reaction rate) increases exponentially with overpotential, which leads to the reactant being progressively depleted at the interface and a concentration gradient in the boundary layer is developed. As an overpotential increases, the rates of electron transfer and mass transport are comparable and both processes control the rate of the overall reaction (the region of the mixed control). As the overpotential increases further, the surface concentration of electro-active species tends to zero and the current density becomes independent of the potential; the reaction rate is wholly determined by the mass transport rate.

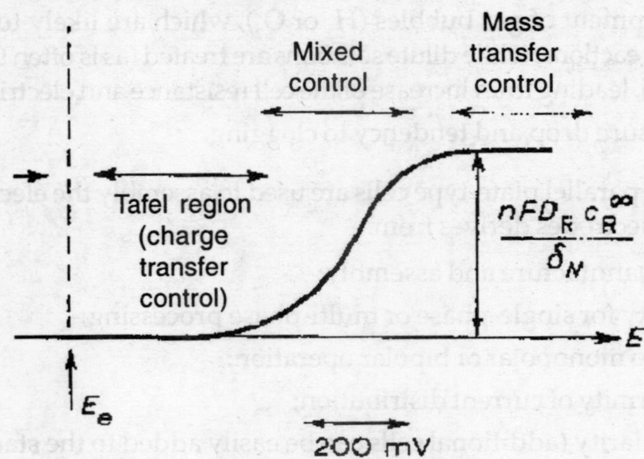

Fig. 2: A Current-potential Curve for the Positive Region

Under the conditions of the complete mass transport control, assuming that migration of the active species in the electric field is negligible (an assumption valid when the concentration of the base electrolyte is high), the limiting current i_L expresses the duty of the reactor:

$$i_L = k_L AnFc^{\infty}$$

where

i_L-limiting current (A); A-electrode surface area (m^2); n-number of participating electrons; F-Faraday constant (96490 C mol^{-1}); $c\infty$-bulk concentration of the reacting species (mol m^{-3}).

Thus, assuming the reactor being a batch or a plug-flow type reactor, the observed change of the concentration of the reactive species, obtained from the mass balance, is given by:

$$\frac{dc_t}{dt} = i_t /(nFV_R)$$

where

i_t is the instantaneous current at time t; V_R is the working volume of the reactor.

Under the complete mass transport control $i_t = i_L = k_L AnFc_t$. Substituting, the following expression is obtained:

$$\frac{dc_t}{dt} = k_t Ac_t / V_R$$

From the last equation the importance of the mass transfer and of the highly developed electrode surface area in electrochemical reactors operating with direct electrochemical oxidation of pollutants is evident. One way to fulfil this requirement is to operate with three-dimensional electrodes. This solution, however, has also several drawbacks, *e.g.*:

1. The difficulty to obtain a uniform potential and current distribution;
2. An easy entrapment of gas bubbles (H_2 or O_2), which are likely to evolve simultaneously with the main reactions, while dilute solutions are treated (as is often the case during treatment of wastewater), leading to an increase of the cell resistance and electrical energy consumption;
3. The high pressure drop and tendency to clogging.

Consequently often parallel plate-type cells are used to assembly the electrochemical reactor. The preference to use plate electrodes derives from:

1. Ease of their manufacture and assembly;
2. Their versatility for single-phase or multi-phase processing;
3. Adaptability to monopolar or bipolar operation;
4. Relative uniformity of current distribution;
5. Flexible modularity (additional cells can be easily added to the stack);
6. Facility to incorporate membranes, if required.

Current/mass transfer correlations for the reactors with parallel plate electrodes can be defined from the first principles. The general form of this correlation is:

$$Sh = a\,\mathrm{Re}^{b}\,Sc^{c}\,L^{d}$$

where

Sh-Sherwood number ($Sh = k_L d_e/D$);

Re-Reynolds number ($Re = vd_e/H$); Sc-Schmidt number ($Sc = H/D$);

L-length of the electrode in direction of flow (m).

The symbols in the dimensionless groups are: d_e-equivalent diameter of flow channel (m); D-diffusion coefficient of electroactive species ($m^2\ s^{-1}$); v-mean linear velocity of electrolyte ($m\ s^{-1}$); H-kinematic viscosity of the electrolyte ($m^2\ s^{-1}$).

The *a*, *b*, *c* and *d* are coefficients, whose values differ in function of the hydrodynamic regime and the cell geometry.

It can be shown (Pletcher and Walsh, 1993) that the limiting current and the Sherwood number are related via the mass transfer coefficient, so we have:

$$Sh = i_L d_e / nFADc^{\infty}$$

As a direct consequence of this relation, the duty of the parallel plate reactor is supposed to increase exponentially with turbulence. In particular, for the plug-flow reactor the duty would be a function of the volumetric flow rate of electrolyte, with the exponent equal to *b*.

Table 1 reports some values of the coefficients of the mass transport correlation for electrolyte flow past a flat plate in a rectangular channel (Picket, 1979).

Table 1: Mass Transport Correlation Coefficients for Electrolyte Flow Past a Flat Plate in a Rectangular Channel

$Sh=aRe^bSc^cLe^d$				*Conditions*
a	*b*	*c*	*d*	
2.54	0.3	0.3	0.3	Fully developed laminar flow; $58<Re<2000$; $2850<Sc<5140$ $0.17<L_e<12.5$; S/B 00.167 or 0.75
1.85	1/3	1/3	1/3	Theoretically derived for fully developed laminar flow assuming infinitely wide electrodes; $ReSc\ (d_e/L)>10^4$ and $L_e<35$
1.85x	1/3	1/3	1/3	x corrects the correlation for a finite width between the electrodes. $ReSc\ (d_e L)>10^4$ and $L_e<35$
0.023	0.8	0.3	-	Fully developed turbulent flow; $2000<Re<20000$ and $L_e>12.5$
0.145	2/3	1/3	1/4	Fully developed turbulent flow. $2000<Re<20000$ and $L_e<7.5$

The following graph (Fig. 3) gives an example of the performance of the eletrochemical batch reactor with parallel plate anodes operating with a complete recycle under different volumetric flow rate conditions (Ricci, 2002).

Recently an increase of the number of commercial electrochemical reactors based upon the parallel plate geometry is witnessed. Particularly the filterpress arrangement achieves great attention (Picket and Wilson, 1982; Schwager *et al.*, 1980; Stork and Hutin, 1981; Trinidad and Walsh, 1996), due to a possibility of realising high mass transfer rates, enhanced by an implementation of the promoters of turbulence. The enhancement of the mass transport rate can be of a factor of two, in function of the geometry of the promoters (Ralph *et al.*, 1996). However, whilst the expanded plastic mesh turbulence promoters do provide a simple and cost effective method of enhancing mass transport, they increase

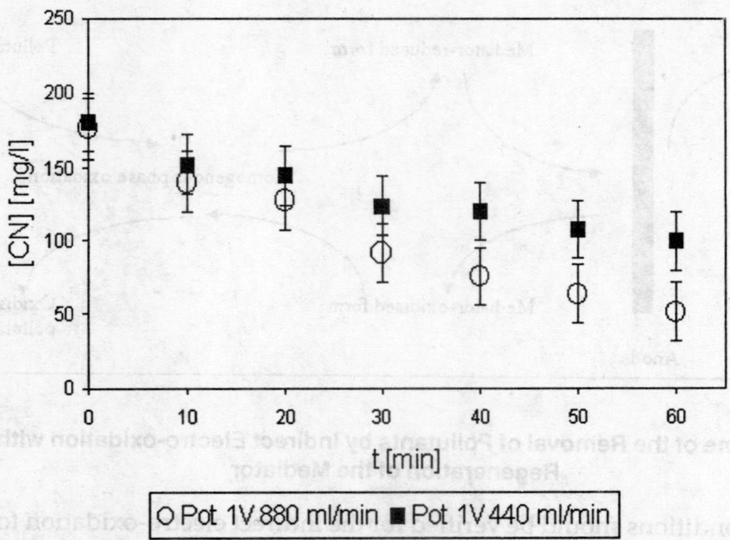

Fig. 3: Time Trend of Cyanide Concentration During Electrolysis of Copper Complex Cyanide Solution

the pressure drop over the reactor, thereby increasing the pumping costs, and can shield much of the electrode surface producing a non-uniform potential distribution. Furthermore they can also give rise to gas blinding and cause flow segregation. As a consequence the mass transfer rates can even be decreased with respect to those for the unmodified channels, as reported by Wragg and Leontaritis (1997).

If the electrodes working potentials are close to those of water electrolysis, differences in local current/voltage distributions, existing over the electrode surface area, can result in gas evolution (O_2 and H_2). This, in turn, can produce a modification of the hydrodynamic boundary layer at the electrode surface, as the rising bubbles generate radial momentum, which locally destroys the layer and induces mass renewal by bringing fresh solution to the edge of the hydrodynamic boundary layer.

Electrochemical Reactors and Indirect Electro-oxidation

In the process of indirect electro-oxidation of pollutants an electrochemically generated redox reagent is used as a chemical reagent (or catalyst) to convert pollutants to less harmful products, or even to CO_2 and H_2O. The redox reagent acts as an intermediator to transport electrons between the pollutant and the electrode. The general scheme of this reaction is:

$$C \rightarrow C^+ + e^-$$

$$C^+ + R \rightarrow O + C$$

where

C-mediator, R-pollutant in the reduced form, O-pollutant in the oxidised form.

Fig. 4 depicts schematically the electrode and bulk reactions involved.

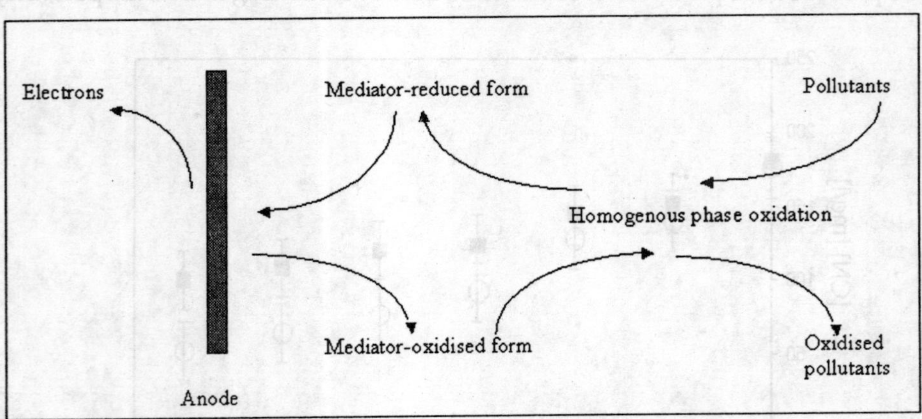

Fig. 4: The Scheme of the Removal of Pollutants by Indirect Electro-oxidation with an Anodic Regeneration of the Mediator

The following conditions should be verified for the indirect electro-oxidation to proceed with a high efficiency:

1. The rate of production of the mediator should be high;

2. the rate of the reaction of the mediator with a given pollutant should be higher than the rate with other compounds eventually present;
3. the potential at which electrogeneration of the mediator occurs should be distant (lower) from the potential of solvent electrolysis, to prevent lost of energy for an unproductive process;
4. the electrode should not be prone to adsorption of the pollutant, to avoid occupation of the active sites, otherwise free for the reaction of the generation of the active form of the mediator.

In fact, indirect electrochemical processes offer a promising alternative solution for destruction of pollutants in industrial wastewaters. They can be particularly attractive if conducted in undivided electrochemical reactors, which can operate at lower cell voltages, and are simpler and cheaper to engineer than their counterparts incorporating ion-permeable membranes, separating anolyte and catholyte. Especially for treatment of low concentrations of pollutants in industrial wastewaters, indirect electrochemical oxidation, mediated in bulk solution by a redox couple with electrochemical regeneration of its oxidised form, is generally preferred to direct electro-oxidation. The most often applied mediators are listed in Table 2 (Rajeshwar *et al.*, 1994).

Table 2: Redox Mediators Applied in Indirect Electro-oxidation of Pollutants and their Standard Potentials

Mediator couple	*Standard Reduction Potential (vs SHE)*
Ag (I/II)	1.98
Co (II/III)	1.82
Ce (IV/III)	1.44
Cl (0/-I)	1.36
Fe (II/III)	0.77

Among the redox couples indicated in Table 2 the chlorine species are of particular importance, due to the presence of chloride in several types of industrial wastewater and the lack of need for its recovery from the purified effluents. This last advantage makes this mediator particularly competitive with Ag and Co and offers a possibility to conduct the process at low operation costs.

Indirect electro-oxidation of pollutants in wastewater by electrogenerated chlorine has been used widely (Szpyrkowicz *et al.*, 1994; Szpyrkowicz *et al*, 1995; Do and Yeh, 1995; Comninellis and Nerini, 1995, Bonfanti *et al.*, 2000), because of the low cost of chloride and the relatively high solubility (ca. 0.1 M-Boxall and Kelsall, 1992) and strongly oxidising properties of active chlorine.

The influence of hydrodynamic conditions in a mediated electro-oxidation of pollutants is much more complex than during direct electro-oxidation. The dependence of the reaction rates on stirring is not straightforward and cannot be obtained from the first principles.

Electrochemical reactors, which apply the indirect electro-oxidation via a Cl_2/Cl^- couple, operate at potentials at which two reactions generally can occur at the anode:

$$2H_2O \rightarrow O_2 + 4H^+ + 4e^-$$

$$2Cl^- \rightarrow Cl_2 + 2e^-$$

Whereas aqueous chlorine predominates at pH < *ca.* 3.3, at higher bulk pHs, its diffusion away from the electrode is coupled to its hydrolytic disproportionation to form chloric (I) acid (*HOCl*) at pH < *ca.* 7.5 and chloric (I) ions at pH > *ca.* 7.5:

$$Cl_2 + H_2O \rightarrow HOCl + H^+ + Cl^-$$

$$HOCl \rightarrow H^+ + ClO^-$$

Chlorine, chloric (I) acid and chloric (I) ions are collectively often referred to as "active chlorine".

The cathodic reactions involve hydrogen evolution by:

$$2H^+ + 2e^- \rightarrow H_2$$

and/or

$$2H_2O + 2e^- \rightarrow H_2 + 2OH^-$$

The rate of chlorine evolution reaction may be kinetically or transport controlled, depending on the applied current density, chloride ion concentration, and the prevailing hydrodynamic conditions, which may also depend on the rate of anodic gas evolution; hence, these parameters also control the current efficiency of chlorine evolution. For the given experimental conditions the limiting current density can be estimated from the equation:

$$j_L = nFk_L C_{Cl}$$

where

n-number of the electrons per Cl^- oxidation (n = 1), F – the Faraday's constant (96 485 C); k_L – mass transfer coefficient for chloride ions (approximately = 10^{-5} m s^{-1}), C_{Cl} – concentration of chloride ions (mol m^{-3}).

Hydrogen evolution generates a high local pH at the cathode, so that chloric (I) ions predominate in the cathode diffusion layer; in an undivided reactor they can be reduced by the (mass transport controlled) cathodic reaction:

$$ClO^- + H_2O + 2e^- \rightarrow Cl^- + 2OH^-$$

This reaction limits the maximum active chlorine concentration that can be achieved in such reactors to *ca.* 0.1 M (Boxal and Kelsall, 1992). In addition, further loss reactions such as chlorate formation may occur in the bulk solution, but can be assumed to have insignificant rates at ambient temperatures, because the steady state chloric (I) concentration is kept low by homogeneous reactions of oxidation of pollutants.

Chlorate may also be formed by oxidation of chloric (I) acid (Boxall and Kelsall, 1992; Czarnecki and Janssen, 1992), mass transport usually controlling the rate, depending on the anode material:

$$6HOCl + 3H_2O \rightarrow 2ClO_3^- + 4Cl^- + 12H^+ + 3/2O_2 + 6e^-$$

As this last reaction and also the oxygen evolution reaction acidify the anode diffusion/reaction layer, a large pH gradient exists across the anode-cathode gap, with possible kinetic consequences for oxidation of pollutants, as the reduction kinetics of active chlorine species decrease with increasing pH.

The speciation of the different oxidising species generated during the electrolysis and their profiles inside the reactor are function of the bulk solution pH, as is shown in Fig. 5 (a) & (b) for different pHs of the bulk solution.

The oxygen evolution reaction is charge transfer controlled, as is the chlorine evolution reaction at low current densities and high chloride concentrations, but the chloric (I) loss reactions at the anode and the cathode are mass transfer controlled. Thus, increasing turbulence in the reactor promotes the transport of Cl^- ions to, and $C1_2$ (aq) away from, the anode, but also enhances the rate of chloric (I) reduction at the cathode and its oxidation at the anode.

As shown in Fig. 5 (a) & (b), oxidative decomposition of pollutants in the electrochemical reactor under alkaline and acidic conditions involves active chlorine as mediator, constantly renewed by charge transfer reaction at the anode, involving the mass transfer of the chloride ion from the bulk to the electrode, and the transport of the chloric (I) to the bulk. With several oxidisable pollutants present during the electrolysis, fast homogeneous reactions will affect the concentration profiles in the anode diffusion/reaction layer, enhancing the chloride flux to the anode, whereas slow reactions will occur predominantly in the bulk solution.

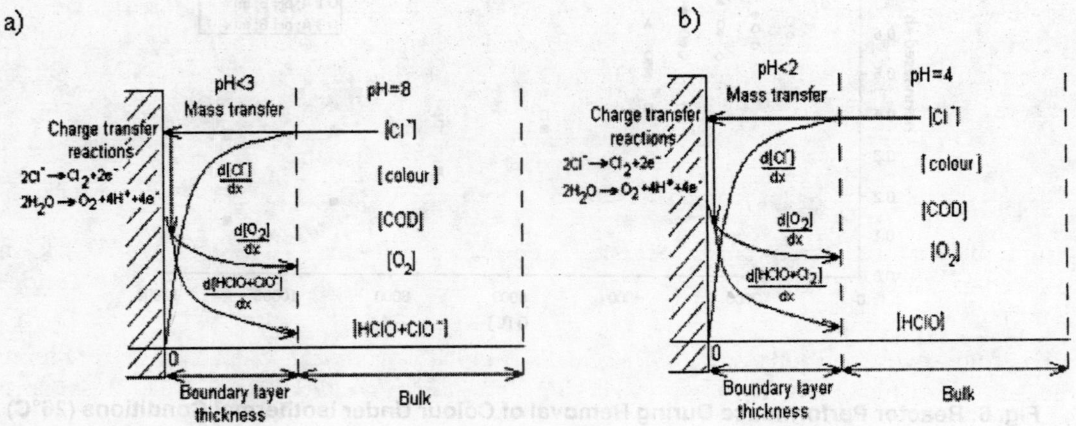

Fig. 5: Schematic Spatial Distribution of Oxidising Species and their Concentration Profiles Inside the Reactor Under Alkaline (a) and Acidic (b) Bulk pH

A mass balance on active chlorine species, formed in the solution of a batch electrochemical reactor of volume V with current efficiency Φ_e at anode of area A_a and mass transport rate coefficient $k_{m,a}$, but consumed by further oxidation of chloric (I) acid at the anode, by reduction of chloric (I) ions at the cathode of area A_c and mass transport rate constant $k_{m,c}$ and by chemical reaction with pollutant R and second order rate constant k, can be represented by:

$$V_R \frac{d['Cl_2']}{dt} = \frac{\Phi_e j_a A_a}{nF} - A_a k_{m,a}[HOCl] - k_{m,c} A_c [ClO^-] - Vk['Cl_2'][R]$$

The anodic current efficiency Φ_e depends on the anode current density j_a, $k_{m,a}$ and chloride concentration. Studies conducted in our laboratory (Szpyrkowicz *et al.*, 2001b), showed that in the absence of pollutants at pH 4, current efficiencies were only 0.1 – 0.2, essentially independent of stirring rate, due to the limited solubility of chlorine and hence its desorption from solution, which would be more restricted in the presence of oxidisable pollutants, when homogeneous phase reaction of their oxidation was operative. At the bulk pH = 8, in quasi-quiescent solutions, current efficiencies were *ca.* 0.65, essentially independent of time for the restricted compositional change produced in an hour. When the solution was agitated, the apparent efficiency of chlorine production decreased from the initial value of 0.84 to 0.50 after 60 minutes, due to enhancement of the rates of loss reactions with increasing active chlorine concentration.

Figures 6 and 7 show the effect of charge respectively, on the decay of colour and of COD, obtained under isothermal conditions (26 °C). The reactor performance was largely dependent on current

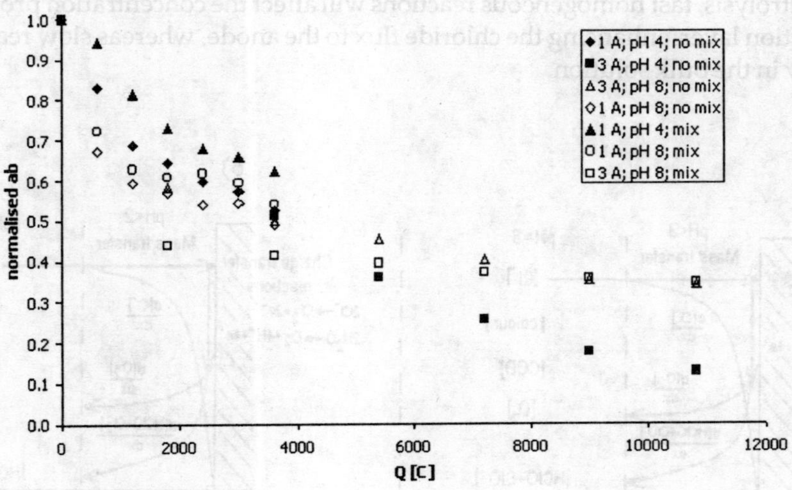

Fig. 6: Reactor Performance During Removal of Colour Under Isothermal Conditions (26°C)

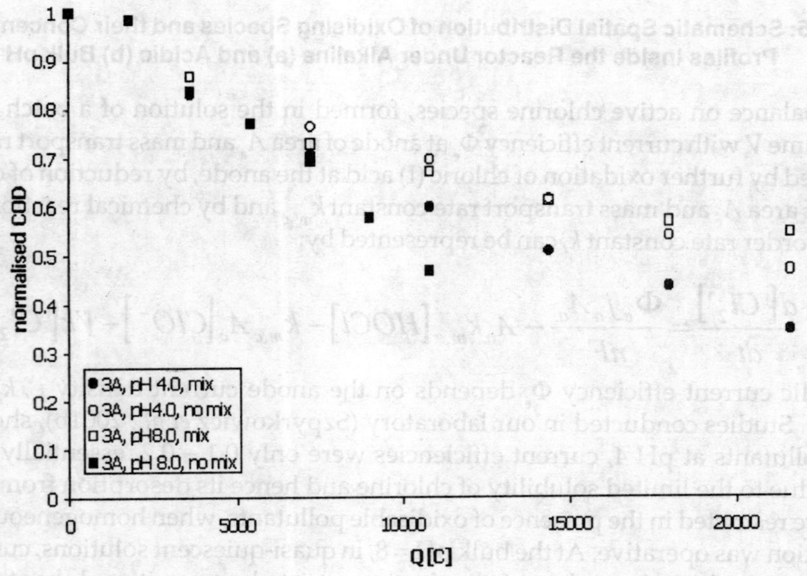

Fig. 7: Reactor Performance for Removal of COD Under Isothermal Conditions (26°C)

density and was influenced by solution stirring and pH; at low current density (1 A dm^{-2}), mixing appeared to have a detrimental effect, due to enhancement of the electrode loss reactions.

In principle, the behaviour of the reactor for such an electrochemical-chemical (EC) process could be modelled, the kinetics of the chemical reaction dictating where its occurs in relation to the electrode surface; fast chemical reactions affect the concentration profiles of mediator species within the electrode diffusion/reaction layer and hence, in this system, the mass transport limited rate of, and current efficiencies for, chloride oxidation (*e.g.* Scott, 1991). Figures 6 and 7 indicate that at least the initial (< 2 kC) rate of the removal of colour and COD could be described by a pseudo-first order reaction rate. The Hatta number was calculated from:

$$Ha = \frac{\sqrt{k_1 D_{Cl_2}}}{k_L}$$

where

k_1 – an observed pseudo-first order reaction rate constant (s^{-1}); k_L – mass transport rate coefficient of chlorine ($m\ s^{-1}$). The value of D_{Cl2} was assumed equal to $10^{-9}\ m^2\ s^{-1}$ (Hikita *et al.*, 1975) and that of k_L to $10^{-5}\ m\ s^{-1}$ (Lin and Zhang, 2001). For the specific case of electro-oxidation of disperse dyes (Szpyrkowicz *et al.*, 2001b) *Ha* was equal to *ca.* 0.1. This value of Ha implies that the chemical reaction occurred both in the electrode diffusion layer and in the bulk solution. At low conversions, increased agitation rates enhanced mass transport rates both of the chloride ions to the anode and chlorinated intermediates away from the anode. However, at longer times, increased adsorption densities of those intermediates tended to partially inhibit Cl_2 evolution, as evident from voltammetric tests shown in Figs. 8 (a) & (b). As a consequence, a sharp decrease in the decolorisation rate after a charge ca. 2 kC had been passed was observed, as shown in Fig. 6. Thus, it can be concluded that under alkaline conditions turbulence initially favours decolorisation, but as intermediates accumulate with time, the benefits of mixing decreases until no difference between quasi-quiescent and agitated conditions is observed.

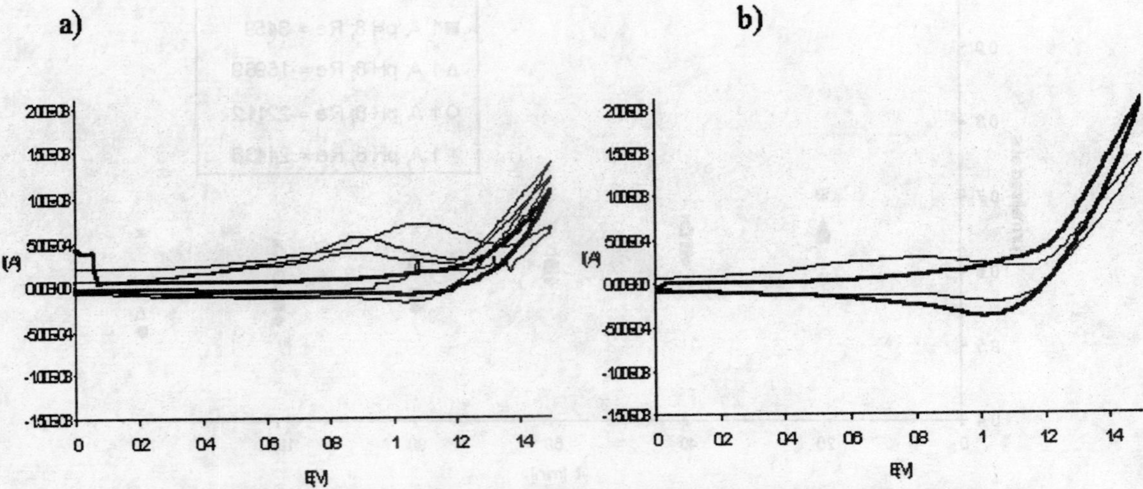

Fig. 8: Cyclic Voltammograms Recorded at the Ti/RhO_x-TiO_2 Electrode in a Solution Containing Disperse Dyes; Supporting Electrolyte 0.1 M $NaClO_4$ and 0.1 N NaCl; (a) pH 8; (b) pH 4.0; Scan Rate 100 mV s^{-1}

As shown in Fig. 7 removal of COD showed different behaviour, with the highest removal rate being observed under alkaline, quasi-quiescent conditions. From the performance of the reactor it can be concluded that the degree of stirring influences the selectivity of the reactor towards the removal of colour and COD. This can be explained by promotion of mineralisation of the intermediates by direct anodic oxidation, for which there was some voltammetric evidence (see Fig. 8). This mechanism can explain far better results obtained by electrochemical chlorine mediated oxidation in comparison to homogeneous phase chlorination under neutral or slightly alkaline conditions (Szpyrkowicz *et al.*, 2001a). It can be hypothesised that under acidic conditions, chlorine mediated oxidation of the dyes is acid catalysed, by analogy to other aromatic compounds (Rebenne *et al.*, 1996).

Figs. 9 and 10 illustrate (Szpyrkowicz *et al*, 2001b) the reactor performance during treatment of dyes-bearing wastewater, in terms of colour and COD removal, respectively, for five agitation rates and in the absence of temperature control, a more common and cheaper practical condition in a batch reactor than isothermal operation. The pH started from about 7, increased to about 9, then stabilised at about 8.5. Quasi-quiescent conditions produced the highest depletion rates of both the colour and COD. The rate of pollutant removal decreased with increasing Reynolds number. Electrolyte temperatures increased from 19° to 26° C at 100 A m^{-2} and 19° to 36° C at 300 A m^{-2}, due to Joule heating, the effect being greater in the absence of additional agitation. The increased temperature caused both the electrode and chemical reaction rates to rise. Hence, when the electrochemical reactor was operated under isoperibolic conditions, with chlorine mediated electro-oxidation being primarily responsible for the destruction of pollutants, it performed better without additional agitation.

Further studies on the influence of stirring during chlorine mediated electro-oxidation of dyes (Moro, 2002), showed that the magnitude of the effect of agitation in an isothermal reactor is comparable with the temperature effects. The pseudo-first order rate constant for electro-oxidation of the Procion

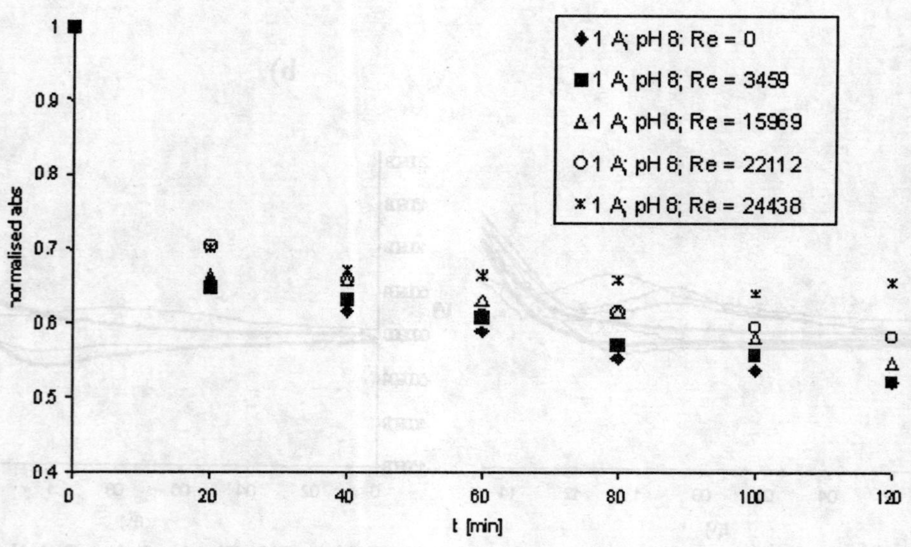

Fig. 9: Performance of the Reactor Under Isoperibolic Conditions During the Removal of Colour

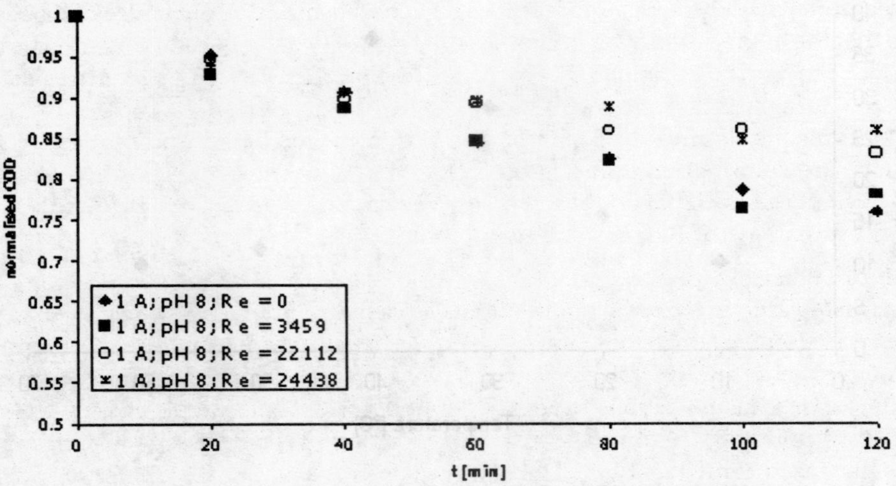

Fig. 10: Performance of the Reactor Under Isoperibolic Conditions During the Removal of COD

Red Rective Dye HEXGL, normalised by the specific surface area of the anode (A/V_R) is shown in Figures 11 and 12, respectively as a function of the agitation (internal recycle) and the temperature of operation.

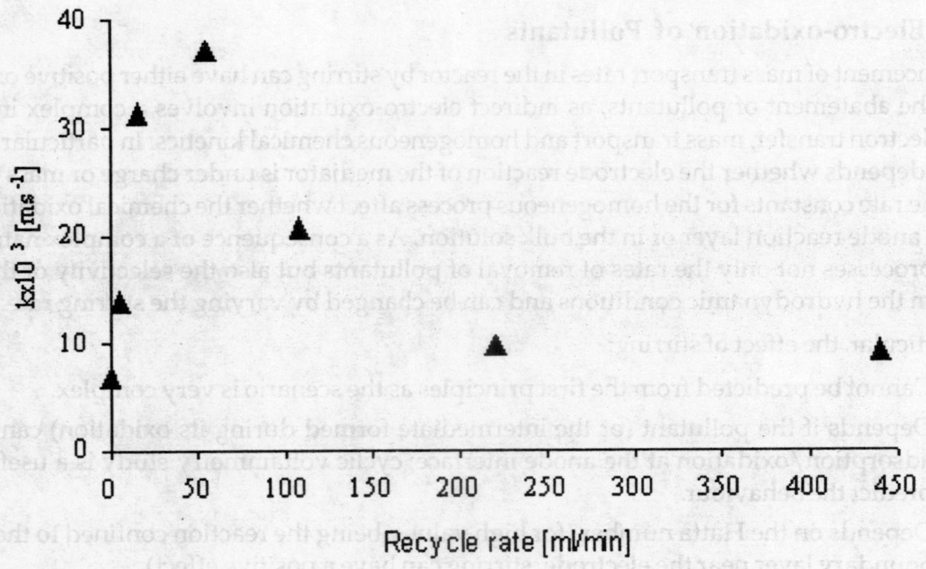

Fig. 11: Variation of the Kinetic Rate Constant (Normalised by the A/V_R Ratio) in Function of the Internal Recycle Rate in the Reactor

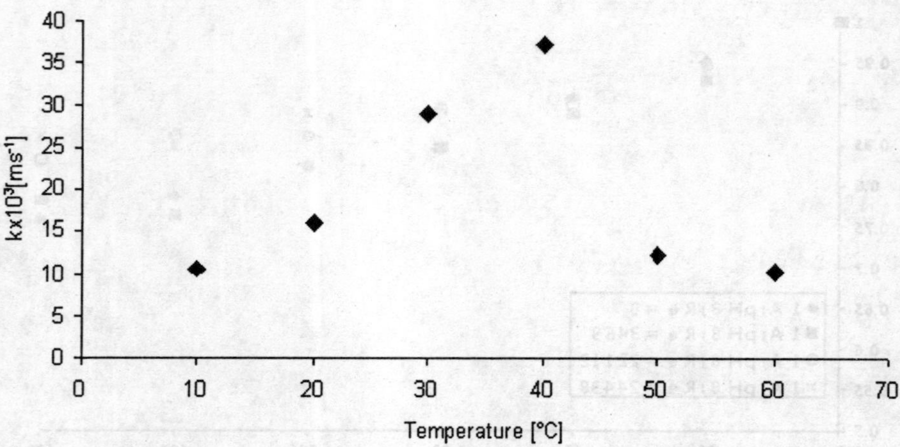

Fig. 12: Variation of the Kinetic Rate Constant (Normalised for the A/V_R ratio) in Function of the Temperature of Operation of the Reactor

Conclusion

Direct Electro-oxidation of Pollutants

Stirring results in enhancing the mass transport rates, which increases the pollutants removal rates; this effect can be predicted from the first principles and is expresses by the correlation of the Sherwood-Gilliland type.

Indirect Electro-oxidation of Pollutants

Enhancement of mass transport rates in the reactor by stirring can have either positive or negative effect on the abatement of pollutants, as indirect electro-oxidation involves a complex interaction between electron transfer, mass transport and homogeneous chemical kinetics. In particular the effect of stirring depends whether the electrode reaction of the mediator is under charge or mass transport control. The rate constants for the homogeneous process affect whether the chemical oxidation occurs within the anode reaction layer or in the bulk solution. As a consequence of a complex nature of the involved processes not only the rates of removal of pollutants but also the selectivity of the reactor depends on the hydrodynamic conditions and can be changed by varying the stirring rate.

In particular, the effect of stirring:

1. Cannot be predicted from the first principles as the scenario is very complex.
2. Depends if the pollutant (or the intermediate formed during its oxidation) can undergo adsorption/oxidation at the anode interface; cyclic voltammetry study is a useful tool to predict the behaviour.
3. Depends on the Hatta number (for high values, being the reaction confined to the reaction boundary layer near the electrode, stirring can have a positive effect).
4. Depends if the reactor operates under isothermal, isoperibolic or adiabatic conditions, as agitation affects the temperature profiles and, consequently, the reaction rates in different zones.
5. Its magnitude, in an isothermal reactor, is comparable with the temperature effects.

REFERENCES

1. Abdo M. S. E. and Al-Ameeri R. S., 1987, Anoxic oxidation of a direct dye in an electrochemical reactor, J. Environ. Sci. Health, A22, 27-42 .
2. Bonfanti F., De Battisti A., Ferro S., Lodi G. and Osti S., 2000, Anodic mineralization of organic substances in chloride-containing aqueous media, *Electrochimica Acta*, 46, 305-314.
3. Boxall, C. and Kelsall, G.H., 1992, Hypochlorite electrogeneration II. Thermodynamics and kinetic model of anode reaction layer, *Electrochemical Engineering, IChem. E. Symp. Series*, I.Chem.E., Rugby, UK, 127, 59-70.
4. Comninellis, Ch., 1992, Electrochemical treatment of wastewater containing phenol. *Trans. IChemE*, 70, Part B, 219.
5. Comninellis, Ch. and Nerini, A., 1995, Anodic oxidation of phenol in the presence of NaCl for wastewater treatment, *J.Appl.Electrochem.*, 25, 23-28.
6. Czarnetzki, L.R. and Janssen, J.J., 1992, Formation of hypochlorite, chlorate and oxygen during NaCl electrolysis from alkaline solutions at an RuO_2/TiO_2 anode, *J. Appl. Electrochem.*, 22, 315-324.
7. Do, J.-S. and Yeh W.-C., 1995, *in situ* degradation of formaldehyde with electrogenerated hypochlorite ion, *J. Appl. Electrochem.*, 25, 483-489.
8. Foti, G., Gandini, D., Comninellis, Ch., Perret, A/, Haenni, W/, 1999, Oxidation of organics by intermediates of water discharge on IrO_2 and synthetic diamond anodes. *Electrochemical and Solid-State Letters* 2, 228.
9. Hikita H., Asai S., Ishikawa H. and Saito Y., 1975, Kinetics of absorption of chlorine in aqueous acid solutions of ferrous chloride, *Chem.Eng.Sci.*, 30, 607 – 616.
10. Ho S. P. Y., Wang C. and Wan C., 1989, Electrolytic decomposition of cyanide effluent with electrochemical reactor packed with stainless steel fiber, *Wat. Res.*, 24, 1317-1321.
11. Lin C. and Zhang J., 2001, Modeling of hydrazodicarbonamide oxidation by chlorine in a gas-liquid-solid system, Chemical Engineering Science, 56, 667 – 671.
12. Moro S., Electrochemical treatment of textile wastewater bearing the dyes, MS Thesis, 2002, University of Venice (in Italian).
13. Murphy, O., Hitchens, G. D., Kaba L. and Verostko C. E., Direct electrochemical oxidation of organics for wastewaters treatment, *Wat. Res.*, 26, 443-451 (1992).
14. Naumczyk, J., Szpyrkowicz, L., De Faveri, M.D. and Zilio-Grandi, F., 1996, Electrochemical treatment of tannery wastewater containing high strength pollutants, *Trans. IChemE*, 74, Part B, 59-68.
15. Picket D. J., Electrochemical Reactor Design, 2nd Edn., Elsevier, Amsterdam (1979).
16. Picket D. J. and Wilson C. J., 1982, Masss transfer in a parallel plate electrochemical cell-the effect of change of flow area and flow cross-section at the cell inlet, *Electrochimica Acta*, 27, 591-594.
17. Pletcher D. and Walsh F. C., 1993, Industrial Electrochemistry, 2nd Ed., , Blackie Academic & Professional, N. Y.
18. Polcaro A. M. and Palmas S., 1997, Electrochemical oxidation of chlorophenols, *Ind. Eng. Chem. Res.*, 36, 1791-1798.

19. Rajeshwar K., Ibanez J. G. and Swain G. M., 1994, Electrochemistry and environment, *J. Appl. Electrochem.*, 24, 1077-1091.
20. Ralph T. R., Hitchman M. L., Millington J. P. and Walsh F. C., 1996, Mass transport in an electrochemical laboratory filterpress reactor and its enhancement by turbulence promoters, *Electrochimica Acta*, 41, 591-603.
21. Rebenne L., Gonzales A. and Olson T., 1996, Aqueous chlorination kinetics and mechanism of substituted dihydroxybenzenes, *Environ. Sci. Technol.*, 30, 2235 – 2242.
22. Ricci F., 2002, Removal of cyanides from industrial wastewater by an electrochemical method, MS Thesis, University of Venice (in Italian).
23. Schwager F., Robertson P. M. and Ibl N., 1980, The use of eddy promoters for the enhancement of mass transport in electrolytic cells, *Electrochimica Acta*, 25, 1655-1665.
24. Scott K., 1991, *Electrochemical Reaction Engineering*, Academic Press, London.
25. Simond, O., Schaller, V and Comninellis, Ch., 1997, Theoretical model for the anodic oxidation of organics on metal electrodes, *Electrochimica Acta*, 42, 2009-2012.
26. Stork A. and Hutin B., 1981, Mass transfer and pressure drop performance of turbulence promoters in electrochemical cells, *Electrochimica Acta*, 26, 127-137.
27. Szpyrkowicz, L., Naumczyk, J., Zilio-Grandi, F., 1994, Application of electrochemical processes for tannery wastewater treatment *Toxicol. Envirn. Chem.* 44, 189-202.
28. Szpyrkowicz, L., Naumczyk, J., Zilio-Grandi, F., 1995, Electrochemical treatment of tannery wastewater using Ti/Pt and Ti/Pt/Ir electrodes *Wat. Res.* 29, 517-524.
29. Szpyrkowicz L., Juzzolino C., Kaul S. N., Daniele S., 2000, Electrochemical oxidation of dyeing baths bearing disperse dyes, *Ind. Eng. Chem. Res.*, 39, 3241-3248.
30. Szpyrkowicz L., Juzzolino C., Kaul S. N., A Comparative study on oxidation of disperse dyes by electrochemical process, ozone, hypochlorite and Fenton reagent, *Water Research*, 2001a, 35, 2129-2136.
31. Szpyrkowicz L., Cherbanski R., Molga E., Kelsall G. H., 2001b, Effect of mixing on mediated electrochemical oxidation of disperse dyes in an undivided electrochemical reactor, ICheaP-5 (Italian Conference on Chemical and Process Engineering), Florence, 73 – 79.
32. Trinidad P. and Walsh F. C., 1996, Hydrodynamic behaviour of the FM01-LC reactor, *Electrochimica Acta*, 41, 493-502.
33. Wragg A.A. and Leontaritis A.A., 1997, Local mass transfer and current distribution in baffled and unbaffled parallel plate electrochemical reactors, *Chemical Engineering Journal*, 66, 1-10.

2

Electrochemical Treatment of Wastewater from a Phosphoric Acid Manufacturing Plant

N.N. RAO, S.N. KAUL** & A. KUMAR***

***National Environmental Engineering Research Institute, Nehru Marg, Nagpur – 440 020 (India)*

**Environmental Biology Research Unit, P. G. Department of Zoology, S. K. University, Dumka-814 101, Jharkhand, India.*

ABSTRACT

Phosphoric acid plant wastewater containing fluoride, phosphate and chemical oxygen demand etc., was treated using electrooxidation & electroflocculation methods. A maximum of 82% F^- and 22.7% COD were removed using Ti/Pt (5) under electrooxidative conditions. Electrooxidation with respect to F^- removal is found mass-transfer limited, and removals below 5-6 mg/L F^- are not achievable. Electroflocculation using Al anode resulted in better removal of F^-, COD & PO_4^{3-}. The various results obtained are discussed in this paper.

Keywords: Phosphoric acid plant wastewater, Ti/Pt anode, Al-anode, fluoride removal

Introduction

Electrochemical methods have been projected to have potential application in wastewater treatment. The electrons and some chemical reagents that can be produced in an electrochemical reactor are clean and powerful redox reagents. Hence, it may help to replace certain conventional chemical reagents, used in typical physicochemical methods of treatment. Thus electrochemical methods may be applied to remove or concentrate dissolved salts, to generate oxidants such as hypochlorite, ozone, hydroxyl radicals, metal redox couples and floc-forming metal ions such as ferrous and aluminium (III) ions. Toxic inorganic ions, organic compounds present in wastewater may be removed through electrochemical routes.

Two methods-electrooxidation and electroflocculation have been applied to treat phosphoric acid plant wastewater. This plant adopted wet process through HCl route for the manufacture of technical grade/food grade phosphoric acid. The major sources of wastewater generation from this plant include: stripped spent brine from solvent recovery section, aqueous effluent from metal removal section and scrubber liquor from dissolution section. The main objective of this paper is to explore the treatability of this wastewater collected from different process sections as well as combined neutralized wastewater. The data obtained may have potential to explore these techniques as tertiary treatment techniques for the removal of F-& PO_4^{3-} from phosphoric acid plant wastewater. The wastewater from this plant is characterized and found to contain significant TDS, COD, fluoride, phosphates, etc.

Experimental Methods

Wastewater sampling has been carried out from the afore-referred major streams. Based on two hourly samples collected and composited for six hours and 24 hours on two different days, detailed analyses for the samples was carried out and the relevant data are presented in Table 1.

Preparation of Ti/Pt

A suspension of K_2PtCl_6 in EtOH was dabbed over the Ti plate and fired in a furnace at 540°C for a period of 2 h., the initially yellow colored K_2PtCl_6 is converted into black Pt deposit during this treatment. This procedure is repeated to attain satisfactory dispersion of Pt on Ti plate. The Ti/Pt anodes that received Pt coating in two and five thermal cycles were used. They are denoted as Ti/Pt (2C) and Ti/Pt (5C), respectively.

All electrochemical experiments were conducted in batch mode in a Plexiglas reactor having 500-ml capacity. For electrooxidation experiments Ti substrate anodes coated with Pt was used in conjunction with a Ti cathode. For electroflocculation experiments, a dissolving Al anode was employed. The reactor contents were thoroughly mixed during the reaction period.

The different amounts of Al^{3+} introduced into wastewater by reaction (7) were calculated as per Faraday's law by keeping the current constant and increasing the time of electrodes dissolution

$$i = \frac{FC}{tn}$$

where

i = DC current, in amperes; t = time in seconds; F = Faraday constant, 96,500 coulombs mol^{-1}; n = Gran equivalent weight of aluminum anode (27/3 = 9); C = Amount of metal ion produced.

The electrochemically treated samples were collected intermittently for requisite analysis of major parameters, *viz.*, F-, PO_4^{3-}.

Table 1: Physico-chemical Characteristics of Different Wastewater Streams

Sl. No.	*Parameters*	*Spent Brine*	*Aqueous Effluent*	*Scrubber Liquor*
1.	pH	0.1	0.2	9.5
2.	Conductivity, mS/cm	13.34	13.16	13.79
3.	Suspended Solids	4970	680	2010
4.	Total Dissolved Solids	201884	6650	147530
5.	COD	2682	1560	6075
6.	BOD	343	155	450
7.	Ammoniacal Nitrogen	22	17	13
8.	Chloride	99868	2080	105060
9.	Phenols	0.5	0.3	0.25
10.	Sulphates	600	180	26688
11.	Phosphates	2600	2000	75
12.	Fluoride	794	116	24.6
13.	*Heavy Metals*			
	Cr	14.676	2.498	BDL
	Pb	1.257	0.300	0.242
	Fe	95.70	109.50	0.158
	Mn	0.870	0.172	BDL
	Cu	0.043	4.573	0.054
	Zn	3.780	9.104	BDL

All values, except pH and Conductivity, are reported in mg/L

Results and Discussion

Electrooxidation using Ti/Pt (2C)

The end-pit wastewater was electrooxidized using Ti/Pt (2C) anode and Ti cathode. The applied voltage was varied from 4.0 to 10 V (current charged from 0.02 to 1.72 A) during a total electrolysis period of 2 h. This treatment caused reduction in fluoride (9.08 mg/L to 5.6 mg/L), in COD (1980 mg/L to 756 mg/L) and a total TDS reduction of 16770 mg/L. However, the Ti/Pt anode was found to dissolve forming tiny flocs of $TiO_2 . H_2O$. The observed removals were mainly due to electroflocculation. The Ti/Pt (2C) anode was unstable in the chosen environment. It was found that Pt coating on Ti plate was non-uniform and islands of Pt deposits were found leaving Ti substrate open for dissolution.

Subsequently, Ti/Pt (5C) anodes were tested under the similar experimental conditions. No dissolution of Ti substrate was found in this case indicating the Pt deposit was uniform. The Ti/Pt (5C) anodes were used to treat thickener overflow, stripped spent brine, neutralized spent brine and the results of treatment are presented in Table 2. It is evident that electrolysis of the above wastewater samples resulted in considerable reduction of fluoride, chemical oxygen demand, while phosphate removals were poor. In the case of thickener overflow with initial pH of 6.4, 81.7% F^- removed, 22.7% COD removal were found at an applied current of 0.6 Adm^{-2} over a period of six hour electrolysis. Stripped spent brine (pH 0.4) was having higher F^-, & COD initially. This showed marginal decrease in F^-, while COD is reduced by 22% upon electrooxidation.

Table 2: Electrooxidation of Wastewater from Phosphoric Acid plant

Anode: Ti/Pt (25 cm²) **Reaction Time: 6h;**
Cathode: Ti (20cm² **Volume of wastewater: 0.4L**

Sample	*TOF, pH 6.4, 0.6Adm⁻²*			*SSB, pH 0.4, 2.0Adm⁻²*			*NSB, 2.0Adm⁻²*		
Parameter	*Before*	*After*	*%R*	*Before*	*After*	*%R*	*Before*	*After*	*%R*
Fluoride	44.8	8.2	81.7	802	742	7.5	25.4	9.8	61.4
Phosphate	-	-	-	-	-	-	193.3	187.2	3.10
COD	2220	1716	22.7	4440	3432	22.7	3120	2808	10.0

TOF: Thickener Overflow; SSB: Stripped Spent Brine; NSB: Neutralized Spent Brine

Neutralization of spent brine to pH 7.0 and applying current caused considerable F-removal, while COD and PO_4^{3-} removals were a little low. Further, the fluoride removal from neutralized spent brine is higher at lower applied currents (Fig. 1). Evidently, the fluoride reduction occurs in two rate zones at least at lower applied currents 0.25 or 0.5A. A major portion of F^- is removed during first one or two-hour electrolysis before it reached slow rate of removal. At higher applied current, *i.e.* 1.0 A, fluoride is removed at a slow rate only.

It may be emphasized here that F^- levels below 5-6 mg/L could not be attained using the electrolytic method, however long are period of electrolysis time to attain these levels. Apparently, a mass transfer limited situation is reached due to progressively declining F^- levels. This was further substantiated by

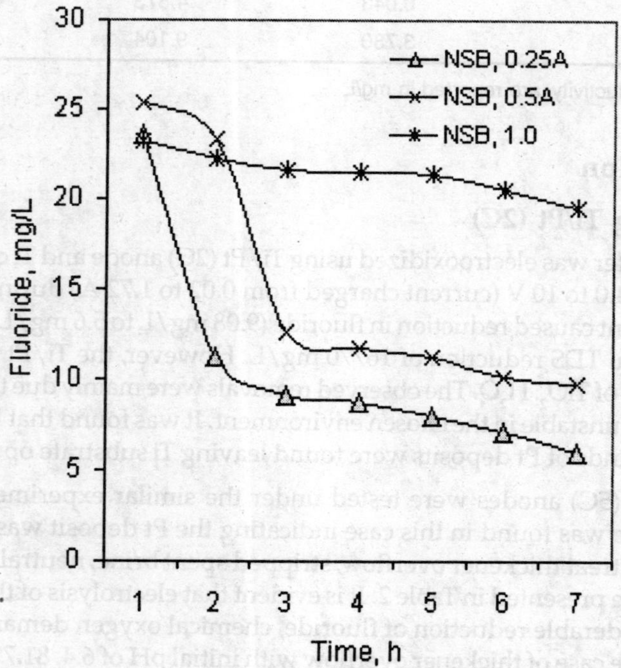

Fig. 1: Electrooxidative Fluoride Removal from Neutralized Spent Brine at Different Current Densities

the fact that combined neutralized waste with initial load of 15 mg/L F^- showed 65% reduction in fluoride level in 3h electrolysis period, while the same combined neutralized waste treated with PAC + PE 156 with initial F^- level of 5.80 mg/L showed no significant reduction in F^-.

Consideration of the following reactions can help to understand the chemistry occurring at anode and cathode in the electrochemical cell:

At Anode

$$2F^- \rightarrow F_2\uparrow + 2e^- \quad (1)$$

$$2H_2O \rightarrow O_2\uparrow + 4H^+ \quad (2)$$

$$2Cl^- \rightarrow Cl_2\uparrow + 2e^- \quad (3)$$

At Cathode

$$2H_2O + 2e^- \rightarrow H_2\uparrow + 2^-OH \quad (4)$$

$$Ca^{2+} + 2e^- \rightarrow Ca \quad (5)$$

The calcium deposited at cathode may from Ca $(OH)_2$ using up OH^- built up at the cathode. Indeed the titanium cathode collected a white precipitate, which flaked off to settle at bottom during electrolysis of Neutralized Spent Brine wastewater. However, at lower pH as is the case with stripped spent brine, the Ti cathode turned grayish black with the deposit of Ca; which turned white after exposing to air.

Fluoride may be expected to be removed by reaction (1) mainly, although the Ca $(OH)_2$ precipitate formed at cathode could also coagulate F^- to some extent (Eq. 6).

$$Ca\,(OH)_2 + F^- \rightarrow Ca\,(OH)_2.\,F\downarrow \quad (6)$$

Fluoride removal according to reaction (1) may face competition from O_2 and Cl_2 evolving reactions (2) & (3), more so at higher applied currents. The attainment of limiting F^- levels can only be due to diffusion-controlled nature of reaction (1) at lower levels of F^- (*i.e.* 5-6 mg/L). No such limiting situation can be expected if F^- were to be removed by coagulation with Ca $(OH)_2$ formed at cathode. Therefore, reaction 6 is of little consequence. The cathode was observed to collect Ca $(OH)_2$ continuously with reactions (2) and (3) occurring at anode as complimentary reactions.

The Cl_2 gas evolved at anode may react with $^{\cdot}OH$ forming HOCl, which is an excellent oxidizing agent capable of causing COD reduction. Phosphate could be removed as adsorbed species on the insoluble hydroxides formed at cathode. Reactions (1), (3), (5) and (6) explain the observed pollutant removal under electrooxidative conditions.

Electroflocculation

Because electrolysis using Ti/Pt anodes was ineffective to bring down F^- levels below 5-6 mg/L, electroflocculation using aluminium anode was tried on combined neutralized wastewater. Insoluble aluminium hydroxide flocs were produced electrochemically by dissolving aluminium anode at constant current (See. reactions (7), (8) and (9).

$$2Al \rightarrow 2Al^{3+} + 6e^- \text{ (anodic)} \quad (7)$$

$$6H_2O + 6e^- \rightarrow 3H_2\uparrow + 6\,(OH)^- \text{ (cathodic)} \quad (8)$$

$$2Al + 6H_2O \rightarrow 3H_2\uparrow + 2Al\,(OH)_3 \text{ (overall)} \quad (9)$$

The results of electroflocculation experiments were conducted on combined neutralized wastewater. As a special case, the combined neutralized wastewater was treated using polyaluminium chloride + anionic polyelectrolyte (Magnafloc 156) and this treated wastewater, denoted as final effluent (FE), was also subjected to electroflocculation experiments. The results are presented in Table 3. The cell and electrode specifications along with working pH of wastewater and dose of Al^{3+} (calculated from Faraday's law) are also included in this table. It is evident that electroflocculation experiment on combined neutralized waste at pH 8.7 showed that at 201.3 mg/L Al^{3+} dose, about 90% fluoride, 46% COD and 56% phosphate removal efficiencies were attained. Further, the observed removals of individual species is a function of pH and dose of Al^{3+} for the same amount of Al^{3+}, the removals are better at higher pH in this case pH 8.7. The removals also increase with increasing dose of Al^{3+}, although this increase is not proportionate. The observed removals of fluoride, COD, phosphate may be attributed to combined action of electrochemically formed Al $(OH)_3$ flocs and complimentary cathodic reactions. The final effluent obtained after treating combined neutralized wastewater with PAC + PE 156 showed further reduction in F^-, COD, and PO_4^{3-} levels upon electroflocculation (see Table 3). With these removals, the final effluent consists of F^-, COD, and PO_4^{3-} levels within the desired control range. The removal fluoride could be assumed to be through adsorption over flocs, while better phosphate removals (compared to electrooxidation) may be due to its ready removal $AlPO_4$.

Table 3: Electroflocculation of Combined Neutralized Wastewater and Final Effluent* from Phosphoric Acid Plant

Anode: Al (15.12 cm²) **Constant current: 0.2A**
Cathode: Ti (20cm²) **Volume of wastewater: 0.2L**

Sample	*CNW, pH 6.4 12 min., Al^{3+} 67.1 mg/L*			*CNW, pH 8.7 12 min., Al^{3+} 67.1 mg/L*			*CNW, pH 8.7 24 min., Al^{3+} 134.2 mg/L*			*CNW, pH 8.7 36 min., Al^{3+} 201.3 mg/L*			*FE (PAC+PE156) 36 min., Al^{3+} 201.3 mg/L*		
Parameter	*Before*	*After*	*%R*	*Before*	*After*	*%R*	*Before*	*After*	*%R*	*Before*	*After*	*%R*	*Before*	*After*	*%R*
Fluoride	13.00	4.85	62.0	12.40	2.85	77.0	12.40	1.56	87.50	12.55	1.26	90.0	3.00	0.65	78.3
Phosphate	8.98	8.08	10.0	9.00	6.38	29.0	5.03	4.41	8.80	8.80	3.88	56.0	4.50	3.07	31.80
COD	3978	2262	43.1	3978	2301	42.1	3978	2067	48.0	3978	2145	46.0	250	131	47.60

CNW = Combined Neutralized Wastewater; Final Effluent: Wastewater treated using PAC + PE 156

Conclusions

The phosphoric acid plant wastewater can be treated for the removal of F^-, COD, and PO_4^{3-} using electrochemical methods. The electrooxidation method using Ti/Pt (5C) anodes is capable of reaching F-levels upto 5-6 mg/L only, as this process is mass transfer limited below this concentration range. Electroflocculation shows promise as a technique to remove F^-, COD, and PO_4^{3-} levels substantially. The electroflocculation of wastewater collected after secondary treatment may be employed successfully to bring down F^-, COD, and PO_4^{3-} levels within the standards specified by a Pollution Control Board.

REFERENCES

1. J. S. Do and M.L. Chen, 1994, Decolourization of Dye containing solutions by electrocoagulation, J. Appl. Electrochem. 24, 785.

2. O. Groterud and L. Smoczynski, 1986, Phosphorus removal from water by means of electrolysis, Water Res. 20, 667.

3. S. Lech and Z Beata, 1993, Wastewaters treatment by direct electrocoagulation. Int. Conf. Mod. Electrochem. Ind. Prot. Environ. Krakow, Poland.

3

Application of Electrochemical Process for Industrial Wastewater Treatment and Environmental Protection

M. RAGHAVAN

Central Electrochemical Research Institute, Karaikudi

Introduction

Electrochemistry has much to contribute towards a cleaner environment. At present there exists pollution control and clean technologies which are based on electrochemical principles.

An area where electrochemistry is able to offer a clean environmental advantage over alternative energy sources is electric transport based on batteries. The philosophy of clean technology is in increasing order of desirability.

Examples, which illustrate this, fall into the following areas:

1. Non-carbon Materials for Pollution Abatement.
2. Environmental Protection in Transport Sector.
3. Fuel cells & Metal air cells for Power Generation.
4. Wastewater Treatment through Electrochemical Technologies.

Non-carbon Materials for Pollution Abatement

The degradation in quality of the environment is termed as air pollution, characterized by a diminished amount of oxygen. It is caused by the presence of gaseous pollutants such as carbon dioxide, carbon monoxide, nitric oxide, nitrogen dioxide, sulphur dioxide etc. Among them carbon dioxide (CO_2) is primarily responsible for the global warming and related environmental hazards. Man-made emission of CO_2 from coal, liquid fuels and metallurgical industries is the main source of these noxious gases. 24 billion tons of CO_2 are emitted to the atmosphere annually, of which 80% is coming from developed countries. USA with a share of 4.8% in population emits 22% of CO_2 while India with 16.2% of world population emits only 3% of CO_2. Interestingly it is also observed that the emission of pollutants is proportional to the GNP of countries. Atmospheric CO_2 level has increased from 280 to 360 ppm during the last 200 years and expected to reach 600 ppm by the year 2050.

Aluminium industry is considered to be environmentally polluting due to the generation of CO_2 during electrolysis. The metal is manufactured by the electrodeposition of pure alumina dissolved in molten cryolite with consumable carbon anode at 960°C. Considering the Indian aluminium sector, 7.0 lakh tons of primary metal is produced by consuming 3.5 to 4.0 lakh tons of high quality carbon with the consequential emission of 1.4 million tons of CO_2 to the atmosphere. Use of non-carbon materials instead of carbon is the best option for pollution abatement in these industries. The exclusion of carbon from the process automatically eliminates CO, CO_2, CF_4, CF_6 and SO_2 mitigates the harmful effects of these pollutants.

Central Electrochemical Research Institute, Karaikudi has done pioneering work (1) in the area of 'green technology' and developed cermet anodes for aluminium electrolysis. These cermet anodes are non-consumable substitutes for consumable carbon. The green technology incorporating cermet anodes liberates pure oxygen instead of CO_2. Nearly 7.0 lakh tons of concentrated oxygen will be available as coproduct. Hence by this novel process, the emission of CO_2 is not only eliminated but the environmental hygiene is also enhanced by the addition of eco-friendly oxygen.

This improvement is considered as 'quantum leap' in the 115 years old aluminium technology. No doubt, the green technology will revolutionize the aluminium technology and envisaged to save 25% overall energy and Rs. 7000 million per annum. Similarly any industry now using carbon can switch over to non-carbon materials so as to limit the generation of carbonaceous pollutants thereby enhancing the environmental hygiene.

Environmental Protection in Transport Sector

Throughout the present century, internal combustion-engine ("motor") vehicles have brought many economic and social benefits. While goods vehicles have oiled the wheels of commerce, private motor cars have given individuals virtually unlimited mobility and, consequently, great freedom in choosing where to live, work and play. Indeed, the motor car has become a major, symbol of increasing prosperity throughout the world, but such transportation benefits have been achieved not without some cost The vehicles derive their power from the combustion of fossil fuel and the resulting exhaust gases are inflicting serious damage to air quality in both regional and global environments (2).

In terms of regional air pollution, the three major harmful emissions from automobiles are: carbon monoxide, nitrogen oxides (which combine with water to form corrosive nitric acid and, subsequently, "acid rain") and hydrocarbons (which react with nitrogen oxides in the presence of sunlight to form ozone, *i.e.* photochemical smog, a lung irritant).

Apart from generating acid rain, photochemical smog and airborne toxins, vehicular emissions also aggravate global environmental problems by the production of large quantities of carbon dioxide. On average, a car will produce nearly four times its own weight of CO_2 every year. This gas is generally considered to be the main cause of the observed rise in the average global temperature. A worldwide temperature increase of only a few degrees may spell disaster for the planet in the form of rising oceans, as well as more frequent and severe hurricanes, droughts and floods. At present, road transport is a major contributor to the carbon dioxide budget. For example, vehicles are responsible for 20, 24 and 26% of all emissions of carbon dioxide in the UK, the USA and Australia, respectively. The problem is particularly sensitive in the USA as the transportation sector of this one country alone accounts for approximately 5% of the world's total output of greenhouse-gas emissions.

The growing awareness of the escalating deterioration in the urban air quality and the possible destructive consequences of global warming have sharpened the debate on how road transport, and in particular the motor car, can be made more "environment friendly". The suggested measures include; the fitting of catalytic converters to remove most of the pollutants; direct curbs by governments on private-car use, together with substantial investments in public transport; increase in fuel efficiency (via lean-burn gasoline and diesel engines, the development of gas turbines, the introduction of ceramic components, etc); alternative, cleaner fuels (*e.g.*methanol, ethanol, compressed natural gas or propane to reduce ozone levels). All these approaches, however, are not without other problems. However, on 29th March 1996, the California Air Regulatory Board's (CARB) 1990 ZEV mandate was softened or rather postponed due to an independent panel's assessment that advanced batteries could not be made available until the year 2000 and also as a result of strong pressure from auto manufacturers and oil suppliers.

Faced with continual violation of federal air quality standards and also due to the severe air pollution in Los Angeles (which is 10 times worse than in any other US city), the State of California, USA, took the lead in early 1990 by adopting the Clean Air Act for the promotion of EV. The legislation originally mandated that 2% of all the new light duty vehicles sold in the state shall be ZEV by 1998 (30, 000 EVs), 5% in 2001 (75, 000) rising to 10% in 2003 (1,50,000).

The ZEVs must not emit directly any of the regulated pollutants, *viz.*, carbon monoxide, nitrogen oxides and non-methane organic gases. In effect, ZEVs have to be electric vehicles powered by batteries and/or fuel cells. Under separate US federal legislation, which applies to States that do not follow California's program, automobile manufacturers must, between model years 1994 and 1996, reduce tailpipe emissions of nitrogen oxides and total hydrocarbons by 60% and 39%, respectively, in all of their light-duty vehicles sales. The EPA may require a further 50% reduction in emissions in 2003.

Although battery-powered electric vehicles are clean alternatives for urban conditions consideration must be given to the indirect emissions associated with generating the electricity required to run such vehicles. In a separate analysis, it was estimated that the introduction of 1.5 million ZEVs in the Los Angeles area would lower ozone concentrations by about 4%, *i.e.* to a level nearly 20% below that projected for the region by 2010.

The impact of battery-powered electric vehicles on greenhouse gases – carbon dioxide, nitrous oxide and methane – is less clear-cut and is particularly dependent on the source of power. In summary, therefore, electric vehicles with non-fossil power offer the best opportunity for reducing, or eliminating, emissions of greenhouse gases from the road transportation sector (3) (Table 1).

Table 1: Comparison of Emissions from Various Fuels

Per Kilometer Drive Type of Fuel	*ROGs**	*NOx*	*CO*
Gasoline	0.156	0.25	2.125
Methanol	0.125	0.169	0.875
LPG	0.213	0.281	0.625
CNG	.119	0.275	0.063
EVs	0.004	0.044	0.005

*ROGs Reactive Organic Gases

It could be further argued that, ultimately, there is no alternative to electric vehicles that take their energy from nuclear power stations (either directly or as electricity stored in batteries and/or indirectly via hydrogen in fuel cells) if the world wants to maintain its mobility. But such an outcome would extend the environmental debate on motor vehicles into entirely new dimensions. When evaluating the prospects for the various candidate batteries, it is necessary to take account of both technical factors (performance, life, safety, etc) and commercial factors (development and production costs, market readiness, product liability, etc.). This situation is especially complex for EVs because the anticipated market is being driven not by consumer demand but principally by environmental pollution concerns, and by the fact that the time-scale of advanced battery development and production is long and the associated cost is high.

The principal technical criteria to be met by an EV battery have been listed by the United States Advanced Battery Consortium (USABC) (3) (Table 2). Sixteen individual criteria are defined, with both mid-term and long-term goals. Many of these are interactive and need to be specified in greater detail than a simple list allows. For example, specific energy is dependent not only on discharge rate, but also on temperature and cut-off voltage. Also, the battery mass requires precise definition with respect to what elements are included. Further, the cycle life depends upon the entire cycling history of the battery, *i.e.*, charge-discharge rates, temperature profile, extent of charge-discharge. Often, it is found that the cycle life in an EV under urban driving conditions is less than that in laboratory simulations. These interactive effects make it difficult to be quantitative when assigning performance levels to the individual battery parameters. Similarly, it is often difficult to assess the wider validity of published laboratory values for these parameters.

Despite the above complications, a semi-quantitative view of five key performance criteria for each of the main candidate batteries has been assembled in Table 3 (4). The principal attributes, both positive and negative, of the most promising EV battery types are now summarized in Table 4 (3). In the medium term, it is expected that the advanced batteries with the best technical prospects for demonstrating a range of at least 160 km (100 miles) in an electric car or van, and being available in pilot-plant quantities for fleet trials, are lithium-ion, nickel/metal hydride, sodium/nickel chloride, and zinc/air.

Aside from the technical performance parameters of the battery, a key commercial consideration is its cost. From the outset, it is very difficult for any battery to compete with the comparatively low cost of lead/acid. For example, nickel/cadmium batteries, although an established industry, are about four times the price of lead/acid counterparts, while nickel/metal hydride batteries are currently even more expensive, although the higher cost of this battery system is offset to some degree by their longer cycle life. Both the initial capital cost and the running cost per km are important. The latter takes account of the possible daily range, the cycle life, and the electricity tariff. It is a reasonable assumption that all serious battery developers have calculated these costs, have extrapolated them to large-scale production rates, and have become convinced that it will be possible to achieve an acceptably low value. Others may not agree with this conclusion; so much depends upon the assumptions made. Some of the extrapolations are very significant, *e.g.*, for lithium-ion batteries from US$ 1000-3000 per kWh at the pilot scale to less than US$ 200 per kWh for production scale. Some authors have made attempts to gather together comparative production costs for different advanced batteries, but there is a tendency for them all to arrive at values around US$ 100-200 per kWh, no doubt to meet the USABC target. The problem of the high expense of purchasing advanced batteries may be resolved to some degree by battery leasing.

Table 2: Principal Attributes of Candidate EV Traction Batteries

Battery	*Principal Attractions*	*Limiting Factors*
Lead/acid	1. Established industry 2. Lowest cost battery 3. Sealed (maintenance-free)	1. Low specific energy 2. Moderate cycle life (~500) 3. Poor low-temperature performance
Nickel/iron	1. Established, robust battery 2. Long cycle life	1 Moderate specific energy 2. Gassing (high maintenance) 3. High cost
Nickel/cadmium	1. Established industry 2. Long cycle life	1. Moderate specific energy 2. Poor performance above 35°C 3. Highly toxic component 4. High cost
Nickel/zinc	1. Reasonable specific energy 2. Good all-round candidate battery	1. Short cycle life (~300)
Nickel/metal Hydride	1. Reasonable specific energy 2. High power 3. Long cycle life 4. Tolerance to overcharge and Over discharge 5. Sealed (maintenance-free)	1. High self-discharge 2. Poor charge-acceptance at high ambient temperature 3. High cost
Metal/air	1. Long EV range before re-plating or mechanical recharging	1. Complexity 2. High volume (low energy density) 3. Low overall electrical efficiency 4. Re-plating technology not yet established
Zinc/bromine	1. Reasonable specific energy 2. Demonstrated in Evs 3. Low–cost components	1. Low power output 2. Shunt-current problems 3. Safety concerns (Br_2 release)
Sodium/sulfur	1. High specific energy 2. Low-cost electrode materials 3. No self-discharge	1. High operating temperature 2. Corrosion of cell components 3. No overcharge capability 4. Safety concerns (fire)

Contd...

Table 2–Contd...

Battery	*Principal Attractions*	*Limiting Factors*
Sodium/nickel chloride	1. High specific energy	1. High operating temperature
	2. Discharged-state assembly	2. Low specific power at 80% DoD
	3. Long cycle life	3. Cost?
	4. No self-discharge	
	5. Successful fleet trials	
Lithium-ion	1. High specific energy and power	1. Early stage of development
	2. Long cycle life	2. Accurate control of charge voltage necessary
		3. Safety?
		4. High cost.
Lithium-ion Sulfide	1. Reasonable specific energy	1. High operating temperature
	2. Good power output	2. Only a bipolar battery will meet specifications for Evs
		3. No pilot production despite 20 + years of research

The data in Table 4 (3) show that no battery system has all the required desirable attributes. It is likely that, progressively, development work will cease on the least promising candidates and only one or two battery types will be commercialized. A further problem, common to all new batteries, is that a developer will be unlikely to set up a full-scale production plant until firm orders have been received from one or more EV manufactures, while potential EV manufactures will be reluctant to place an order for batteries until they are certain that they can sell their vehicles. In addition, EV manufactures will require firm prices, delivery dates and performance warranties, which the battery maker will be hesitant to provide when based solely on the experience gained in operating a pilot plant. These are generic problems which have yet to be resolved

Fuel Cells and Metal/Air Batteries for Power Generation

The increase of atmospheric pollution may provoke irreparable damage to the Earth, particularly the emissions of gases that are related to the greenhouse effect and oxides of nitrogen and sulphur. The production of energy to sustain the human activities, is one of the greatest responsible for this fact.

Ninety per cent of world energy is provided by burning fossil fuels and the consequent result is to add more carbon dioxide to the already burdened atmosphere and so increase global warming with direct impact on the climate. Base on the global climatic models, we can estimate that Earth will suffer a rise in the average surface temperature between 1.5°C and 4.5°C if the concentration of greenhouse gases duplicate; moreover, if the actual tendency does not change, this temperature increase will occur before the year 2050. These alterations would have a bad impact on human activity (4).

The necessity of reducing the gaseous products of the combustion lead to the need for alternative technology, concerning the conservation of energy resources and more efficient burning of petroleum and coal, the use of alternative fuels with lower or no emissions of greenhouse gases, the use of satisfactory renewable energy sources, and so on.

Fuel cells and metal/air batteries are a set of energy generators, which appear to hold considerable potential for future efficient and cleaner, generation of electricity. These systems are presently being developed for stationary power plants and for transport applications.

Table 3: USABC Objectives for EV Battery Packs

USABC Objective	*Mid-term*	*Commercialization*	*Long-term*
		Primary objectives	
Specific energy, Wh / kg (C_3/3 discharge rate)	80 (100 desired)	150	200
Energy density Wh/L (C_3/3 discharge rate)	135	230	300
Specific power, W/ kg (80% DoD* per 30 s)	150 (200 desired)	300	400
Specific power (regen.), 20% DoD/ 10s, W /kg	75	150	200
Power density, W/L	250	460	600
Recharge time (20 → 100% SoC'), h	<6	6 (4 desired)	3 to 6
Fast recharge time, minutes	(40 → 80%SoC) 15<	<30 @150 W/kg 20→70% Soc) (<20@ 270 W/kg[1] desired)	<15 (40→
End-of-life	20% degradation of power & capacity specification	20% degradation of power & capacit specification	20%degradation of power & capacity specification
Calendar life, years	5	10	10
Life, cycles	600 @ 80% DoD	1000 @ 80% DoD 1600 @ 50% DoD 2670 @ 30% DoD	1000 @ 80% DoD
Life, urban miles	100 000	100 000	100 000
Ultimate cost, US$/kWh	<150	<150 (75 desired)	<100
Operating environment, oC	-30 to +65	20% loss at Extremes of –40 & +50 (10% desired)	-40 to +85
Continuous discharge in 1 h (no failure), % rated energy capacity	75	—	75
		Secondary Objectives	
Efficiency, % (C_3/3 discharge rate, 6-h charge, at end-of-life)	75	80	80
Off-tether pack energy loss			
-Self-discharge, %	<15 (in 48 h)	12 days: cumulative loss <25% with some Performance loss @ extreme temperature limits	<15 (in 1 month)
-thermal loss	<3.2 W/ kWh (<15%in 48 h)	3 days: <15% with no performance loss	<3.2 W/ kWh (<15%in 48 h)
Maintenance	Zero		zero

*DoD = depth-of-discharge; SoC = state-of-charge.

Table 4: Status of Battery Systems in Competition with Lead/acid for EVs

System	*Specific Energy (Wh/kg)*	*Peak Power (W/kg)*	*Energy Efficiency (%)*	*Cycle Life*	*Self-Discharge (% per 48h)*	*Cost (US$/kWh)*
Acidic Aqueous Solution						
Lead/acid	35-50	150-400	>80	500-1000	0.6	120-150
Alkaline Aqueous Solution						
Nickel/Cadmium	40-60	80-150	75	800	1	250-350
Nickel/Iorn	50-60	80-150	65	1500-2000	3	200-400
Nickel/Zinc	55-75	170-260	70	300	1.6	100-300
Nickel/metal Hydride	70-95	200-300	70	750-1200	+6	200-350
Alumin ium/air	200-300	160	<50	?	?	?
Iron/air	80-120	90	60	500+	?	50
Zinc/air	100-220	30-80	60	600+	?	90-120
Flow						
Zinc/bromine	70-85	90-110	65-75	500-2000	?	200-250
Vanadium redox	20-30	110	75-85			400-450
Molten Salt						
Sodium/sulfur	150-240	230	85	800+	0*	250-450
Sodium/Nickel Chloride	90-120	130-160	80	1200+	0*	230-345
Lithium/Iron (FeS)	100-130	150-250	80	1000+	?	110
Organic/lithium						
Lithium-ion	80-130	200-300	>95	1000+	0.7	200

The main advantage of fuel cells is to produce electricity and heat with negligible emissions of nitrogen oxides, carbon monoxide and dioxide, and other airborne pollutants. A study of the World Fuel Cell Council realized by Arthur D Little in 1991 indicates that fuel cells could reduce energy use and associated carbon dioxide emissions by 40-60% in those sectors that use over 90% of energy, and the emissions of harmful oxides of nitrogen would be reduced by 50-90% compared even to conventional technologies that employ pollution control equipment in a wide range of both stationary and transportation applications.

A further advantage of fuel cells and metal/air batteries over all other methods of energy conversion is that chemical energy is converted directly into electrical energy. The conventional methods of energy conversion, apart from hydroelectric, involved the intermediate conversion of heat to mechanical energy, the efficiency of which is Carnot limited, thus the efficiencies of fuel cells are higher than those of other energy conversion methods (Table 5) (5).

Other advantages of fuel cells are the modular construction. Cells can be made in any shape or size, which allows them to be used for a wide range of applications, adaptability to operate over any desired temperature range and simple operation compared to other energy conversion devices.

An overview of the state of the art and short-term perspectives of fuel cells and metal/air batteries and analysis of their environmental benefits and drawbacks are presented.

The climatic conditions of the Earth are determined, in large scale, by the atmosphere compositions. The Earth's atmosphere acts as a filter that controls the exchange of energy between the sun, the Earth and space. Certain gases in the atmosphere, like carbon dioxide, methane, nitrous oxide, troposphere ozone and chlorofluorocarbons (CFC), can act as greenhouse gases, by absorbing infrared radiation and retaining heat, thus contributing to the calculated global mean surface warming.

Table 5: Efficiencies of Energy Conversion Devices (5)

Energy Generator	*Efficiency (%)*
Thermoelectric	10
Thermionic	22
Photovoltaic	18
Magnethohydro dynamics	60
Fuel cells	60
Gas turbines	30
Internal combustion engines	30

The main greenhouse gas is carbon dioxide, which is responsible for 49% of the effect. The emissions result chiefly from burning of fossil fuels in the electricity generation, transports and industry (5 Gton of C per year), and deforestation (0.5 to 2 Gton of C per year) (4). Table 6 shows the percentage contribution to carbon dioxide emissions from human activities between 1980-85 (6).

Table 6: Percentage Contributions to CO_2 Emissions from Human Activities (1)

Sector	*% contributions to CO_2 emissions*
Desflorestation	23%
Electricity	22.5%
Transport	22.5%
Industry	16%
Others	16%

Between the years 1960 and 1985, the atmosphere concentration of carbon dioxide increase from 315 ppm to 345 ppm. Measurements of carbon dioxide in bubbles of air trapped in the Antarctic ice have shown that for 10,000 years the concentration remained constant at 270 ppm and only started rising 100 years ago. Actually the rate of increase of the concentration is 1.5 ppm per year (4).

Methane arises from agricultural animals, paddy fields and oil and gas wells. Recent observation shows clearly that methane emissions have been rising since the beginning of the last century, with a more rapid rise in the last few decades. This is partly because changing agricultural practices, waste disposal and mining have produced significant amounts of methane. In 1983 the concentration of methane in the atmosphere was 1.68 ppm and is increasing by 1.1% per year since 1951 (4).

The *chlorofluorocarbons* are chemical products with application in air refrigeration, refrigerators, solvents, packing and expulsants in aerosols. In 1985 the atmosphere concentration of CFC-11 and CFC-12 in four different places in the planet was respectively 0.21 to 0.32 ppb and 0.37 to 0.39 ppb. Although its concentration is lower than that of carbon dioxide, CFC is responsible for 14% of the actual greenhouse effect because its retention efficiency is 10 000 times greater.

The emissions of *nitrous oxide* are resulting chiefly from burning of fossil fuels and biofuels. In 1985 the concentration in the atmosphere was of 300 ppb. Actually the rate of increase of the concentration of nitrous oxide is 0.25% per year (4). Table 7 shows greenhouse gases contributions to global warming of the Earth (6).

Table7: Greenhouse Gas Contributions to Global Warming (6)

Greenhouse Gas	*% Contributions to Global Warming*
Carbon dioxide	49%
Methane	18%
CFC-11 & CFC-12	14%
Nitrous oxide	6%
Others	13%

Oxides of Nitrogen and Sulphur

The emissions of nitrogen oxides, NO_x, arise from two sources, the oxidation of nitrogen in the combustion air and the oxidation of the nitrogen contained in the fuel. Sulphur oxides, SO_x, arise from oxidation of the sulphur in the fuel and all the sulphur is normally converted to the oxide. Sulphur occurs naturally in coal and oil.

The principal transmitters are refineries, chemical product factories, thermoelectric centers and automobiles (solely NO_x). As an example, a 1000 MW oil thermoelectric centre emits 7 ton of SO_2, 1 ton of NO_x, and 0.1 ton of particles per hour, and an automobile circulating at 90 km/h mist, per km, approximately 4 g of CO, 3 g of NO_x and 0.7 g of hydrocarbons (6).

The combustion installations of the chemical industry, paper paste, cork and various units of small and medium dimensions, and the energy production sector, set almost exclusively by thermoelectric centres, are the greatest responsible for the total emissions of sulphur oxides; their contributions are respectively 42% and 38%. The main contributor to the total emission of nitrogen oxides is the transport sector, with 61%.

There are strong indications that sulphur dioxide (SO_2) emissions increased by about a factor of two between 1950 and 1970 in most of Europe and the same general trend also holds for nitrogen oxides (NO_x). The emissions of sulphur dioxide have generally leveled off since 1970, while the nitrogen oxides increase with the number of motor vehicles.

Fuel Cells

Fuel cells (FCs) are electrochemical devices that directly convert fuel energy into electricity without the need for a thermal cycle. They are essentially galvanic cells in which the electrodes only collect and convey electrical charges, but (unlike in the Volta pile and all other electric cells and batteries) they do not participate in the electrochemical reaction, since they are chemically and electrochemically inert conductors (amorphous carbon, sintered nickel oxide, etc.).

In a FC, such electrodes are porous to allow fluid reactants to be continuously adducted to the reaction zones; the contact surface between the anode and the electrolyte for the fuel, and that between the cathode and the electrolyte for the oxidant.

In principle, all those exothermic reactions between fluid reactants that can be subdivided into two electrochemical semi-reactions could be used in a FC. In practice, only the oxidation reactions of

H_2 to form H_2O and of CO to form CO_2 are considered, because the reaction products at the FC working temperature are non-aggressive and gaseous: therefore, involve no problems in the choice of structural materials, and are easily eliminated.

If we take the H_2 oxidation as an example, the theoretical cell voltage is:

$$E=-\Delta G/nF = 474609/4x96516= 1.229V$$

where

ΔG is the Gibbs free energy of the reaction, expressed in joules, n is the number of electrons participating in the reaction, per molecule, and F is the Faraday number.

Table 8: The Sectorial Emissions of SO_2 and NO_x in Portugal During 1987 (3)

Sector	SO_2		NO_x	
	Ton	%	Ton	%
Combustion: Electricity production	83877	38	15834	14
Oil refinery	14318	7	2343	2
Industry combustion	91823	42	13534	12
Production processes	19815	9	13402	11
Transports	8489	4	70748	61
TOTAL	218322	100	115861	100

In practice, polarization effects reduce the voltage available at the electrodes to nearly half of its theoretical value: $V_{oc} = 0.71$ V/cell is the open-circuit voltage, with current density i=200 mA/cm^2. To realize useful voltages it is therefore necessary to pile up several cells (as in any electrical battery): this is the fuel-cell stack. Electrode surfaces of about one square foot, with current intensities in the range of 200 A are very common when using the present technology, and stacks of some hundreds of cells are also commonly used.

A pre-fuel reformer must convert any commercial fuel (gasoline, LPG, methane, naptha, methanol, etc.) into a hydrogen-rich and/or carbon-monoxide-rich gaseous mixture to feed the fuel-cell stack. Moreover, since fuel cells are electrochemical devices, they produce direct current: an inverter must therefore be set after the stack, when a.c. current is needed. The inverter subsystem is obviously not necessary for transportation applications.

Advantages and Drawbacks

The main advantages of FC generators are: high electrical-conversion efficiency, modularity and very low environmental impact.

Among the devices that produce electricity using fossil fuels, FC are the only ones that produce no NO_x since this mixture only forms well above 1000°C, as in burner flames, but not in FCs, which all operate below this temperature.

Fuel Cells obviously produce the same stoichiometric amount of CO_2 as conventional generators do per amount of fuel used. However, since their conversion efficiency is higher, the rate of CO_2 production per kWh produced is lower.

FC exhausts do not contain SO_2 since FCs do not accept sulphurated fuels. Finally, FCs produce negligible noise since they have neither moving parts nor intermittent irreversible reactions (like combustion or explosion). They can have high power-to-weight ratios (up to 0.7 kW/kg) and power density (up to 1 kW/litre).

So simple is the principle and so many are the advantages, that one might wonder why FCs are not yet commonly adopted. Indeed, they were conceived by Sir Humphrey Davy in 1801, only one year after the invention of the Volta pile, and first realized by William Grove in 1839, as a laboratory prototype. From then on, a century-and-a-half long development took place to bring that prototype to the threshold of industrialization. The reason is that the kinetics of the oxidation and reduction semi-reactions is unfavourable at room temperature, and this hindrance can be overcome either by increasing the operation temperature (high-temperature FCs, 600-950°C) or by adding suitable materials stability and in the second case catalyst lifetime, and in both cases current density, have raised century-lasting problems that could be solved only recently, with very mature technologies.

One further consideration may be added. FCs are typical high-added-value, low-material-cost devices (in spite of precious catalysts, which only cover 0.5 per cent of the cost of the present commercial prototypes, and 5 per cent of the final expected asymptotic value for mass-production would make FC generators competitive, at least until environmental impact was not considered as a bottle-neck, and economy and performances were the only comparison grounds for powering devices.

Taking into account that the automobile industry is a typical mass-production industry, it is deemed worthwhile to draw the attention of automobile designers to the FC technology and performances.

Metal/Air Batteries

Metal/air cells are a semi-fuel cell, a hybrid between a battery and a fuel cell. The system combines *in situ* a metal anode, which dissolves in an alkaline or neutral electrolyte, and a porous diffusion cathode, in which air is continuously admitted.

These systems can be electrically or mechanically recharged. Electrical recharging can be performed either in the battery itself or in a special recharging or electrolysis cell. Mechanical recharging consists of the replacement of the metal anode plate. If the battery is mechanically recharged, it will be necessary to remove the reaction products by re-circulation of the electrolyte or by replacement periodically of the electrolyte.

The first metal/air cell mentioned in the literature was a zinc/air cell constructed by Smee in 1840 (7). In 1878, Maich (8), modified a Leclanche cell by replacing the conventional manganese dioxide cathode by a mixture of platinum and carbon powder. Ever since the work of Maich, various metals have been studied for use in this type of system in different applications. Among others, zinc, iron and aluminium have been considered for electric vehicles applications. It is, however, in the field of electric vehicles that most of the research has been conducted.

Advantages and Drawbacks

The main characteristics of electrochemical power sources that must be considered in electric vehicles applications are specific energy, specific power and cost.

The main advantage of metal/air batteries is primarily their high specific energy in comparison with other types of batteries. This is achieved through not requiring incorporating positive active components within the cell (Table 9).

The operational costs of metal/air batteries are low. Air is freely available and inexhaustible. The metals that have been used, such as aluminium, zinc, iron, are cheap and abundant.

Table 9: Metal/Air Systems

Metal	*Electrochemical Equivalent (Ah/g)*	*Theoretical Cell Voltage (V)*	*Theoretical Specific Energy (Wh/g)*	*Operating Voltage (V)*
Li	3.86	3.4	13.1	2.4
Al	2.98	2.7	8.1	1.6
Mg	2.20	3.1	6.8	1.4
Zn	0.82	1.6	1.3	1.2
Fe	0.96	1.3	1.2	1.0

The major limitation of metal/air batteries, that have prevented their more wide-spread applications, is their low specific power. This limitation arises from the air performance electrode, which suffers from considerable polarization losses on discharge, largely due to mass transport limitations. However, advances have been made in reducing polarization by improving the physical electrode properties.

The metal/air systems produce electricity without emissions of airborne pollutants. Thus, the use of large population of metal/air electric vehicles may greatly reduce the air pollution and associated health impacts in urban areas. However, even though metal/air electric vehicles do not themselves emit exhaust gases, their running will incur emission from metal industries, which can be resolved with non-carbon molten salt technology.

The air emissions from the metal production plants are principally due to fossil-fuel generated electricity, which emissions can be reduced by using alternatives electricity generators, such as, fuel-cells. Consequently, it is important to the implementation of metal/air electric vehicles, that metal production plants develop their technology to reduce global impact on the environment.

Wastewater Treatment through Electrochemical Technologies

Many electrochemical processes for environmental pollution control involve the direct reaction of species at electrode surfaces, while others involve production of active species at the electrode and further reaction with the targeted pollutants. The term direct electrolysis shall be used here in connection with those processes in which an electron transfer reaction to or from the undesired pollutant occurs at the surface of an electrode. Likewise, the term indirect electrolysis refers here to processes in which a dissolved redox reagent either exists in or is generated from the electrolyte or from the electrode phase in order to participate in a targeted reaction. The direct and indirect, electrolytic processes include the oxidation and reduction of organic and inorganic pollutants. Indirect processes also include processes of electrocoagulation, electroflotation, and electroflocculation.

Positive Features of Electrochemical Approaches to Pollution Control (9)

Electrochemical techniques offer many distinctive advantages:

Environmental Compatibility

The main reagent used is the electron, which is a clean reagent, and usually there is no need for adding chemicals.

Versatility

Electrochemical processes involving direct or indirect oxidation and reduction can generate neutral, positively, or negatively charged inorganic, organic, or biochemical species. They can also deal with solid, liquid, or gaseous pollutants and can induce the production of precipitates, gaseous species, pH changes, or charge neutralization. The products from the electrolysis of pollutants often are even useful. In addition, various kinds of reactor and electrode materials, shapes and configurations can be utilized. Frequently, the same reactor can be used for different electrochemical reactions with only minor changes. In addition, point-of-use production of chemicals is facilitated by electrochemical techniques (*e.g.*, for water disinfection). Last, volumes of fluid from microliters to millions of liters can be treated.

Energy Efficiency

Electrochemical processes often have lower temperature and pressure requirements than those of equivalent non-electrochemical counterparts (*e.g.*, incineration, supercritical oxidation). The applied potentials can be controlled and electrodes and cells can be designed to minimize power losses due to poor current distribution and voltage drops.

Safety

Electrochemical processes are safe because of the mild conditions usually employed, and the small quantity and innocuous nature of the added chemicals.

Selectivity

The applied potential in many cases can be controlled to selectively to avoid production of by-products.

Amenability to Automation

The electrical variables used in electrochemical processes (I, E) are particularly suited for facilitating data acquisition, process automation, and control.

Cost Effectiveness

The required equipment and operations are normally simple and, if properly designed, can also be made relatively inexpensive.

The Process Treatment of Industrial Effluents: Metal Recovery by Electrodeposition (10)

The use of electrochemical techniques for the treatment of a wide range of industrial effluents and process streams is increasing rapidly and applications include the removal and recovery of dissolved metals, the destruction of dissolved organic, and the treatment of waste gases. The methods are often used in conjunction with other methods because there may be several types of contaminating species present including suspended particulate, microorganisms, and dissolved organic and inorganic compounds. In many applications, specific technologies have limitations in their effectiveness unless they are coupled to other established or developing technologies. In the case of electrodeposition, for low metal ion concentrations the efficiency and cost effectiveness start to fall. Thus a method such as ion exchange, which can be more effective with lower metal ion concentrations, can be utilized to recover, concentrate and recycle the metal ions back to the electrodeposition. Several methods have been researched and developed for the removal of dissolved metals in medium to low concentrations from wastewaters including adsorption, ion exchange, extraction, precipitation, membrane separation and biochemical treatment (10) (Table 10).

Table 10: Alternative Technologies for the Removal of Metal Ion Solutions (10)

Methods	*Quality of Treated Effluent*	*Cost of Effluent Treatment (a)*	*Selectivity for Individual Metals*	*Feasibility of Metals Recovery for Saleable Product*	*Relative Costs of Metals Recovery*	*Other Factors*
Precipitation *e.g.* as hydroxide or sulfide	Moderate good	Low-moderate (depends on soid-liquid separation)	None	Poor	High	Additional treatment to destroy complexing anions
Ion exchange	Very good	High	High or moderate (depends on pH etc.)	Good	Moderate	Resin may be affected by oxidising agents
Electrolysis	Moderate–high (suitable for closed loop operation)	High	High (with control of potential) depending upon standard potentials	Very good	Low (included in a)	Susceptible to other ions and organics
Reverse osmosis	Medium high	Depends on desired effluent quality	None	Good for in-plant solution recycle	Low (included in a)	Restriction on pH of effluents
Evaporation	Minimal effluent	High	None	Good for in-plant solution recycle	Low	
Liquid-liquid extraction	Low-high (organic as well as residual metals in raffinate)	Moderate to high (depends on solvent loss)	High or moderate (depends on pH and extractant)	Good	Low for in-plant recycle otherwise moderate	Continuous countercurrent operation
Flotation	Moderate	Low-moderate	None	Poor	High	Relative
Electrodialysis	Moderate	High	Poor	Good for recycling	High	Relatively high effluent concentration
Cementation	Poor	Low	Poor	Good	Low	High effluent concentration

These techniques compete with one another, and can complement electrochemical methods, which are finding even wider practical applications. Notably the removal of the metal by electrodeposition, at source, is seen to be attractive due to the potential for a one-step clean method of metal recycle. The electrochemical approach will not always provide the primary solution to the problem, but can be part of an integrated waste management scheme. The recovery of metals from solution by electrodeposition can conveniently be divided into two areas depending upon the metal ion concentration to be treated.

1. In the metal ion concentration range 1-3 g dm^{-3} of solution feed, two-dimensional cathodes are generally used to maintain effluent concentration levels in the range 0.1-0.5g dm^{-3}. Thus the removal of a metal from a process stream to a pre-determined concentration enables the recycling of either solution or metal to the process.

2. At approximate concentrations up to 100 mg dm^{-3} (100 ppm) of metal ion feed solution, three-dimensional electrodes are used preferably, especially if effluent levels of <1 mg dm^{-3} (1 ppm) are required. In the case of certain precious metals, two-dimensional electrodes are used to treat lower concentration solutions when power requirements are not crucial. The removal of metal to very low concentrations enables the resulting spent liquor to be recycled, discharged, or passed to further treatment stages for other species. Electrodeposition cells can operate continuously or in a batch mode, and can produce either compact metal for reuse or resale, powdered or flakes metal, or concentrates for recycling. The metal may be produced as an alloy as an alternative to pure metal or occasionally as an insoluble metal compound (hydroxide, oxide and sulfide). The choice of the metal form depends largely on the use or destination of the material. It may be recycled directly within the process operation, or to another process, or reprocessed either in house or sold for refining or scrap.

There are approximately 32 metals that can be recovered effectively by electrodeposition from aqueous solutions, which is only partly indicated in the appropriate standard electrode potentials shown in Table 11.

The simplest way of recovering the metal is by direct reduction of the metal cation:

$$M^{z+} + z\,e \rightarrow M \qquad (1)$$

In certain cases the metal is not deposited from a free ion but from a complex with an inorganic or organic ligand *e.g.*

$$CuCl_3^{2-} + e \rightarrow Cu + 3Cl^- \qquad (2)$$

$$Cd.EDTA^{2-} + 2e \rightarrow Cd + EDTA^{4-} \qquad (3)$$

The chemistry of the system can have a significant bearing on the performance of the electrodeposition and the ability to recover the metal.

Applications of Metal Deposition Systems

In the treatment of dilute solutions of metal ions the attraction of a clean one-step recovery of the metal value is a compelling argument for its adoption. The parallel plate, tank electrolyser is widely used for the electrowinning of metals from high concentrations of ions, but is generally unsuitable for low concentrations of the order of 1 kmol m^{-3}. The major limiting factor is that the poor hydrodynamics and thus mass-transfer rates produce limiting currents, which are low and uneconomic. Thus for wastewater applications considerable development into cell technology has occurred to improve the performance. Available cell designs can be put into two general categories (Table 12):

1. Direct metal recovery;
2. Indirect recovery as a metal ion concentrator (concentrator cells).

Metal ion removal or recycling by electrodialysis techniques includes the following:

1. Rinsing wastewater from electroplating processes can be desalted and reused while the concentrate is recycled into the plating bath. In addition, this electrodialytic process facilitates the regulation of concentration required in the rinse tanks.
2. The OH ions produced at the cathode of an electrodialytic cell are used for the precipitation and removal of insoluble radioactive metal ion hydroxides.

Table 11: The Electromotive Series: Standard Electrode Potentials

Electrode Reaction				*E°, V at 25ºC*
$S_2O_8^{2-}$	$+ 2e^-$	→	$2SO_4^{2-}$	+2.01
Co^{+++}	$+ 2e^-$	→	Co^{++}	+ 1.82
H_2O_2	$+ 2H^+ + 2e^-$	→	$2H_2O$	+1.78
Au^+	$+ e^-$	→	Au	+1.69
2Cl	$+ 2e^-$	→	$2Cl^-$	+1.36
$½O_2$	$+ 2H^+ + 2e^-$	→	H_2O	+1.23
Pd^{++}	$+ 2e^-$	→	Pd	+0.99
Ag^+	$+ e^-$	→	Ag	+0.80
Fe^{+++}	$+ e^-$	→	Fe^{++}	+0.77
Cu^+	$+ e^-$	→	Cu	+0.52
Cu^{++}	$+ 2e^-$	→	Cu	+0.34
$2H^+$	$+ 2e^-$	→	H_2	+0.00
Pb^{++}	$+ 2e^-$	→	Pb	-0.13
Sn^{++}	$+ 2e^-$	→	Sn	-0.14
Mo^{++}	$+ 2e^-$	→	Mo	-0.20
Ni^{++}	$+ 2e^-$	→	Ni	-0.25
Co^{++}	$+ 2e^-$	→	Co	-0.28
Cd^{++}	$+ 2e^-$	→	Cd	-0.40
Cr^{+++}	$+ e^-$	→	Cr^{++}	-0.41
Fe^{++}	$+ 2e^-$	→	Fe	-0.44
Zn^{++}	$+ 2e^-$	→	Zn	-0.76
Mn^{++}	$+ 2e^-$	→	Mn	-1.19
Al^{+++}	$+ 3e^-$	→	Al	-1.66
Mg^{++}	$+ 2e^-$	→	Mg	-2.36
Na^+	$+ e^-$	→	Na	-2.71

Table 12: Categories of Cell Designs

Deposited Metal	*Reusable Anode Substrate*	*Metal on a Combustible*	*Metal Ion Concentrate*	*Metal Flake or Powder*	*Particulate Metal*
Parallel plate 1. Recowin 2. Retec cell Concentric cylinder 3. Cyclone cell Continuous belt electrodes	Chemelec cell in electroplating Foam electrodes	Packed-porous cathodes 1. Retec cell 2. Enviro-cell	All three-dimensional *e.g.* 1. HAS reactor 2. Environ-cell 3. ER cell	Rotating or vibrating cells RCE cells (batch or continuous) Goecomet reactor 1. CEER cell	Moving beds Circulating beds Fluidised beds

3. Sulfate ions become concentrated during the electroless deposition of copper on printed circuit boards and undergo undesired side reactions. They can be removed by electrodialysis, and sulfuric acid is produced on the anode of the cell, while the OH ions produced on the cathode are used to neutralize the plating solution.
4. Silver ions are concentrated from spent photographic fixing solutions by electrodialysis and sent to an electrolytic cell for metal recovery.

Advantages of using electrodialytic techniques (most of which are also common to other IEM-based separation processes) include the following:

1. Essentially no extra reagents are required for the processes.
2. Some ionic dissolved substances that cannot be separated by conventional methods can be removed by electrodialysis.
3. Electrolyte solutions can be concentrated up to ~20% (or more in some cases).
4. Diluates can be obtained with concentration down to 100 ppm (or less in some cases).
5. Concentrate/diluate concentration ratios of ~100 can be obtained.
6. Nonelectrolytes can be either separated by exclusion or not affected by the process.
7. IEMS are rather durable (typically several years).
8. High current efficiencies can be attained (80-90% are common, 100% in optimal cases).
9. Unlike ion exchange resins, IEMs do not require a periodic regeneration step.
10. Multivalent and complexed ions can be rejected.
11. Modular equipment adds versatility to the processes. Up to a hundred or more units cells can be stacked, and only one anode and one cathode are required.
12. Continuous processes with relatively high flow rates can be designed.
13. On-site, on-demand production of some substances can be done with attendant reduction in transportation and storage costs.
14. No phase changes are required.
15. Hybrid processes can be designed to optimize the benefits obtainable from two or more separation processes.

For example, a combined electrodialysis-conventional dialysis process has been used for the separation of excess acid from a metal ion solution in an electroplating plant. Ion-exchange materials can be used as packing between membranes to increase both the contact area and the conductivity of the medium. This approach is called electrodialysis-ion exchange, EDIX. The electrical resistance in the ion exchange material packed bed can be decreased by applying a suitable binder in an approach called electrochemical ion exchange (EIX), see later. Larger decontamination factors can be obtained with EDIX than with EIX.

Uses of EIX include the following:

Separation of Radionuclides

The concentration of Cs in a feed solution has been drastically reduced by a factor of 5, 000 using EIX. Also, the radioactivity of a wastewater containing a mixture of radionuclides (predominantly Co and Co) was decreased by a factor of >99%. Likewise, cleanup factors of >11 and 70 were obtained for

Pu and Am containing solutions; potential applications in the treatment of Tc and Tu solutions have been demonstrated. When radioactive and metal ion species are removed from a waste stream by a cation exchange resin, they can be migrated toward the cathode, where they can be deposited.

Water Polishing

High-purity water (15 Mohm cm resistivity) can be obtained by an electrochemical deionization process (EDP) called electrodiaresis polishing (EDIP) based on the passage of pretreated water (for example, by reverse osmosis) through ion exchange resins in an electrochemical cell with compartments separated by an alternate CEM/AEM arrangement. The resin beds increase the electrical conductivity between the electrodes, facilitating treatment of low-conductivity media. In addition, they promote better mass-transfer characteristics and longer residence times. Large flow rates may promote resin crushing as well as too short residence times.

Recycling of Chromate Ions

The EDIP process just described can also be used for the removal and concentration of anions from an effluent stream. In particular, chromate ions from a simulated plating waste stream can be removed and concentrated up to a factor of ~8 without carrying over other metal residues in the form of cations. Here, the inlet and cathodic compartment is filled with cation exchange resin. The two membranes are AEMs.

Nitrate Removal

Nitrates found in wastewaters, in nuclear processing salt cake residues, or in groundwater can be absorbed by an anion exchange resin in a three-compartment electrochemical cell (catholyte, anion exchange resin, and anolyte compartments). Under the influence of an electrical field, the nitrate ions migrate toward the anode while OH ions produced at the cathode (from water reduction) replace them in the resin, regenerating it. The nitrate containing anolyte is then sent to a tortuous path electrochemical cell for its conversion into N_2, O_2, and H_2O. Alternatively, the nitrate ions can be routed to the anolyte compartment to be combined with protons forming nitric acid.

Other Applications

These include the removal of toxic metal ions, the recovery of precious metals, and corrosive anion removal.

In these categories both two-and three-dimensional electrodes are used and a range of reactors used for metal deposition is shown in Table 13, virtually all of which have seen some commercial application.

The performance of reactors for wastewater treatment is based on several criteria; energy consumption, space time yield and the specific investment cost for a unit volume of solution treated. This can be quantified in the normalized space velocity, Nsv, *i.e.* the volume of process solution whose concentration can be reduced ten-fold in a unit time in a unit volume of solution treated. For a batch reactor, operating with amass transfer limited current:

$$Nsv = 3600\, I\, (CE) / (C_I - C_o)\, V_r\, nF \log (Ci/C_o)$$

where

I is the total current, C_i and C_o are the initial and final concentrations of the active species and V_r is the active reactor solution volume.

Table 13: Commercial Reactors for Electrodeposition

Reactor	Cathode	Method of Product Removal	Operating Range (ppm)
Chemelec cell (BEWT Water Engineers Ltd)	Vertical mesh (or plate) in an electrolyte with fluidized glass beads	Discontinuous by manual scraping or re-use as anodes in electroplating	750
Concentric cell (Wilson Process Systems)	Inner surface of a cylindrical foil	Discontinuous-brass cathode may be furnace refined in the case of gold	100
Rotating electrode cell	Rotating cylindrical foil, or static cylindrical foil with rotating anode	Discontinuous by manual scraping or flexing	1-100
Eco-cell (Steetley Engineering Ltd)	Outer surface of a rotating cyclinder (divided cell)	Continuous via powder formation	1-100
Reconwin (Ecotec)	Vertical plates, air agitation	Manual stripping of sheet	750
CEER cell (Finishing Services Ltd)	Vertical plate (divided cell)	Discontinuous metal powder or flake	$>10^3$
Retec cell (Eltech)	Vertical metal (or carbon) foam electrodes in a tank	Discontinuous, cathode may be steel which may be chemically stripped or carbon (for precious metals) cathode may be re-used	0.1-100
Haraeus cell	Rotating Zn band	Continuous by manual peeling	1-100
SCADA Systems	Carbon fibre	Discontinuous by stripping	0.1-100
ER cell (Electro Cell AB)	Packed bed of carbon particles within a parallel plate and frame (divided cell)	Discontinuous via leaching	0.1-100
Enviro-cell (Deutsche Carbone)	Contoured packed bed of carbon (divided cell)	Discontinuous via leaching or vacuum removal of bed	0.1-100
FBE reactor (Billiton Research by)	Fluidised bed of metal particles in a tube-and-shell type geometry (divided cell)	Discontinuous via with-drawal of grown particles	0.1-100
HAS reactor	Packed bed of carbon particles (divided cell)	Discontinuous by leaching	0.1-100
Porocell (EA Technology)	Carbon fibre high surface area	Incineration of carbon substrate	0.1-100
Geocomet	Rotating tubular or impact rod	Continuous as powder	1-100

The Process Treatment of Industrial Effluents: Organic and Inorganic Species

The electrochemical methods for the treatment of wastewater, which contain organic and inorganic species and hazardous substances, and of flue and waste gases include direct and indirect oxidation, photo electrochemical oxidation, electroreduction, electrosorption and electroflotation.

Treatment of Organic Compounds

Process for the removal of organic species from solutions includes:

1. Chemical oxidation, using ozone, UV etc
2. Chemical reduction, *e.g.* catalytic dehalogenation
3. Biological treatment

4. Concentrating methods, *e.g.* membrane or adsorption
5. Incineration

Incineration is a high cost method when dealing with dilute wastewater due to high transportation and fuel costs. Methods for concentrating solutions depend on further treatment steps and have limited applications. Chemical oxidation methods are useful although they are not generally used for one major class of toxic compounds, *i.e.* halogenated organics. Chemical reduction methods are expected to be particularly suitable for concentrated waste streams. Biological methods using special microorganisms are at a relatively early stage of development.

The electrochemical treatment or destruction of organics in wastewater is potentially a powerful method of pollution control, offering a low temperature alternative to incineration and a clean process. In many cases no other chemicals are released into the wastewaters. Owing to the nature of the process streams and the implications of the electrochemistry, the methods which are applied either partially or completely detoxify or decompose the organic species. Complete destruction often means the oxidation of the organic species to carbon dioxide. This oxidation can be achieved either directly or indirectly at the anode, or indirectly using appropriate anodically generated reagents.

In certain cases the complete destruction of an organic molecule by anodic oxidation is not an appropriate or easy process to perform and a partial transformation may be more desirable. The complete oxidation of large organic molecules requires a large number of electrons and energy consumption can be relatively high, as will the cost of the cell hardware. Ideally, current efficiencies for the process should be high, but more importantly any products from the treatment should be either non-toxic, readily biodegradable, or in sufficiently low concentrations to satisfy environmental legislative requirements.

Direct Electrolysis of Pollutants

Pollutants capable of undergoing direct electrochemical oxidation or reduction at an electrode can in principle be removed from water streams or reservoirs by the application of appropriate potentials in electrochemical reactors. Here, oxidation or reduction processes occur directly on inert electrodes without the involvement of other substances. Unfortunately, rather than the removal of unwanted material being the dominant electrode process, side reactions (particularly solvent breakdown) almost always occur; for example:

$$2H_2O \rightarrow O_2 + 4H + 4e \quad (4)$$

$$2H_2O + 2e \rightarrow H_2 + 2OH- \quad (5)$$

While these reactions do have a deleterious effect on the overall process efficiency, clever schemes can be devised to exploit the production of gases as well as the pH changes associated with Reactions (4) and (5) to bring about useful reaction chemistry. For example, the production of H_2 (g) has been used to facilitate the flotation and removal of unwanted substances in the electrocoagulation-electroflotation process. Likewise, the generation of hydroxyl ions at the cathode [Reaction (5)] serves to immobilize the electrolytically generated Cr (III) as the insoluble hydroxide, in the electrochemical reduction of the (very toxic) Cr (IV) to Cr (III). Finally, the production of H_3O^+ has been used in the field of electrokinetic or electro-osmotic remediation of soil to increase acidity in soils, solubilize pollutant species, and so forth.

Other Applications

Anodic oxidation has been researched as a pollution control device for several applications and for several wastewaters.

1. The spent wash from distillery effluent containing high biological oxygen demand (BOD) and dark-brown colour;
2. Wastewater containing alcohol and sugars;
3. From the production of amino ether;
4. Nitration waste sulfuric acid emanating from the manufacture of trinitrotoluene.

A significant area for the application of anodic oxidation is the treatment of dyes and surfactants. The consumption of water in dyeing operations is large, for example in the processing of 1.0 tonne of cloth, ca.100-400m^3 of water can be used. There are several methods for the treatment of waters, which contain dyes and surfactants, which include:

1. Coagulation, by lime, or by aluminium of iron salts. This treatment involves increasing the degree of mineralization of the effluent, a large consumption of reagents and the production of a wet solid.
2. Extraction, with solvents such as diethyl ether and chloroform, which requires a large consumption of the solvent.
3. Biological methods.
4. Adsorption on activated carbon although the adsorbent can rapidly become de-activated.
5. Oxidation using chlorine or ozone, which require large amounts of reagents, *e.g.* up to 9 kg of dye.

Treatment of Dyes

The following classes of dyes have been considered in the electrochemical treatment of wastewater (11): anthraquinone, thiazines, oxazines, phenazines, triphenylmethanes, xanthenes, vat and azo dyes. Purification through electrolysis is usually obtained by electro-oxidation with non-dissolving anodes or electrocoagulation, using dissolving anodes. Direct reduction has been proposed for the decolourisation of the dyes although the energy costs are significantly greater than anodic amine. The degradation products from the oxidation of azo dyes are typically CO_2, N_2 and Na_2SO_4, with the possible formation or aromatic esters, phenols, aliphatic carboxylic acids, cyclic and aliphatic hydrocarbons and aromatic amines etc. The oxidation of the azo groups takes place first followed by the further oxidation of the decomposition products. The oxidation rate is dependent upon the anode material. A range of coated titanium metal anodes formed from ruthenium oxide, cobalt oxide, manganese dioxide and lead dioxide and graphite have been successfully employed. In alkaline solution nickel or an alloy of nickel with titanium is suitable.

In the presence of chloride ions the oxidation of azo dyes is largely through the generation of 'active' chlorine, although direct anodic oxidation does play a part. The energy consumption of the process, using electrochemically generated active chlorine is in the range 25-35 kWh kg^{-1}. Several reactor designs have been used in the treatment of dyes; packed bed anodes of lead spheres, carbon felts and carbon fibers.

Soluble metal anodes are applied in an electrocoagulation process in which the resultant metal hydroxides adsorb the dye. Steel has been used mainly as the anode material. The rate of anode consumption is in the region of 5-200 g m^{-3} of effluent. There is some evidence for the partial direct oxidation of the dye at the anode. The electricity costs for complete removal of azo dyes at a concentration of 0.1 g dm^{-3} are between 30-100 kWh kg^{-1}. The cell design for coagulation must incorporate a froth collector and a flotation chamber.

Electrochemical reduction is attractive in the case of metal chelate complexes where the toxic metal ion may be eliminated as solid metal or the valence state is changed to that more suitable to ion exchange.

Chloride and Chlorinated Compounds

An important issue in the electrochemical treatment of wastewaters and related streams is the presence of halogens, and in particular chloride. Electrochemical oxidation of chloride ions will readily produce chlorine, which may react, either as free chlorine or as monatomic chlorine, with the dissolved organics species of any oxidation intermediates, to produce halogenated compounds. These halogenated compounds are generally more toxic than their unchlorinated counterparts and thus would make anodic oxidation unsuitable. Materials, which are poor electrocatalysts for chlorine evolution, are needed, *e.g.* tin oxide, although at sufficiently high concentrations of chloride, chlorine may be generated. Thus the potential formation of halogenated organics cannot be excluded in the use of anodic oxidation. Tin oxide has been used in the oxidation of wastewater from the chlorine-bleaching step of a pulp plant, which contained both organic and halogenated organic species (12). Both the COD and TOC were reduced and an adsorbable organic carbon was also removed.

Indirect Electrolysis of Pollutants

The ideal here is to use an electrochemically generated redox reagent as a chemical reactant (or catalyst) to convert pollutants to less harmful products. The redox reagent acts then as an intermediary for shuttling electrons between the pollutant substrate and the electrode. These redox reagents can be electrochemically generated either in a reversible or irreversible manner. These two strategies are discussed in turn next. The general scheme is:

$$C \rightarrow C^{+} + e \qquad (6)$$

$$C^{+} + R \rightarrow O + C \qquad (7)$$

For the oxidation case and

$$C + e \rightarrow C^{-} \qquad (8)$$

$$C^{-} + O \rightarrow R + C \qquad (9)$$

Indirect electrolytic processes can be viewed as chemical on-off switches, since they stop when the current supply is discontinued. Among the principal requirements for obtaining high efficiencies in the indirect electrolytic processes are the following:

1. The potential at which the intermediate species, C, is produced must not be near the potential for O_2 or H_2 evolution, since then a large portion of the current will not be employed in the desired reaction.
2. The rate of generation of C must be large.
3. The rate of reaction of C with the pollutant R (or O) must be higher than the rates of any competing reactions (for example, $C+H_2O \rightarrow O_2$+products.).

4. Adsorption of the pollutant (or any other species) must be minimized since electron exchange of C at the electrode often becomes sluggish as a result.

Electrocoagulation offers several advantages, including the following:

1. The smallest charged colloids can be treated, since they can move more easily than their larger counterparts within an electrical field; this facilitates coagulation. Furthermore, such motion avoids the need for mechanical agitation, which may destroy precipitates as soon as they are formed.
2. The amount of required chemicals is much lower. For example, in conventional lime-neutralization processes, water hardness is increased.
3. As mentioned previously, a smaller amount of sludge is produced, due to the higher content of dry solids. For example, conventional addition of ferric chloride followed by lime or sodium hydroxide produces up to 30 liters of sludge for every liter of removed oil. In addition, the sludge produced by the electrochemical treatment is more hydrophobic, which leads to more compact residues; this also leads to shorter decantation times.
4. No mixing of chemicals is required.
5. The durability of the electrodes translate to low "down times" for maintenance or replacement.
6. Organic matter removal (including non-biodegradable organic) is more effective; this facilitates subsequent biological treatment.
7. Coagulant dosing as well as required overpotentials can be easily calculated and controlled.
8. These processes are suitable for small water treatment plants.
9. A pH increase is normally observed, which aids in the removal of heavy metal ions by their precipitation as hydroxides1 or by adsorption into other flocs or precipitates (*e.g.* in acid mine wastewater)
10. Often, pH control is not necessary, unless this parameter acquires extreme values; this facilitates process design and operation (*e.g.*, in oil removal, lignin destruction).
11. High current efficiencies (~90%) can be achieved in well-designed systems.
12. Short contact times are required. For example, an 8-min treatment suffices for decolorization of typical textile wastewater.
13. Operating costs are much lower when compared with most of the conventional technologies.

Conclusion

Electrochemical technology may be used for a range of pollution control and emission free processes and the treatment of a wide variety of environmental pollutants is already at a mature stage of development. In fact, commercialization efforts have been underway during the past few years. In many cases, such as the removal of toxic metals from waste streams, electrochemistry offers a viable process that is easy to control and operate. The long-term problem of emission from road transport could be effectively combated by the use of battery propelled vehicles.

REFERENCES

1. C.O.Augustin, U. Sen; Proc. International Conference on Aluminium (INCAL – 98) New Delhi Vol.II 1998, p. 173.

2. DAJ. Rand, Proc. International Conference Ni-Cd Batteries up date-92 Munich Oct.5.6.1992.
3. DAJ. Rand, Batteries for Electric Vehicles, Society of Automotive Engineers, Inc.Warrendale, pa.
4. CE Document, COM (88) 656/2, 'The Problem of Greenhouse Effect and the Community', 1989.
5. B.V. Tilak, R.S.Yeo, and S. Srinivassan, 'Comprehensive Treatise of Electrochemistry;Vol.3 Electrochemicsl Energy Conversion and storage', Ed.J.O'M.Bockris, Brian E.Conway, Ernest Yeager and Ralph E White, Plenum Press, New York, 1981; 39-122.
6. UNEP Industry and Environment, April-May-June 1990; 2-8.
7. A.Smee: Phil.Mag.III, 16, 1840; 315.
8. L.Maiche, French pat-127, 069, 1878.
9. K.Rajeshwar, J.Ibanez, Environmental Electrochemistry. Academic Press-1997; 361-497.
10. Keith Scott. Electrochemical Processes for Clean Technology, The Royal Society of Chemistry-1995.

4

Fenton-Type Treatment: State of the Art

CARLO MERLI*, ELISABETTA PETRUCCI, ANNA DA POZZO, MIMMA PERNETTI

* *Corresponding author. Fax: (+39) – 0644585622*
E-mail: carlo.merli@ingchim.ing.uniroma1.it

ABSTRACT

The different currently used Fenton-type treatment, either chemically or electrochemically generated, are reviewed. A particular attention was devoted to the traditional Fe^{++}/H_2O_2 chemical process and to the indirect electrochemical oxidation which uses *in situ* generated hydrogen peroxide. Mechanisms and experimental conditions employed for the optimisation of each technology are reported and moreover theoretical and practical advantages and main limitations are discussed. Both technologies can be considered as a simple and environmentally safe approach for removing pollutants from wastewater. Although electrochemical process requires higher cost it offers a remarkable potentiality mostly when total removal are needed. Therefore the treatment efficiency on real wastewater and the process scale-up, so far scarcely studied, deserve further investigation.

Introduction

In the last years particular attention has been devoted to the different methods to remove organic and inorganic substances from both industrial and civil wastewater, in order to attain the standard limits of water quality. High-contaminated wastewater can be treated by physical methods, biodegradation and chemical processes.

Amongst the chemical treatment, oxidation processes are the most interesting, leading to degradation of persisting pollutants to non-toxic products or to more easily biodegradable intermediates. The oxidation may be induced by chemical agents or electrochemical, photochemical phenomena.

Advanced Oxidation Processes (AOPs) (5, 6), mainly employing radical species for organic substrate oxidation, include (Chiron *et al.*, 2000):

1. *Photochemical processes*, employing artificial light source or natural sunlight; they are slow processes, leading to uncompleted degradation;
2. *Ozonation processes*, efficient but expensive (Hoigné and Bader, 1976);
3. *In situ generation of free hydroxyl radicals*, avoiding the addition of external agents. These processes are: Fenton's Reagent, novel vacuum light, high power ultrasound, electrochemical treatment and wet air oxidation.

This work will focus on Fenton's Reagent process, one of the most interesting and most simple methods to generate hydroxyl radicals.

Fenton's Reagent

Hydrogen peroxide is one of the strongest chemical oxidants, whose high oxidative power may be further increased through the combination with UV radiation (Maillard *et al.*, 1992, Venkatadri and Peters, 1993) or metal salts. In particular, the aqueous mixture of ferrous ion and hydrogen peroxide is widely employed as one of the most effective oxidants of organic substrates. This combination is known as Fenton's reagent, after the scientist who first used it for the oxidation of malic acid (Fenton, 1894). Only forty years later a mechanism, in which hydroxyl radicals was involved as the actual oxidant species, was proposed by Haber and Weiss (Haber *et al.* 1934). Their initial hypothesis has been subsequently proved by many techniques including ESR spectroscopy.

Under acidic conditions the reaction is (Mohanty *et al.*, 1993):

$$H_2O_2 + Fe^{2+} \rightarrow Fe^{3+} + HO^{\bullet} + HO^{-} \qquad k=51\ M^{-1}s^{-1}$$

resulting in the formation of hydroxyl radicals, remarkably unstable but provided with a high oxidative potential (Table 1). They react with organic substances by hydrogen abstraction and electrophilic addition to double bounds. Reacting with molecular oxygen, they initiate a non-selective sequence of oxidative reactions (dehydrogenation and/or hydroxylation), resulting in demolition of organic pollutants and conversion to CO_2 and inorganic. Hydroxyl radicals may also react with aromatic rings where halogens are attached, thus generating phenol homologues.

Table 1: Redox Potential of the Strongest Oxidants

Oxygen	$O_2 + 4\,H^+ + 4\,e^- \Leftrightarrow 2\,H_2O$	1.23
Hydrogen peroxide	$H_2O_2 + 2\,H^+ + 2\,e^- \Leftrightarrow 2\,H_2O$	1.78
Ozone	$O_3 + 2\,H^+ + 2\,e^- \Leftrightarrow O_2 + H_2O$	2.07
Hydroxyl radical	$HO^{*} + H^+ + e^- \Leftrightarrow H_2O$	2.8
Fluorine	$F_2 + 2\,e^- \Leftrightarrow 2\,F^-$	2.87

Different Fenton's Reagent processes are known:

Chemical

1. Classical
2. Photo-assisted, where the classical FR is enhanced by coupling with UV irradiation (Zepp *et al.*, 1992, Lu *et al.*, 1997, Pignatello, 1992).

Electrochemical

1. Anodic
2. Cathodic
3. Alternative electrochemical processes, with "in situ" production of Ferrous ion and Hydrogen Peroxide addition (Hwang *et al.*, 1999, Cha *et al.*, 1999, Pratap *et al.*, 1998).

Chemically Generated Fenton's Reagent

The use of the Fenton's reagent has been proven effective for the treatment of a wide range of well known organic pollutants such as pentachlorophenol, trihalomethanes (Tang *et al.*, 1997), aromatic amines (Casero *et al.*, 1997), ethylene glycol; a few commercial azodyes and textile effluents (Kang *et al.*, 1997; Szpyrkowicz *et al.*, 2001; Kang *et al.*, 2002) as well as wastewater and landfill leachates (Bae *et al.*, 1997, Kang and Hwang, 2000), nitrophenols (Goi *et al.*, 2002), surfactants (Kitis, 1999) and pesticides (Scott *et al.*, 1995; Pignatello *et al*, 1995). This process has proven interesting also for the employment of low-cost materials and no special equipment (Venkatadri and Peters, 1993). Fenton's reaction has been rarely used for the treatment of inorganic pollutants and one of the rare case is arsenic removal from drinking water (Balarama Krishna *et al.*, 2001).

The initial reaction:

(1) $H_2O_2 + Fe^{2+} \rightarrow Fe^{3+} + HO^{\bullet} + HO^{-}$ $k=51\ M^{-1}s^{-1}$

can be followed by different reactions, depending on the presence / absence of other compounds:

In absence of organic substrate and in H_2O_2 excess:

(2) $HO^{\bullet} + H_2O_2 \rightarrow H_2O + HO_2^{\bullet}$ $k=3.3\bullet10^{7}\ M^{-1}s^{-1}$

Hydro peroxide radicals ($HO_2^{\bullet}$) thus formed, though less oxidant than $HO^{\bullet}$ (E°=-0.30 V), may compete with them and react (in presence of H^{+} ions) with iron bivalent ions:

(3) $Fe^{2+} + HO_2^{\bullet}\ (+ H^{+}) \rightarrow Fe^{3+} + HO_2^{-}\ (+ H^{+}) \rightarrow Fe^{3+} + H_2O_2$

This reaction results in ionisation of hydro-peroxide radicals and formation of H_2O_2, thus dissipating hydroxyl radicals and iron bivalent ions.

In absence of organic substrate and in Fe^{2+} excess, the following parallel reaction may occur:

(4) $Fe^{2+} + HO^{\bullet} \rightarrow Fe^{3+} + HO^{-}$ $k=3{\cdot}10^{8}\ M^{-1}s^{-1}$

Resulting in ionisation of hydroxyl radical and dissipation of both Fe^{2+} ions and $HO^{\bullet}$ radicals.

In *presence of organic substrate* (indicated with R-H) the following reactions may occur:

Hydroxylation

(5) $HO^{\bullet} + R\text{-}H \rightarrow HR\text{-}OH$ $k=10^{9}\text{-}10^{10}\ M^{-1}s^{-1}$

Where the chain reaction is terminated.

Extraction of an Hydrogen Atom

(6) $HO^{\bullet} + R\text{-}H \rightarrow H_2O + R^{\bullet}$ $k=10^7\text{-}10^{10}\ M^{-1}s^{-1}$

Which can be followed by another reaction:

(7) $R^{\bullet} + H_2O_2 \rightarrow ROH + HO^{\bullet}$

These last two reactions are both propagation steps, as radicals are regenerated.

The reactions (2), (3), (4) and in some cases also the hydroxylation (5) are termination steps, as the following:

(8) $2HO^{\bullet} \rightarrow H_2O_2$ $k=5.5 \cdot 10^9\ M^{-1}s^{-1}$

(9) $R^{\bullet} + HO^{\bullet} \rightarrow R\text{-}OH$

(10) $2R^{\bullet} \rightarrow$ Inert dimmer
(if substrate is organic)

The reactions (4), (5), (8), (9) consume $HO^{\bullet}$ radicals, thus reducing the oxidation efficacy of the system. Moreover the reaction (4) may compete with (6), therefore it is necessary to minimise the (Fe^{2+}) / (substrate) ratio in order to prevent the termination reaction (4) and the consequent dissipation of both $HO^{\bullet}$ and Fe^{2+}.

The initiation reaction (1), having a low kinetic constant, results to be the limiting step of the whole process.

From the first reaction (1) it can be observed that bivalent iron ions activate the hydrogen peroxide decomposition in HO^- and $HO^{\bullet}$ but, unlike classical catalysis, they are consumed in the reaction and transformed to trivalent ions. Trivalent iron ions produced in Fenton's reaction (1) can be regenerated to bivalent ions through different chemical reduction reactions, depending on the oxidant agent:

Hydrogen Peroxide (H_2O_2)

(11) $Fe^{3+} + H_2O_2 \rightarrow Fe^{2+} + HO_2^{\bullet} + H^+$ $k=7 \cdot 10^{-6}\ M^{-1}\ s^{-1}$

Hydro-peroxide Radical ($HO_2^{\bullet}$)

(12) $Fe^{3+} + HO_2^{\bullet} \rightarrow Fe^{2+} + O_2\text{-} \uparrow + H^+$

Organic Substrate

(13) $R^{\bullet} + Fe^{3+} \rightarrow R^{+} + Fe^{2+}$

Which is followed by the other reaction:

(14) $R^+ + H_2O \rightarrow R\text{-}OH + H^+$

At usual operating conditions (25°C, pH 3.5), the reactions (11), (12), (13) are extremely slow. Comparing the two kinetic constants of (1) and (11): respectively 51 $M^{-1}s^{-1}$ and $7 \cdot 10^{-6}\ M^{-1}s^{-1}$, it can be concluded that bivalent iron ions consumption is faster than their regeneration, resulting in limitation of hydroxyl radicals production along the process.

The most *significative parameters* influencing this oxidative treatment are:

pH

H^+ is essential for H_2O_2 decomposition to hydroxyl radicals: in acidic solutions, the energy required for electron transfer from Fe^{2+} to H_2O_2 and velocity of peroxide decomposition are minimized. Moreover at pH higher than 4, Fe^{2+} ions may transform in Fe^{3+}, which easily form complexes, dissipating bivalent ions. On the other hand, too strong acidic pH may reduce the treatment efficiency, slowing down the Fenton's reaction (1). In conclusion, pH should be rigorously controlled and kept between 2 and 4.

Temperature

Temperature has two opposite effects on the process: high temperatures promote the reaction of $HO^\bullet$ production (kinetic effect); on the other hand, low temperatures are required, being the process exothermic (ΔH =-98.30 kJ/mol) (thermodynamic effect). Moreover, high temperatures have negative effects on hydrogen peroxide stability. Therefore the process is generally carried out at ambient temperature and in any case always below 40-50°C.

Substrate Nature and Concentration

It was found that $HO^\bullet$ radicals are more reactive towards unsatured hydrocarbons, such as Trichloroethylene (TCE) and substituted phenols, due to double bonds, richer in electrons. Depending on substrate nature, the oxidation may follow different reactions, developing many intermediates and by-products. Substrate concentration influences the quantity of Fenton's Reagent to be added, which must be optimised in order to prevent undesired reactions.

Concentration of Hydrogen Peroxide and Iron Bivalent Ions

On one side it is important to provide a minimal amount of iron bivalent ions, in order to prevent their competition with substrate in the reaction with $HO^\bullet$ radicals (4).

On the other side an excess of hydrogen peroxide though providing a (positive) large amount of $HO^\bullet$ radicals, it may cause other parallel (undesired) reactions.

The careful dosage of the reactive and optimisation of (Fe^{2+}) / (H_2O_2) ratio is therefore of crucial importance.

Procedures for Reagent Addition

Subsequent additions of both reagents may enhance the treatment efficiency mostly when higher concentration are to be removed.

Reaction Time

The complete oxidation of contaminants requires short reaction times. Most treatment are carried out for reaction times between 5 minutes and 2 hours, though more complex substrates may require even some days.

Other Ions Interference

Fenton's Reaction is strongly influenced by the presence of inorganic anions, such as phosphates and chlorides. Anions probably react with iron ions, producing complexes which inhibit hydroxyl radicals. Cl^- may even react with $HO^\bullet$, thus consuming them.

The advantages of Chemical Fenton's treatment are:

1. High removal yields for substances difficult to be degraded,
2. Simple realization, easy introduction in existing plants, low capital costs,
3. Low operating costs, as Fenton's Reagent is completely miscible in water, easy storable and no dangerous at all,
4. No affection by wastewater constituents, thus suitable to treatment of dye wastewater, high molecular weight PAHs, PCE, where other oxidation processes fail.

It must be stressed that this is the simplest way of producing hydroxyl radical because no special reactants or apparatus are required. Furthermore, iron and hydrogen peroxide are non-toxic chemicals so that the whole system appear to be easy to handle, environmentally safe and cost-effective. Nevertheless, the disadvantages of this treatment are:

1. Continuous pH adjusting, necessary to achieve the best operating conditions,
2. High hydrogen peroxide consumption, resulting in necessity of optimized concentration,
3. Rapid iron ions dissipation and slow regeneration,
4. Difficult iron ions disposal, as Fenton Reagent is a homogenous system. A combination of coagulation, flocculation and precipitation is often required.

Cost Analysis

The main costs, depend on the specific wastewater, are due to reagents, sludge treatment, energy operators, maintenance and plant amortization. However the main cost is generally hydrogen peroxide (66.6%), iron salts and sludge disposal.

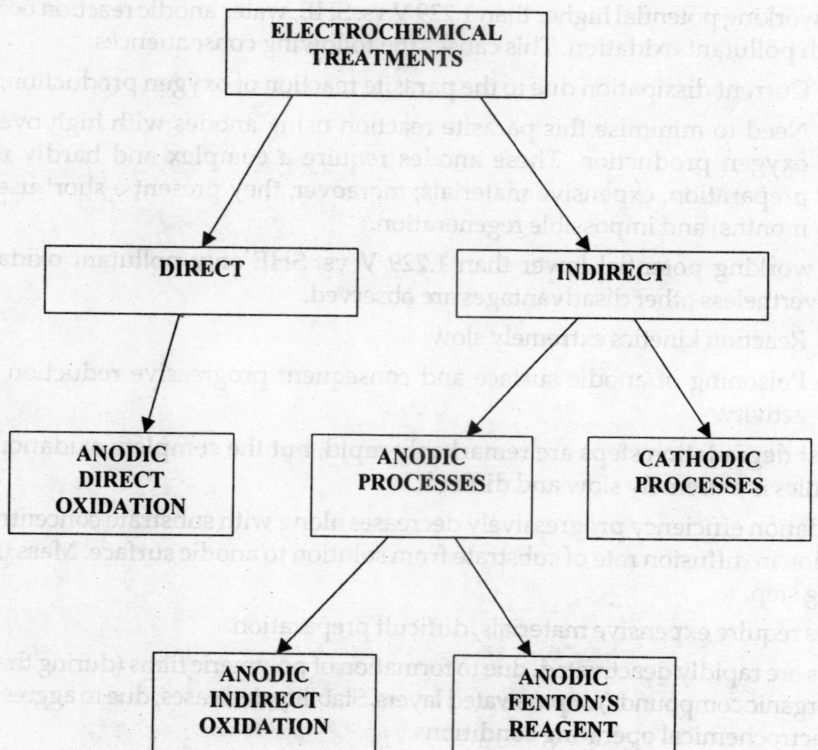

The global cost is low if compared with other treatment with similar efficiency, such as H_2O_2/UV, requiring complex equipment, or Ozonation, requiring expensive on-site ozone production and storage.

Electrochemically Generated Fenton's Reagent

Electrochemical processes offer promising solutions for environmental problems. Some of their most attractive characteristics are: environmental compatibility (the main reagent, the electron, is "clean" and it often does not require the addition of extra reagents), versatility, energy efficiency, suitability to automation and cost-effectiveness.

It is possible to classify electrochemical treatment according to the procedure employed for conversion reactions of pollutants to less harmful products or to more degradable intermediates. In particular, oxidation processes can occur directly on inert electrodes without involving other substances ("direct" processes) or through an electrochemically generated redox reagent, which acts as an intermediary [as a chemical reactant (or catalyst)], shuttling electrons between the substrate and the electrode ("indirect" processes). In order to better understand this classification, a scheme with the different processes is shown.

Anodic Direct Oxidation

In direct oxidation processes the substrate is oxidized at the anode, by direct electron transfer (Brillas *et al.*, 1998). Anodic direct oxidation, though allowing degradation of many organic and inorganic compounds, presents several disadvantages:

1. The process requires high anodic potentials (or high cell currents), with consequent high energy consumption.. In particular:
   (i) At working potential higher than 1.229 V vs. SHE, water anodic reaction occurs together with pollutant oxidation. This causes the following consequences:
      (a) Current dissipation due to the parasite reaction of oxygen production;
      (b) Need to minimise this parasite reaction using anodes with high overvoltage for oxygen production. These anodes require a complex and hardly reproducible preparation, expensive materials; moreover, they present a short useful life (few months) and impossible regeneration.
   (ii) At working potential lower than 1.229 V vs. SHE only pollutant oxidation occurs. Nevertheless other disadvantages are observed.
      (a) Reaction kinetics extremely slow;
      (b) Poisoning of anodic surface and consequent progressive reduction of electrode activity.
2. The first degradation steps are remarkably rapid, but the complete oxidation to CO_2 and inorganics is extremely slow and difficult.
3. Degradation efficiency progressively decreases along with substrate concentration, due to reduction in diffusion rate of substrate from solution to anodic surface. Mass transfer is the limiting step.
4. Anodes require expensive materials, difficult preparation.
5. Anodes are rapidly deactivated, due to formation of polymeric films (during the treatment of some organic compounds) or passivated layers. Stability decreases, due to aggressive chemical and electrochemical operating conditions.

Anodic Indirect Processes

Special anodes allow hydroxyl radicals production according to the following reaction:

(15) $H_2O \rightarrow HO\cdot + H^+ + e^-$

Substrates are then successfully degraded following the reactions (6) and (7). This treatment presents some disadvantages:

1. High anodic potential are required to generate $HO^\bullet$ free radicals ($E > E_0 = +2.74$ V vs. SHE) or $HO^\bullet$ radicals chemisorbed on the electrode surface ($E > E_0 = +0.208$ V). This causes the following consequences:
   (i) High energetic consumption
   (ii) Requirement of special anodes, with expensive materials and complex preparation
2. As substrate concentration decreases, $HO^\bullet$ radicals concentration becomes the controlling factor, attaining zero some Mm from anodic surfaces and decreasing along time, as electrode ages.
3. Anodes useful life is very short (1 or 2 months), due to aggressive chemical and electrochemical operating conditions.

Anodic Fenton's Reagent

It is possible to produce hydroxide peroxide at the anode, in order to realize an electrochemical "anodic" Fenton. At the anode water can be directly oxidized to H_2O_2 (electrochemical potential: E= 1, 77 V):

(16) $H_2O \rightarrow ½\, H_2O_2 + H^+ + e^-$

The presence of Fe^{2+} ions, in combination with H_2O_2, might lead to bulk generation of $HO^\bullet$ radicals according to the Fenton reaction (1), quite effective at low pH values (about 2) reached at the anodic compartment of the cell. This reaction can be achieved together with direct substrate oxidation at anode.

Nevertheless, this process presents some disadvantages typical of other anodic processes, such as expensive materials and rapid deactivation. Above all, with anodic generation of Fenton's reagent, electro-regeneration of ferrous ions is not possible.

Cathodic Fenton

Chemical and anodic Fenton system present two main limitations:

1. The regeneration of Fe^{2+} via reaction ($R + Fe^{3+} \rightarrow Fe^{2+}$ + product) is less efficient than electrochemical reduction of Fe^{3+}.
2. A continuous supply of H_2O_2 is required.

In fact, in a Cathodic Fenton process (Alvarez Gallegas *et al.*, 1998, Brillas *et al.*, 1998, Do *et al.*, 1984, Hsiao *et al.*, 1993, Oturan *et al.*, 2000, Oturan *et al.*, 2000, Oturan *et al.*, 2001, Panizza *et al.*, 2001, Ponce *et al.*, 1995, Pulgarin *et al.*, 1994, Qiang *et al.*, 2002, Sudon *et al.*, 1986, Tomat *et al.*, 1979, Tomat *et al.*, 1984, Tzedakis *et al.*, 1989), H_2O_2 can be electrochemically generated by reduction of dissolved oxygen in acidic solution at selected cathodes, following the reaction:

(17) $O_2 + 2\, H_3O^+ + 2e^- \rightarrow H_2O_2 + 2\, H_2O$

Oxygen is constantly sparged in the solution and it progressively adsorbs on the electrode. The reduction yield depends on the medium and on oxygen solubility: being apolar, O_2 is more soluble in organic solvents than in water; at 25 °C, 1 atm, its saturation concentration is 1mM. The limiting step of this reaction is the transfer of the first electron, which generates $O_2^{\bullet-}$, strongly alkalinic (following Brœnsted definition):

(18) $O_2 + e^- \rightarrow O_2^{\bullet-}$ $E_0 = -0.33$ V vs NHE

If electrons are supplied, it is possible to generate (electrochemically) all the other reduced forms of oxygen (HO_2^-, H_2O_2, $HO^\bullet$, $HO_2^\bullet$, H_2O, HO^-) through different reactions.

In aqueous solution $O_2^{\bullet-}$ is rapidly converted to O_2 and HO_2^-

(19) $2\,O_2^{\bullet-} + H_2O \rightarrow O_2 + HO_2^- + HO^-$ $E_0 = 0.2$ V vs NHE

On inert electrode and with protons excess a rapid chemical reaction can occur:

(20) $O_2^{\bullet-} + H_3O^+ \rightarrow HO_2^\bullet + H_2O$

Hydroperoxide radical is a weak acid ($pK_a = 4.89$), which can give homolythic disproportion:

(21) $HO_2^\bullet + HO_2^\bullet \rightarrow H_2O_2 + O_2$ $k = 8.6.\ 10^5\ M^{-1}\ s^{-1}$ $K_{(25°C)} = 10^{25}$ atm $E_0 = 1.44$ V vs NHE

Hydroperoxide radical may react with $O_2^{\bullet-}$ (at pH lower than 5):

(22) $HOO^\bullet + O_2^{\bullet-} \rightarrow HOOH + O_2 + H_2O$ $E_0 = 1.44$ V vs NHE (a pH 0)

In this Fenton's system the reduction of Fe^{3+} to Fe^{2+} also occurs at the cathode:

(23) $Fe^{3+} + e^- \rightarrow Fe^{2+}$ E_0=+0.771 V vs. SHE

The H_2O_2 generated at the cathode (17) can be coupled with Fe^{2+} in order to produce the Fenton reagent (1). The cathodic regeneration of Fe^{2+} in acidic solution allows providing this reagent continuously to the system, thus preventing the rapid decrease in irons ions, which is observed in chemical Fenton method.

The results attained from studies carried on until now prove that Electrochemical Fenton method, eventually associated to U.V. irradiation, is one of the most promising electrochemical systems for wastewater treatment, worthy of further studies for large scale applications.

Though a few researchers have studied H_2O_2 electro-generation in acidic solutions, there exist contrasting results. Sudoh (Sudoh *et al.*, 1986) investigated the decomposition of aqueous phenol by electro-generated Fenton's reagent, obtaining the highest current efficiency (85%) with a cathodic potential of –0, 6 Volt vs. a saturated Ag/AgCl electrode (SSE) and pH 3. Tzedakis (Tzedakis *et al.*, 1989) reported that the electrolysis of an oxygen-saturated H_2SO_4 solution (0, 6 M), using a stirred mercury pool electrode yielded a current efficiency of 55% at a cathodic potential of –0, 30 V vs. SCE. Chou (Chou *et al.*, 1999) used the electro Fenton process to remove aqueous chlorophenols at cathodic potential of –0.6 V vs. SCE and reported that the current efficiency increased with decreasing pH. Hsiao and Nobe (Hsiao and Nobe, 1993) studied the oxidative hydroxylation of phenol and chlorobenzene using electro-generated Fenton's reagent, thus attaining an optimal cathodic potential of –0.55 V vs. SCE and a H_2O_2 generation favoured at low pH values.

The *significative parameters* influencing the Electro-Fenton process are:

pH

H^+ is essential in Fenton treatment. At pH higher than 4, Fe^{2+} ions may transform in Fe^{3+}, which easily form complexes and precipitating hydroxides, thus dissipating bivalent ions. Therefore, pH should be rigorously controlled and kept at 2.

Stirring

This is important to enhance oxygen mass transfer coefficient and to ensure reagent mixing.

Electrolysis Voltage

Voltage influences current efficiency for production of hydrogen peroxide. The potential (or current density) must be fixed in order to allow reduction reaction of both Fe^{3+} and O_2.

Temperature

Low temperatures are required, being the process exothermic (ΔH =-98.30 kJ/mol). Moreover, low temperatures promote oxygen solubility. Therefore the process is generally carried out at ambient temperature, 20°C.

Substrate Nature and Concentration

It was found that $HO^{\bullet}$ are more reactive towards unsatured hydrocarbons, such as Trichloroethylene (TCE) and substituted phenols, due to the double bond, richer in electrons. Moreover, the initial substrate concentration may influence the degradation efficiency.

Concentration of Hydrogen Peroxide and Iron Bivalent Ions

It is important to provide a minimal amount of iron bivalent ions, nevertheless it is necessary to prevent their competition with substrate in the reaction with $HO^{\bullet}$ radicals (4). Therefore the (Fe^{2+}) / (substrate) and (Fe^{2+}) / (H_2O_2) ratio should be optimised.

The Electrochemical Fenton shows some disadvantages:

1. Limiting diffusion of oxygen from solution to cathode surface and low oxygen solubility.
2. Need to add a bivalent iron salt to the substrate solution.
3. "Expected" current yields for substrate removal lower than the first ones in cathodic processes.
4. Accumulation of H_2O_2 at cathode/solution interface, resulting in partial decomposition.
5. Possible competition between high concentration protons and oxygen, resulting in gaseous hydrogen release.

The last two problems reduce the current efficiency of H_2O_2 production. Therefore, in acidic solutions, cathodic potential and solution pH are two essential factors to control the current efficiency.

The main advantages of electro-Fenton processes include:

1. Small amounts of reagents (Fe^{2+}) are required to attain substrate removal in relatively short times (thus reducing problems due to limiting oxygen diffusion and solubility (disadvantage 1.).
2. Work potential is lower than the anodic one (E < +0.682 V vs. SHE for oxygen reduction in acidic solution, E < +0.774 V vs. SHE for Fe^{3+} reduction). Moreover, it should be possible to

fix the potential in order to minimize undesired reactions, causing current dissipation or damages to electrodes.

3. $HO^{\bullet}$ are produced in bulk solution, thus eliminating mass transfer problems (of radicals from electrode to solution and substrate from solution to electrode)
4. The oxidative efficiency should not decrease along with pollutant concentration, as mass transfer problems are reduced.
5. $HO^{\bullet}$ concentration depends on solution conditions, easy to control, and not from electro-catalytic cathode activity.
6. This system employs low cost cathodes, with inexpensive materials, easy preparation, long useful life, not exposed to aggressive conditions, specially if compared with anodic potential.
7. H_2O_2 can be continuously generated on-site whenever needed, thus eliminating acquisition, shipment and storage.
8. Dilute H_2O_2 solutions ensure safety during material handling.
9. Production process can be simply carried out at ambient pressure and temperature.
10. Fe^{2+} can be electrochemically regenerated at the cathode, thus minimizing iron sludge production and not requiring chemical precipitation.
11. Oxygen or air sparging enhances solution mixing.
12. Reagent dosage is not required.

Cost Analysis

The most relevant cost is represented by electric energy consumption (66%), lower expenses are due to oxygen and iron salts (30%).

Electro-Fenton treatment results to be more expensive than Chemical Fenton, also taking into account operating and construction costs. However, Electro-Fenton process may offer new potentiality for the complete oxidation of persistent substances.

Perspectives

Both Chemical and Electrochemical Fenton's processes represent two interesting technologies for wastewater treatment, due to their versatility and high efficiency. Nevertheless Electro-Fenton, being more recent, leaves more potentialities for further study and improvement. In particular, process enhancement and plant design are the most important aspects to be further developed.

Process

In order to develop the process in view of large scale application, it is necessary to optimize the most significative parameters, such as pH, oxygen concentration, voltage and to study the chemical engineering problems, such as mass transfer. Moreover the process may become attractive and cost-effective if valuable by-products are generated. As a matter of fact, it is possible to employ anodic compartment to produce gaseous oxygen, eventually recycled in the cathodic compartment:

(24) $2H_2O \rightarrow 4H^+ + O_2 + 4e^-$ $E_0 = 0,987$ V vs. SCE

Hence protons are generated, which can be electro-statically driven to the catholyte and partially compensate protons consumption during H_2O_2 synthesis.

Plant

In order to enhance the suitability to large scale the following features should be further studied and developed:

1. Cathodes with high specific surface for oxygen adsorption and reduction, in order to minimise mass transfer limitation.
2. Enhancement of cathode electrocatalythic activity, employing for instance a preliminary anodization treatment, in order to increase the yield of oxygen reduction.
3. Cells promoting permeation of oxygen in the electrode, employing a forced convection flux for the electrolyte.
4. Recycling of oxygen at the solution exit and reintegration at the cathodic surface.
5. Small superpressure, in order to enhance the oxygen partial pressure and its solubility in the medium.

REFERENCES

1. Alvarez-Gallegos A., Pletcher D., 1998. The removal of low level organics via hydrogen peroxide formed in a reticulated vitreous carbon cathode cell. Part 1: the electrosynthesis of hydrogen peroxide in aqueous acidic solutions, Electrochimica Acta 44, 853-861.
2. Alvarez-Gallegos A., Pletcher D., 1998. The removal of low level organics via hydrogen peroxide formed in a reticulated vitreous carbon cathode cell. Part 2: the removal of phenols and related compounds from aqueous solutions, Electrochimica Acta 44, 2483-2492.
3. Bae J., Kim S., Chang H., 1997. Treatment of landfill leachates; ammonia removal via nitrification and denitrification and further COD reduction via Fenton's treatment followed by activated sludge. Wat. Sci. Tech. 36, 341-348.
4. Balarama Krishna M.V., Chandrasekaran K., Karunasagar D., Arunachalam J., 2001. A combined treatment approach using Fenton's Reagent and zero valent iron for the removal of arsenic from drinking water. J. Haz. Mat. B84, 229-240.
5. Beltran F.J., Encinar J.M., González J.F., 1997. Industrial Wastewater Advanced Oxidation. Part 2. Ozone combined with Hydrogen Peroxide or UV Radiation. Wat. Res. 31 (10), 2415-2428.
6. Beltran F.J., Gonzalez M., Gonzalez J.F., 1997. Industrial Wastewater Advanced Oxidation. Part 1. UV Radiation in the presence and absence of Hydrogen Peroxide. Wat. Res. 31 (10), 2445-2414.
7. Brillas E., Mur E., Sauleda R., Sanchez L., Peral J., Domenech X., Casado J. 1998. Aniline Mineralization by AOP's: anodic oxidation, photocatalysis, electrofenton and photoelectrofenton processes. Applied Catalysis. B. 16. 31-42.
8. Brillas E., Sauleda R., Casado J., 1998. Degradation of 4-chlorophenol by anodic oxidation, electro-Fenton, photoelectro-Fenton, and peroxi-coagulation processes. J. Electrochem. Soc. 145, 759-765.
9. Casero I., Sicilia D., Rubio S., Pérez-Bendito D., 1997. Chemical Degradation of Aromatic Amines by Fenton's Reagent. Wat. Res. 31 (8), 1985-1995.
10. Chiron S., Fernandez-Alba A., Rodriguez A., Garcia-Calvo E., 2000. Pesticide chemical oxidation: state of the art", Wat.Res. 34 (2), 366-377.

11. Chou S., Yao-Hui Huang, Shen-Nan Lee, Gaw-Hao Huang, Chihpin Huang, 1999. Treatment of high strength hexamine containing wastewater by electro-fenton method. Wat.Res. 33 (3), 751-759.

12. Comninellis Ch., Pulgarin C., 1991. Anodic oxidation of phenol for wastewater treatment. J. Appl. Electrochem. 21, 703-708.

13. Comninellis Ch., Pulgarin C., 1993. Electrochemical exidation of phenol for wastewater treatment using tin dioxide anodes. J. Appl. Electrochem. 23, 108-112.

14. Do J.S., Chen P., 1994. *in situ* oxidative degradation of formaldehyde with hydrogen peroxide electrogenerated on the modified graphite. J. Appl. Electrochem. 24, 936-942

15. Fenton H.J.H., 1894. Oxidation of Tartaric Acid in presence of Iron. J. Chem. Soc. 65, 899-910.

16. Goi A., Trapido M., 2002. Hydrogen peroxide photolysis, Fenton reagent and photo-Fenton for the degradation of nitrophenols: a comparative study. Chemosphere 46, 913-922.

17. Haber F. and Weiss J., 1934. The Catalytic Decomposition of Hydrogen Peroxide by Iron Salts Proc. R. Soc., A, 147.

18. Hoigné J. and Bader H., 1979. Ozone. Sci. Eng., 1, 357-372.

19. Hsiao Y. L., Nobe K., 1993. Hydroxylation of chlorobenzene and phenol in a packed bed flow reactor with electrogenerated Fenton's reagent. J. Appl. Electrochem. 23, 943-945.

20. Hsiao Y. L., Nobe K., 1993. Oxidative reactions of phenol and chlorobenzene with *in situ* electrogenerated Fenton's reagent. Chem Eng Commun 126, 97-110.

21. Kang S.F., Chang H.M., 1997. Coagulation of Textile Secondary Effluents with Fenton's Reagent. Wat. Sci. Tech. 36 (12), 215-222.

22. Kang S.F., Liao C.H., Chen M.C., 2002. Preoxidation and coagulation of textile wastewater by the Fenton process. Chemosphere 46 (6), 923-928.

23. Kang Y.W., Hwang K.Y., 2000. Effects of reaction conditions on the oxidation efficiency in the Fenton process, Wat. Res. 34 (10), 2786-2790.

24. Kitis M., Adams C. D., Daigger G.T., 1999. The effects of Fenton's Reagent pretreatment on the biodegradability of nonionic surfactants. Wat. Res. 33 (11), 2561-2568.

25. Kwon B.G., Lee D.S., Kang N., Yoon J., 1999. Characteristics of *p*-chlorophenol oxidation by Fenton's reagent. Wat. Res. 33 (9), 2110-2118.

26. Lu M.C., Chen J.N., Huang C.P., 1997. Effects of Inorganic Ions on the Oxidation of Dichlorvos Insecticide with Fenton's Reagent. Chemosphere 35 (10), 2285-2295.

27. Maillard C., Guillard C., Pichat P., 1992. Comparative Effects of the TiO_2-UV, H_2O_2-UV, H_2O_2-Fe++ Systems on the Disappearance Rate of the Benzamide and 4-hydroxibenzamide in Water. Chemosphere 24 (8), 1085-1094.

28. Mohanty N.R., Wei I.W., 1993. Oxidation of 2, 4-Dinitrotoluene Using Fenton's Reagent: Reaction Mechanisms and Their Practical Applications. Hazardous Waste & Hazardous Materials 10 (2), 171-183.

29. Oturan M.A., 2000. An ecologically effective water treatment technique using electrochemically generated hydroxyl radicals for *in situ* destruction of organic pollutants: Application to herbicide 2, 4-D. J.Appl. Electrochem. 30, 475-482.

30. Oturan M.A., Oturan N., Lahitte C., Trevin S., 2001. Production of hidroxyl radicals by electrochemically assisted Fenton's reagent. Application to the mineralization of an organic micropollutant, pentachlorophenol. J. Electroanalysis Chem. 507, 96-102.

31. Oturan M.A., Peiroten J., Chartrin P., Acher A.J., 2000. Complete destruction of p-nitrophenol in aqueous medium by electro-fenton method. Environ. Sci. Technol. 34, 3474-3479.

32. Panizza M., Cerisola G., 2001. Removal of organic pollutants from industrial wastewater by electrogenerated Fenton's reagent. Wat.Res 35 (16), 3987-3992;

33. Pignatello J.J., 1992. Dark and Photoassisted Fe^{3+}-Catalysed Degradation of Chlorophenoxy Herbicides by Hydrogen Peroxide. Environ. Sci. Technol. 26, 944-951.

34. Pignatello J.J., SUN Y., 1995. Complete Oxidation of Metolachlor and Methyl Parathion in Water by the Photoassisted Fenton Reaction. Wat. Res. 29 (8), 1837-1844.

35. Ponce De Leon C., Pletcher D., 1995. Removal of formaldehyde from aqueous solutions via oxygen reduction using a reticulated vitreous carbon cathode cell. Journal of applied electrochemistry 25, 307-314.

36. Pratap K, Ann T. Lemley, 1998. Fenton electrochemical treatment of aqueous atrazine and metolachlor. J. Agric. Food Chem. 46, 3285-3291.

37. Pulgarin, C., Adler, N., Peringer, P., and Comninellis, C., 1994. Electrochemical detoxification of a 1, 4-benzoquinone solution in wastewater treatment. Water Res., 28, 887.

38. Qiang Z., Jih-Hsing Chang, Chin-Pao Huang, 2002. Electrochemical generation of hydrogen peroxide from dissolved oxygen in acidic solutions. Wat.Res. 36, 85-94.

39. Scott A.M., Hickey W.J., Harris R.F., 1995. Degradation of Atrazine by Fenton's Reagent: Condition Optimization and Product Quantification. Environ. Sci. Technol. 29, 2083-2089.

40. Sudoh M., Kodera T., Sakai K., Zhang J. Q., Koide K., 1986. Oxidative degradation of aqueous phenol effluent with electrogenerated Fenton's reagent. J. Chem Eng. of Japan 19 (6), 513-517.

41. Szpyrkowicz L., Juzzolino, Santosh Kaul N., 2001. A Comparative Study on Oxidation of Disperse Dyes by Electrochemical Process, Ozone, Hypochlorite and Fenton Reagent. Wat. Res. 35 (9), 2129-2136.

42. Tang W.Z., Tassos S., 1997. Oxidation Kinetics and Mechanisms of Trihalometanes by Fenton's Reagent. Wat. Res. 31 (5), 1117-1125.

43. Tomat R., Rigo A., 1979. Oxidation of polymethylated benzenes promoted by $HO^{\bullet}$ radicals", J. Appl. Electrochem. 9, 301-305.

44. Tomat R., Rigo A., 1984. Electrochemical oxidation of toluene promoted by $HO^{\bullet}$ radicals. J. Appl. Electrochem. 14, 1-8.

45. Tzedakis T., Savall A., Clifton M. J., 1989. The electrochemical regeneration of Fenton's reagent in the hydroxylation of aromatic substrates: batch and continuous processes. J. Appl. Electrochem. 19, 911-921.

46. Venkatadri R., Peters R.W., 1993. Chemical Oxidation Technologies: Ultraviolet Light/Hydrogen Peroxide, Fenton's Reagent and Titanium Dioxide-Assisted Photocatalysis. Hazardous Waste & Hazardous Materials 10 (2), 107-149.

47. Weiss J., 1935. The Catalytic Decomposition of Hydrogen Peroxide on different Metals. Trans. Farad. Soc. 3L, 1547-1557.

48. Yao-hui Huang, Shanshan Chou, Ming-Ging Perng, Gaw-Hao Huang and Sheng-Shung Cheng., 1999. Case study on the bioeffluent of petrochemical wastewater by electro-fenton method. Wat. Sci.Tech. 39 (10-11), 145-149.

49. Zepp R.G., Faust B.C., Hoigné J., 1992. Hydroxyl Radical Formation in Aqueous Reactions (pH 3-8) of Iron (II) with Hydrogen Peroxide: the Photo-Fenton Reaction. Environ. Sci. Technol. 26, 313-319.

5

Electrochemical Methods for Pollution Abatement

DE BATTISTI ACHILLE, FERRO SERGIO, MARTINEZ-HUITLE, CARLOS A., URGEGHE, CHRISTIAN

Dipartimento di Chimica dell'Università degli Studi di Ferrara; via L. Borsari, 46; I-44100 Ferrara; Italia

ABSTRACT

The intensification of industrial activities since the second half of XIX century and during all the XX century has inevitably caused severe environmental pollution with dramatic consequences in atmosphere, waters, soils. The consequent restrictions imposed by new legislations require effective initiatives for pollution abatement not only in gaseous emissions and industrial aqueous effluents, but also adequate decontamination in soils. Typically, in the latter case different classes of pollutants may have accumulated during long periods of uncontrolled waste disposal and reclamation may represent a serious technological problem. In all these cases universal strategies of reclamation can hardly be found, due to the very different features of pollution phenomena. As far as organic pollutants in wastewaters are concerned, biological abatement may be sometimes impossible, due to the biorefractory character of substrates. Use of chemicals, like ozone or chlorine dioxide is not always effective and also transportation and storage of reactants may be a significant inconvenience for safe processing. In these cases the electrochemical incineration, direct or mediated, may be an interesting alternative. Basically, the electrochemical incineration proceeds under the action of strong oxidants as well as the chemical incineration, but the in-situ electrogeneration allows better efficiency of the abatement of the organic substrates and avoids the need for transportation and storage.

The direct electrochemical incineration has been studied with particular interest since the end of the eighties-beginning of the nineties (*e.g.*: (1) and literature therein). The influence of the nature of the electrode material on the efficiency of the process has been widely discussed and better performance has been found at high-oxygen-evolving electrodes, like PbO_2 (2) and SnO_2-based electrodes (1). A mechanism considering the different stabilization exerted by the electrode material on electrosorbed hydroxyl radicals has been proposed by Ch. Comninellis (1). The above interpretation has been widely confirmed by recent data on Highly-Boron-Doped Diamond Electrodes (3, 4). The mediated electrochemical incineration, based on species like Cl^-, $S_2O_8^-$, Ce (III), may also be an interesting alternative. Cl^--mediated mineralization has been shown to give good results at low-oxygen-overvoltage electrodes, like Pt (5, 6). On the other hand, the CerOx process, based on the action of electrogenerated Ce (IV) supplies an interesting alternative to the direct incineration, basing on a volume-, rather than surface reaction (*e.g.*: (7).

Another important application of electrochemisty in environmental remediation is the electrokinetic reclamation of polluted soils (e.g.: (8) and literature therein). In this case migration of charged pollutants, as well as electroosmotic transport of neutrals have been proved to supply a very effective way for decontamination, not only of soils, but also sludges.

REFERENCES

1. Comninellis Ch., 1994, Electrochim. Acta 1994. 39 (11-12), 1857-62.
2. Savall A., Tahar N. B., Manaranche C., Elaoud S.C., Abdelheidi R., 1999, Trans. SAEST. 34 (3-4), 93.
3. Gherardini, L., Michaud, P. A., Panizza, M., Comninellis, Ch., Vatistas, N., 2001, Journal of the Electrochemical Society. 148 (6), D78-D82.
4. Panizza, M., Michaud, P. A., Cerisola, G., Comninellis, Ch., 2001, Electrochem.Comm. 3 (7), 336-339.
5. Boscolo Boscoletto, A., Gottardi, F., Milan, L., Pannocchia, P., Tavan, M., Amadelli, R., De Battisti A., 1994, J. Appl. Electrochem. 24 (10), 1852-8.
6. Comninellis Ch., Nerini A., 1995, J. Appl. Electrochem. 25 (1), 23-8.
7. Nelson N.J., 2002, Platinum Metals Review. 46 (1), 18-23.
8. Page M.M., Page L.P., 2002, J. Environm. Technol. 208.

6

Laboratory Study on Electrokinetic Extraction of Heavy Metals from Contaminated Soil

ASOK ADAK & MANAS BANDYOPADHYAY

Indian Institute of Technology, Kharagpur - 721302, West Bengal

ABSTRACT

Electrokinetic Extraction is an innovative technology for the remediation of fine grained soils like clay contaminated with heavy metals where other technologies like Soil Washing, Immobilization, Metlcap method etc. are not suitable due to low permeability of fine grained soils. Laboratory studies were conducted with synthetically contaminated soils under different test conditions as such variation of voltage, variation of head, different soil pH, and different initial concentrations of heavy metals-Zn, Cd, Cu, and Ni). From the study it has been found that about 80% of heavy metals can be extracted in the processing time of about 225 hours. Use of surfactant can help to increase the removal efficiency. Also the increase of processing time increases the removal efficiency.

***Key Word:** Electrokinetic Extraction, Soil Washing, Immobilization, Metlcap method, Surfactant*

Introduction

The release of heavy metals in the environment has become a great concern because of their accumulation in the environmental compartments (soil, water and air) that would impart toxic effects on biotic life. The main anthropogenic sources of heavy metals are various industrial point sources, including present and former mining activities, foundries and smelters, and diffuse sources such as piping, constituents of products, combustion by-products, traffic, etc.

There are many soil remediation methods like soil washing, soil flushing, immobilization, electrokinetic extraction, etc. The electrokinetic remediation is the soil remediation method in which low direct current is applied through the soil and water is passed from anode to cathode. The electrokinetic method relies upon the advection generated by electro-osmotic flow and migration of ions in the electric field. Electro-osmosis is a phenomenon in which pore fluid in a porous medium moves due to application of direct current. Figure 1 represents a schematic diagram of the ion flow during electro-osmosis. The hydraulic potentials maintained constant at both ends. When a direct current is applied across the soil specimen an acid front is generated at the anode by the electrolysis reaction. A corresponding basic front appears at the cathode. In tests with an open electrode configuration, this acid front advances toward the cathode in time by the imposed electrical, chemical and hydraulic potential gradients.

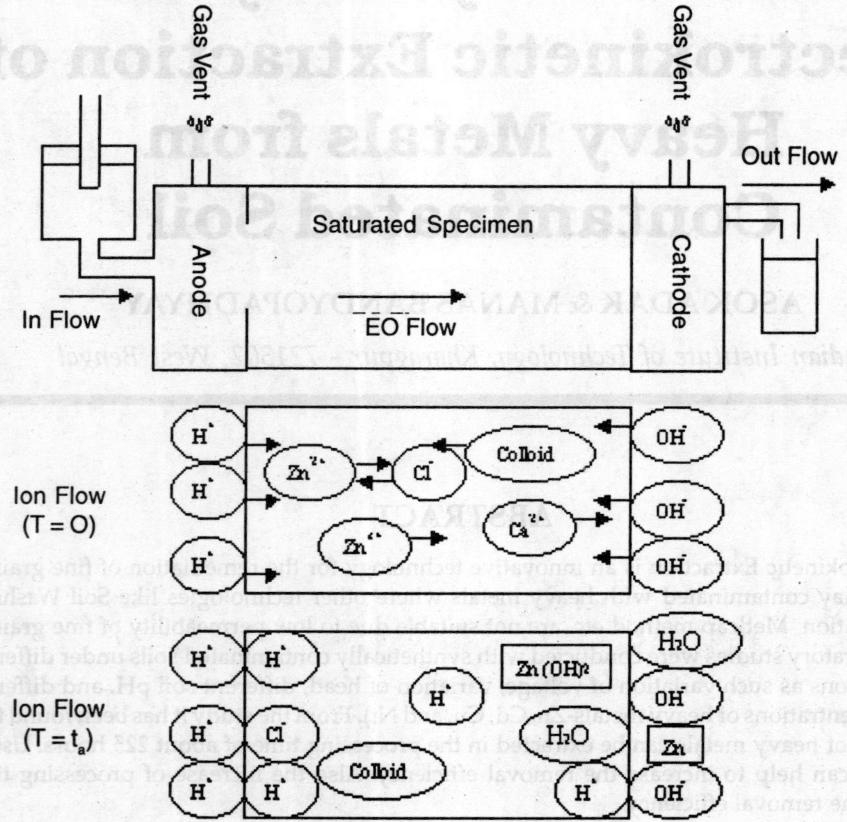

Fig. 1: Schematic of Electrokinetic Soil Processing and Ion Flow

The acid front eventually flushes across the specimen and neutralizes the base generated at the cathode. The movement of the acid front by migration (electrical potentials), diffusion (chemical potentials) and advection (hydraulic potentials) will cause the desorption of cations and other species from clay surfaces and facilitates their release into the pore fluid. This phenomenon together with the concurrent electro-osmotic flow would then constitute the mechanism for removing contaminants from soils.

In the absence of all but traces of electrolyzable solutes in the pore fluid (*i.e.* metal ions, Cl^-, organic matter, pollutants, dissolved oxygen etc.) and for the case of inert electrodes the following primary electrolysis reactions occur at the anode and cathode inserted in a saturated soil mass.

$$2H_2O - 4e^- \rightarrow O_2 + 4H^+ \text{ (anode)} \quad (1)$$

$$4H_2O + 4e^- \rightarrow 2H_2 + 4OH^- \text{ (cathode)} \quad (2)$$

Secondary reactions may exist depending upon the concentrations of available species, *e.g.*

$$H+ + e^- \rightarrow \tfrac{1}{2} H_2 \quad (3a)$$

$$Zn^{2+} + 2e^- \rightarrow Zn \quad (3b)$$

$$Pb^{2+} + 2e^- \rightarrow Pb \quad (3b)$$

The production of H^+ ions at the anode decreases the pH by the Eq. 1 reaction and the Eq. 2 reaction increases the OH^- concentration at the cathode, increasing the pH. For the same quantity of electricity twice as many water molecules are electrolyzed at the cathode than at the anode, producing a chemical gradient of molecular water. Thus as a consequence of reactions in Eq. 1 and 2, in addition to the migration of existing anions and cations in the pore fluid of the specimen under the electric field, two supplemental ionic species, H+ and OH-, are generated that can have a significant influence on local conductance. Figure 1 shows this changing chemistry across the cell.

In electro-osmosis, the flow rate, q_e (cm3/s), is conventionally related to potential gradient i_e (V/cm), the current I (amp) by an empirical relationship as given below.

$$q_e = k_e i_e A = k_i I \quad (4)$$

Where

k_e is coefficient of electro-osmotic permeability (cm2/V-s); k_i is coefficient of water-transport efficiency (cm3/amp-s) and A is cross-sectional area of the specimen; k_e and k_i are considered to be constants for a specified soil (Mitchell 1976).

In the last five decades electro-osmosis has been investigated and used for different applications, such as improving stability of excavations (Chappell and Burton 1975); increasing pile strength (Butterfield and Johnston 1980); stabilization of fine-grained soils (Mitchell and Wan 1977); dewatering of sludges (O'Bannon *et al.* 1976); ground-water-lowering and barrier systems (Gray 1969); chemical grout injection (Esrig 1967); removal of metallic objects from the ocean sea bottom (Esrig and Henkel 1966); decreasing pile penetration resistance (Thompson 1971); increasing petroleum production (Amba *et al.*, 1964); determination of volume change and consolidation characteristics of soils (Banerjee and Mitchell 1980); and separation and filtration of materials in soils and solutions (Yukawa *et al.* 1971). Jacobs and Mortland (1959) demonstrated that Na^+ and K^+ could be leached out of the clays by electro-osmosis. Krizek *et al.* (1976) showed that the soluble-ions content substantially increased in the effluent in electro-osmotic consolidation of polluted dredging. Segal *et al.* (1980) discovered that heavy metals concentration and organic carbon content substantially increased in the effluent after electro-osmotic consolidation sludges. Shmakin (1985) notes that the technique has been used in the Soviet Union since the early 1970s as a method for concentrating metals and exploring minerals in deep soil deposit. Shmakin (1985) mentions its use in the USSR in prospecting for Cu, Ni, Co and Au. Runnel and Larson (1986) demonstrated that $CuSO_4$ at a concentration of 617 ppm could be moved from silty sand by electro-osmosis.

Materials and Methods

Equipments

The pH of the soil was determined with the help of digital pH-meter (Syberscan 510 pH). Electronics balance (sartorious) and ordinary was used for weighing chemicals and soil. Zinc concentration was measured by Atomic Absorption Spectrophotometer (Shimadzu AA-6650). The voltage and current were measured by multimeter. For supplying direct current to the electric cell an adopter was made with the capacity of 39 volt.

Laboratory Setup

Electrokinetic test specimens were assembled as presented in Figure 2. Inert perforated graphite electrodes were selected to prevent introduction of other corrosion products that might cause complicated electrochemistry due to electrode-electrolysis products. Two sheets of Whatman50 filter papers were placed at the ends of specimen. The electrodes were held in place with polyacrylate end caps connected tightly with the cell. Water was supplied from anode side from constant head reservoir. Holes were made into the top of each end cap to allow venting gaseous electrolysis products.

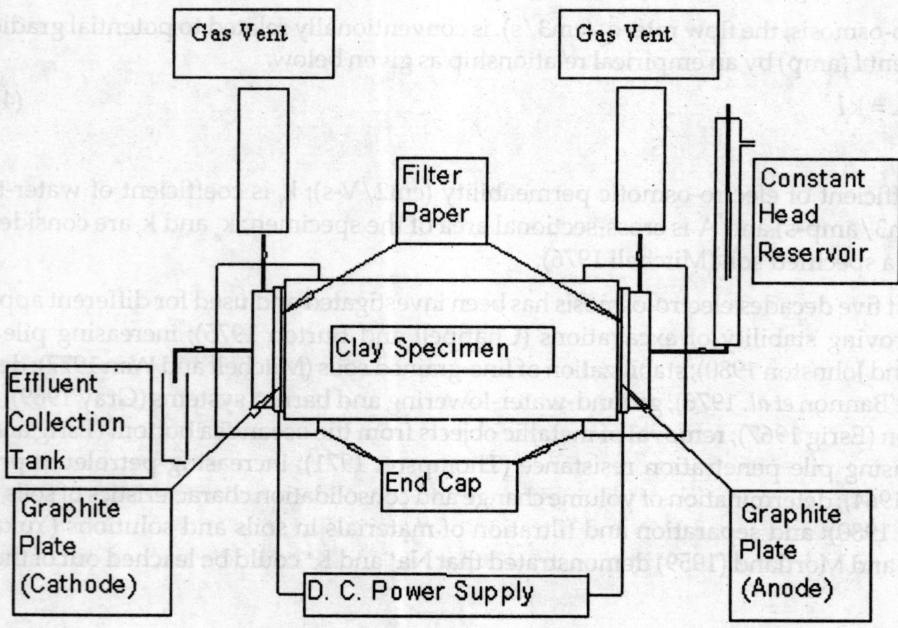

Fig. 2: Schematic of Electrokinetic Cell and Test Setup

Specimen Preparation

The sample preparation for the test – 1 is described as below:

1. 1 kg of soil was taken
2. Soil mixed with 20 ppm zinc solution (Standard AAS Zinc Solution)
3. Resulting slurry stirred for 30 minutes
4. Supernatant allowed to evaporate
5. The electrokinetic cell filled with the soil

Testing

Specimens were kept in a horizontal configuration. The electric current was applied to the cell through a constant DC power supply. The voltage was kept about 20 Volt for the first test. The effluent was analyzed for pH. After the selected period of current application, the specimen was promptly removed from the cell and sliced into four/five sections. Each sliced was analyzed for pH and metal concentration. A summary of testing program and processing data is given below:

Table 1: Summaries of Testing Program and Processing Data

Parameter (unit)	*Test Number*				
	Test – 1	*Test – 2*	*Test-3*	*Test – 4*	*Test-5*
Area (cm²)	28.26	28.26	28.26	28.26	28.26
Length (cm)	7	7	8.5	8	8.4
Head of Water (cm)	7.5	21	60	60	60
Initial pH of the Soil	6.1	6.1	5.15	5.85	2.70
Initial Zn Conc. (ppm)	20	20	130	120	120
Initial Cd Conc. (ppm)	0	0	0	130	130
Initial Cu conc. (ppm)	0	0	0	120	120
Initial Ni Conc. (ppm)	0	0	0	60	60
Porosity	0.56	0.58	0.57	0.59	0.55
Duration (hr)	223.5	194	205.5	146.5	129.5
Total Effluent Volume (ml)	306.6	194	552	472.8	344.7

Results and Discussion

Electrical Potential Gradients and Flow

Figure 3 presents the development of the net average electrical potential gradient across the cell electrodes and the Figure 4 represents the total flow in time. The following observations are made.

1. Under the test conditions, the average cell electrical potential gradients increased gradually from an initial value of about 2.7 V/cm to 3.2 V/cm for the test-1. For the test-1, the applied voltage was about 24 V while for the other tests it was about 39 V. As the duration of test increased, the resistance of the soil increased since the metal ions gradually started moving out the soil. The current also gradually decreased. Since the resistance increased and current decreased, the voltage across the cell increased.

2. The initial flow rate was quite high and then nonlinearly decreased in time.
3. The flow rate depended on the porosity, tortuosity of soil, head of water, and potential gradient across the soil. The porosity for the all test was more or less same (0.55). The head of water for the test-1 was 7 cm, for test-2 was 21 cm and for other tests was 60 cm.

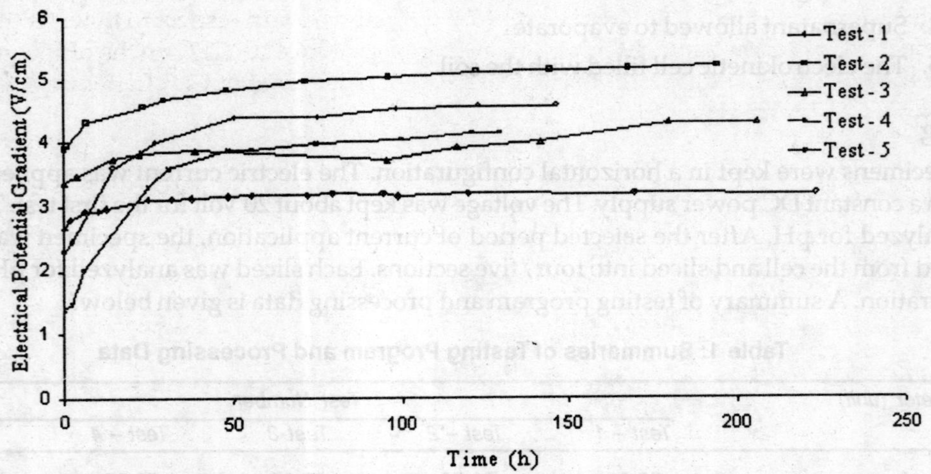

Fig. 3: Change of Electrical Potential Gradient in Time

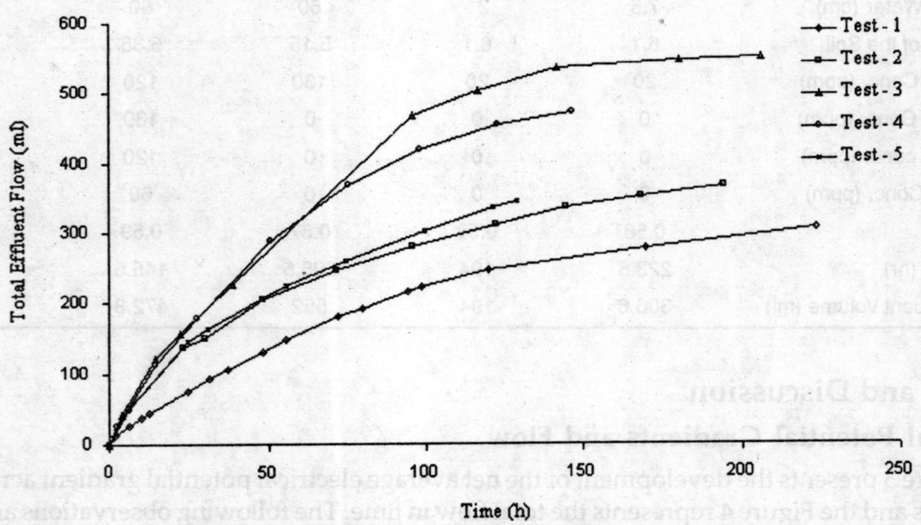

Fig. 4: Total Flow Volume in Time

From the data obtained it has been seen that as the voltage, current and head of water increase from test to test, the flow rate and total flow volume increase. For the test-1, the flow rate was 7.2 ml/h and the maximum value was 11.47 ml/h for the test-3. For the last three tests, the head of water was same (60 cm) but for the test-3, the electrical potential gradient was more than that of test-4 and test-5. Hence the flow rate was more for the test-3 than others.

Figure 5 presents the change in the electro-osmotic co-efficient of permeability k_e. The tests indicate that k_e gradually decreased in time from a value of 2.52×10^{-5} cm^2/V-s to 1.65×10^{-6} cm^2/V-s. The time-dependant change in k_e demonstrate that significant changes occurred in the overall cell resistance and hence the chemistry across the cell during the process.

Effluent pH

Figure 6 presents a plot of the effluent pH at the cathode with respect to time. At the start of the experiment, the pH at the cathode compartment rose to a value of 10-12. Then the pH decrease with the decreased in flow rate. For the test-5, the initial pH of the soil was about 2.7. In this case, the maximum effluent pH was found to be 9.8.

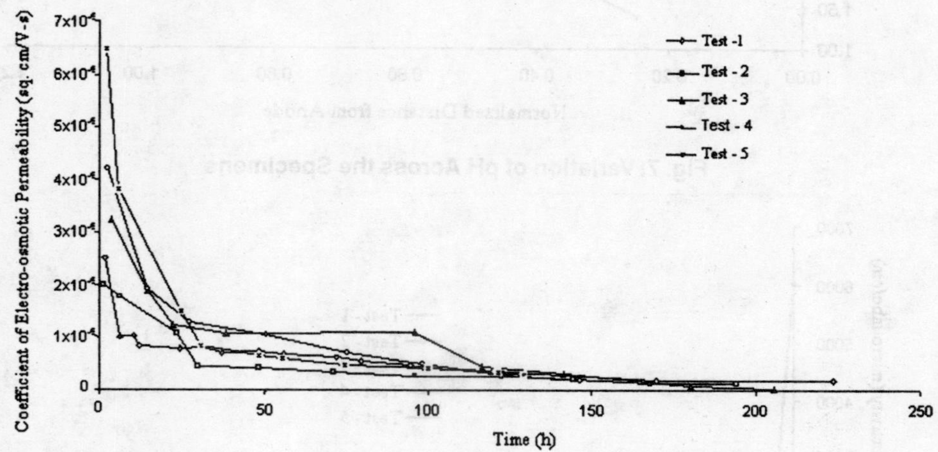

Fig. 5: Change of Co-efficient of Electro-osmotic Permeability in Time

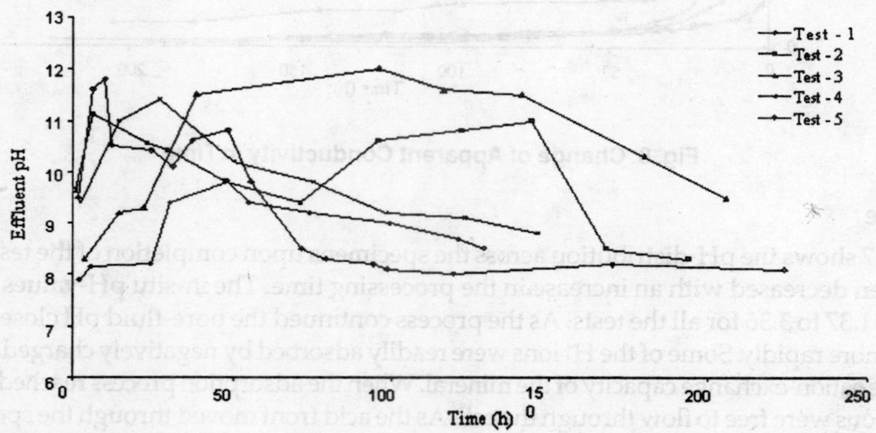

Fig. 6: Change in Effluent pH in Time

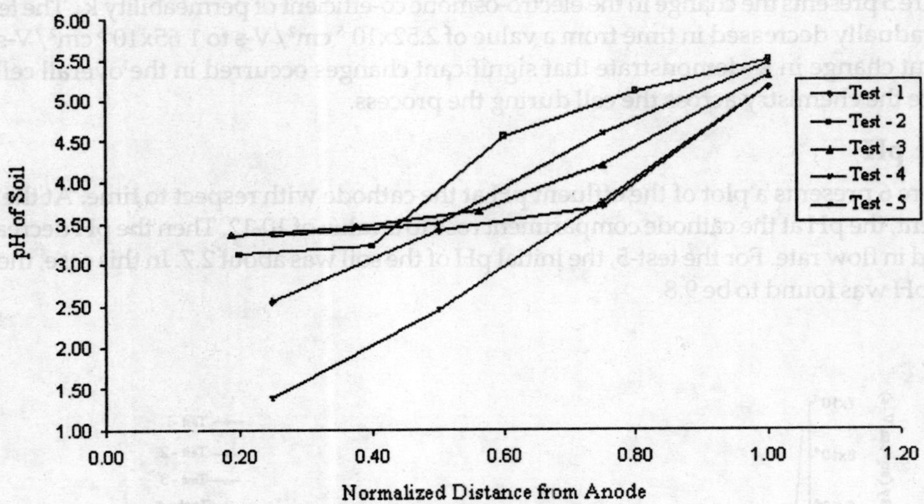

Fig. 7: Variation of pH Across the Specimens

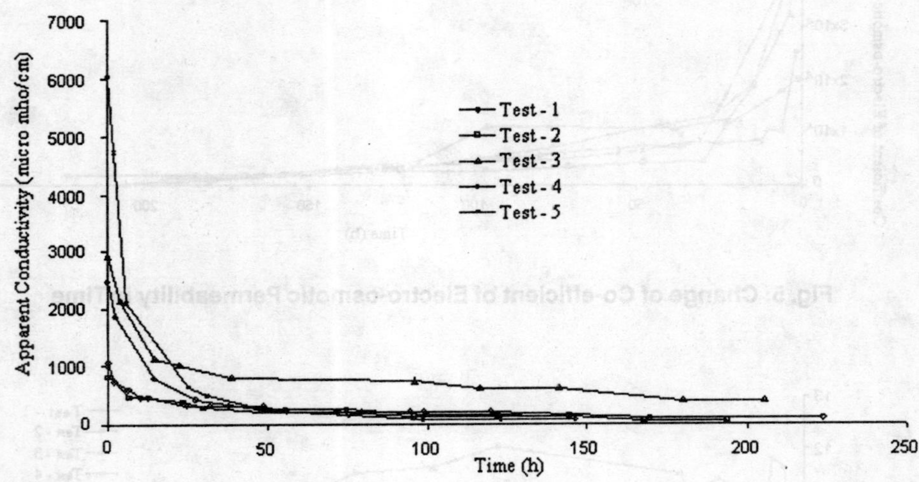

Fig. 8: Change of Apparent Conductivity in Time

pH Profile

Figure 7 shows the pH-distribution across the specimens upon completion of the tests. The pH of the specimen decreased with an increase in the processing time. The in-situ pH-values at the anode were about 1.37 to 3.36 for all the tests. As the process continued the pore-fluid pH close to the anode decreased more rapidly. Some of the H^+ ions were readily adsorbed by negatively charged clay surfaces to satisfy the cation-exchange capacity of the mineral. When the adsorption process reached equilibrium, any excess ions were free to flow through the cell. As the acid front moved through the specimen it was dispersed by both advection and diffusion. The pore-fluid with an H^+ ion concentration lower finally met and neutralized the OH^- ions generated at the cathode. Thereby, the OH^- ion concentration in the vicinity of the cathode was eventually lowered.

Conductivity

Conductivity is inversely proportional to the resistance offered to the current flow. Conductivity is directly related to the mobility and concentration of the ions charge. The resistance would change due to variation in pore sizes (porosity), tortuosity in the porous medium and variation in pore-fluid and electrolyte concentrations. Conductivity is measured in siemens (mho) per distance between the measured points in a medium. Conductivity rises from a very low value (~ 1Ms/cm) for deionized, distilled water to orders of magnitude higher for a fluid containing electrolyte depending on their type and concentrations. In electro-kinetic soil processing, conductivity can either be estimated from the electrical potential drop across the electrodes and the current across the electrodes (apparent conductivity), or it can be directly measured by probes in the soil or the pore fluid with a conductivity meter. The apparent conductivity (K_a) calculated from the electrical potential drop across the electrodes, is defined as:

$$K_a \text{ (siemens/cm)} = I_t \text{ (amp) } L \text{ (cm) } / V_t \text{ (volt) } A \text{ (cm}^2\text{)}$$

where

I_t is current; V_t is voltage difference between the current-generating electrodes; L is specimen length and A = cross-sectional area of the specimen.

Figure 8 presents the change in apparent conductivity across the electrodes in these experiments. In all the tests the decrease in conductivity reflected an increase in resistance offered and an increase in voltage was required since there was decrease in current. In the first two tests distilled water was used as influent and the Zn concentration was only 20 ppm. For the other tests tap water was used as influent and the heavy metal concentration was more. For this reason the conductivity for the last three tests was more than that of previous two. In the test-5 the soil ph was 2.7, which gives the highest conductivity.

Efficiency of Metal Removal

Figure 9 presents the ratio of final to initial Zn concentration in soil for all the tests. Figure 10 to Fig. 12 present the ratio of final to initial Cd, Cu and Ni concentration in soil for test-4 and test-5. It is noted that Zn, Cd, Cu, Ni were removed along the anode section and these metals were accumulated

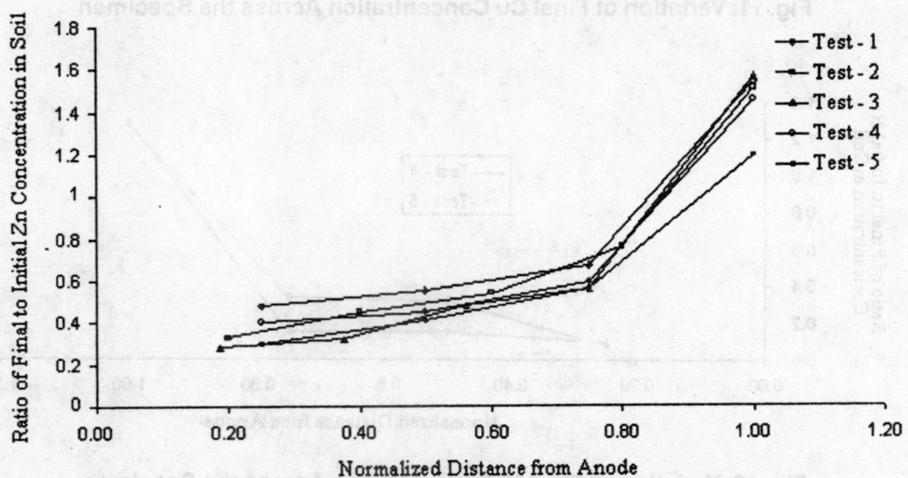

Fig. 9: Variation of Final Zn Concentration Across the Specimens

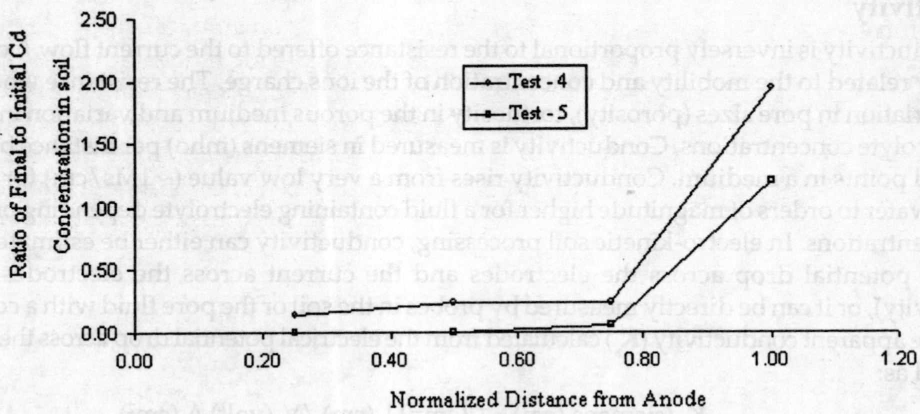

Fig. 10: Variation of Final Cd Concentration Across the Specimens

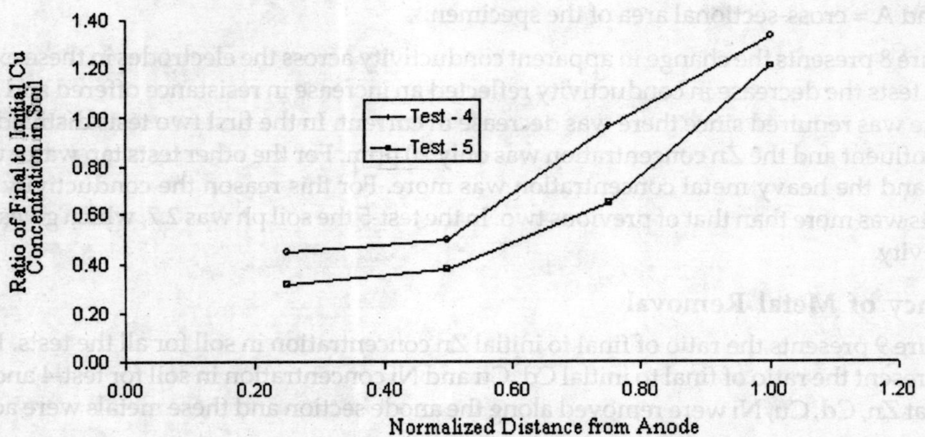

Fig. 11: Variation of Final Cu Concentration Across the Specimen

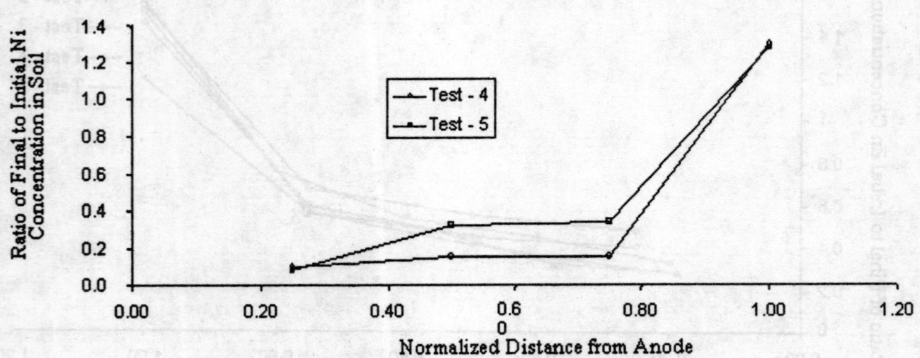

Fig. 12: Variation of Final Ni Concentration Across the Specimen

at the cathode section. It has been seen that increase in water head, voltage and current increased the removal efficiency of the heavy metal. The initial soil pH was also an important factor for the removal of the metals from soil. Use of anionic surfactant also increased the efficiency of the process. The maximum efficiency for Zn removal was obtained for the test where surfactant (Sodium Lauryl Sulphate) was used as influent. Surfactants would washout some of the heavy metals. In the test-5 the Cd, Cu and Ni removal from acidic soil was more than that of test-4. The precipitation of the heavy metals in the vicinity of the cathode caused the increase in heavy metal concentration at the cathode zone. The precipitation of metal ions in the vicinity of the cathode could be prevented by a cathode acidification arrangement.

Conclusions

The following conclusions can be drawn based on the results of the study, which investigated heavy metal removals from saturated less permeable soil like clay.

1. The acid front generated at the anode in electro-osmosis flushes across the specimen, ultimately decreasing the pH of the effluent.
2. In response to the advection of the acid front across the cell, significant variations occur in chemistry and conductivity across the cell.
3. The coefficient of electro-osmotic permeability decreases in time as a response to an increase in voltage across the cell and/or the decrease in flow.
4. The removal of Zn, Cd, Cu and Ni is due to the flushing of acid generated at the anode across the specimen resulting in desorption of metal ions together with their migration and advection by electro-osmotic flow.
5. In short-duration tests in which the acid front did not completely flush across the specimen, heavy metals were removed from the acidified anode section while these accumulate at the cathode section.
6. For acidic soil (*i.e.* pH is low) the removal efficiency of heavy metals is more than the soil having higher pH.
7. The soil pH after the treatment is a matter of concern. At the anode side the pH is about 2 and at cathode side it is 5-7. The lime treatment is required after the electrokinetic process to recover from low pH condition.
8. In the process the effluent treatment is required since its pH is high and contain heavy metals. However, it is found that the metal concentration in the effluent in not significant.

REFERENCES

1. Acar, Y. B. and Alshawabkeh, A. W. 1993, "Principals of Electrokinetic Remediation." *Environmental Science and Technology*, 27 (13): 2638-2648.
2. Alshawabkeh, A. N., Young, A. T. and Bricka, M. R. 1999, "Practical Aspects of In-situ Electrokinetic extraction." *Journal of Environmental Engineering*, ASCE, 125 (1): 27 – 34.
3. Anderson, W. C. 1993, "Soil Washing/Soil Flushing, Innovative Remediation Technology." *American Academy of Environmental Engineers Annapolis*, Vol. 3.
4. Avar, Y. B., Gale, R. j and Li, H. 1992, "Phenol Removal from Kaolinite by Electrokinetics." *Journal of Geotechnical Engineering*, ASCE, 118 (11): 1137 – 1852.

5. Das, S., 2001, "Laboratory Studies on Removal of Chromium from Contaminated Soil." Department of Civil Engineering, IIT, Kharagpur.

6. Hamed, J., Avar, Y. B. and Gale, R. J. 1991, "Pb (II) Removal from Kaolinite by Electrokinetics." ASCE, *Journal of Geotechnical Engineering*, 117 (2): 241 – 269

7. Li, A., Cheung, K. A. and Reddy, K. R., 2000, "Cosolvent-Enhancement Electrokinetic Remediation of Soils Contaminated with Phenanthrene." *Journal of Environmental Engineering*, ASCE, 126 (6): 527 – 533.

8. Li, L. Y. and Li, R. S., 2000, "The Role of Clay Minerals and the Effect of H^+ ions on Removal of Heavy Metal (Pb^{2+}) from Contaminated Soils." *Journal of Canadian Geotechnique*: 296 – 307.

9. Li, R. S. and Li, L. Y., 2000, "Enhancement of Electrokinetic Extraction from Lead-spiked Soils", ASCE, *Journal of Environmental Engineering*, 126 (9): 849 – 856.

10. Osman, S. Md. 1995, "Biosorption of Heavy Metals from Water Environment by Macro Fungi V. volvacea", Department of Civil Engineering, IIT, Kharagpur.

11. Segall, B. A., O'Aannon, C. E. and Matthias, J. A. 1980, "Electro-osmosis Chemistry and Water Quality", *Journal of Geotechnical Engineering*, ASCE, 106 (10): 1148 – 1152.

12. Verma, B. and Shukla, N. P., 1999, "Electrokinetic Separation of Chromium from Chrome Tannery Wastewater", *Indian Journal of Environment Health*, 41 (1): 43 – 48.

13. Zagury, G. J., Dartiguenave, Y. and Seteir, J., 1999, "Ex-situ Electroreclamation of Heavy Metal Contaminated Soils: Pilot Scale Study", *Journal of Environmental Engineering*, ASCE, 125 (10): 1049 – 1053.

7

Degradation of Organic Compounds: The Role of Electro-Fenton Process

T.S.N. SANKARA NARAYANAN

National Metallurgical Laboratory, Madras Centre, CSIR Complex, Taramani, Chennai - 600 113

ABSTRACT

The present study explores the utility of electro-fenton process for the degradation of o-chlorophenol from aqueous solution. Extent of degradation and reduction in COD was studied as a function of applied current, electrolysis time concentration of o-chlorophenol and concentration of ferrous ion, at 27°C. The extent of degradation is largely a function of applied current and electrolysis time and under optimized conditions it is possible to achieve efficiencies higher than 70%. From the study it can be concluded that electro-Fenton process is highly useful for the degradation of o-chlorophenol from aqueous solution.

Introduction

Fenton's reagent (Fenton, 1894), a mixture of H_2O_2 and a ferrous salt, has been proved as an effective and simple oxidant for various types of organic contaminants such as chlorobenzene, nitrophenols and aromatic amines (Buxton *et al.*, 1988; Ewa, 1991; Sedlak and Andren, 1991; Kochany and Bolton, 1992; Zhu *et al.*, 1996; Casero *et al.*, 1997). Fenton's reagent is not stable. Once H_2O_2 and Fe^{2+} are mixed, several reactions take place simultaneously. These reactions produce hydroxyl radicals ($HO^{\bullet}$), hydroperoxyl radicals ($HO_2^{\bullet}$), Fe^{3+}, and O_2 [Equations (1) to (6)]. The basic reaction mechanism of the Fenton system was determined by Haber and Weiss (1934) and further extended by Barb *et al.* (1951).

$$Fe^{2+} + H_2O_2^{\bullet} \rightarrow Fe^{3+} + HO^- + HO^{\bullet} \quad (1)$$

$$Fe^{2+} + HO^{\bullet} \rightarrow Fe^{3+} + HO^- \quad (2)$$

$$H_2O_2 + HO^{\bullet} \rightarrow H_2O + HO_2^{\bullet} \quad (3)$$

$$Fe^{2+} + HO_2^{\bullet} \rightarrow Fe^{3+} + HO_2^- \leftrightarrow H_2O_2 \quad (4)$$

$$Fe^{3+} + HO_2^{\bullet} \rightarrow Fe^{2+} + H^+ + O_2 \quad (5)$$

$$Fe^{3+} + H_2O_2 \rightarrow Fe^{2+} + HO_2^{\bullet} + H^+ \quad (6)$$

Although hydroxyl radicals, hydroperoxyl radicals, hydrogen peroxide, and oxygen are all oxidants, hydroxyl radicals have the strongest oxidation capability, and are considered responsible for oxidizing various organic compounds. The mixture of organics and Fenton's reagent results in additional reactions, [Equations (7) to (12)] (Walling, 1975). Hydroxyl radicals react strongly with the dissolved organic compounds, H_2O_2, and Fe^{2+} ions in solution. They also react with other hydroxyl radicals themselves. The reaction rates of hydroxyl radicals are much faster than the generation of these radicals, which gives them a very short life span. Oxidation of the organics initially depends on the selectivity of the hydroxyl radicals for the target organic compounds and then on the selectivity and availability of these radicals to propagate the reaction.

$$RH + HO^{\bullet} \rightarrow R^{\bullet} + H_2O \text{ (chain propagation)} \quad (7)$$

$$R^{\bullet} + H_2O_2 \rightarrow ROH + HO^{\bullet} \text{ (chain propagation)} \quad (8)$$

$$R^{\bullet} + HO^{\bullet} \rightarrow ROH \text{ (non-chain termination)} \quad (9)$$

$$2R^{\bullet} \rightarrow \text{Product (dimer) (non-chain termination)} \quad (10)$$

$$R^{\bullet} + Fe^{3+} \rightarrow Fe^{2+} + \text{Product} \quad (11)$$

$$R^{\bullet} + Fe^{2+} \rightarrow Fe^{3+} + RH \text{ (chain termination)} \quad (12)$$

Application of Fenton's reagent in the destruction of organic compounds has been limited by the slurry system because it produces a substantial amount of $Fe(OH)_3$ sludge, which requires further separation and disposal. Another drawback of Fenton's reagent is that the overall oxidation process is considerably slowed down after conversion of Fe^{2+} to Fe^{3+}, as the reduction of Fe^{3+} to Fe^{2+} by hydrogen peroxide [reaction (6)] is much slower than reaction (1) (Barb *et al.*, 1951). The other possible source of regeneration of Fe^{2+} via reaction between Fe^{3+} and $R^{\bullet}$ [reaction (11)] is less efficient. Besides, a continuous supply of H_2O_2 is essential.

In recent years, there is a great interest in the development of practical electrochemical methods, such as anodic oxidation and indirect electrooxidation, for the destruction of toxic and biorefractory organic pollutants for wastewater treatment. Many studies have been carried out on the direct anodic oxidation of wastewater containing organic pollutants (Scharifian and Kirk, 1986; Stucki *et al.*, 1991; Comninellis and Pulgarin, 1991, 1993; Feng *et al.*, 1995; Tennaloon *et al.*, 1996; Rajeshwar and Ibanez, 1998; Johnson *et al.*, 1999; Gandini *et al.*, 2000), in which adsorbed $OH^{\bullet}$ are produced due to the oxidation of water at the anode. Several anode materials have been tested, but the complete mineralization of the pollutants has only been obtained using high oxygen overvoltage anodes such as SnO_2 (Comninellis and Pulgarin, 1993), PbO_2 (Belhadj and Savall, 1998; Bonfatti *et al.*, 1999) and boron-doped diamond (Foti *et al.*, 1999; Perret *et al.*, 1999). Besides direct oxidation, organic pollutants can also be treated by an indirect electrolysis by *in situ* generation of strong oxidants such as hypochlorite ions (Panizza *et al.*, 2000; Do and Yeh, 1995; Comninellis and Nerini, 1995; Dziewinski *et al.*, 1996;

Yang, 1999), ozone (Tatapudi and Fenton, 1993), and hydrogen peroxide (Sudoh *et al.*, 1986; Do and Chen, 1994). Among these processes, the *in situ* generation of hydrogen peroxide received considerable attention since it can be coupled with ferrous ions to produce Fenton's reagent. This process is termed as "electro-Fenton process". The electro-Fenton process can be divided into three types. The first type uses Fe^{2+} and electrogenerated H_2O_2 (Sudoh *et al.*, 1986; Hsiao and Nobe, 1993a, 1993b; Brillas *et al.*, 1996). The second type (Anodic Fenton treatment [AFT)] utilizes H_2O_2 and electrogenerated Fe^{2+}, which is produced via the oxidation of iron, the sacrificial anode (Pratap and Lemley, 1994; Huang *et al.*, 1999; Wang and Lemley, 2002; Saltmiras and Lemley, 2002). The final type (Fered-Fenton process) applied H_2O_2 and electrogenerated Fe^{2+} produced via the reduction of ferric sulphate or ferric hydroxide sludge (Chou *et al.*, 1999; Huang *et al.*, 2001a). Several authors have studied the removal of organic pollutants, such as formaldehyde (Do and Chen, 1993), aniline (Brillas *et al.*, 1996, 2002), phenol (Alvarez-Gallegos and Pletcher, 1999), 2, 4-dichlorophenoxyacetic acid (Brillas *et al.*, 2000), pesticides (Aaron and Oturan, 2001), pentachlorophenol (Oturan *et al.*, 2001) and atrazine (Ventura *et al.*, 2002) by the first type of electro-Fenton process. The present work aims to study the degradation of o-chlorophenol by first type of electro-Fenton process, which uses Fe^{2+} and electrogenerated H_2O_2.

Chlorophenols have been detected during the manufacture of pesticides, bleaching of industrial wastewater, and chlorination of drinking water (Ahlborg and Thunberg, 1980; Karlsson *et al.*, 1995; Smeds *et al.*, 1995). Because of their numerous origins, they can be found in industrial wastewaters, soils and surface waters. In fact, US EPA has listed several of them among the 65 priority pollutants (Keith and Telliard, 1979). The chlorinated byproducts, produced by chlorination of industrial wastewater and drinking water, have led to several health risks to human beings, and give bad odour and taste (Minero *et al.*, 1995; Muftikian *et al.*, 1995). The conventional destructive technologies that are usually employed to degrade these compounds include biological treatments, incineration, adsorption on activated carbon, air stripping, etc. (Jardin *et al.*, 1997). However, these conventional methods are associated with serious limitations. Biological treatments are affected by the toxicity of these substances. Incineration presents considerable emission of other hazardous compounds whereas adsorption or air stripping requires a post-treatment to remove the pollutants from the newly contaminated environment. Removal of these refractory compounds becomes difficult and expensive when a very low level of these compounds is dealt with (Wu *et al.*, 1992).

Degradation of chlorophenols has received much attention from several researchers. Brillas *et al.* (1998b) compared the efficiency of anodic oxidation, electro-Fenton, photoelectro-Fenton and peroxicoagulation processes towards the degradation of 4-chlorophenol. Kwon *et al.* (1999) adopted the classical Fenton oxidation to study the characteristics of p-chlorophenol oxidation. Lin and Ma (2000) used ultrasound/Fenton method for the oxidation of o-chlorophenol. Oturan *et al.* (2001) studied the mineralization of pentachlorophenol by electro-Fenton process. Benitez *et al.* (2000) have employed advanced oxidation processes for the decomposition of 4-chlorophenol, 2, 4-dichlorophenol, 2, 4, 6-trichlorophenol and 2, 3, 4, 6-tetrachlorophenol. Lu (2000) and Huang *et al.* (2001b) employed Fenton-like processes, using various iron oxides, for the degradation of 2-chlorophenol. Jung (2001) explored the possible use of ozonation method for the destruction of 2-chlorophenol. Sauleda and Brillas (2001) utilized ozonation catalyzed with Fe^{2+} and UVA light for the mineralization of 4-chlorophenol. The degradation of o-chlorophenol was not studied by electro-Fenton process and hence attempted in this study.

Experimental

All chemicals used were of analytical reagent grade and they were used without any purification. Sodium sulphate solution (0.05 M) was used as the electrolyte solution. The pH of the sodium sulphate solution was adjusted to 3.5 using sulphuric acid. The temperature of the electrolyte was maintained at 27°C. For the electrolytic generation of hydrogen peroxide, oxygen gas was purged into the electrolyte solution at a constant flow rate of 3.75 l/min. A rectangular stainless plate served as the cathode whereas a platinum gauze electrode served as the anode. In order to generate the Fenton's reagent, ferrous ion was added to the electrolyte solution. A D.C. power source was used to supply the required current for the electrogeneration of hydrogen peroxide. The concentration of o-chlorophenol was in the range of 40-160 ppm. The effect of applied current, concentration of ferrous ions, electrolysis time and concentration of o-chlorophenol, on the extent of degradation was studied. The amount of chloride ions generated as a result of degradation of o-chlorophenol was analyzed by Volhard's method. COD was also determined to correlate the extent of degradation and the nature of intermediates formed.

Results and Discussion

Effect of Applied Current

Fig. 1 shows the effect of applied current on the percentage degradation of o-chlorophenol along with the percentage decrease in COD after electrolysis. It is evident from Fig. 1 that the extent of degradation of o-chlorophenol increases with increase in applied current in the range of 200 to 600 mA. The applied current is the driving force for the reduction of oxygen leading to the generation of hydrogen peroxide at the cathode. Higher applied current increases the quantum of hydrogen peroxide produced, thus increasing the number of hydroxyl radicals in the electrolyte medium (reaction 1), which are highly reactive and responsible for the degradation of o-chlorophenol. However, the increment in percentage degradation is decreased with increase in applied current from 200 to 600 mA. This could be attributed to the competing hydrogen evolution reaction, which decreases the efficiency of the process. Panizza and Cerisola (2001) and Qiang *et al.* (2002) have also observed a decrease in current efficiency due to the secondary reaction of hydrogen evolution at higher applied potentials.

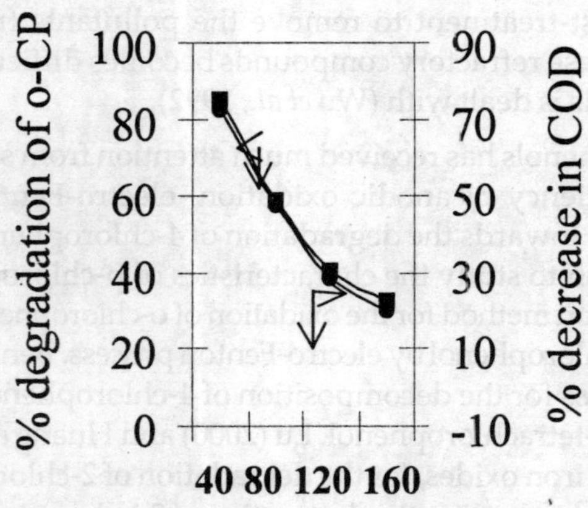

Fig. 1: Effect of Applied Current on the Percentage Degradation of o-CP and the Percentage Decrease in COD (Concentration of o-CP: 80 ppm; Concentration of ferrous ion: 0.5 ppm; Electrolysis time: 60 mins)

The electrogeneration of hydrogen peroxide following the reduction of oxygen was studied on several electrodes, such as, mercury pool, graphite, vitreous carbon or carbon-PTFE-O_2-fed cathodes (Tomat and Rigo, 1972; Matsue *et al.*, 1981; Tzedakis *et al.*, 1989; Oturan *et al.*, 1992; Hsiao and Nobe, 1993a, 1993b; Brillas, *et al.*, 1996, 1998a, 1998b). Carbon is widely used as a cathode material for hydrogen peroxide generation as it exhibits a range of electrochemical activities towards oxygen reduction, high overpotential for hydrogen evolution and low catalytic activity for hydrogen peroxide decomposition. Stainless steel cathode is used in the present study. As far as the activity of cathode is concerned, no deleterious effect is observed with the stainless steel cathode, except some surface staining due to the adsorption of the degradation products formed in the electrolyte, which requires solvent cleaning.

Effect of Electrolysis Time

The percentage degradation of o-chlorophenol along with the percentage decrease in COD as a function of electrolysis time is shown in Fig. 2. Though the extent of degradation increases with increase in electrolysis time, a saturation level is reached after some time. It should be noted that during the entire duration of the experiment oxygen is continuously purged into the electrolyte solution, which implies a continuous generation of hydrogen peroxide. Moreover, the electro-Fenton process enables the generation of ferrous ions in the electrolyte solution by electrochemical reduction of ferric ions at the cathode. Hence the reason for saturation in extent of degradation of o-chlorophenol should be due to the decrease in availability of ferrous ions due to some competing reactions.

It is well known that ferrous and ferric ions in water readily form stable complexes with organic substances (Martell and Smith, 1977; Potter, 1993). In fact, such an attribute is validated by the observation of a dark brown coloration of the electrolyte medium, which could be predicated to the formation of iron-organic complexes. Eisenhauer (1964) and Kwon *et al.* (1999) also made similar observations. Kwon *et al.* (1999), using elemental analysis, atomic absorption spectroscopy and FTIR

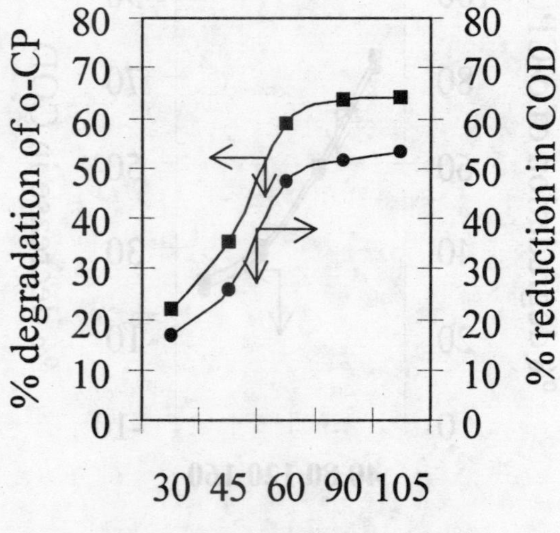

Fig. 2: Effect of Electrolysis Time on the Percentage Degradation of o-CP and the Percentage Decrease in COD (Concentration of o-CP: 80 ppm; Concentration of ferrous ion: 0.5 ppm; Applied current: 600 mA)

spectroscopy, have confirmed the unavailability of iron, due to complexation, for further reactions. The results of the present study supports the hypothesis made by Eisenhauer (1964) and Kwon *et al.* (1999) on the potential scavenging effect by the iron-organic complexes.

Effect of Concentration of o-chlorophenol

Fig. 3 depicts the percentage degradation of o-chlorophenol along with the percentage decrease in COD for varying concentration of o-chlorophenol in the range of 40-160 ppm. For a given set experimental parameters, it is obvious to expect a decrease in the extent of degradation of o-chlorophenol when its concentration in the electrolyte solution is increased. This implies the requirement of additional treatment time when the concentration ofo-chlorophenol is higher in the electrolyte solution. When the treatment time is increased, obviously, being an electrolytic process, the increase in energy consumption significantly affects the economics of the process. Hence it is logical to recommend electro-Fenton process for the degradation of o-chlorophenol, when it is present in the medium at very low concentrations (less than 100 ppm). Alverez-Gallegos and Pletcher (1999) have also recommended the use of electrochemical treatment process when the concentration of organics and dyes are in the range of 25-100 ppm.

Effect of Ferrous Ions Concentration

The effect of variation in ferrous ion concentration on the percentage degradation of o-chlorophenol along with the percentage decrease in COD is shown in Fig. 4. The extent of degradation increased with increase in concentration of ferrous ions. Since hydrogen peroxide is continuously electrogenerated at the cathode, increase in ferrous ion concentration is likely increase the generation of hydroxyl radicals, which is the major reactive species for the degradation of o-chlorophenol. However, ferrous ions in the electrolyte solution, when present in excess, could consume the hydroxyl radicals

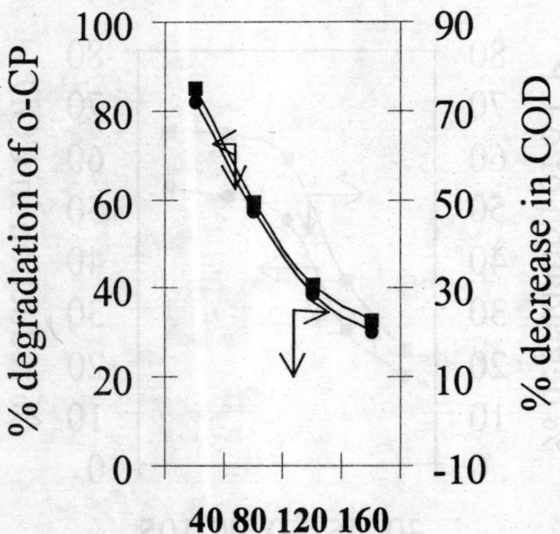

Fig. 3: Effect of Concentration of o-chlorophenol on the Percentage Degradation of o-CP and the Percentage Decrease in COD (Concentration of ferrous ion: 0.5 ppm; Applied current: 600 mA; Electrolysis time: 60 mins)

according to reaction (2) and affect the extent of degradation. Besides, the ferrous/ferric ions readily form stable complexes with o-chlorophenol. Hence it is preferable to use a limited concentration of ferrous ions.

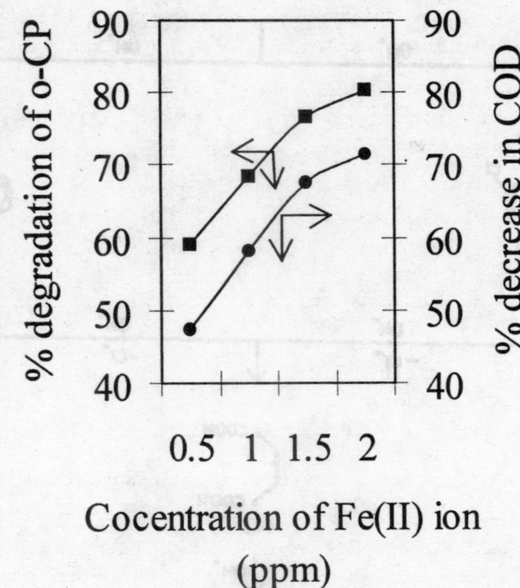

Fig. 4: Effect of Concentration of Fe (II) Ions on the Percentage Degradation of o-CP and the Percentage Decrease in COD (Concentration of o-CP: 80 ppm; Applied Current: 600 mA; Electrolysis time: 60 mins)

Mechanism of Degradation of o-chlorophenol

A general mechanism for the degradation of o-chlorophenol by electro-Fenton process is proposed in Fig. 5. The degradation process involves dehydrogenation and hydroxylation reactions of o-chlorophenol with $OH^{\bullet}$. The attack of $OH^{\bullet}$ on o-chlorophenol could occur at *ortho* or *meta* or *para* positions which could produce 6-chloro-1, 2-dihydroxybenzene or 6-chloro-1, 3-dihydroxybenzene or 6-chloro-1, 4 dihydroxybenzene, by the addition of $OH^{\bullet}$ in the respective positions of o-chlorophenol and abstraction of a hydrogen atom from the corresponding position. Huang *et al.* (1995) found that $OH^{\bullet}$ reacts with phenol in the *ortho* and *para* positions.

Ma and Lin (1998), based on the formation of 2-chloro-p-benzoquinone as the major intermediate, suggest the preferential attack of $OH^{\bullet}$ on *para* position during the decomposition of o-chlorophenol. Several authors have also noted the formation of chlorinated benzoquinone as the major intermediate during the decomposition of chlorophenols using chemical processes (Kotronarou *et al.*, 1992; Serpone *et al.*, 1992; Minero *et al.*, 1995; Stefan *et al.*, 1996). Brillas *et al.* (1998b) and Sauleda and Brillas (2001) proposed the attack of $OH^{\bullet}$ on the *ortho* position during the degradation of p-chlorophenol. Based on the observations made in earlier studies, it is believed that the attack of $OH^{\bullet}$ on o-chlorophenol is most likely to occur at *ortho* and/or *para* positions, respectively yielding 6-chloro-1, 2-dihydroxybenzene and/or 6-chloro-1, 4-dihydroxybenzene. These compounds upon further oxidation, with loss of Cl^-, might yield a variety of intermediates. Jung (2001) proposed the formation of hydroxyhydroquinone, muconic acid, tartaric acid, oxalic acid, maleic acid, and hydroxymalonic acid as the intermediates

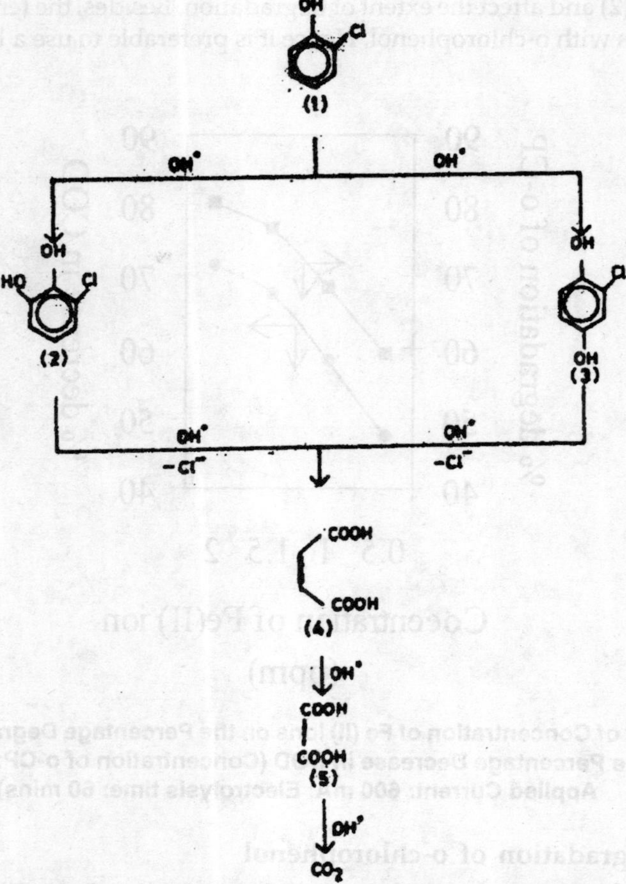

Fig. 5: The degradation mechanism of o-chlorophenol by electro-Fenton process (Key for parent compound and intermediates: (1) o-chlorophenol; (2) 6-chloro-1,2-dihydroxbenzene; (3) 6-chloro-1,4-dihydroxybenzene; (4) maleic acid; (5) oxalic acid)

during the degradation of o-chlorophenol by ozonation. Brillas *et al.* (1998b) and Sauleda and Brillas (2001) have identified maleic acid, its trans-isomer, fumaric acid, and oxalic acid as intermediates of degradation of p-chlorophenol by electro-Fenton process. Hence it is obvious to expect the formation of carboxylic acids as the major by-products during the degradation of o-chlorophenol. These carboxylic acids are finally mineralized to CO_2 via oxalic acid.

Conclusions

From the study it is evident that the degradation of o-chlorophenol by electro-Fenton process is a function of applied current, electrolysis time, concentration of o-chlorophenol and concentration of ferrous ions in the medium. As far as the activity of cathode is concerned, no deleterious effect is observed with the stainless steel cathode, except some surface staining due to the adsorption of the degradation products formed in the electrolyte, which requires solvent cleaning. The extent of degradation increases with increase in applied current in the range of 200 to 600 mA. Though the extent of degradation increases with increase in electrolysis time initially; a saturation level is reached after some time following the decrease in availability of ferrous ions due to the formation of iron-organic complexes in the medium. For a given set experimental parameters the extent of degradation of o-chlorophenol decreases when its concentration is increased. Hence the use of electro-Fenton process can be recommend for the degradation of o-chlorophenol only at very low concentrations (less than 100 ppm). The extent of degradation increased with increase in concentration of ferrous ions. The ferrous ion concentration in the electrolyte solution, when present in excess, consumes the hydroxyl radicals. Besides, the excess ferrous/ferric ions readily form stable complexes with o-chlorophenol and affect the extent of degradation. Hence it is preferable to use a limited concentration of ferrous ions. A general mechanism for the degradation of o-chlorophenol by electro-Fenton process is proposed. The degradation process involves dehydrogenation and hydroxylation reactions of o-chlorophenol with $OH^{\bullet}$. These compounds upon further oxidation, with loss of Cl–, might yield a variety of intermediates. Carboxylic acids are found to be the major by-products during the degradation of o-chlorophenol. These carboxylic acids are finally mineralized to CO_2 via oxalic acid.

Acknowledgement

The author expresses his sincere thanks to Dr. S.P. Mahrotra, Director, National Metallurgical Laboratory, Jamshedpur and Dr. S. Srikanth, Deputy Director, NML Madras Centre, for their constant support and encouragement to carry out this research work. The help rendered by Mr. P. Sattanathan and Mr. G. Magesh, Masters degree chemistry students of Anna University, Chennai, in generating some experimental data is gratefully acknowledged.

REFERENCES

1. Aaron, J.J. and Oturan, M.A., 2001, New photochemical and electrochemical methods for the degradation of pesticides in aqueous media: Environmental applications, Turk. J. Chem., 25, 509-520.
2. Ahlborg, U.G. and Thunberg, T.M., 1980, Chlorinated phenols: Occurrence toxicity metabolism and environmental impact, CRC Crit. Rev. Toxicol., 7, 1-35.
3. Alverez-Gallegos, A. and Pletcher, D., 1999, The removal of low level organics via hydrogen peroxide formed in a reticulated vitreous carbon cathode cell. Part 2: The removal of phenols and related compounds from aqueous effluents, Electrochim. Acta, 44, 2483-2492.
4. Barb, W.G., Baxendale, J.H., George, P. and Hargrave, K.R., 1951, Reaction of ferrous and ferric ions with hydrogen peroxide, Trans. Faraday Soc., 47, 462-500, 591-616.
5. Benitez, F.J., Beltran-Heredia, J., Acero, J.L. and Rubio, F.J., 2000, Contribution of free radicals to chlorphenols decomposition by several advanced oxidation processes, Chemosphere, 21, 1271-1277.
6. Belhadj Tahar N. and Savall A., 1998, Mechanistic Aspects of Phenol Electrochemical Degradation by Oxidation on a Ta/PbO2 Anode, J. Electrochem. Soc., 145, 3427-3434.
7. Bonfatti, F., Ferro, S., Levezzo, F., Malacarne, M., Lodi, G. and De Battisti, A., 1999, Electrochemical Incineration of Glucose as a Model Organic Substrate: I. Role of the Electrode Material, J. Electrochem. Soc., 146, 2175-2179.
8. Brillas, E., Mur, E. and Casado, J.., 1996, Iron (II) catalysis of the mineralization of aniline using a carbon-PTFE-O2-fed cathode, J. Electrochem. Soc., 143, L 49-53.

9. Brillas, E., Mur, E., Sauleda R., Sanchez, L., Peral, J., Domenech, X., Casado, J.., 1998a, Aniline mineralization by AOPs: anodic oxidation, photocatalysis, electro-Fenton and photoelectro-Fenton processes, Appl. Catal. B:Environ., 16, 31-42.

10. Brillas, E., Sauleda, R., Casado, J., 1998b, Degradation of 4-chlorophenol by anodic oxidation, electro-Fenton, photoelectro-Fenton, and peroxicoagulation processes, J. Electrochem. Soc., 145, 759-765.

11. Brillas, E., Calpe, J.C. and Casado, J., 2000, Mineralization of 2, 4-D by advanced electrochemical oxidation processes, Wat. Res., 34 (8, 2253-2262.

12. Brillas, E. and Casado, J., 2002, Aniline degradation by electro-Fenton and peroxi-coagulation processes using a flow reactor for wastewater treatment, Chemosphere, 47, 241-248.

13. Buxton, G.V., Greenstock, C.L., Helman, W.P. and Ross, A.B., 1988) Critical review of rate constants for reaction of hydrated electrons, hydrogen atoms and hydroxyl radicals (OH./O.-) in aqueous solution, J. Phys. Chem. Ref. Data, 17 (2, 513-886.

14. Casero, I., Sicilia, D., Rubio, S. and Perez-Bendito, D., 1997) Chemical degradation of aromatic amines by Fenton's reagent, Water Res., 31., 1985-1995.

15. Chou, S., Huang, Y.H., Lee, S.N., Huang, G.H. and Huang, C., 1999, Treatment of high strength hexamine-containing wastewater by electro-Fenton method, Water Res., 33 (3, 751-759.

16. Comninellis, Ch. and Pulgarin, C., 1991, Anodic oxidation of phenol for wastewater treatment, J. Appl. Electrochem., 21, 703-708.

17. Comninellis, Ch. and Pulgarin, C., 1993, Electrochemical oxidation of phenol for wastewater treatment using SnO_2 anodes, J. Appl. Electrochem., 23, 108-112.

18. Comninellis, Ch. and Nerini, A., 1995, Anodic oxidation of phenol in the presence of NaCl for wastewater treatment, J. Appl. Electrochem., 25, 23-28.

19. Do, J.S. and Chen, C.P., 1993, *in situ* oxidative degradation of formaldehyde with electrogenerated hydrogen peroxide, J. Electrochem. Soc., 140, 1632-1637.

20. Do, J.S. and Chen, C.P., 1994, *in situ* oxidative degradation of formaldehyde with hydrogen peroxide electrogenerated on the modified graphite, J. Appl. Electrochem., 24, 936-942.

21. Do, J.S. and Yeh, W.C., 1995, *In situ* degradation of formaldehyde with electrogenerated hypochlorite ion, J. Appl. Electrochem., 25, 483-489.

22. Dziewinski, J., Marczak, S. and Smith, W., 1996, Electrochemical destruction of mixed wastes, Chemtech, 30-33.

23. Eisenhauer, H. R., 1964, Oxidation of phenolic wastes, J. Water Pollut. Control Fed., 36, 1117-1127.

24. Ewa, L.K., 1991, Degradation of aqueous nitrophenols and nitrobenzene by means of the Fenton reaction, Chemosphere, 22, 529-536.

25. Feng, J., Houk, L.L., Johnson, D.C., Lowery, S.N. and Carey, J.J., 1995, Electrocatalysis of anodic oxygen-transfer reactions: the electrochemical incineration of benzoquinone, J. Electrochem. Soc., 142, 3626-3632.

26. Fenton, H.J.H. (1894, Oxidation of tartaric acid in presence of iron, J. Chem. Soc., 65, 899-910.

27. Foti, G., Gandini, D., Comninellis, Ch., Perret, A. and Haenni, W., 1999, Oxidation of organics by intermediates of water discharge on IrO_2 and synthetic Diamond anodes, Electrochem. Solid-State Lett., 2, 228-230.

28. Gandini, D., Mahe, E., Michaud, P.A., Haenni, W., Perret, A. and Comninellis, Ch., 2000, Oxidation of carboxylic acids at boron-doped diamond electrodes for wastewater treatment, J. Appl. Electrochem., 30, 1345-1350.

29. Haber, F. and Weiss, J., 1934, The catalytic decomposition of hydrogen peroxide by iron salts, Proc. Roy. Soc. London, Part A., 147, 332-351.

30. Hsiao, Y.L. and Nobe, K., 1993a, Hydroxylation of chlorobenzene and phenol in a packed bed flow reactor with electrogenerated Fenton's reagent, J. Appl. Electrochem., 23, 943-946.

31. Hsiao, Y.L. and Nobe, K., 1993b, Oxidative reactions of phenol and chlorobenzene with *in situ* electrogenerated Fenton's reagent, Chem. Eng. Commun., 126, 97-110.

32. Huang, C.P., Shu, C.S., Chen, C.W. and Dong, C., 1995, Electrochemically generated Fenton's reagent for the removal of chlorophenols in water, Proceedings of the 1st Workshop on Drinking Water Management and Treatment Technologies.

33. Huang, Y.H., Chou, S.S., Peng, M.G., Huang, G.H. and Cheng, S.S., 1999, Case study on the bioeffluent of petrochemical wastewater by electro-Fenton method, Water Sci. Technol., 39 (10-11, 145-149.

34. Huang, Y.H., Chen, C.C., Huang, G.H. and Chou, S.S., 2001a, Comparison of a novel electro-Fenton method with Fenton's reagent in treating a highly contaminated wastewater, Water Sci. Technol., 43 (2, 17-24.

35. Huang, H.H., Lu, M.C. and Chen, J.N., 2001b, Catalytic decomposition of hydrogen peroxide and 2-chlorophenol with iron oxides, Wat. Res., 35 (9, 2291-2299.

36. Jardin, W.F., Morales, S.G. and Takiyama, M.M.K., 1997, Photocatalytic degradation of aromatic chlorinated compounds using TiO_2: toxicity of intermediates, Water Res., 31, 1728-1732.

37. Johnson, S.K., Houk, L.L., Feng, J., Houk, R.S. and Johnson, D.C., 1999, Electrochemical incineration of 4-chlorophenol and the identification of products and intermediates by mass spectrometry, Environ. Sci. Technol., 33, 2638-2644.

38. Jung, O.J., 2001, Destruction of 2-chlorophenol from wastewater and investigation of by-products by ozonation, Bull. Korean Chem. Soc., 22 (8, 850-856.

39. Karlsson, S., Kaugare, S., Grimvall, A., Boern, H. and Savehed, R., 1995, Formation of 2, 4, 6-trichlorophenol and 2, 4, 6-trichloroanisole during the treatment of and distribution of drinking water, Water Sci. Technol., 31 (11, 99-103.

40. Keith, L.H. and Telliard, W.A., 1979, Priority Pollutants: A prospective view, Environ. Sci. Technol., 13, 416-424.

41. Kochany, J. and Bolton, J.R., 1992, Mechanism of photodegradation of aqueous organic pollutant. 2. Measurement of primary rate constants for reaction of OH radicals with benzene and some halobenzenes using the EPR spin-trapping method following the photolysis of H_2O_2, Envrion. Sci. Technol., 26 (2, 262-265.

42. Kotronarou, A., Mills, G. and Hoffmann, M.R., 1992, Decomposition of parathion in aqueous solution by ultrasonic irradiation, Envrion. Sci. Technol., 26 (10, 1460-1462.

43. Kwon. B.G., Lee, D.S., Kang, N. and Yoon, J., 1999, Characteristics of p-chlorophenol oxidation by Fenton's reagent, Water Res. 33 (9, 2110-2118.

44. Lin, J.G. and Ma, Y.H., 2000, Oxidation of 2-chlorophenol in water by ultrasound/Fenton method, J. Environ. Engg., February, 130-137.

45. Lu, M.C., 2000, Oxidation of chlorophenols with hydrogen peroxide in the presence of goethite, Chemosphere, 40, 125-130.

46. Ma, Y.S. and Lin, J.G., 1998, Removal of 2-chlorophenol from wastewater using the ultrasonic/Fenton process, J. Chinese Inst. Environ. Engg., 8 (3, 215-226.

47. Martell, A. E. and Smith, R. M., 1977, Critical Stability Constants, Vol. 3. Plenum Press, New York.

48. Matsue, T., Fujihira, M., Osa, T., 1981, Oxidation of alkylbenzenes by electrogenerated hydroxyl radical, J. Electrochem. Soc., 128, 2565-2569.

49. Minero, C., Pelizzetti, E., Pichat, P., Sega, M. and Vincenti, M., 1995, Formation of condensation products in advanced oxidation technologies: The photocatalytic degradation of dichlorophenols on TiO_2, Envrion. Sci. Technol., 29 (9, 2226-2234.

50. Muftikian, R., Fernando, Q. and Korte, N., 1995, A method for the rapid chlorination of low molecular weight chlorinated hydrocarbons in water, Water Res., 29 (10) 2434-2439.

51. Oturan, M., Pinson, J., Bizot, J., Deprez, D., Terlain, B., 1992, Polyhydroxylation of salicylic acid by electrochemically generated OH radicals, New J. Chem., 16, 705-710.

52. Oturan, M.A., Oturan, N., Lahitte, C. and Trevin, S., 2001, Production of hydroxyl radicals by electrochemically assisted Fenton's reagent: Application to the mineralization of an organic micropollutant, pentachlorophenol, J. Electroanal. Chem., 507, 96-102.

53. Panizza, M., Bocca, C. and Cerisola, G., 2000, Electrochemical treatment of wastewater containing polyaromatic organic pollutants, Water Res., 34, 2601-2605.

54. Panizza, M. and Cerisola, G., 2001, Removal of organic pollutants from industrial wastewater by electrogenerated Fenton's reagent, Water Res., 35 (16, 3987-3992.

55. Perret, A., Haenni, W., Skinner, N., Tang, X.-M., Gandini, D., Comninellis, Ch., Correa, B. and Foti, G., 1999, Electrochemical behavior of synthetic diamond thin film electrodes, Diamond Rel. Mater., 8, 820-823.

56. Potter, F. J., 1993) The oxidation of phenols with Fenton's reagent: A kinetic study, Ph.D. Dissertation, Vanderbilt University, Nashville, TN, U.S.A.

57. Pratap, K. and Lemley, A.T., 1994, Electrochemical peroxide treatment of aqueous herbicide solutions, J. Agric. Food Chem., 42, 209-215.

58. Quiang, Z., Chang, J.H. and Huang, C.P., 2002, Electrochemical generation of hydrogen peroxide from dissolved oxygen in acidic solutions, Water Res., 36, 85-94.

59. Rajeshwar, K. and Ibanez, J., 1998, Environmental Electrochemistry: Fundamentals and Applications in Pollution Abatement, Academic Press, San Diego, CA, Chapter 5.

60. Saltmiras, D.A. and Lemley, A.T., 2000, Degradation of ethylene Thiourea (ETU) with three Fenton treatment processes, J. Agric. Food Chem., 48, 6149-6157.

61. Sauleda, R. and Brillas, E., 2001, Mineralization of aniline and 4-chlorophenol in acidic solution by ozonation catalyzed with Fe2+ and UVA light, Appl. Catal. B: Environ., 29, 135-145.

62. Scharifian, H. and Kirk, D.W., 1986, Electrochemical oxidation of phenol, J. Electrochem. Soc., 133, 921-924.

63. Sedlak, D.L. and Andren, A.W., 1991, Oxidation of chlrobenzene with Fenton's reagent, Envrion. Sci. Technol., 25, 777-782.

64. Serpone, N., Terzian, R., Colarusso, P., Minero, C., Pelizzetti, E. and Hidaka, H., 1992, Sonochemical oxidation of phenol and three of its intermediate products in aqueous media: Catechol, hydroquinone and benzoquinone. Kinetic and mechanistic aspects, Res. On Chem. Intermediates, 18, 183-202.

65. Smeds, A., Franzen, R. and Kronberg, L., 1995, Occurrence of some chlorinated enol lactones and cyclopentene-1, 3-diones in chlorinated waters, Envrion. Sci. Technol., 29 (7): 1839-1844.

66. Stefan, M.I., Hoy, A.R. and Bolton, J.R., 1996, Kinetics and mechanism of the degradation and mineralization of acetone in dilute aqueous solution sensitized by UV photolysis of hydrogen peroxide, Environ. Sci. Technol., 30 (7, 2382-2390.

67. Stucki, S., Kotz, R. and Carcer, B (1991, Electrochemical wastewater treatment using high overvoltage anodes. Part II. Anodes performance and applications, J. Appl. Electrochem., 21, 99-104.

68. Sudoh, M., Kodera, T., Sakai, K., Zhang, J.Q. and Koide, K., 1986, Oxidative degradation of aqueous phenol effluent with electrogenerated Fenton's reagent, J. Chem. Eng. Jpn.., 19, 513-518.

69. Tennaloon, C.L.K., Bhardwaj, R.C. and Bockris, J.O.M., 1996, Electrochemical treatment of human wastes in a packed bed reactor, J. Appl. Electrochem., 26, 18-25.

70. Tatapudi, P. and Fenton, J., 1993, Synthesis of ozone in a proton exchange membrane electrochemical reactor, J. Electrochem. Soc., 140, 3527-3530.

71. Tomat, R. and Rigo, A., 1976, Electrochemical production of OH radicals and their reaction with toluene, J. Appl. Electrochem., 6, 257-261.

72. Tzedakis, T., Savall, A. and Clifton, M.J., 1989, The electrochemical regeneration of Fenton 's reagent in the hydroxylation of aromatic substrates: batch and continuous processes, J. Appl. Electrochem., 19, 911-921.

73. Ventura, A., Jacquet, G., Bermond, A. and Camel, V., 2002, Electrochemical generation of the Fenton's reagent: Application to atrazine degradation, Water Res., (in press).

74. Walling, C., 1975, Fenton's reagent revisited, Acc. Chem. Res., 8, 125-131.

75. Wang, Q. and Lemley, A.T., 2002, Oxidation of diazinon by anodic Fenton treatment, Water Res., (in press).

76. Wu, J.M., Huang, H.S. and Livengood, C.D., 1992, Ultrasonic destruction of chlorinated compounds in aqueous solution, Envrion. Progress, 11 (3): 195-201.

77. Yang. C.H., 1999, Hypochlorite production on Ru-Sn binary oxide electrode and its application in the treatment of dye wastewater, Can. J. Chem. Eng., 77, 1161-1168.

78. Zhu, W.P., Yang, Z.H. and Wang, L., 1996, Applications of ferrous-hydrogen peroxide for the treatment of H-acid manufacturing process wastewater, Water Res., 30 (12): 2949-2954.

8

Removal of Cyanides by Electroxidation

LIDIA SZPYRKOWICZ, FRANCESCO RICCI, SALVATORE DANIELE

Department of Environmental Sciences, University of Venice, Dorsoduro 2137, 30123 Venice, Italy

Introduction

In copper plating operations an electrolytic bath containing copper and potassium cyanides is often preferred to a sulphate bath due to the brightness of the metal deposits and their good adherence. In this case the wastewaters which are generated during the rinsing of worked pieces bear copper cyanide complexes. Conventional treatment of these wastewaters consists in oxidation of the cyanide by chlorine gas or hypochlorite and the subsequent removal of cuprous and cupric ions by precipitation as hydroxides. This method requires careful control of pH and a dose of hypochlorite in excess of the stoichiometric quantity required by the reaction. The resulting sludge is difficult to handle as it contains copper hydroxides, which can be subject to easy dissolution giving rise to a copper-rich leakage. Electrochemical oxidation, particularly if conducted in a reactor with plate electrodes, enabling the cathode to be reused, can be an interesting alternative.

Application of electrochemical reactors for the treatment of industrial wastewaters gains an increasing attention (Bergmann *et al.*, 1992; Buso *et al.*, 1997a; Szpyrkowicz et. al., 1995; Rao *et al.*, 2001; Szpyrkowicz et. al., 2001a). Use of electrochemical oxidation for the destruction of cyanides as an alternative to chemical oxidation has been studied since the early seventies. Most of the studies were focused on the destruction of simple cyanides and on the mechanism of the reaction. It has been shown that direct oxidation is slow on platinum (Tamura *et al.*, 1974) and graphite (Arikado *et al.*, 1976), exhibiting low current efficiency. Better results were obtained using Ni anodes (Kelsall *et al.*, 1991) or PbO_2-coated (Hine at al., 1986) or a reticulated three-dimensional electrode (Tissot *et al.*, 1994).

The reaction pathways and by-products of cyanide decomposition depend strongly on the OH^- concentration of the medium. It is generally accepted that in alkaline conditions cyanide complexed with copper can be electrochemically oxidized giving carbon dioxide and nitrogen gas (Tan *et al*, 1985; Hwang *et al*, 1987):

$$Cu\,(CN)_2^{(n-1)} + 2n\,OH^- = Cu^- + nCNO^- + nH_2O + 2ne^- \quad (1)$$

$$CN^- + 2OH^- = CNO^- + H_2O + 2e^- \quad (2)$$

$$2CNO^- + 4OH^- = 2CO_2 + N_2 + H_2O + 6e^- \quad (3)$$

As can be seen from reaction (2) the direct oxidation of cyanide to CNO^- in alkaline media occurs with consumption of 2 moles of alkalinity for 1 mole of reacted cyanide, producing N_2 and CO_2 as final products (Hwang *et al.*, 1987).

At lower pH the presence of oxalate and of a brown polymer $(HCN)_x$, called "azulmin" (Yoshimura *et al.*, 1980; Hwang *et al.*, 1987) can also be detected.

Several researches have observed that the rate of cyanide decomposition can be favourably influenced by the presence of copper in the solution (El-Ghaoui *et al.*, 1982a; Ho *et al.*, 1990, Szpyrkowicz *et al.*; 1998) postulating that the film of copper oxides deposited on the anode exhibits electrocatalytic properties. The mechanism of this process was studied by Wels and Johnson (1990) in cyanide decomposition experiments carried out on the copper oxide film obtained by electrodeposition from an $Na_3Cu\,(CN)_4$ solution. They postulated that anodic conversion of Cu^{III} to Cu^{II} provides a favourable mechanism for the oxygen transfer step, which occurs concomitantly with electron transfer in the oxidation of CN^- to CNO^-. The oxygen transfer step occurs via generation of hydroxyl radical ($^{\cdot}OH$) ads at the Cu^{III} site (S) during anodic discharge of H_2O:

$$S + OH^- = S\,(^{\cdot}OH)_{ads} + e^- \quad (4)$$

$$S\,(^{\cdot}OH)_{ads} + CN^- + OH^- = S + CNO^- + H_2O + e^- \quad (5)$$

The net result of the above reactions is reaction (2)

In addition to the above reactions cyanate can also decompose producing ammonium (Kelsall *et al.*, 1991):

$$CNO^- + 3H_2O = CO_3^{-2} + NH_4^+ + H_2O \quad (6)$$

The above reaction is believed to be autocatalytic with respect to carbonate ions, which can be also formed independently of reaction (3) by absorption of atmospheric CO_2 in the alkaline cyanide solution.

The problem of removing copper simultaneously to the destruction of its organic complexes, particularly with cyanides, has gained less attention. El-Ghaoui *et al.* (1982b) who studied the removal of cyanides from copper complexed solutions in a packed-bed glass tower, found the process feasible. However, no data on copper removal were presented.

The simultaneous removal of copper and cyanide by electrochemical process proved feasible by Hwang *et al.* (1987), who examined different end products of cyanide decomposition in function of the pH. No information on the energy consumption is however, given. In the electrolysis of the copper cyanide using a two concentric cylindrical electrodes under the conditions of pH equal to 11, Tan *et al.* (1985) have eliminated 90 % of copper from the solution: 60 % was recovered as metallic copper at the cathode, 30% was found in the anodic film. Szpyrkowicz et. al., (2000a; 2000b), removed copper from copper complex cyanides solution with the 80 % initial faradic efficiency.

Most of experimental studies performed with the objective to test on application of different materials as anodes in the electrochemical destruction of cyanides lack information on the voltammetric behaviour of these materials. Thus *e.g.* it is not possible to compare the data on the potential at which electrooxidation occurs.

There are other anode materials, proved electrocatalytic for oxidation of organic pollutants, whose activity towards cyanides is worth to be explored. In our previous works we have tested several electrode materials for the destruction of organic pollutants. The following anode materials proved to perform well for such compounds as dyes and Thiourea Dioxide during direct electrochemical oxidation (Szpyrkowicz *et al.*, 200a; 2001b): Ti/Pt, Ti/PbO_2, Ti/70% Pt + 30% Ir, Ti/75% MnO_2 + 25% RuO_2, Ti/RhO_x + TiO_2, Ti/70% PdO + 30% Co_3O_4. Because of the catalytic properties of these electrode materials it seemed useful to investigate their catalytic activity towards direct oxidation of cyanides. Thus the purpose of this paper was to study the performance of several electrode materials which were never yet applied for such a process and to compare their behaviour with a copper oxide film, deposited in situ, during electrooxidation of copper complexed cyanide solution. The investigation was carried out mainly by cyclic voltammetry under different experimental conditions.

The paper presents also the results of the performance of the tests of the exhaustive electrolysis in a batch reactor with plate electrodes.

Materials and Methods

The electrochemical investigation was carried out in a classical three-electrode call. The working electrodes were prepared by connecting a piece of the tested materials with a copper wire to the external electronic circuit. This electrode system was sealed in an epoxy resin and an approximately square surface of about 2 mm² area was exposed to the electrolytic solution. The active surface area was evaluated from the peak current recorded by linear sweep voltammogramms in a 1 mM $K_4Fe(CN)_6$ + 0.1 M $NaClO_4$ solution, using the following equation (Bard and Faulkner, 1980):

$$i_p = 2{,}68\times10^5(n^{3/2}\times A\times D^{1/2}\times v^{1/2}\times C^0) \qquad (7)$$

where

i_p – the peak current (A); *n* – number of electrons (n=1); *D* – diffusion coefficient (D=0, 63x10^{-6} cm^2s^{-1}); *v* – scan rate (Vs^{-1}); C^o – bulk concentration (mol cm^{-3}); *A* – area (cm^2).

The active surface area was evaluated for each electrode from the slope of i_p vs $v^{1/2}$ plot.

The counter electrode was a Pt spiral, while the reference electrode was an Ag/AgCl saturated with KCl. Voltammetric measurements were performed with an Autolab PGSTAT20 potentiostat.

The measurements were carried out in solutions containing 0.1 M $NaClO_4$ without and with KCN, over the concentration rage of cyanide 0.03-0.05 g/l and at different (9.4 < pH = 13.0), adjusting it by an addition of a concentrated solution of NaOH.

Exhaustive electrolysis was carried out in a batch electrochemical reactor of 0.6 l working volume, equipped with stainless steel anode and cathode, as shown in Fig. 1. Experiments of exhaustive electrolysis were performed at pH 13.0, defined in a previous study (Szpyrkowicz et. al., 2000a, 2000b) as optimal for the removal of cyanides by direct electrooxidation. The reactor was under potentiostatic control. Isothermal conditions were obtained by using a thermostatic bath (Haake Fisons, mod. K15). The experiments presented in this study were performed under the temperature of 20°C ± 1. All the parameters were determing following Standard Methods (APHA, 1989).

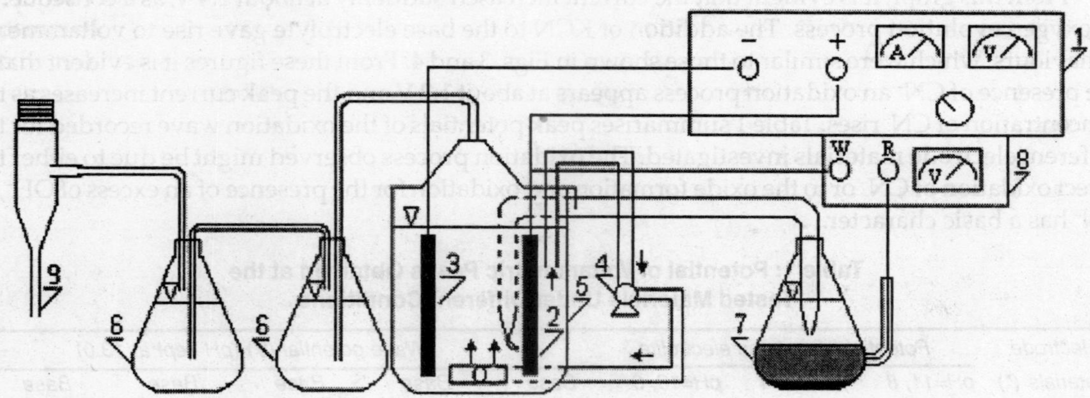

Fig. 1: The Experimental Set-up of the Batch Reactor with Hydraulic Stirring (1-potentiostat DASA; 2-anode; 3-cathode; 4-peristaltic pump; 5-electrochemical reactor; 6-NaOH trap; 7-reference electrode SCE; 8-voltmeter; 9-vacuum pump)

Results and Discussion

Preliminary cyclic voltammogramms were performed in solutions containing the base electrolyte over the positive potential region. All electrode materials employed displayed a similar behaviour, as those shown in Fig. 2.

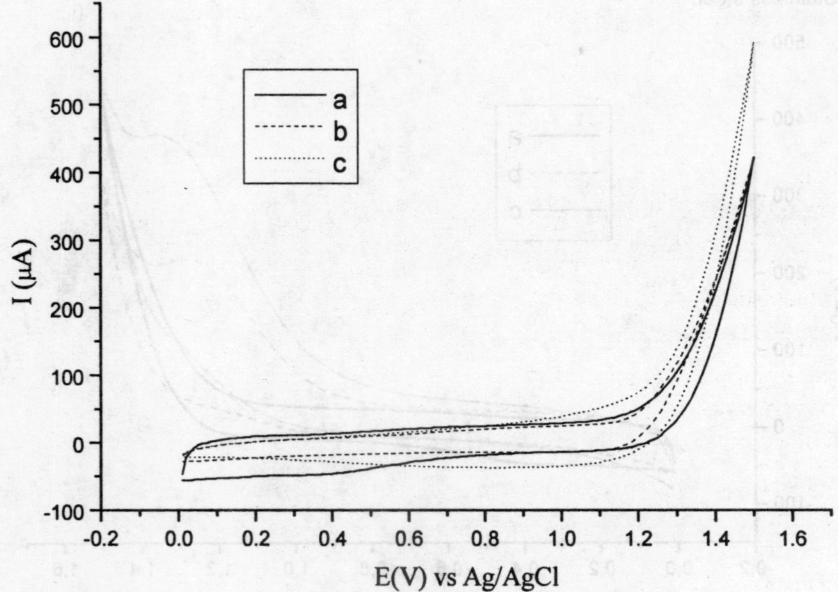

Fig. 2: Cyclic Voltammetry Registered in a Solution of $NaClO_4$ (12.24 g/l) (a) a Ti/Pt electrode; (b) a Ti/70% Pt + 30% Ir electrode; (c) a Ti/RhO_x + TiO_2 electrode

From this graph it is evident that the current increased suddenly at about 1.4 V, as a consequence of oxygen evolution process. The addition of KCN to the base electrolyte gave rise to voltammetric behaviours, which were similar to those shown in Figs. 3 and 4. From these figures it is evident that in the presence of CN^- an oxidation process appears at about 1.2 V and the peak current increases as the concentration of CN^- rises. Table 1 summarises peak potentials of the oxidation wave recorded for the different electrode materials investigated. The oxidation process observed might be due to either the direct oxidation of CN^- or to the oxide formation and oxidation for the presence of an excess of OH^-, as CN^- has a basic character.

Table 1: Potential of Voltammetric Peaks Obtained at the Tested Materials Under Different Conditions

Electrode Materials (*)	*Potential (V) in base electrolite*			*Wave potential (V) (pH kept at 13.0)*				
	pH=11, 8	*pH=12, 4*	*pH=13, 0*	*Base Electrolyte+ CN 0, 03g/l*	*Base Electrolyte +CN 0, 05g/l*	*Base Electrolyte*	*Base Electrolyte+ CN 0, 03g/l*	*Base Electrolyte +CN 0, 05g/l*
1	1.17	1.17	1.17	1.35	1.29	1.17	1.09	1.29
2	0.97	0.99	1.10	1.31	1.17	1.50	1.50	1.41
3	1.01	1.01	1.01	1.12	1.15	1.03	0.85	0.92
4	0.94	0.94	0.94	1.10	1.15	1.06	1.05	1.06
5	0.93	1.12	1.05	1.18	1.38	1.15	1.44	1.50
6	0.86	0.86	0.86	1.04	1.11	0.89	0.87	0.91
7	0.92	0.97	0.95	1.09	1.18	0.96	0.97	1.01

(*) 1: Ti/Pt; 2: Ti/PbO_2; 3: Ti/70% Pt + 30% Ir; 4: Ti/75% MnO_2 + 25% RuO_2; 5: Ti/RhO_x + TiO_2; 6: Ti/70% PdO + 30% Co_3O_4; 7: Stainless steel.

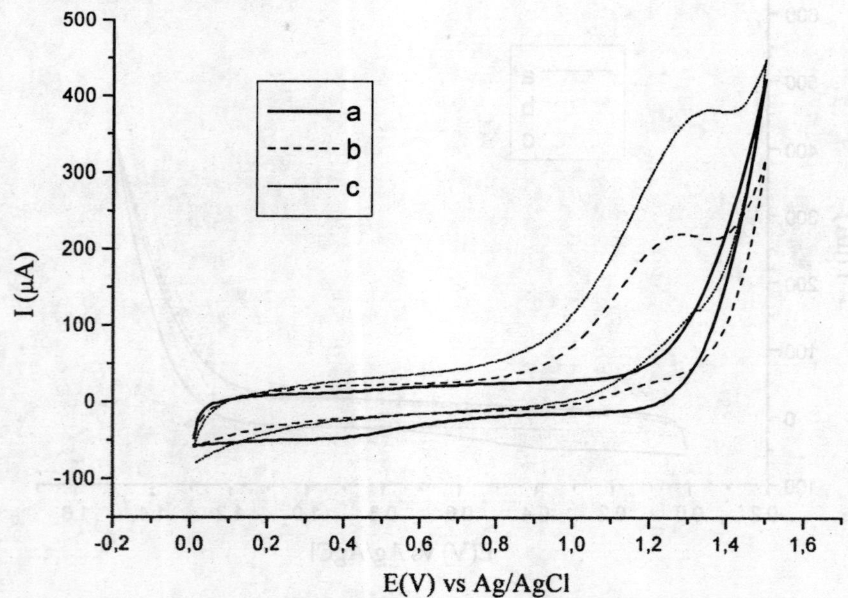

Fig. 3: Cyclic Voltammetry Registered with a Ti/Pt Electrode in: (a) base electrolite $NaClO_4$ (12.24 g/l); (b) $NaClO_4$ (12,24 g/l) + CN^- (0,03 g/l); (c) $NaClO_4$ (12,24 g/l) + CN^- (0,05 g/l); scan rate: 0,1 Vs^{-1}

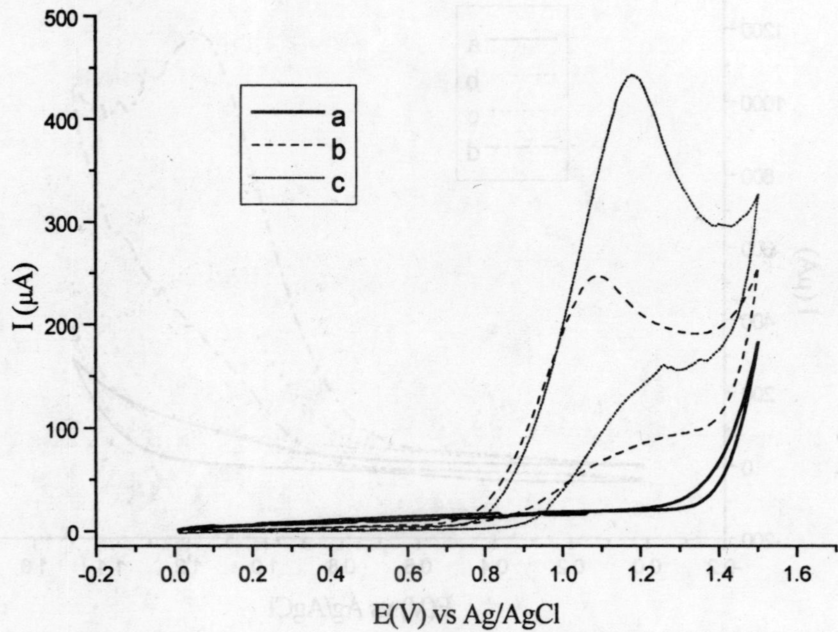

Fig. 4: Cyclic Voltammetry Registered with a Stainless Steel Electrode in: (a) base electrolyte $NaClO_4$ (12.24 g/l); (b) $NaClO_4$ (12,24 g/l) + CN^- (0,03 g/l); (c) $NaClO_4$ (12,24 g/l) + CN^- (0,05g/l); scan rate: 0,1 Vs^{-1}

In order to ascertain which process was responsible for such waves, a series of cyclic voltammogramms were run in solutions containing the base electrolyte made alkaline with NaOH, but without CN^-. Figures 5 and 6 show typical cyclic voltammograms thus obtained. Under these conditions oxidation waves also appear, whose peak potentials are slightly less positive than those observed in the presence of cyanides, as can be seen from data included in Table 1. These results confirm that the hydroxide ions are also responsible for the process observed in Figs. 3 and 4, which probably overlap with the direct oxidation of CN^-. On the basis of the recent literature reports, it may also be that the waves shown in Figs. 3 and 4 are due to a complex CE (chemical reaction preceding the electron transfer process) mechanism of the type:

$$CN^- + H_2O = HCN + OH^- \quad (8)$$

$$4\,OH^- = O_2 + 2\,H_2O + 4e^- \quad (9)$$

This aspect has been further investigated by performing a series of measurements in a solution buffered at constant pH equal to 13.0, to which the increasing amounts of KCN were added. Under these conditions all electrode materials behave in a similar manner. Fig. 7 shows cyclic voltammogrammes recorded for the Ti/Pt anode, which can be considered as typical and representative of the voltammetric pictures of other tested materials: 1: Ti/Pt; 2: Ti/PbO_2; 3: Ti/70% Pt + 30% Ir; 4: Ti/75% MnO_2 + 25% RuO_2; 5: Ti/RhO_x + TiO_2; 6: Ti/70% PdO + 30% Co_3O_4; 7: Stainless steel.

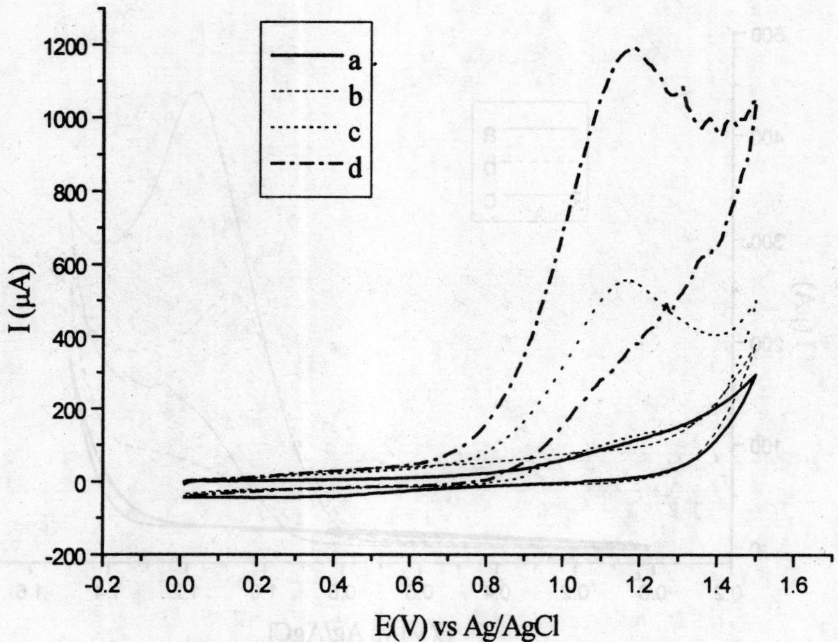

Fig. 5: Cyclic Voltammetry Registered with a Ti/Pt Electrode in Base Wlectrolyte $NaClO_4$ (12.24 g/l): (a) pH=9,4; (b) pH=11,8; (c) pH=12,4; (d) pH=13; scan rate: 0,1 Vs^{-1}

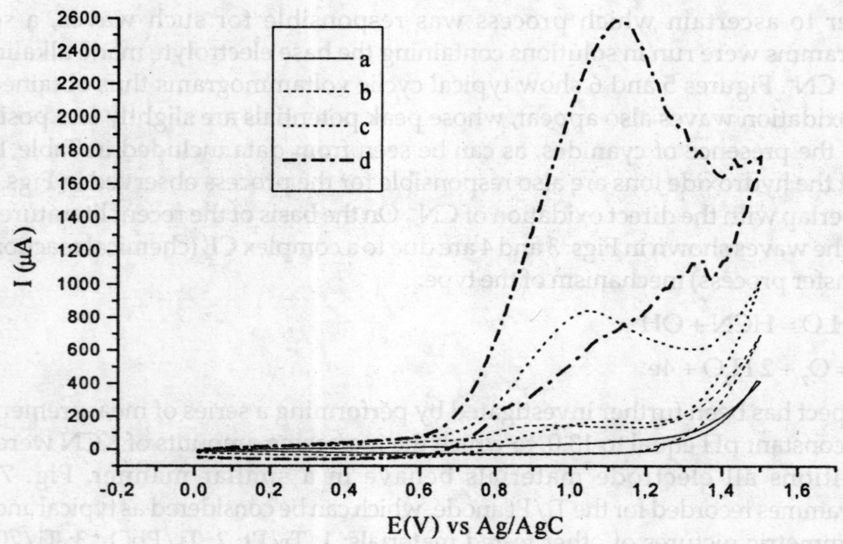

Fig. 6: Cyclic Voltammetry Registered with a $Ti/RhO_x + TiO_2$ in Base Electrolyte $NaClO_4$ (12.24g/l): (a) pH=9,4; (b) pH=11,8; (c) pH=12,4; (d) pH=13; scan rate: 0,1 Vs^{-1}

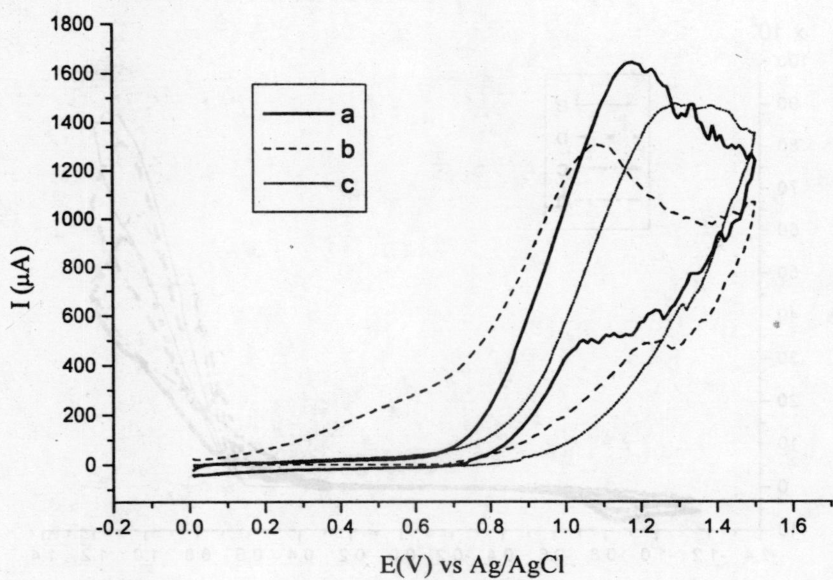

Fig. 7: Cyclic Voltammetry Registered with a Ti/Pt Electrode in: (a) base electrolite $NaClO_4$ (12.24 g/l); (b) $NaClO_4$ (12,24 g/l) + CN^- (0,03 g/l); (c) $NaClO_4$ (12,24 g/l) + CN^- (0,05 g/l); scan rate: 0,1 Vs^{-1}

Starting from the solution at the pH 13.0 an addition of KCN led, at a lower concentration, to a decrease of the peak current; a further increase of KCN quantity led to a rise of the peak current with respect to the base electrolyte. Because the pH was kept constant, the latter behaviour can be explained as due to a partial passivation effect of the electrode surface due to an adsorption of the CN^- oxidation product. This hypothesis is partially supported by the slight shift of the waves towards less positive values. This phenomenon seems less evident at higher CN^- concentrations.

Results similar to those reported above were also obtained for a stainless steel as an electrode material. The potential data are also included in Table 1.

Data reported in Table 1 put into evidence, that for the materials: Ti/Pt; Ti/PbO_2; Ti/70% Pt + 30% Ir; Ti/75% MnO_2 + 25% RuO_2; Ti/RhO_x + TiO_2; Ti/70% PdO + 30% Co_3O_4 and Stainless steel, there is essentially only a very small variation in the potential of the voltammetric waves, which can be attributed to oxidation of free cyanides. The potential at which these pollutants can be electrooxidised is approximately 1 V.

Dramatically different results were obtained when cyanides were present as copper complexes. A typical voltammetric picture obtained for the electrolyte containing copper cyanide complexes is depicted in Fig. 8 for the stainless steel electrode. In the presence of CuCN a formation of a new peak, situated in the potential window 0.6-0.7 V, can be noticed. In the backward scan a "crossover" of a current can be observed, which suggests that in this potential window a complex electrochemical phenomena occur, comprising oxidation of copper, with a probable formation of a new surface film, and electrooxidation of cyanides. A "crossover" of the current, which is an evidence of a nucleation phenomenon, suggests that a new film is being formed. The height of the peak is proportional to the concentration of the free cyanides (see Fig. 9), which clearly confirms that these ions are also involved.

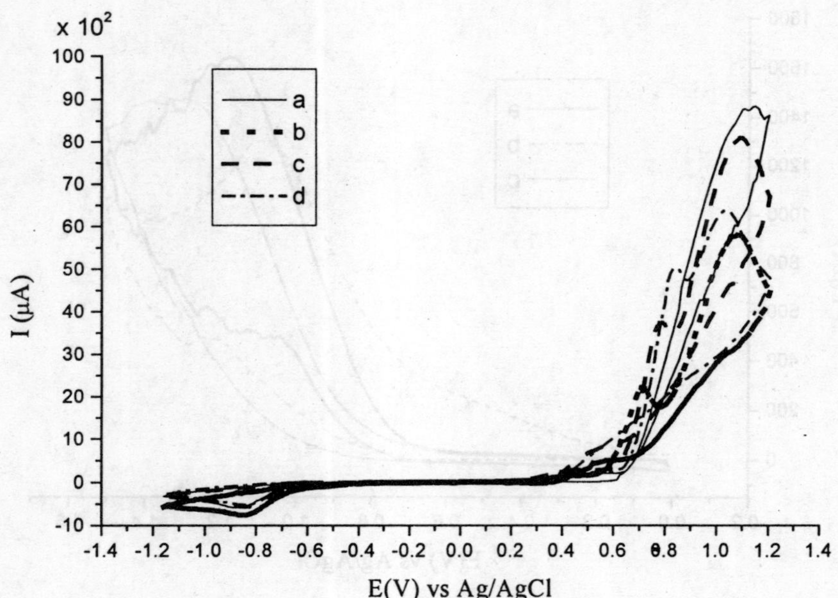

Fig. 8: Cyclic Voltammetry Registered with a Stainless Steel Electrode in: (a) base electrolyte $NaClO_4$ (12.24 g/l); (b) $NaClO_4$ (12,24 g/l) CuCN (0,015 6 g/l); (c) $NaClO_4$ (12,24 g/l) + CuCN (0,0156 g/l) + KCN (0,012 g/l); (d) $NaClO_4$ (12,24 g/l) + CuCN (0,015 6g/l) + KCN (0,023 g/l); scan rate: 0,1 Vs^{-1}

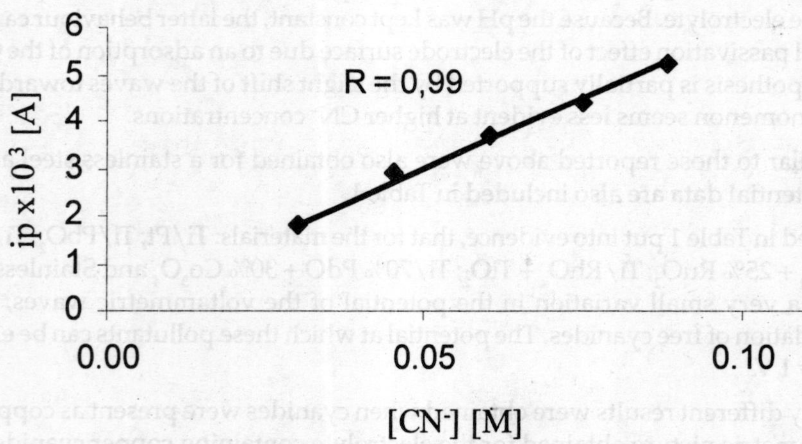

Fig. 9: Dependance of the Peak Current (i_p) on the Concentration of Free Cyanide

The exhaustive electrolysis was performed with a stainless still anode under different hydrodynamic conditions, obtained by mechanic, hydraulic and pneumatic agitation. The results here presented refer to the performance of the reactor under hydraulic stirring.

Fig. 10 shows trend in time of the concentration of cyanides in the reactor operated under the conditions of different anode potential (hydraulic stirring at the same rate for all the tests). As can be

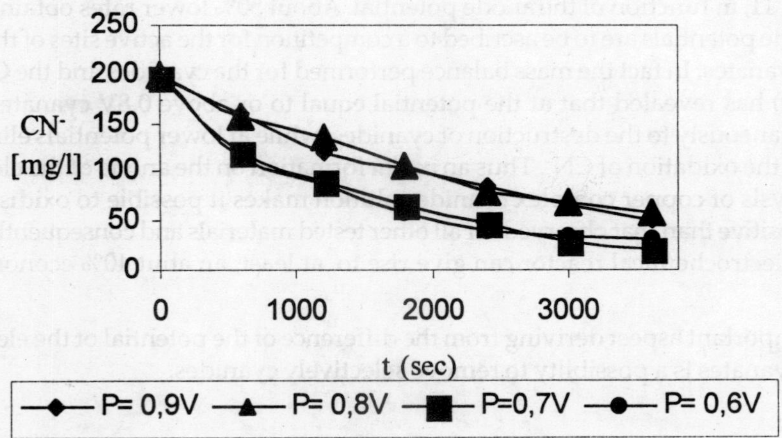

Fig. 10: Destruction of Total Cyanides in Function of the Anode Potential

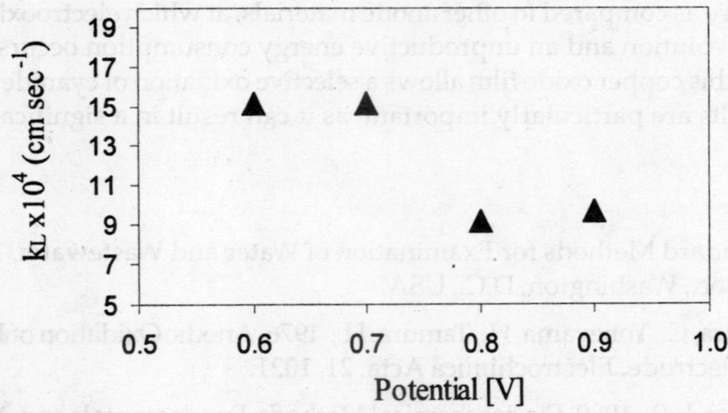

Fig. 11: Variation of the Apparent Mass Transport Coefficient During Electro-oxidation of Cyanides in Function of the Anode Potential

seen, curves obtained for the potential equal to 0.9 and 0.8 V (vs SCE) overlap, as do also the curves relative to the experiments conducted at the potential equal to 0.6 and 0.7V (vs SCE).

The values of the apparent mass transport coefficient were calculated using the following equation (Pletcher and Walsh, 1993):

$$c_t = c_0 \exp\left(-\frac{k_L A}{V_R} t\right) \tag{10}$$

where

c_o and c_t denote cyanide concentration (mole m^{-3}) at the beginning of the electrolysis and after the time t, respectively, A_t is the anode area (m^{-2}) and V_r is the reactor volume (m^3). The k_L values are depicted in Fig. 11, in function of the anode potential. About 50% lower rates obtained for runs at the two higher anode potentials are to be ascribed to a competition for the active sites of the anode between cyanides and cyanates. In fact the mass balance performed for the cyanides and the Chemical Oxygen Demand (COD) has revealed that at the potential equal to or above 0.8V cyanate ions were being oxidised simultaneously to the destruction of cyanides, while at lower potentials elimination of COD is totally due to the oxidation of CN^-. Thus an *in situ* formation on the anode of the electrocatalitic film during electrolysis of copper complex cyanide solution makes it possible to oxidise cyanides at the potential less positive than that observed for all other tested materials and consequently, a potentiostatic control of the electrochemical reactor can give rise to, at least, an abut 40% economy of the energy consumption.

Another important aspect deriving from the difference of the potential of the electro-oxidation of cyanides and cyanates is a possibilty to remove selectively cyanides.

Conclusions

The present study has confirmed that a film deposited *in situ* on the anode during electrolysis of the copper cyanide bearing solutions has electrocatalytic properties towards the oxidation of cyanides. The anode potential at which occurs electrooxidation of cyanide on this *in situ* formed film is equal to 0, 6 V and is less positive as compared to other anode materials, at which electrooxidation proceeds at potentials of oxygen evolution and an unproductive energy consumption occurs. By contrast to an other anode materials this copper oxide film allows a selective oxidation of cyanides, leaving cyanides unreacted. These results are particularly important, as it can result in a significant economy of the current consumption.

REFERENCES

1. APHA, 1989, Standard Methods for Examination of Water and Wastewater, 17th Ed. American Public Health Assoc., Washington, D.C., USA.

2. Arikado, T., Iwakura, C., Yoneyama, H., Tamura, H., 1976, Anodic Oxidation of Potassium Cyanide on the Graphite Electrode. Electrochimica Acta. 21, 1021.

3. Bard, A. J., Faulkner, L.R., 1980, Electrochemical Methods-Fundamentals and Applications, Wiley and Sons, N.Y.

4. Bergmann, H., Hertwig, H., Nieber, F. 1992, Experimental and Theoretical Studies on a New Type of Electrochemical Reactor for Wastewater Treatment, Chem. Eng. Proc., 31, 195.

5. Buso, A., Giomo, M., Boaretto, L., Paratella, A., 1997, New Electrochemical Reactor for Wastewater Treatment: Mtematical Model, Chem. Eng. Proc. 36, 411.

6. El-Ghaoui, E. A., Jansson, R. E., Moreland C., 1982a, Application of Trickle Tower to Problems of Pollution Control. III. Heavy-metal cyanide solutions, J. Appl. Electrochem. 12, 75.

7. El-Ghaoui, E. A., Jansson, R.E., 1982b, Application of Trickle Tower to Problems of Pollution Control. II The Direct and Indirect Oxidation of Cyanide, J. Appl. Electrochem. 12,.

8. Hine, F., Yasuda, M., Iida, T., Ogata, Y., 1986, On the oxidation of cyanide solutions with lead dioxide coated anode, Electrochimica Acta, 31, 1389.

9. Ho, S. P., Wang, Y. Y., Wan, C.C., 1990, Electrolytic decomposition of cyanide effluent with an electrochemical reactor packed with stainless steel fiber, Wat. Res. 24, 1317.

10. Hwang, J.Y., Wang, Y.Y., Wan, C.C., 1987, Electrolytic oxidation of cuprocyanide electroplating wastewaters under different pH conditions, J. Appl. Electrochem. 17, 684.
11. Kelsall, G. H., Savage, S., Brandt, D., 1991, Cyanide Oxidation at Nickel Anodes, J. Electrochem. Soc., 138, 117.
12. Pletcher D., Walsh F.C., 1990, "Industrial eletrochemistry" secon editino Blackie Academic & Professional N.Y.
13. Rao, N.N., Somasekhar, K. M., Kaul, S.N. and Szpyrkowicz, L., 2001, Electrochemical Oxidation of Tannery Wastewater, *Journal of Chemical Technology and Biotechnology*, 76, 1124 – 1131.
14. Szpyrkowicz, L, Naumczyk, J., Zilio-Grandi F., 1995, Electrochemical treatment of tannery wastewater using Ti/Pt and Ti/Pt/Ir electrodes, *Wat. Res.* 29, 517.
15. Szpyrkowicz, L, Zilio Grandi F, Kaul S.N., Rigoni-Stern S,. 1998, Electrochemical treatment of copper cyanide wastewaters using stainless steel electrodes, Wat. Sci. Techn. 38, 268.
16. Szpyrkowicz, L., Kaul S. N., E. Molga, De Faveri M., 200a, Comparison of the Performance of a Reactor Equipped with a Ti/Pt and an SS Anode for Simultaneous Cyanide Removal and Copper Recovery, *Electrochimica Acta*, 46, 381 – 387.
17. Szpyrkowicz, L., Zilio Grandi F., Kaul S. N., Polcaro A. M., 2000b, Copper Electrodeposition and Oxidation of Complexed Cyanide from Wastewaters in an Electrochemical Reactor with a Ti/Pt anode, *Ind. Eng. Chem. Res.*, 39, 2132-2139.
18. Szpyrkowicz, L., Iuzzolino C., Kaul, S.N., 2001a, A Comparative study on Oxidation of Disperse Dyes by Electrochemical Process, Ozone, Hypochlorite and Fenton Reagent, *Wat. Res.*, 35, 2129-2136.
19. Szpyrkowicz, L., Juzzolino, C., Daniele, S., De Faveri, M., 2001b, Electrochemical Destruction of Thiourea Dioxide in an Undivided Parallel Plate Electrodes Batch Reactor, *Catalysis Today*, 66, 519-527.
20. Tamura, T., Arikado, T., Yoneyama H., Matsuda Y., 1974, Anodic oxidation of potassium cyanide on platinum electrode, Electrochimica Acta 19, 273.
21. Tan T. C., Teo W. K., Chin D-T., 1985, Electrochemical destruction of complex cyanide, Chem. Eng. Commun, 38, 125.
22. Tissot P., M. Fragniere, 1994, Anodic oxidation of cyanide on a reticulated three-dimensional electrode, J Appl. Electrochem. 24, 509.
23. Wels, B., Johnson, D.C., 1990, Electrocatalysis of Anodic Oxygen Transfer Reactions: Oxidation of Cyanide at Electrodeposited Copper Oxide Electrodes in Alcaline Media, J. Electrochem. Soc., 137, 2785.
24. Yoshmura S., Katagiri A., Deguchi Y., Yoshizawa S., 1980, Studies of the Anodic Oxidation of the Cyanide Ion in the Presence of the Copper Ion. IV. The Kinetics and Machanism of the Decomposition of the Intermediate Tetracyanocuprate (II) Ion, Bull. Chem. Soc. Jpn, 53, 2437.

9

Destruction of Cyanide in Aqueous Waste by Electrochemical Oxidation

M. NISHA PRIYA & K. PALANIVELU

Centre for Environmental Studies, Anna University, Chennai - 25

ABSTRACT

A laboratory study was carried out by electrochemical oxidation to destroy cyanide in aqueous waste. The electrode used in this study was a triple oxide coated titanium based mesh type anode and a carbon cathode. Direct and indirect methods were both carried out at alkaline conditions and indirect oxidation method in the presence of chloride was found to be more efficient. The cyanide concentration in the aqueous waste collected from an electroplating unit was reduced from 128 mg/L to 0.04 mg/L with a energy consumption of 387.5 KWh/KgCN$^-$ within 4.6 hours by indirect oxidation method. The degradation of cyanide followed first order kinetics with respect to cyanide concentration (first order reaction rate K=1.381min^{-1}).

Keywords: Cyanide, Electrochemical Oxidation, Chloride, Ti/TiO$_2$-RuO$_2$-IrO$_2$ anode, plating wastewater treatment.

Introduction

Cyanides are used in a number of chemical synthesis, electroplating industry and in metallurgical processes. Cyanides are highly toxic and must be destroyed from waste prior to discharge. The most common method for treating free or simple cyanide is alkaline chlorination. However chlorination of cyanide results in highly toxic intermediates (*e.g.* cyanogen chloride) and if organic material is present chlorinated VOC's are formed. These compounds together with the residual chlorine create additional environmental problems. The method also suffers the draw back of generating a large quantity of sludge, which contains heavy metal after chemical decomposition and requires subsequent precipitation process (Hwang *et al.*, 1985, Ho *et al.*, 1990). Consequently, there is a growing need for alternative methods for destroying cyanide.

The electrochemical method is an alternative process for destruction of cyanide at the anode and collecting heavy metal at the cathode simultaneously without generating a sludge problem. (Tan *et al.*, 1985). Complex cyanide and concentrated cyanide solutions can also be handled. This method also has an advantage of the non-toxic products that are formed. The oxidation of cyanide at different conditions gives different products. Direct oxidation of cyanide at pH 5.2-6.8 yields $C_2O_4^{2-}$ and ammonia, at pH 7.0-11.7 yields ammonia and CO_3^{2-} and at pH 12 yields CO_2 and N_2 (Hwang *et al.*, 1987). Indirect oxidation of cyanide yields CO_2 and N_2 as end products. (El-Ghaoui *et al.*, 1982) The process is especially suitable for the electroplating industries where DC supply and other plating equipments can be adopted for waste treatment process. (Bakir *et al.*, 1999).

Studies conducted on electrochemical destruction of copper cyanide using plate stainless steel electrodes showed efficient degradation of cyanide at anode with simultaneous recovery of copper at the cathode. (L.Szpyrkowicz *et al.*, 1998).

Degradation of cyanide using graphite raschig rings as electrodes was also found to be efficient in continuous or recirculating batch modes. During the treatment the rate of removal decreased as the concentration decreased with time. More energy was used to reduce the cyanide content from 30mg/L to zero (70 KW h/kg CN^-) than to reduce the cyanide concentration from 300 to 30mg/L (8.75 KW h/kg CN^-) (Bakir Ogutveren *et al.*, 1999). Studies were also carried out using packed-bed electrochemical reactor with stainless steel fiber as electrodes to decompose cyanide containing effluent. (Ho *et al.*, 1990). The study showed that copper ion has an accelerating effect on the cyanide oxidation.

Thus electrochemical route has been investigated in the study, since no expensive chemicals are required, no sludge is produced, complex cyanide can be degraded and high current efficiencies can be obtained. The aim of this present work was to investigate the electrochemical oxidation of cyanide using titanium substrate insoluble anode (TISA) coated with TiO_2-RuO_2-IrO_2.

Experimental Details

The laboratory scale experiments were carried out in galvanostatic condition in a batch process. The experimental set up (Fig. 1) consists of undivided electrolytic cell of 500 ml capacity. The solution was constantly stirred with a magnetic stirrer. The electrodes used were titanium based mesh type anode coated with TiO_2-IrO_2-RuO_2 (supplied by M/S Titanium equipment and anode manufacturing company Ltd., Chennai, India) and a carbon cathode. The dimensions of the electrodes were 10 × 5 cm. The surface area of the anode was 28cm^2 (neglecting the space contributed by the voids) and that of the cathode is 50 cm^2. The electrodes were suspended into the electrolytic cell using a stand, which also bears a thermometer to check the temperature of the solution. The electrodes were connected to a DC supply (Transistor power supply-613) which supplies current in the range of 0-2A. The concentrations of cyanide in the samples were determined using cyanide-ion-selective electrode (WTW PMX 3000). The sample pH (using model pH 197 WTW) and temperature were frequently monitored during the degradation process.

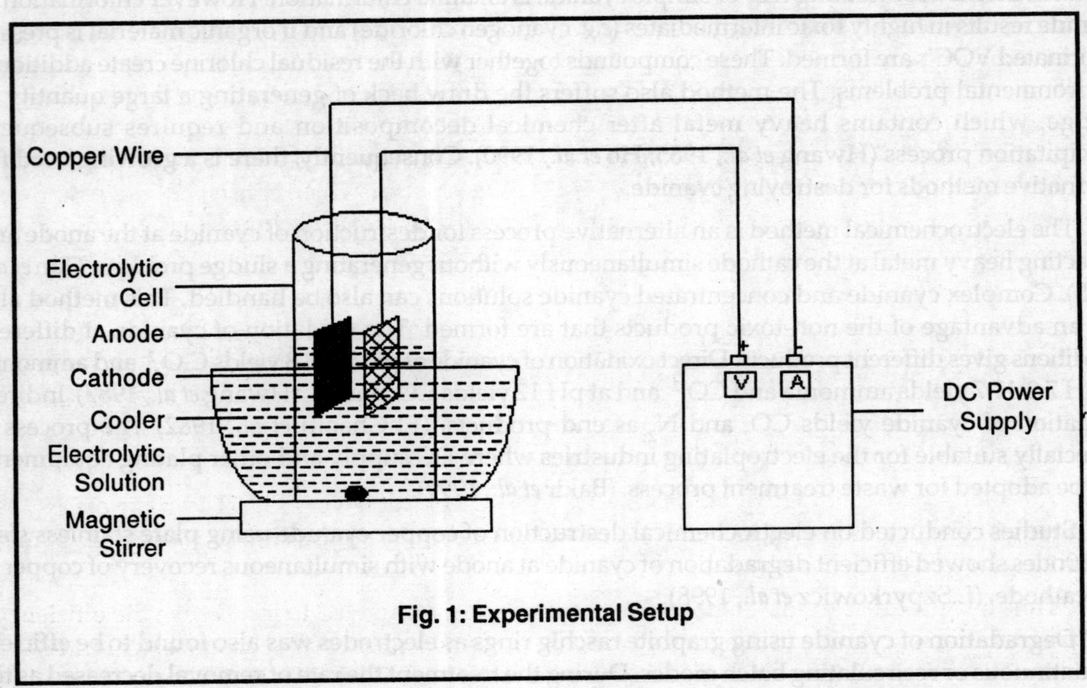

Fig. 1: Experimental Setup

Results and Discussions

Preliminary Study

Initially a preliminary study was conducted with 100 ppm of cyanide using direct oxidation at pH 12 and with the current density of 5.4 A/dm^2. The cyanide degradation using indirect oxidation was also carried out in the presence of 500mg/L and 1000mg/L of chloride and the efficiencies of both the treatment methods were compared.

The energy consumption for 100 ppm of cyanide using direct oxidation was calculated to be 3330 KWh/kg CN^- with 92.33% CN^- removal, which took about 160 minutes whereas the energy consumption decreased with indirect oxidation with 99.8% cyanide removal for 500mg/L chloride to 189 KWh/kg CN^-, which took about 100 minutes. The power requirement in presence of 1000mg/L of chloride further reduced to 114 KWh/Kg CN^- with 99.9% degradation in 60 minutes.

Thus on comparing the electrochemical degradation of the direct and indirect oxidation of cyanide, the indirect oxidation method proved to degrade cyanide faster with lesser energy consumption. Thus indirect oxidation method was chosen for further studies.

Effect of pH

pH effect on the degradation of cyanide was carried out in alkaline conditions, since cyanide has the risk of forming poisonous HCN^- in acidic conditions. The pH was adjusted to 9, 10, 11, 12 with dil. H_2SO_4 and dil. NaOH.

From the experiments carried out with various pH, it was observed that the pH showed no significant effect on the degradation of cyanide. Though there was some difference in the rate of degradation in the initial stages of degradation, the time taken for 99% removal of cyanide was similar in all the four cases. Hence the cyanide degradation can be carried out at any alkaline pH.

Effect of Electrolyte Concentration

One of the important factors for the electrochemical process which consist of heterogenous ion transfer reactions is the conductivity of the solution to be treated. Consequently a supporting electrolyte such as NaCl was added to increase the conductivity of the solution.

The electrolyte used was chloride prepared from sodium chloride. The supporting electrolyte concentration was varied as 250mg/L, 500mg/L, 750mg/L and 1000mg/L and the other conditions were maintained at pH 12, current density 5.4 A/dm^2 and the cyanide concentration 200 ppm. The results obtained for the degradation of 200 ppm CN^- with various chloride concentration are represented in the Fig. 2 which shows that the chloride concentration had positive effect on the degradation.

The energy consumption for various electrolyte concentration were 189.5 KW h/kg of CN^-, 132.9 KW h/kg of CN^-, 94.99 KW h/kg of CN^-, 75.99 KW h/kg of CN^- for 250mg/L, 500mg/L, 750mg/L, 1000mg/L of chloride respectively.

Thus it was observed that increase in chloride concentration led to increase in the efficiency of the process. Since an optimum level of chloride has to be chosen with lesser chemical (chloride) usage which yields good efficiency, a chloride concentration of 500mg/L was chosen as optimum concentration for further studies.

However, the energy consumption and the degradation of cyanide varies depending on the chloride present in the effluent.

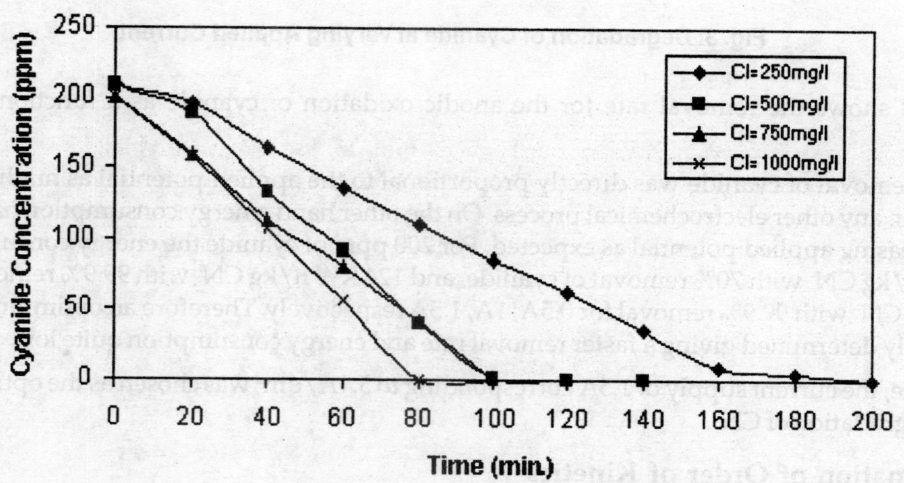

Fig. 2: Effect of Electrolyte Concentration (Chloride) (ph = 12, Current Density, 5.4A/drn2, Cyanide = 200 ppm)

The reaction of chloride can be explained as follows:-

At the anode: $2\,Cl^- \rightarrow Cl_2 + 2e^-$

At the cathode: $2H_2O + 2e^- \rightarrow H_2 + 2OH^-$

At the bulk: $Cl_2 + H_2O \rightarrow HCl + HOCl$

$HOCl \rightarrow H^+ + OCl^-$

The generated hypochlorite ion oxidises the cyanide (Hwang *et al.*, 1987).

Effect of Applied Current

Effect of applied current on the degradation process was studied, the current supply was varied as 0.5A, 1A, and 1.5A corresponding to 1.8A/dm², 3.6A/dm² and 5.4A/dm² keeping the pH–12, cyanide concentration 200 ppm and chloride concentration at 500mg/L.

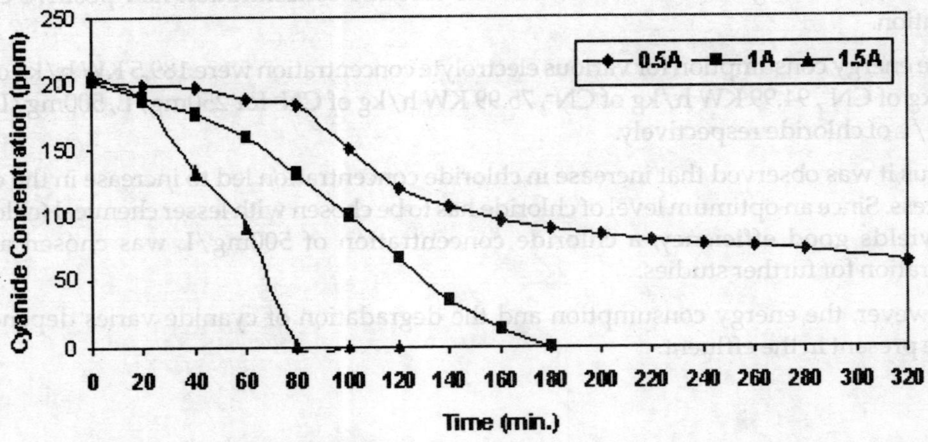

Fig. 3: Degradation of Cyanide at Varying Applied Current

Fig. 3 shows the removal rate for the anodic oxidation of cyanide as a function of applied potential.

The removal of cyanide was directly proportional to the applied potential as might have been expected in any other electrochemical process. On the other hand, energy consumption has increased with increasing applied potential as expected. For 200 ppm of cyanide the energy consumption was 145 KW h/kg CN^- with 70% removal of cyanide, and 124 KW h/kg CN^- with 99.9% removal and 132 KW h/kg CN^- with 99.9% removal for 0.5A, 1A, 1.5A respectively. Therefore an optimum point must be carefully determined giving a faster removal rate and energy consumption quite low.

Hence, the current supply of 1.5A corresponding to 5.4A/dm² was chosen as the optimal current for the degradation of CN^-.

Determination of Order of Kinetics

The rate of order of kinetics was studied for the degradation process at optimal conditions by ploting $\log_{10}(C_t/C_i)$ versus time 't' where C_t is the cyanide concentration at time 't' and C_i is the initial CN^- concentration. The order of kinetics was found to confirm first order reaction as a straight line was obtained on plotting $\log_{10}(C_t/C_i)$ versus time t, Fig. 4.

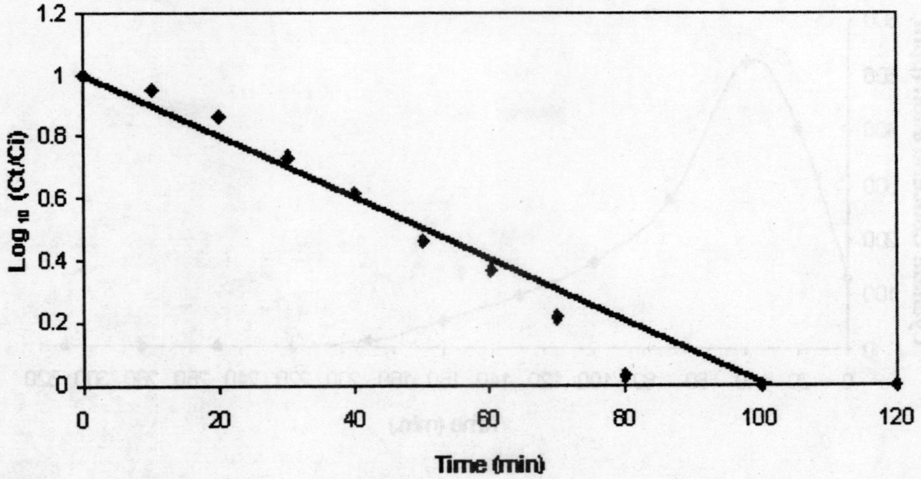

Fig. 4: First Order Reaction Plot for Cyanide Degradation (Cyanide concentration = 200 ppm, Current applied = 1.5 A, pH = 12, Chloride concentration = 500 mgl)

From the graph obtained, the k value was calculated as 1.381 min^{-1}.

Application of the Treatment Method

The treatment method developed was applied to the cyanide containing effluent collected from a silver-electroplating unit.

The sample was found to contain significant amount cyanide, sulphate, chloride, total dissolved solids and suspended solids. The chloride was found to be 832 mg/L. Hence no chloride was added externally. The pH of the solution was 10.3 which also required no adjustment.

The cyanide containing wastewater was treated by electrolytic method, and the results obtained are graphically represented in Fig. 5.

During the degradation process, the cyanide concentration in the solution was observed to initially increase and then decrease, the reason may be due to the breakdown of metal cyanide complex originally present in the solution as per the equation:

$$[M(CN)_y]^{(y-z)-} \rightarrow M^{z+} + yCN^-$$

This is followed by the oxidation of the free cyanide ion to cyanide ion at the anode. The reaction took about 4.3 hours for completion with a energy consumption of 387.5 KWh/Kg CN^-.

After the degradation process the cyanide sample was analysed and it was found to show decrease in cyanide concentration from 128mg/L to 0.04mg/L and in chloride concentration from 832mg/L to 244mg/L confirming the degradation of cyanide and also the conversion of chloride to hypochlorite. The sample characteristics before and after degradation is given in the Table 1 as follows.

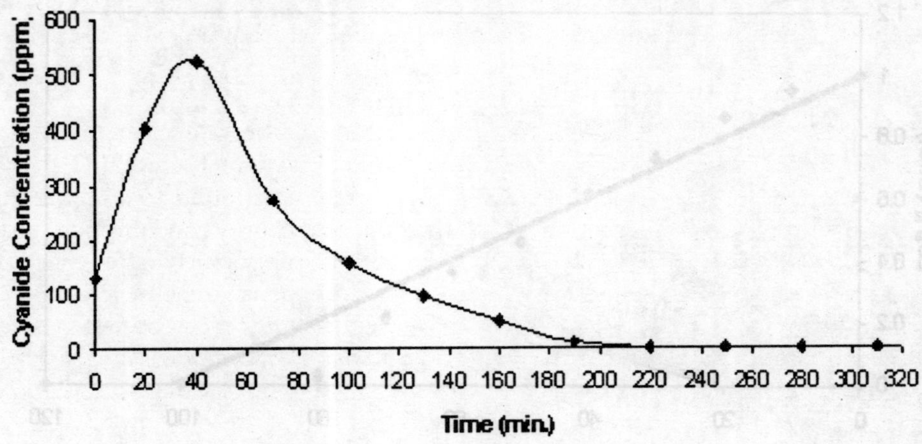

Fig. 5: Degradation of Cyanide Containing Industrial Effluent

Table 1: Characteristics of Cyanide Containing Effluent Before and After Treatment

Sample Characteristics	*Before Treatment*	*After Treatment*
CN^-	128 mg/L	0.04 mg/L
pH	10.33	10.2
Conductivity	6.79 mS/cm	7.02 mS/cm
Cl^-	832 mg/L	244 mg/L
Sulphate Concentration	820 mg/L	750 mg/L
TDS	5672g/L	5796 g/L

During the degradation process, there was a deposition on the cathode surface leading to its colour change. The deposition was due to the attraction of the positively charged metal ions present in the effluent, by the cathode. The metal ions deposited on the cathode was dissolved using dil. HNO_3. The dissolved metal ions were tested for silver by adding sodium chloride solution to the dissolved metal which formed silver chloride white precipitate confirming that the deposited metal was silver. The deposition of silver on the cathode was found to not affect the degradation efficiency of the electrodes. Thus, this process efficiently decomposes cyanide at the anode and remove silver at the cathode simultaneously.

Cost Evaluation

The operational cost of the process is contributed by the usage of NaCl and the energy consumption. Sodium chloride is not generally added to the treatment process because chloride is present in most of the raw effluent from electroplating units in a considerable amount. Even if the addition of NaCl is required it can be done in very less amount i.e 500 mg/L which will cost about 12 paise per litre which is negligible. (500g of NaCl = Rs.60)

To degrade the cyanide of 200ppm using the process at a current density of 5.4A/dm^2 for 140 minutes, the energy consumption was 132.8 KWh/Kg CN^- which will cost about Rs.396.

Conclusions

It was found in the work that the removal of cyanide from wastewater can be carried out effectively using titanium based mesh type anode coated with triple oxides namely RuO_2-IrO_2-TiO_2 and a carbon cathode. It was also found that indirect oxidation method was found to be more efficient in the treatment of cyanide wastes in terms of degradation time and energy consumption. The degradation of cyanide followed first order reaction. This method has advantages such as low energy consumption and no sludge formation. This method also does not require expensive chemical usage. The electrodes used in this study has large surface area, which provides better mass transfer conditions, and is simple to set up. Thus it can be concluded that anodic oxidation of cyanide is quite an efficient method and can be applied for the removal of cyanide from the wastewater successfully along with electrodeposition of associated metal ion at the cathode.

REFERENCES

1. Bakir Ogutveren U., Koparal S., 1998, "Removal of Cyanide by Anodic Oxidation" Journal Wat. Sci. Tech. Vol. 40, No. 5 pp. 1851-1856.
2. El–Ghaoui, E.A., Jansson, R.E and Moreland, C., 1982, "Application of Trickle Tower to Problems of Pollution Control III, Heavy–metal cyanide solutions". Journal Appl. Electrochem. Vol. 12, pp. 75–80.
3. El-Ghaoui, E.A., Jansson, R.E.W. and Moreland C., 1982, "Application of the trickle lower to problems of pollution control". II, The direct and indirect oxidation of cyanide'. J. Appl. Electrochem. 12, 669-673.
4. Ho. S.P., Wang. Y.Y and Wan C.C.,1990, "Electrolytic decomposition of cyanide effluent with an electrochemical reactor packed with stainless steel fiber", Journal Water Res. Vol. 24 No.11, pp. 1317–1321.
5. Hwang J.Y., Wang Y.Y and Wan C.C., 1987, "Electrolytic oxidation of cuprocyanide electroplating wastewaters under different pH conditions", Journal Appl. Electrochem. Vol.17, pp. 684–694.
6. Kenneth F. Cherry, 1982, "Treatment of common metal wastes", Plating waste treatment. Ann. Arbor Science Publications, Michigan, pp. 69-74.
7. Szpyrkowicz L., Zilio-Grandi F., Kaul S.N. and Rigoni-Stern S., 1998, "Electrochemical Treatment of copper cyanide wastewaters using stainless steel electrodes" Journal Wat. Sci. Tech. Vol.38, No.6, pp. 261-268.
8. Tan T.C., Teo W.K. and Chin D.T., 1985, "Electrochemical destruction of complex cyanide", Chem. Eng. Commun. Vol.38, pp. 125-133.

10

Electrocatalytic Treatment of Waste: Studies on Discoloration of an Industrial Azo Dye Effluent

S.S. VAGHELA, A.D. JETHVA, M.S. GOHIL, S. ADIMURTHY, P.M. GOUR, V.R.K.S. SUSARLA, G. RAMACHANDRAIAH & P.K. GHOSH

Central Salt and Marine Chemicals Research Institute, Bhavnagar - 364 002

ABSTRACT

A textile dye effluent containing chiefly reactive azo dyes has been treated electrochemically for discoloration and COD (chemical oxygen demand) reduction at different current densities, flow rates and dilution. Experiments have been carried out in a thin electrochemical reactor under single pass conditions using a dimensionally stable catalytic anode (DSA) and a stainless steel cathode. In the case of 50% diluted effluent, the discoloration has been successfully achieved upto 85-99% at 10-40 mA/cm^2 at 5 ml/min flow rate or 50-88% at 30-40 mA/cm^2 at high (10-15 ml/min) flow rates. The % COD reduction has been observed to be maximum (81%) at 39.9 mA/cm^2 or above when the average solution-anode (Ct) contact time was as high as 21.7 s/cm^2 (5 ml/min) and decreased as Ct declined at a given current density. The cyclic voltammetric study on the effluent at a glassy carbon electrode has confirmed the indirect oxidative decomposition of dissolved dye at the anode. A probable electrochemical mechanism mediated by active OCl–ion operating in the discoloration and COD reduction processes has been suggested. The apparent pseudo-first-order rates, k'_{obs} for the color removal and k''_{obs} for COD reduction and the corresponding second order rate constants have been discussed.

Introduction

Industrial sector needs huge quantities of pure water for direct or indirect use in their processes. On the contrary, it generates more or less an equal volumes of wastewater. The discharge of these wastes is increasingly becoming more and more difficult for them to dispose in recent years. This is because of increased public concern and ecological awareness which led to strict norms for keeping the environment clean and safe for all living organisms. This noble cause and the industrial demand for huge quantity of pure water, have paved way to look in to the treatment of these wastewaters for their safe reuse.

Dye manufacturers and the coloring industries release wastewaters in huge quantities having the dye[1-4] to the extent of 0.001-0.7% (w/v) often with dissolved inorganic salts, dispersing agents, surfactants, harmful metals like chromium in the form of coordination compounds and organics washed out from the materials. Reports also suggest that the dyehouse effluents contain 0.1-2% (w/v) dye[5-8] resulting in 2-9% which is equivalent to nearly 50, 000 tones of total global annual dye production[9-11]. This situation would get further aggravated in the ensuing years as the demand for cotton and other fabrics would shoot up exponentially[9]. The pH of these discharges is reported[12-14] to vary between 2 to 12, depending upon the dye and its application.

The deleterious aspects of the dye effluents are their unacceptable color and high COD content. Further, the dye components are hardly degradable by physico-chemical or biological methods and become highly difficult as the textile dyes are constantly being replaced by every passing day with modern dyes, resistant to chemical, photochemical and biological degradation[2,9,15]. These untreated discharges once released may deteriorate the soil, the water resources and the environment[16-18]. They reported to reduce light penetration in water streams or reservoirs which ultimately affect the growth of plants and other forms of wildlife.

The conventional biochemical oxidative destruction of the wastewaters often results in colored water unfit for its reuse. Methods such as coagulation, absorption, chemical oxidation, reverse osmosis, ultra filtration and photo decomposition etc. tried earlier, single or in combination, are found to be ineffectual[10-21] due to their cost, regeneration or reusability and secondary pollution. Chemicals such as hypochlorite, ozone and hydrogen peroxide, in the absence and in the presence of UV light and hydrogen peroxide with ferrous ions, have been used for the pretreatment of dye-bearing wastes[22-26].

Electrochemical process is yet another rewarding method because of intrinsic high electrolytic conductivity of the textile wastewaters. However, only a few studies[27-31] have been conducted so far. The electro-oxidation of synthetic textile wastewaters having partially soluble disperse dyes employing different anode materials with 0.1 M NaCl as supporting electrolyte have been claimed[27] to give 39% of COD reduction and about 90% discoloration in a 40 min electrolysis. Lin and Chen[22] and Lin and Peng[28] have studied the dye effluents by electro-coagulation methods using steel electrodes. Under the best conditions of current density (92.5 A m^{2-}) the efficiency of COD removal was 60% when polyaluminium chloride (PAC) was added and was only 50% without addition of PAC. Gaur *et al*[29] have succeeded to reduce 88% of color and only 18.6% of COD of an effluent from a Saree printing unit in two successive cycles in an electrochemical cell. The investigations[30] on a reactive cellulosic azo dyeing and finishing processing units employing Ti/Pt anode and stainless steel 304 cathode in presence of different additives showed 92% COD reduction with 94% discoloration. The textile wastewaters were treated at Ti/RuO_2, Ti/Pt and Ti/Pt/Ir electrodes[31] and found the effective direct and/or indirect oxidation of organics bringing the levels of COD to 85-92% and the organic-C by ~85% in a 60 min duration at 6 A/dm^2.

In the present paper, as a sample case, the electrochemical discoloration and COD reduction of an orange red color effluent containing a mixture of reactive azo (0.4-0.5%) dyes corresponding to one of the local leading textile industry has been studied under single pass flow conditions employing a dimensionally stable catalytic anode *vs* a stainless steel cathode mounted in a thin flow cell. The % of discoloration based on the absorption spectral data and the % of COD removal by the analytical methods have been studied at various possible current densities and flow rates. The observed rates for discoloration and COD reduction have been worked out and the second order rate constants with respect to applied current density per unit area have been presented here. A probable mechanism for the hypochloride mediated oxidative decomposition of the dye has been discussed.

Experimental Section

Dye Effluent. The effluent containing a mixture of reactive azo dyes *viz.* Red 2 (CI 18200), Red 20 (CI 25810), Reactive Orange 4 (CI 18260) and Reactive Orange 13 (CI 18270) was obtained from M/s Atul Industries Limited, Atul, a leading textile industry situated in Gujarat, India. The *p*H, TDS, chemical oxygen demand (COD), Hazen value and the NaCl composition as provided by the company for the orange-blood red effluent were 7.3, 17.6% (w/v), 6, 015 ppm, 75, 000, 16.7% (w/v), respectively. The above effluent was studied at different 0, 25 and 50% dilutions with deionized water.

Instrumentation. Discoloration and COD reduction in the dye effluent were carried at a dimensionally stable catalytic anode (DSA)[32] against a stainless steel cathode in an undivided flow cell under Galvanostatic conditions. An Aplab (India) model L 1288 DC regulated power supply was employed as a constant current source. The absorption spectra were recorded on a Shimadzu UV-160 spectrophotometer coupled to a temperature controlling unit employing 1 cm quartz cuvettes. The solution pH was recorded using Adair Dutt digital pH meter, sensitive to 0.01 pH units.

The ^1H-NMR spectra of the dried residues of the treated and untreated effluent by electrochemical and chemical methods were recorded in D_2O using Bruker Avance DPX-200-FT NMR-200 MHz spectrometer. The IR spectra of the same were recorded using a Perkin Elmer Spectrum (GX FT-IR system) while the elemental analysis was made on a Perkin Elmer CHNS/O 2400.

Cyclic voltammetric responses of the effluent samples were recorded with EG & G PAR models 273 A Potentiostat/Galvanostat coupled with a three electrode cell assembly and Gateway 2000 (4DX2-66) computer. A glassy carbon electrode (0.0314 cm^2) and its potentials were measured with reference to Ag/AgCl (0.222 V *vs* NHE) in 3 M NaCl. The platinum wire separated from the analytical solution by a vycor tip bridge served as a counter electrode[33].

Flow Reactor and Methods

A rectangular (18 cm × 15 cm × 5 cm) PVC cell (Figure 1), consisting of an expanded DSA (10.5 cm × 6 cm) with an effective area of 50.16 cm^2 as anode[32] and a thin stainless steel plate (11 cm × 8 cm) as cathode served as a flow reactor. The internal cell volume was measured to be equivalent to 90 ml of the effluent. Additionally, the cell had a provision with one inlet at the bottom and one outlet at the top as shown in Figure 1. The dye effluent was allowed to flow from the bottom of the cell at a regulated rate. A current was applied across the two current-carrying electrodes while the effluent passed steadily through the cell such that the average solution-electrode contact time (CT) per unit area varied between 5-22 s/cm^2. The cell potential across the cathode and the anode and the pH of the solution at the outlet were monitored.

The chemical oxygen demand of the original and the treated dye effluent samples were evaluated by the standard method involving potassium dichromate[29].

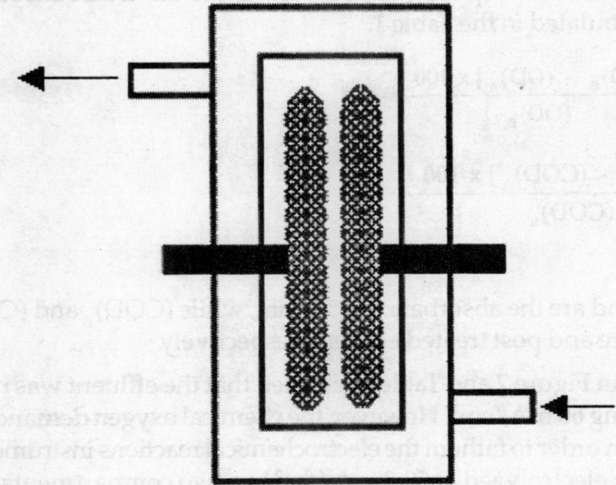

Fig. 1: A View of the Flow Reactor

The effluent after the electrochemical treatment under a given set of experimental conditions was collected at the outlet of the flow reactor and analysed. The % discoloration from the absorption spectral data and the % COD removal from the COD data were calculated comparing with the corresponding data of the original dye effluent.

Results and Calculations

Physical Properties

A 100 ml effluent was evaporated and dried. The residue weighed 17.62 g which possessed 16.88 g of non-combustible inorganic matter as sodium chloride and 0.74 g of combustible or organic dye indicating the dye content in the effluent supplied as 0.74% w/v. Measurements showed its chemical oxygen demand (COD) as 5957 ppm while its absorption spectra (showed in the latter part of the paper) recorded between 200-800 nm exhibited a characteristic broad or a group of overlapped bands of the azo dye in the visible region between 550-600 nm and an intense band in the UV region due to excess NaCl. The intensity of the band at 550-660 nm which centered around 576 nm, slightly decreased to about 4.5% and 8.1% when the effluent was diluted by 25% and 50%, respectively with deionized water while that of the band in the UV region changed linearly. However, the COD values decreased commensurately to the extent of dilution. The COD for 25% and 50% diluted effluents was estimated to be 4470 and 2985 ppm, respectively.

Electrochemical Treatment

Original Effluent

The effluent was passed through the flow cell (Figure 1) at a current density selected in the range between 0-100 mA/cm^2 while fixing the flow rate at 5 ml/min. The electrolysis between 0-30 mA/cm^2 was less effective as the color and the COD value of the effluent changed negligibly after the treatment. However, at higher current densities, the color and COD of the treated effluent changed remarkably despite the copious evolution of chlorine by the side reaction. The spectral changes in the effluent before and after the treatment at different current densities are shown in Figure 2 (a-e). The percentage of discoloration and the COD reduction corresponding to these solutions related to those of the original effluent were calculated with the help of equations 1 and 2 and the absorbance at 576 nm and the COD data, respectively, are tabulated in the Table 1.

$$\%\text{Discoloration} = \frac{[(OD)_B - (OD)_A] \times 100}{(OD)_B} \quad (1)$$

$$\%\text{COD} = \frac{[(COD)_B - (COD)_A] \times 100}{(COD)_B} \quad (2)$$

where

$(OD)_B$ and $(OD)_A$ and are the absorbance at 576 nm, while $(COD)_B$ and $(COD)_A$ are the chemical oxygen demand of the pre and post treated effluent, respectively.

The data presented in Figure 2 and Table 1 revealed that the effluent was nearly colorless at high current densities exceeding 60 mA/cm^2. However, the chemical oxygen demand (COD) of the solution reduced to 42-53% only. In order to fathom the electrochemical reactions instrumental for these changes, the dye effluent was also electrolyzed at 59.8 mA/cm^2 in a two compartmental flow cell[32], especially built with an ion-exchange membrane wherein a cation exchange membrane was placed between the two electrodes at a distance of 2-3 cm on either side. The effluent was passed through the anode or cathode compartment while deionized water was circulated through the other during the experiment. The flow of the counter ion across the membrane from higher concentration to the lower one supports the reaction at the electrode. The color and the COD of the dye became less when it was passed through the anode compartment as effectively as in the case of single compartment cell. There was no appreciable change in color intensity and COD value when they were passed through the cathode compartment implying that the azo dye present in the effluent is self active or reactive with the species produced in the anodic reaction with the solution.

Table 1: The Percentages of Color Removed and COD Reduced in the 0, 25 and 50% Diluted Effluent at High Current Densities. Flow of the effluent in each case was adjusted at initial stages to 5 ml/min.

Sl. No	*mA/cm²*	*% Dilution*	*% Color removed*	*% COD reduced*
1.	39.9	0 25 50	87.1 94.6 99.2	40.1 63.1 80.8
2.	59.8	0	92.7	42.8
3.	79.7	0	98.0	42.5
4.	99.7	0	98.8	52.8

Dilution Effect

To study further effects of flow rate and applied current at low current densities, the original effluent was diluted by 25 and 50% with deionized water and studied independently. The absorption spectra of these solutions collected after the treatment in the flow cell at 39.9 mA/cm^2 along with those of original effluent pre and post electrolysis under identical conditions are compared in Figure 2B wherein the absorbance at 576 nm in the case of 50% dilute solution tends to zero. The % data of color removed and the COD reduction for the 25 and 50% diluted solutions are given in Table 1. The data indicated that the % of color removed increases nominally by 7.5 and 12.1 while the COD reduction was affected to an extent of 23 and 40.7% in the case of 25 and 50% diluted effluent, respectively.

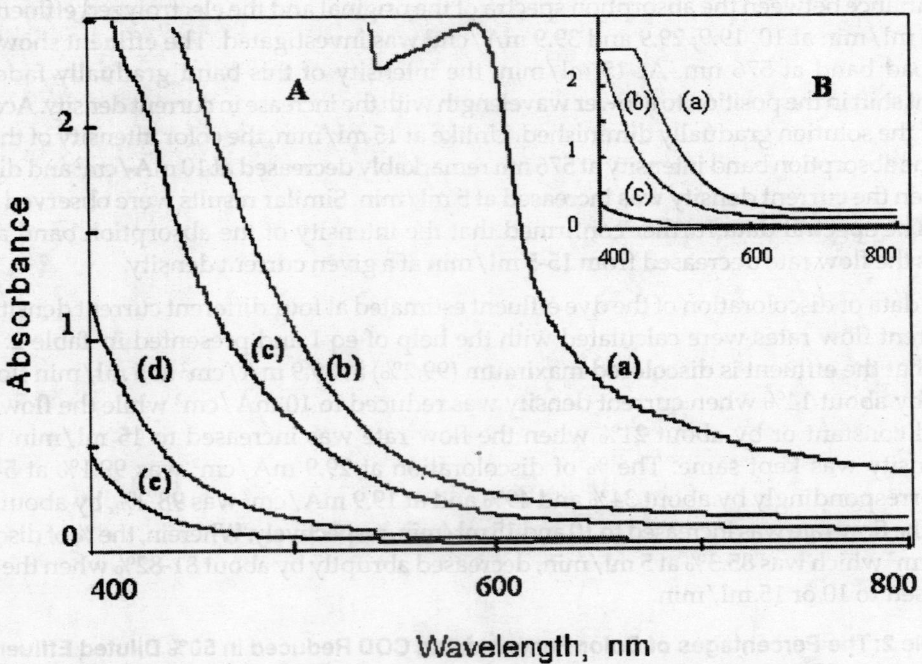

Fig. 2: Absorption Spectra of (A) the effluent after the treatment at (a) 0; (b) 39.9; (c) 59.8, (d) 79.8, (e) 99.7 mA/cm²; (B) the diluted effluent after the treatment at 39.9 mA/cm². (a) 0%, (b) 25%, (c) 50% dilution. The flow rate in all cases was 5 ml/min.

Further, the liberation of elemental chlorine was less in the case of 25% and negligible in the case of 50% diluted effluent as compared to that in the original. Thus, the 50% diluted solution was considered suitable for understanding further the relative effects of applied current density, flow rate, pH, electrical conductance on the percentages of color removed and COD reduced and to evolve a possible mechanism for the said effluent treatment by electrochemical method.

Effect of Current Density and Flow Rate

The 50% diluted effluent was passed through the electrochemical flow reactor at three different flow rates *viz.* 5, 10 and 15 ml/min to get an average solution residence time (R_t, eq 3) in the cell between 20-4 min. The electrolysis was carried out by varying the current densities between 9-40 mA/cm² at a given flow rate. During the experiments, the potential drop at the two current carrying electrodes was steady and it varied between 3 and 4 V when the magnitude of the current changed. The treated solution was collected at the outlet of the flow reactor under the single pass conditions and then studied by absorption spectra and evaluated for their chemical oxygen demand. The pH of the diluted effluent was 8.1 ab initio which decreased to 6.0-7.6 depending on the current density and the flow rate studied here.

$$R_t = \frac{\text{Cell capacity}}{\text{Flow rate}} \quad (3)$$

The variance between the absorption spectra of the original and the electrolyzed effluent flowing at 5 and 15 ml/min at 10, 19.9, 29.9 and 39.9 mA/cm^2 was investigated. The effluent showed single intense broad band at 576 nm. At 15 ml/min, the intensity of this band gradually faded with a concomitant shift in the position to a lower wavelength with the increase in current density. Accordingly, the color of the solution gradually diminished. Unlike at 15 ml/min, the color intensity of the effluent as well as the absorption band intensity at 576 nm remarkably decreased at 10 mA/cm^2 and diminished further when the current density was increased at 5 ml/min. Similar results were observed at 10 ml/min flow. The spectral data further confirmed that the intensity of the absorption band at 576 nm declined as the flow rate decreased from 15-5 ml/min at a given current density.

The % data of discoloration of the dye effluent estimated at four different current densities and at three different flow rates were calculated with the help of eq 1 and presented in Table 2. The data indicated that the effluent is discolored maximum (99.2%) at 39.9 mA/cm^2 at 5 ml/min flow rate. It decreased by about 14% when current density was reduced to 10 mA/cm^2 while the flow rate was maintained constant or by about 21% when the flow rate was increased to 15 ml/min while the current density was kept same. The % of discoloration at 29.9 mA/cm^2 was 99.1% at 5 ml/min, reduced correspondingly by about 34% and 49% and at 19.9 mA/cm^2 was 98.2%, by about 63% and 80% when the flow rate was increased to 10 and 15 ml/min, respectively. Wherein, the % of discoloration at 10 mA/cm^2 which was 85.3% at 5 ml/min, decreased abruptly by about 81-82% when the flow rate was increased to 10 or 15 ml/min.

Table 2: The Percentages of Color Removed and COD Reduced in 50% Diluted Effluent (COD 2978 ppm) as a function of current density and flow rate or solution-electrode contact time, Ct.

Sl. No	*mA/cm^2*	*% Color Removed At*			*% COD Reduction at*		
		5 (ml/min)	*10 (ml/min)*	*15 (ml/min)*	*21.5 (s/cm^2)*	*10.8 (s/cm^2)*	*7.2 (s/cm^2)*
1.	10.0	85.3	3.8	3.1	26.1	12.7	1.3
2.	19.9	98.2	34.8	18.5	39.7	16.7	13.4
3.	29.9	99.1	65.4	50.2	53.0	24.8	14.0
4.	39.9	99.2	88.0	78.5	80.8	32.2	20.8

Cell voltage was 3-4 V at 9-41 mA/cm^2; In all cases, the pH of the original solution was 8.10 while that of the electrolysed solution varies between 6.00-7.60 depending on the current density applied.

Figure 3 shows the variation in the % of discoloration as a function of the current density applied at the anode. It is obvious from Figure 3 that the maximum discoloration of the above said effluent under the present set of experimental conditions is possible at both low current density between 10-30 mA/cm^2 when the flow rate is regulated at 5 ml/min at high current densities greater than 30 mA/cm^2 and with a flow rate of 10 ml/min or more. The data in Figure 3 at 10 ml/min and 15 ml/min indicated that the % of discoloration at 10 mA/cm^2 or less are ineffectectual. This implies that 10 mA/cm^2 is the threshold current from where the discoloration occurs proportionately at moderately high flow rates. The plot of flow rate *vs* the % discoloration shown in Figure 4 also indicated that the % of discoloration decreases linearly with increase in the flow rate only at current densities greater than 30 mA/cm^2. It (% of discoloration) decreased linearly with flow rate at 2.1 units/ml/min at 39.9 mA/cm^2 while at 4.9 units/ml/min at 29.9 mA/cm^2. This denotes that the process is to be carried out at high current densities and at high flow rates for quick and better results as far as discoloration is concerned.

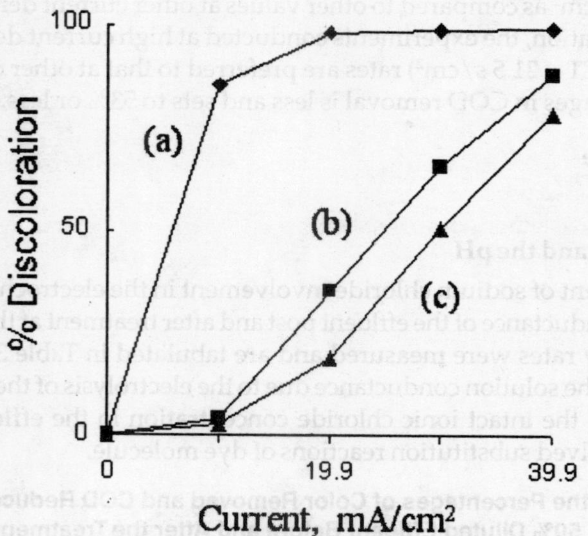

Fig. 3: Plot of Applied Current, mA/cm² *vs* % of Discoloration of Effluent at Different Flow Rates (a) 5 ml/min; (b) 10 ml/min; (c) 15 ml/min.

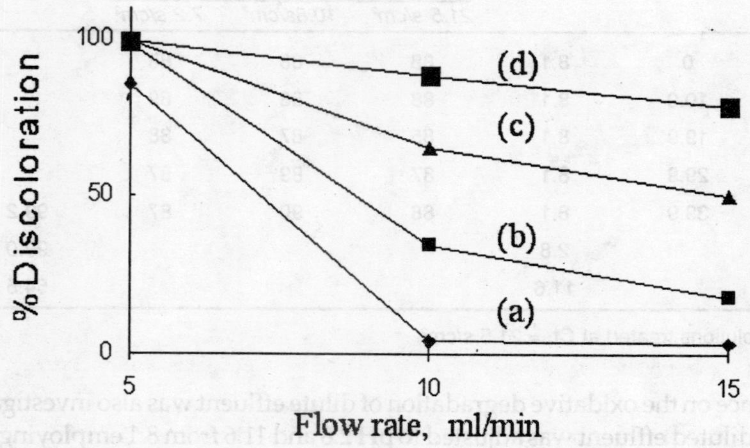

Fig. 4: Plot of Flow Rate *vs* the % of Discoloration at Different Current Densities (a) 10, (b) 19.9, (c) 29.9; (d) 39.9 mA/cm²

The COD data were measured simultaneously for the electrolyzed solutions at different current densities at different flow rates. The % data of COD removal in the effluent as a function of current and the average solution-electrode contact time (CT, eq 4) in seconds are presented in Table 2. The data revealed that the % COD removal was maximum (81% at 39.9 mA/cm²) when the contact time (CT) was 21.5 s/cm². It gradually decelerated to 26% as the current density decreased to 10 mA/cm² and further plummeted to 21% with the increase in CT. The % of COD removal was maximum (81%) at 39.9

mA/cm^2 at CT = 21.5 s/cm^2 as compared to other values at other current densities. This confirms that as in the case of discoloration, the experiments conducted at high current densities (39.9 mA/cm^2) at low flow (5 ml/min or CT = 21.5 s/cm^2) rates are preferred to that at other current densities at other flow rates since the changes in COD removal is less and sets to 53% or less.

$$Ct = \frac{\text{Residence time}}{\text{Anode area}} \tag{4}$$

Electrical Conductance and the pH

To measure the extent of sodium chloride involvement in the electrochemical decomposition of the dye, the electrical conductance of the effluent post and after treatment at the above said four current densities and three flow rates were measured and are tabulated in Table 3. The data did not show considerable change in the solution conductance due to the electrolysis of the dissolved dye. This may partly be accounted for the intact ionic chloride concentration in the effluent and suffices for the absence of chlorine involved substitution reactions of dye molecule.

Table 3: The Data of the Percentages of Color Removed and COD Reduced, and the Electrical Conductance of the 50% Diluted Effluent Before and After the Treatment at Different Current Densities, Solution-electrode Contact Times (CT) and pH

Sl. No.	*mA/cm²*	*pH*	*Electrical Conductance (m mhos/cm)*			*% Color Removed*[a]	*% COD Reduced*[a]
			21.5 s/cm²	*10.8s/cm²*	*7.2 s/cm²*		
1.	0	8.1	88	88	88		
2.	10.0	8.1	88	88	89		
3.	19.9	8.1	85	87	88		
4.	29.9	8.1	87	89	87		
5.	39.9	8.1	88	90	87	99.2	80.8
		2.8				99.0	27.4
		11.6				54.5	50.7

[a] for the solutions treated at Ct, = 21.5 s/cm^2

pH influence on the oxidative degradation of dilute effluent was also investigated. In these studies, the pH of 50% diluted effluent was adjusted to pH 2.8 and 11.6 from 8.1 employing minimum quantities of concentrated hydrochloric acid and sodium hydroxide solutions, respectively and electrolyzed in the flow cell at 39.9 mA/cm^2 with 5 ml/min flow rate. The percentage of color removed and COD reduced are presented in the Table 3. The color removal at pH 2.8 was 99%, which is equivalent to that of pH 8.1, but it reached only to 45% at the pH 11.6. The inhibition in the desired reaction at higher pH > 11 may be attributed for the stabilization of the species responsible for the decomposition of dye molecules. On the contrary, the % of COD reduced steeply decreased at both the pH, 50.7 at pH 2.8 and 27.4 at pH 11.6 against 80.8 at pH 8.1. This asserts that the reaction of electro-decomposition of dye is purely controlled by species generated at the electrode whose formation and stability is controlled by pH of the solution.

Chemical Treatment

The decomposition of the dye molecules, the 50% diluted effluent was subjected to chemical treatment. In these studies, sodium hypochlorite (4%) and hydrogen peroxide (30%) were used as oxidizing agents to substitute the electrode.

Sodium Hypochlorite

50 ml of the diluted effluent was reacted variously with 1, 2, 5, 15 and 20 ml of 4% NaOCl solution. The color of the solutions faded slowly with time. The absorption spectral data of these solutions when recorded after 48 h of mixing with the hypochlorite solution exhibited respectively a negligible (<1) value, 5.9, 90.2, 98.1, 97.7% decrease in the color. Consequently, the COD in these solutions was found to be decreased. The % of COD reduced in the latter three solutions are 65, 85 and 90% which are comparable to the quantity of hypoclorite employed and in good agreement with the results in Table 2.

Hydrogen Peroxide

A 10 ml of the effluent was mixed with 1 ml of 30% H_2O_2 and allowed to react at room temperature in a dark place. There was no much change in the solution color and COD even after its storage for a week. Similar observations were derived when 1 ml concentrated HCl was added along with 1 ml 30% H_2O_2 to the 10 ml effluent. This ascertained the inability of hydrogen peroxide to oxidize the present effluent in its present form like in the electrochemical process and hypochlorite treatment. Therefore no further study was made with the hydrogen peroxide.

Analysis of the Treated Effluent

A 10 ml sample of pre and post electrochemically treated (50% diluted) effluent at 39.9 mA/cm² with 5 ml/min flow rate were evaporated on water bath, dried and finally powdered. The residues were studied without further treatment by FT-IR, ^{1}H-NMR and analyzed for %C. Similarly, a 25 ml of the 50% diluted effluent was treated with 10 ml of 4% hypochlorite solution for 48 h and evaporated on hot water bath and its IR spectrum were recorded after drying. The %C found in the electrochemically treated sample was found to be 1.29 versus 1.61 in the untreated solution indicating a 0.32% loss in the total dye carbon composition after the oxidation.

In D_2O, the extracted dye sample from untreated effluent exhibited two multiplets between 7.17-778 ppm and 8.35-8.66 ppm which are probably due to distinguishable aromatic protons[34]. Besides, a sharp peak at 3.59 ppm for N-methyl protons and peaks at 1.13, 1.79, and 2.25 ppm due to aliphatic protons were also observed. The sample extracted after the electrolysis reproduced the peaks corresponding to aromatic protons between 7.17-7.78 ppm, N-methyl protons at 3.62 ppm and aliphatic protons at 1.15, 1.46 and 1.82. However, the multiplet between 8.35-8.66 corresponding to aromatic protons of second kind disappeared and a new multiplet of low intensity, appeared between 7.97 and 8.15 ppm in its place, tentatively suggesting the structural and conformational changes in the dye molecules anodic oxidation.

The IR spectra of the solid residues isolated from the effluent before and after the electrochemical and chemical treatment were recorded. Considerable changes were observed in the region between 400-700 cm–^{1} and 1000-1700 cm–^{1}. Two bands at 619 cm–^{1} and 676 cm–^{1} probably due to C-Cl stretch present in all the dye molecules were also observed. The former band appeared at 622 cm–^{1} while the later one tend to disappear in treated effluents. Similarly, the band which appeared at 1197 cm–^{1} in original effluent appeared as broad and relatively low intensive peak in treated samples, indicating that the corresponding group is the potential site for oxidation. However, the bands at 1400, 1429 and 1465 cm^{-1} were absent in electrochemically treated sample and concomitantly a new sharp peak appeared at 1384 cm^{-1}. On the contrary, the bands at 1400, 1429 and 1465 cm^{-1} were observed as two intense peaks at 1430 and 1458 cm^{-1} in hypochlorite treated sample, indicating that the corresponding functional groups are sensitive for oxidation. However, an intense band at 1384 cm^{-1} was observed in electrochemically treated sample replacing the three bands at 1400, 1429 and 1465 cm^{-1}. Besides, an additional peak at 970 cm^{-1} appeared in the chemically treated sample may be assigned for the ClO_3^- ion which are generally present in the hypochlorite solution. Peak identical to 970 cm^{-1} in hypochlorite treated sample was absent in electrochemically treated sample which ascertains that ClO_3^- did not form as the side product during the progress of electrolysis. Since the band intensities in the region 505-760 cm^{-1} were not intensified due to C-Cl stretch[35] in treated samples, the possibility of formation of chlorine substituted byproducts which may lead to secondary pollution, was ruled out in both electrochemical and chemical treatment of the effluent.

Cyclic Voltammetry

The redox behavior of the dye effluent was studied at glassy carbon electrode (0.0314 cm^2) against Ag/AgCl (3 M NaCl). The responses of 0.1 M NaCl before and after the addition of 0.2 and 0.4 ml of 50% diluted effluent in the potential range of 1.5 V to-1.5 V are recorded and compared. No major difference between the electrode response before and after the introduction of the effluent into the cell was noted owing to electro-inactive nature of the dyes (Red 2, Red 20, Reactive Orange 4 and Reactive Orange 13) said above in the potential range investigated here. However, at anodic potentials between 1.0 and 1.5 V, a considerable decrease in the anodic currents was observed and it further decreased with the hike in the dyes composition denoting the adsorption of dye molecules over the anode surface prior to its decomposition. The large anodic currents at and above 1.0 V may be due to the displacement of chlorine and oxygen from the medium. The above results confirm the indirect oxidation of dye molecules mediated by both active chlorine and oxygen as discussed in the latter part of this paper.

Calculation of Kinetic Constants

The rates of discoloration and COD reduction in the effluent depend on dye concentration (Dye) and the oxidizing strength of the anode *i.e.* the concentration of active species (discussed in the latter part of this paper) denoted as (O) by which the color and COD in the effluent are reduced. Thus, the kinetics of the electrochemical effluent treatment at the anode are expressed as:

$$-d\,(\text{color})/dt = k'\,(\text{Dye})\,(\text{O}) \tag{5}$$

$$-d\,(\text{COD})/dt = k''\,(\text{Dye})\,(\text{O}) \tag{6}$$

where

k' and k'' are the second order rate constants for the discoloration and the COD reduction reactions at the anode, respectively. During electrolysis, the value of (O) in eqs. 5 and 6, will remain constant under a given set of experimental conditions but it varies as the applied current density is altered. Presumably, under stationary conditions there is no loss in the oxidizing strength of the anode or (O), the rate of active species (O) is equal to the rate of its consumption in both of the electrode processes and at a given current density, the eqs 5 and 6 are modified as:

$$-d\,(\text{color})\,/\,dt = k'_{obs}\ (\text{Dye}) \qquad (7)$$

$$-d\,(\text{COD})\,/\,dt = k''_{obs}\ (\text{Dye}) \qquad (8)$$

where

k'_{obs} and k''_{obs} are the apparent pseudo-first-order kinetic constants for the electrochemical processes of discoloration and COD reduction, respectively at anode under a given set of experimental conditions.

The values of k'_{obs} and k''_{obs} for the treatment of 50% diluted effluent at different current densities and flow rates were calculated by substituting the data in Table 2 in eqs 9 and 10, respectively and the results are summarized in Table 4. The k'_{obs} value at a given flow rate increased linearly with the hike in the current density except at 5 ml/min where it was almost independent of current density. At a given current density, it decreased to a measly value with the increase in flow rate at 10 mA/cm^2, but it decreased uniformly at 19.9 mA/cm^2 due to regular decrease in the liberation of species reacting with the dye molecules. On the contrary k'_{obs}, increased uniformly with the flow rate at high 29.9 and 39.9 mA/cm^2 maintained status quo in the liberation of chemically active species at the electrode.

$$k'_{obs} = \%\text{Discoloration} \times \text{flow rate} \qquad (9)$$

$$k''_{obs} = \%\text{COD} \times \text{flow rate} \qquad (10)$$

Similarly, the k''_{obs} vales at a given flow rate, increased correspondingly with the increase in current density. Expectedly, the liberation of chemically active species increases with the increase in current density. Unlike the k'_{obs}, it (k''_{obs}) decreased marginally with the increase in flow rate at all current densities indicating that the chemical steps responsible for the discoloration and COD reduction are two independent processes.

The observed rates k'_{obs} and k''_{obs} at a given flow rate were plotted against the applied current density. The plots k'_{obs}, *vs* current density at 10 and 15 ml/min were linear and showed an intercept on current (X) -axis between 8 and 9 mA/cm^2 as the threshold current to be applied for the discoloration. Assuming that there is no accumulation of oxidising species (O) in the reactor and the rate of its production and the rate of its consumption are equal, the value of (O) –mole equivalents of active (O) species produced per mole of electron discharged at the anode–at a given current density and flow was obtained by eq 11. The plots k'_{obs} against (O) at a given flow rate were linear and showed a negative intercept on at high flow (10 and 15 ml/min) rates. From slopes of these plots, the second order rate constants k' was calculated to be 0.29 M^{-1} s^{-1} at 5 ml/min, 7.62 M^{-1} s^{-1} at 10 ml/min and 15.61 M^{-1} s^{-1} at 15 ml/min. Unlike the plots k'_{obs}, *vs* (O), the plots of k''_{obs} *vs* (O) were linear and passed through the origin at all flow rates. The values of k'' obtained from the slopes of these plots remained more or less (1.19 $M^{-1} s^{-1}$ at 5 ml/min, 1.79 $M^{-1} s^{-1}$ at 10 ml/min and 3.57 $M^{-1} s^{-1}$ at 15 ml/min) constant.

$$[O] = \frac{1000 \times \text{Current density} \times \text{time}}{96487 \times \text{flow rate}} \quad (11)$$

Discussion and Conclusions

Azo dyes are synthetic organic colorants[1]. Most of them consist a functional group possessing two doubly bonded nitrogens (–N=N–) attached to a tri valent (sp^2) carbon atom of aromatic or heterocyclic nucleus (R) on one side and an unsaturated molecule of the carboxylic, heterocyclic or aliphatic (R') type on the other side. Dyes are classified as mono-, disazo-, trisazo-and polyazo-dyes depending on the number of azo linkages, and also according to their solubility in water, types of components used in their syntheses, their methods of application or core use. Usually, azo dyes are classified as acid, base/cationic, direct, disperse, mordant and reactive types.

The chromophoric nature of a dye effluent is due to the electronic transitions between the non-bonding electrons of N-atoms of the azo group and the Π-molecular orbital energy levels of conjugal systems attached to it. Partial or complete cleavage of N=N bond of the azo group by an oxidative or reductive process results in the discoloration of the effluent. The reductive cleavage of N=N double bond is a difficult process in aqueous or near aqueous solutions and demands drastic conditions and/or a catalyst[36]. On the other hand, the oxidative cleavage of N=N double bond is relatively easier. It may be achieved either at the anode or by employing a potentially equivalent oxidizing agent. Quintessentially, the chemical oxygen demand (COD) of a dye effluent is the requirement of total oxidizing power capable to annihilate the dye molecule and the organics. In the process of COD reduction, the effluent may also get discolorized and hence these methods are preferred.

In electrochemical methods, the reactions at the anode depends on the nature of dye, the type and composition of organic and inorganic ingredients, and the nature of electrode materials used. Of the two possible ways of anodic reactions, the direct oxidation of the effluent molecules depends on the potentials of the dye and the organic ingredients. If these potentials are greater than the decomposition of water or the chloride which is excessively present in all effluents, then the direct oxidation of the dye and the other organics (other organics are no more discussed in the remaining part of the paper) is extremely difficult unless electrocatalytic electrodes such as platinum are employed, or else, it is done by indirect ways. In all the indirect methods, an electro-active species present already or externally added to the effluent oxidizes itself to its possible highest valent state at the electrode surface and immediately converts back to its original state in the subsequent chemical reactions with the proximate dye molecules. This process continues till the COD of the effluent at the electrode-solution interface, thereby, the bulk solution reaches minimum value.

Most of the dye effluents contain sodium chloride as the major constituent in them. Thus, the method of treating such solutions electrochemically is easy as they involve no addition of chemicals for supporting the electrolysis. Where dye molecules are inactive, the chemical changes at the anode are water (eq 12) and/or the chloride ion oxidations (eq 13) which generate active OH and/or Cl radicals leading to O_2 and/or Cl_2 liberation, respectively. O_2 is relatively weak oxidant and hence not useful in the effluent treatment, while the Cl atom and Cl_2 are equally robust oxidizing agents, hence Cl_2 and OH are considered for the sake of simplicity. The counter reactions at the cathode would be the reduction of only water when no other reducible species are present. Therefore, no further discussion has been made on cathodic reductions.

$$2H_2O \rightarrow O_2 + 4H+ + 4e- \quad (12)$$

$$2Cl- \rightarrow Cl_2 + 2e- \quad (13)$$

Reactions 12 and 13 occur in dilute chloride solutions equally. However, the latter reaction predominantly occurs in concentrated ((Cl–) > 1M) solutions. The gaseous chlorine dissolves in aqueous solutions due to ionization as indicated in eq 13. It is less in acidic solution due to HOCl instability and considerably more in basic solutions due to the ready formation of OCl– (pKa 7.44)[37] ions (eq 15) implying that the basic or neutral pH conditions are more favorable for conducting Cl_2 involved reactions.

$$Cl_2 + 2H_2O \rightarrow HOCl + HCl \quad (14)$$

$$HOCl + OH- \rightarrow OCl- + H_2O \quad (15)$$

The 0, 25 and 50% diluted effluent studied here contains considerable amount of dye (0.74-0.37% w/v) with high COD (5957-2979 ppm) and inorganic (16.88% w/v) salts, mainly chlorides having pH 7.35-8.1 favorable for the study. The dye molecules are electro-inactive in the permissible potential range +1.5 to-1.5 V. Thus, the decomposition that led to the discoloration and COD reduction in the effluent is thus attributed for the anodic oxidation of dye molecules mediated chiefly by OCl–and to a lesser extent by OH radicals as electrocatalysts.

The active oxygen (OH) and the active chlorine (OCl–at pH > 7, HOCl pH < 5) species are taken as the mediators generated at the anode. Electrodes like DSA employed in the present investigation preferentially work as a catalytic electrode to liberate chlorine from inorganic chloride solutions[38].

Earlier investigations on fading of dye color in aqueous solution[39] suggest that the more aggressive oxidants like OCl–and per acids act as electrophiles and initiate oxidation by reaction with the more nucleophilic nitrogen atom of the azo group of the dye common anion. This is probable because, the N-azo-atom are oxidizable relatively at ease. Based on these reports together with the fact that the N (azo dye) -atoms under go easy oxidation, we believe that the high valent chlorine end of OCl–first attacks the electron rich N atom and get itself reduced to Cl–and corresponding N-oxide intermediate which rapidly dissociates to give colorless oxides of type R=O and R′=O and N_2 as seen in Scheme I.

Scheme I

R–N=N–R′ + 2 OCl– → R–N (ClO) =N (ClO) -R′ (intermediate)

R–N (ClO) =N (ClO) -R′ → R–N (O) =N (O) -R′ (intermediate)

R–N (O) =N (O) -R′ → RO + R′O + N_2

As evidenced in Table 1, the treatment of (~50%) diluted effluent is beneficial as far as the maximum reduction in COD, discoloration and the complete utilization of liberated chlorine in the normal current density range ($\leq$ 40 mA/cm^2) at the anode. The observed data (Table 2, Figures 4 and 5) reveal that the highest % of discoloration at low current densities-low flow rates or high current densities-high flow rates while the highest % of COD reduction at high current densities and slow flow rates, indicate that the mechanism of both the processes are different wherein the discoloration is a simple step that needs less electrical energy as compared to the process of COD reduction. The percentages of discoloration and COD reduction in 50% diluted effluent are correlated as a function of applied current density in Figure 8 at 5 ml/min flow. It shows that discoloration is almost independent of current density applied here. However, the % of COD reduction increases linearly with corresponding increase in current density.

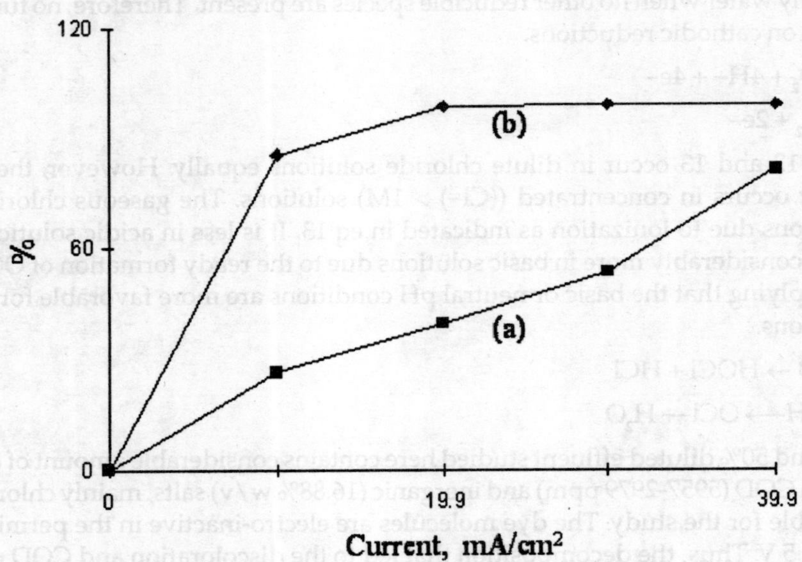

Fig. 5: Effect of Applied Current on % of (a) discoloration, (b) COD reduced at 5 ml/min

The negligible change in the pH of the effluent and the minor change in the electrical conductance of the effluent after the treatment show that the chlorine derivetized products of the decomposed dye are not liberated during the electrolysis. This is further supported by the absence of new characteristic IR (C–Cl) band in electrochemically treated sample. The proximate similarities reveals that the decomposed products in both chemical and electrochemical treatments are nearly the same. However, the decrease in % of COD in acidified effluent (pH 2.8, Table 3) shows the low cell-efficiency at low pH which may be accounted for the low stability of the active chlorine (HOCl) in acidic solution. Similarly, the low performance of the cell at high pH 11.6 is again attributed for the loss of active chlorine (OCl–) due its dispropotionation as shown in eq 16.

$$2OCl^- + 2OH^- \rightarrow ClO_3^- + Cl^- + H_2O \quad (16)$$

Further, the above study inferred that the dye effluent is effectively decomposed at DSA anode mediated by ClO– and OH radicals under single pass condition to disposable products. The kinetic data in Table 4 showed that the pseudo first order rates with respect to discoloration and COD

Table 4: The Apparent Pseudo-first Order Rates (k'_{obs}) of Discoloration and (k''_{obs}) COD Reduction in the 50% Diluted Effluent as a Function of Current Density at Different Flow Rates

Sl. No.	mA/cm²	k'_{obs} × 10² (s⁻¹)			k''_{obs} × 10² (s⁻¹)		
		5 ml/min	10 ml/min	15 (ml/min)	5 ml/min	10 ml/min	15 (ml/min)
1.	10.0	7.11	0.63	0.78	2.18	2.12	0.33
2.	19.9	8.18	5.80	4.63	3.31	2.78	3.35
3.	29.9	8.26	10.90	12.55	4.42	4.13	3.50
4.	39.9	8.27	14.67	19.63	6.73	5.37	5.20

reduction are in the order of 10^2 s–^{1}. Szpyrkowicz *et. al.*[27] who studied the electro-oxidation of pollutants in synthetic textile wastewater containing partially soluble disperse dyes in 0.1 M NaCl solution with different anode materials, have suggested a similar mechanism. They have reported a 39% of COD reduction in a span of 40 min. However, the apparent pseudo-first-order rates for the color removal is only 2.54×10^{-4} s–^{1}. The very low rate of color removal in their investigations may be attributed to the dispersive nature of dye and the low concentration chloride ion. The present study also suggests that the electrochemical method of treatment does not always give rise to secondary pollutants like chlorosubstituted by products.

Acknowledgment

The authors are grateful to Dr. Prakash Bhate, Vice President, Research, M/s Atul Industries Limited, Atul, Gujarat, India for providing the effluent and the details of dyes present in it.

REFERENCES

1. Kirk, I., Othmer, D. F., 1997, *The Encyclopedia of Chemical Technology*, Ed. H. G. Mary, 4th edn, Vol. 3, J. Wiley & Sons: New York, p821.
2. Laing, I. G., 1991, *Rev. Prog. Coloration, 21*, 56-71
3. Koprivanac, N., Jovanovic-Kolar, J., Bosanac, G. and Meixner, J., 1992, *Microchemical* Journal, *46*, 379-384.
4. Koprivanac, N., Bosanac, G., Grabaric, Z and Papic, S., 1993, *Environmental Technology*, *14*, 385-390.
5. Easton, J.R., 1995, *Colour in Dyehouse Effluent*, Ed by Cooper P, Society of Dyers and Colourists, The Alden Press, Oxford, pp 9-21.
6. Shelley, T.R., 1994, *International Dyer* 79: 26-31.
7. Pierce, J., 1994, *Journal of the Society of Dyers and Colourists* 110: 131-133.
8. Gahr, F., Harmanutz, F. and Oppermann, W. Ozonstion, 1994, *Wat Sci Tech* 30: 255-263
9. Neill C O, Hawkes F R, Hawkes D L, Lourenco N D, Pinheiro H M & Delée W, *J Chem Technol Biotechnol* 74 (1999) 1009-1018.
10. Correia, V.M., Stephenson, T. & Judd, S.J., 1994, *Environ Technol* 15, 917-929.
11. Brown, D., 1987, *Ecotoxicology and Enironmental Safety* 13, 139-147.
12. Gurnham, C.F., Ed., 1965, Industrial Waste Control: Academic Press: New York.
13. Lin, S. H., Lin, C. M., 1993, *Water Res.*, *27*, 1743.
14. Szpyrkowicz, I., Zilio, G. F., Canepa, P., 1996, *Toxicol. Environ. Chem.*, 56, 23.
15. Michelsen, D.L., Fulk, L.L, Woodby, R.M. & Boardman, G.D., 1993, In *Emerging Technologies in Hazardous Waste Management III, ACS Symposium Series*, 119-136.
16. Churchley, J. H. 1998, *International Ozone Association*, *20*, 111-120.
17. Clarke, E. A. and Anliker, R., 1984, *Rev. Prog. Coloration*, *14*, 84-89.
18. Loyd, C.K., Boardman, G.D. & Michelsen, D.L., 1992, *Hazard Indust Wastes* 4, 593-601.
19. Rao, B.V. and Sastry, C.A., 1987, *Indian J. Environ Protect*, 7, 441.

20. Rao, B.V. and Sastry, C.A., 1987, *Indian J Environ Protect*, 7, 363.
21. Nowak, K.M. and Winnichi, T., 1986, *Desalination*, 60, 59.
22. Lin, S.H., Chen, M.L., 1997, *Water Res.*, *31*, 868.
23. Sarasa, J., Roche, M. P., Ormad, M. P., Gimeno, E., Puig, A., Ovelleiro, J.L., 1989, *Water Res.*, *23*, 2721.
24. Wu, J., Witeman, M.A., Law, S.E., 1998, *J. Enviorn. Eng.*, March, 272.
25. Kuo, W.G., 1992, *Water Res.*, *26*, 881
26. Carriere, J., Jones, J.P., Broadbent, A.D., 1993, *Ozone Sci. Eng.*, *15*, 189
27. Szpyrkowicz, L., Juzzolino, C., Kaul, S.N., Daniele, S. and De Faveri, M.D., 2000, *Ind. Eng. Chem. Res.*, *39*, 3241-3248.
28. Lin, S.H., Peng, C.F., 1994, *Water Res.*, 28, 277.
29. Gaur, P.M., Prajapati, M.N. and Dasare, B.D., 1993, *Indian J Technol* 31, 663-665.
30. Vlyssides, A.G. & Israilides, C.J., 1998, *J. Envron Sci Health, Part A Toxic/Hazard Subst Environ* A33 (5): 847-862.
31. Szpyrkowicz, N.J. & Z.F. Grandi Z.F., 1996, *Water Sci Technol* 34, 17-24.
32. Vasudevan, D., Vaghela, S.S. and Ramachandraiah, G., 2000, *J. Appl. Electrochem.*, 30, 1302.
33. Prakash, R., Tyagi, B. Ramacandraiah, G., 1997, *Indian J. Chem.* 36, 201.
34. Silverstein, R.M. and Bassler, G.C., 1963, Spectrometric Identification of Organic Compounds, John Wiley and Sons, Inc., New York, p. 83-88.
35. Socrates, G., 1980, *Infrared Characteristic Group Frequencies*, John Wiley & Sons, New York.
36. Chatt, J. Dilworth, J.R. and Richards, R.L., 1978, *Chem. Rev. 78*, 589 and the references there in it.
37. Kumar, K. and Margerum, D.W., 1987, *Inorg. Chem.*, 26, 2706-2711.
38. Vasudevan, D. and Anantharaman, P. N., 1993, 1994, *J. Appl. Electrochem.* 23, 808, 24, 1188.
39. J. Oakes *et. al. J. Chem. Soc. Perkin Trans* 1998, 2, 1857, 1998, 2, 2261, 1998, 2, 2201, 1998, 2, 2693, 1998, 2, 2699.

11

Application of Eletrochemical Reactors for Industrial Wastewater Treatment

ANNA MARIA POLCARO* & LIDIA SZPYRKOWICZ**

** Department of Chemical Engineering and Materials, Piazza d'Armi, 09123 Cagliari, Italy*

*** Department of Environmental Sciences, University of Venice, Dorsoduro 2137, 30123, Venice, Italy*

Electrochemistry offers promising approaches for treatment of effluents and waste: in particular electrochemical techniques could be very useful when toxic materials must be removed from the final stage of the process.

Attractive advantages of electrochemical processes consist in their versatility: processes of direct or indirect oxidation, reduction, phase separation can be realised in order to remove persistent pollutants from industrial wastewater. Moreover, high energy efficiency can be achieved when electrodes and cells are suitably designed to minimising power losses, the cell construction is generally simple and the system variables (electrode potential and cell current) are suitable for facilitating process automation.

However, considering that the electrochemical reaction takes place at the electrode surface, limitations of mass transfer could affect the performance of the electrochemical process, in particular when dilute solutions of pollutants have to be treated. Therefore, the main characteristic of electrochemical reactors, designed in the recent years for industrial wastewater treatment, consists of high specific area of the electrode and large mass transfer coefficients. Reactors based on three dimensional electrodes are contributing to the electrochemical approach to effluent treatment, since their properties lead to a high electrolysis rate even if the solution concentration is very low: indeed such electrodes are characterised by high superficial area and determine turbulent mass transport conditions so that high values of limiting current can be achieved.

Three-dimensional electrodes are now available in many forms. Different types of three-dimensional electrodes have been considered: static electrodes (porous electrodes such as felts and foams, packed bed electrodes), moving electrodes (fluidised bed, vibrated beds, beds with pulsed flow).

Two principal arrangements with respect to the current and electrolyte flow are possible (Fig. 1): they are denoted as flow-through (parallel electrolyte and current flow) and flow-by (perpendicular electrolyte and current flow).

The design of practical system will depend primarily on the objective of the process: three dimensional electrode cells are well suited to remove low level of pollutant when the main objective of the process is the improvement of water quality. Some typical, commercially available cell systems are shown in Fig. 2.

Chemelec cell[1-2] which consists of a series of closely spaced expanded metal electrodes (undivided and with alternate anodes and cathodes), separated by a partially fluidised bed of non conducting spherical particles which leads to an efficient mass transport regime.

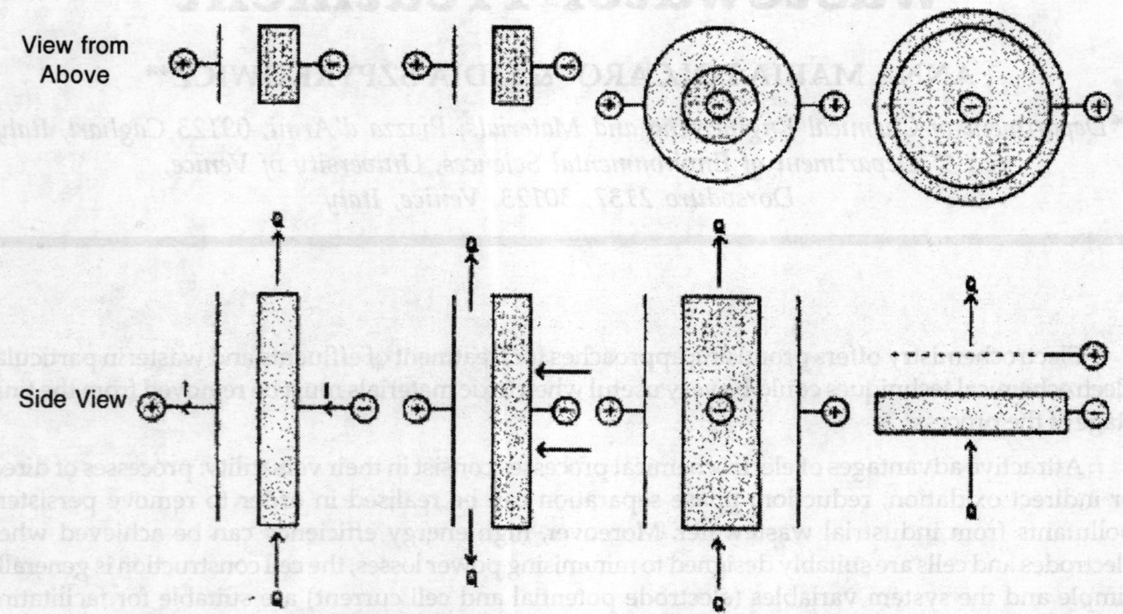

Fig. 1: Geometric Configurations of Porous Tridimensional Cathode Cells

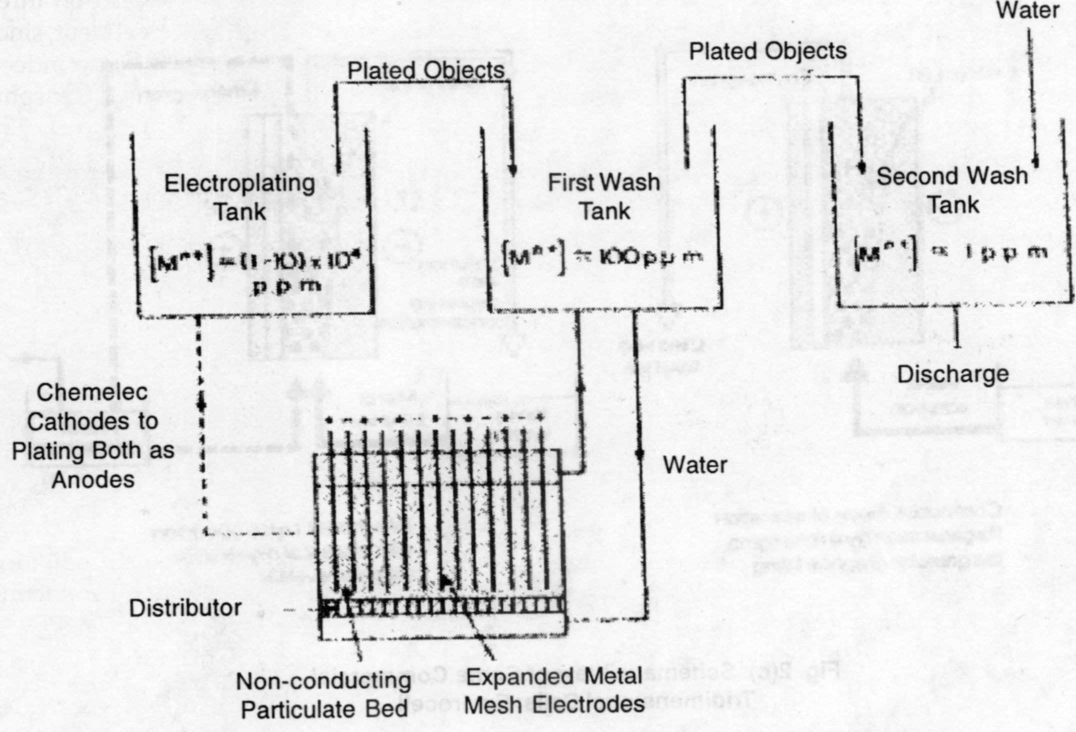

Fig. 2(a): Schematic View of Some Commercial Tridimensional Cells: Chemelec Cell

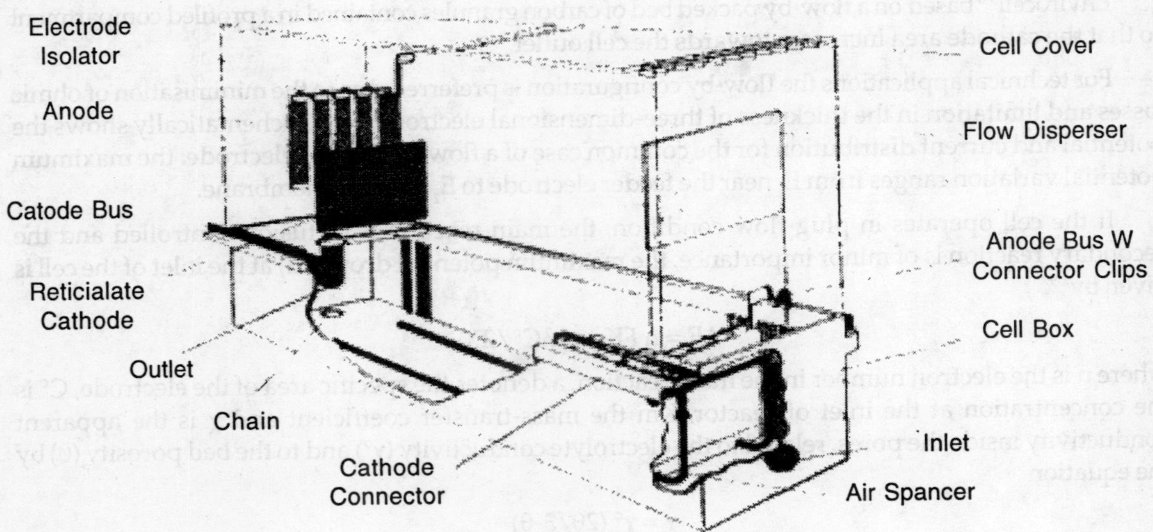

Fig. 2(b): Schematic View of Some Commercial Tridimensional Cells: RETEC Cell

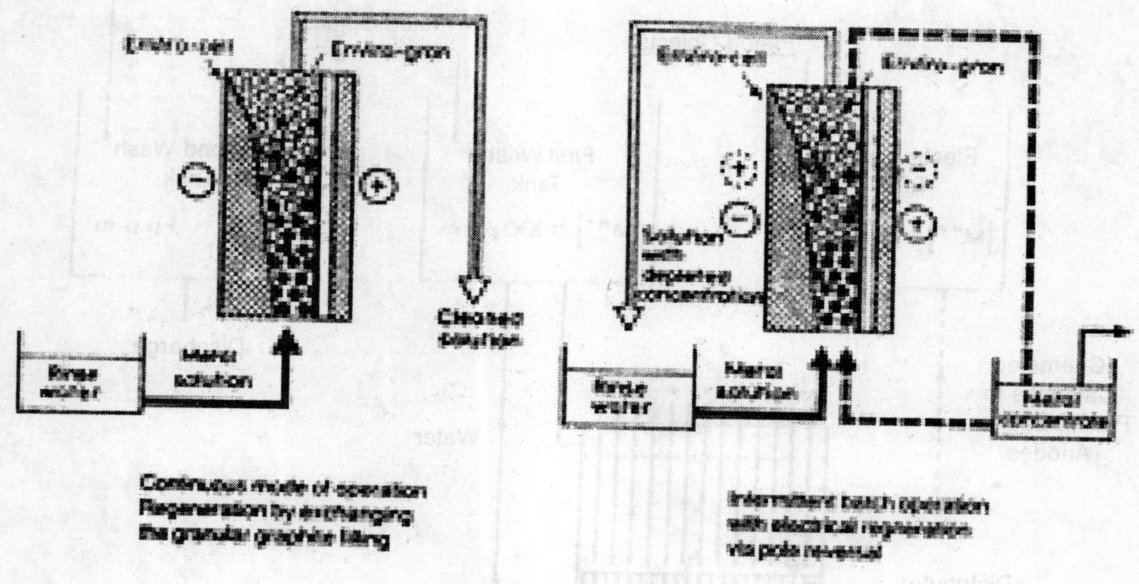

Fig. 2(c): Schematic View of Some Commercial Tridimensional Cells: Envirocell

Retec cell[3-4] based on reticulated carbon or metal foam cathodes, held in place by grooves on the inside tank wall: between each pair of cathode is an inert anode.

Envirocell[5-6] based on a flow-by packed bed of carbon granules contained in a profiled compartment so that the cathode area increases towards the cell outlet.

For technical applications the flow-by configuration is preferred due to the minimisation of ohmic losses and limitation in the thickness of three-dimensional electrode. Fig. 3 schematically shows the potential and current distribution for the common case of a flow-by porous electrode: the maximum potential variation ranges from E_1 near the feeder electrode to E_2 near the membrane.

If the cell operates in plug-flow condition, the main reaction is diffusion controlled and the secondary reaction is of minor importance, the maximum potential drop (ΔE) at the inlet of the cell is given by[7-9]:

$$\Delta E = n\, FK_m\, a\, h^2 C^\circ / 2X$$

where n is the electron number in the main reaction, a denotes the specific area of the electrode, C° is the concentration at the inlet of reactor, Km the mass-transfer coefficient and χ is the apparent conductivity inside the pores, related to the electrolyte conductivity (γ°) and to the bed porosity (θ) by the equation

$$\chi = \chi^\circ (2\theta/3\text{-}\theta)$$

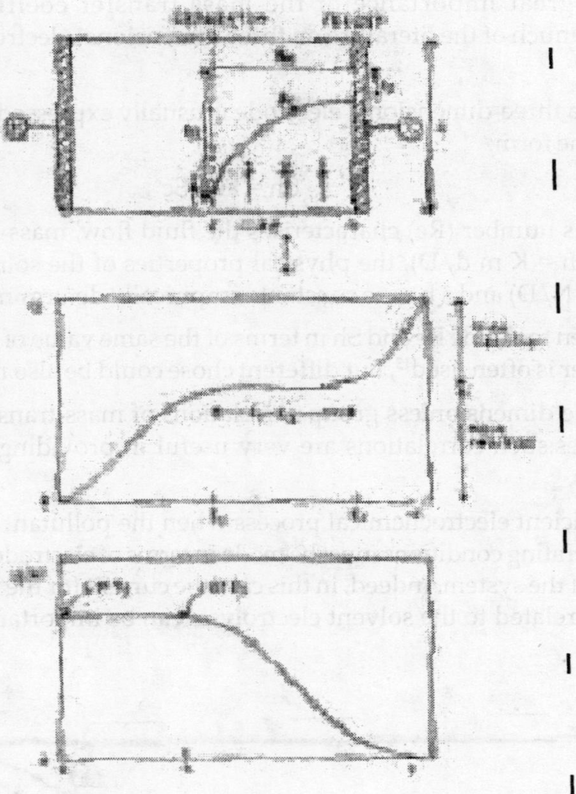

Fig. 3: Potential and Current Distribution in a Flow-by Porous Electrode

The optimal bed dept can be obtained when ΔE* corresponds to the potential region of the limiting current plateau determined in the micro-kinetic polarisation curves.

In the case of dynamic porous electrodes or fluidised bed device, the behaviour is complicated by the electrode movement. Thus, an increase in fluid velocity enhances the average mass-transport coefficient but causes complex potential distributions due to the lowering of particle-particle and particle-feeder contact.

The reactor performance can be expressed by space-time yield (P) defined as the amount of pollutant (dm) removed from a solution volume V in a time interval dt

$$\rho = dm/Vdt$$

According to Faraday' law, dm is proportional to the electrolysis charge dQ= aH i dt, where H is the faradic efficiency of the process.

The optimal process conditions are met when the rate of the electrochemical reaction attains its maximum at the limiting diffusion current density, therefore, the space-time yield can be expressed as:

$$\rho = \eta\, M\, a K_m\, C^\circ$$

Because of the great importance of the mass transfer coefficient in determining the reactor performance, much of the literature on three dimensional electrodes concerns mass-transfer studies[10-12].

Mass-transfer to a three-dimensional electrode is usually expressed by means of dimensionless group correlation in the form:

$$Sh = aRe^{b}Sc^{c}$$

in which the Reynolds number (Re) characterises the fluid flow, mass-transfer is described by the Sherwood number (Sh = K m d/D), the physical properties of the solution are used to define the Schmidt number (Sc= N/D) and a b c are constants empirically determined.

Care must be taken to define Re and Sh in terms of the same value of characteristic length (d): the mean particle diameter is often used[13], but different chose could be also made[14].

Fig. 4 shows some dimensionless group correlations of mass-transfer data for different three-dimensional electrodes:such correlations are very useful in providing a single expression which unites many variables.

To achieve an efficient electrochemical process when the pollutant concentration is very low a careful selection of operating conditions must be made in terms of electrode material, working potential and hydrodynamics of the system. Indeed, in this case the current for the main reaction is very low, so that the side reaction related to the solvent electrolysis can be important and can affect the faradic yield of the process.

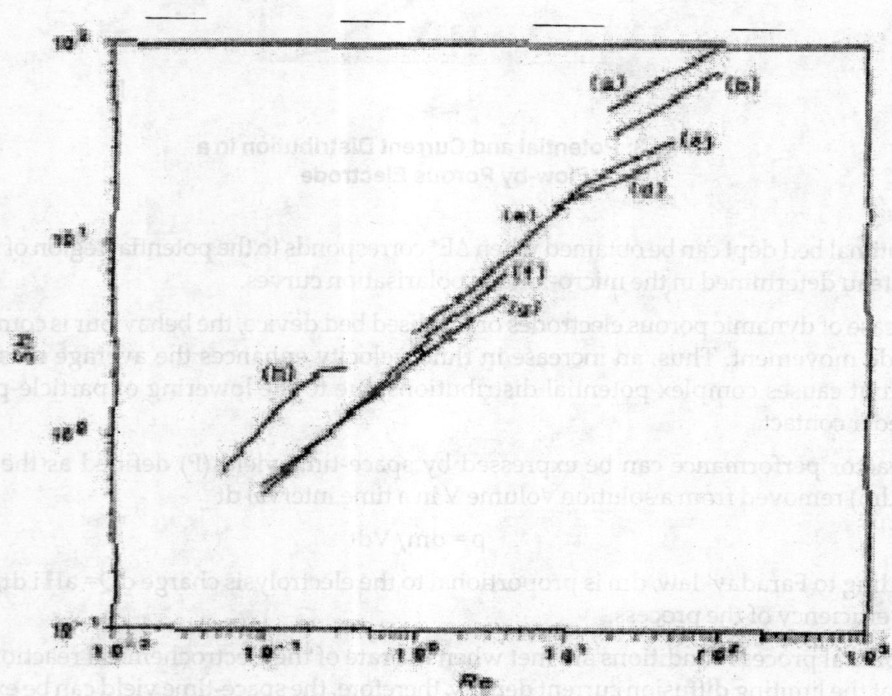

Fig. 4: Re-Sh Correlation for the Various Porous Electrodes Materials

For example, during the cathodic removal of heavy metal ions, if the side reaction is kinetically controlled, a higher efficiency could be achieved by performing the electrolysis at lower cathodic potential, at which the side reactions are less influential.

Nevertheless, if the working potential is out of the range at which metal reduction is under diffusion control, the rate of deposition is not at its maximum value; moreover, when the reversible potential of metal reduction approached, further problems could arise, because of the difficulty in the deposition of the first metal nuclei[15-16].

The lowering of faradic yield due to the side reaction and to the crystallisation overpotential during the electrochemical reduction of copper from very dilute solutions was studied in our laboratory[17]: the initial metal concentration ranged from 10 to 20 ppm and reticulated vitreous carbon (RVC) was selected as cathodic material. RVC, frequently used for three-dimensional electrodes, is an open pore material with a honeycomb structure, and it is available in several porosity grades, from 10 to 100 pores/in (ppi). Among its properties, chemical inertness and stability, low resistance to fluid, and high superficial area: these properties combined with self supporting rigidity and stability in a wide potential range, makes it a versatile electrode material for practical applications.

Three working potential values were tested: E=-0.4 V and E=-0.5 V, at which metal deposition is under diffusion control, and E=-0.25 V at which the effect of oxygen reduction is limited, but deposition does not occur at its maximum. Three cathodic reactions were considered: copper deposition, hydrogen evolution from water, dissolved oxygen reduction and the reaction rate was expressed as a combined function of the electrode kinetics (Tafel approximation)

$$r = k \exp(-A E) C_s$$ and mass transport

$$r = k_m (C - C_s)$$

where

C and C_s are the concentration of species in the bulk and at the interface, k is the rate constant and k_m is the mass transfer coefficient.

A simple mathematical model has been performed that was in general agreement with the experimental data to quantitatively interpret the influence of dissolved oxygen during electro-deposition of copper from extremely dilute solution (Fig. 5).The results suggest some conclusions: an increase in electrolyte flow rate enhances the current intensity for copper deposition, as a logical consequence of mass transfer control. Nevertheless, only a small effect on the faradic efficiency was observed, because the main secondary reaction (oxygen reduction), is under mixed control.

As the overpotential decreases, an increase in faradic efficiency is calculated; nevertheless, the range of lower cathodic potentials, at which problems of slow growth of the crystal nuclei could arise, should be avoided in order to obtain reliable measurements of the behaviour of the electrode.

As was already noticed, the main effect on faradic efficiency is connected to the presence of oxygen in the electrolyte. An initial purging of the solution, although not easy achievable in practical application, does not seem sufficient to determine appreciable benefits in terms of faradic efficiency. A better effect could be obtained by minimising the contact surface area between electrolyte and air, so that re-absorption of oxygen is limited during the electrochemical treatment.

Much of literature[18-19] and most reported applications of three dimensional electrodes deal with the removal of heavy metal ions, but some studies also concern the electrochemical oxidation of organic pollutants. Although the best results for electrochemical combustion of organic solutes have

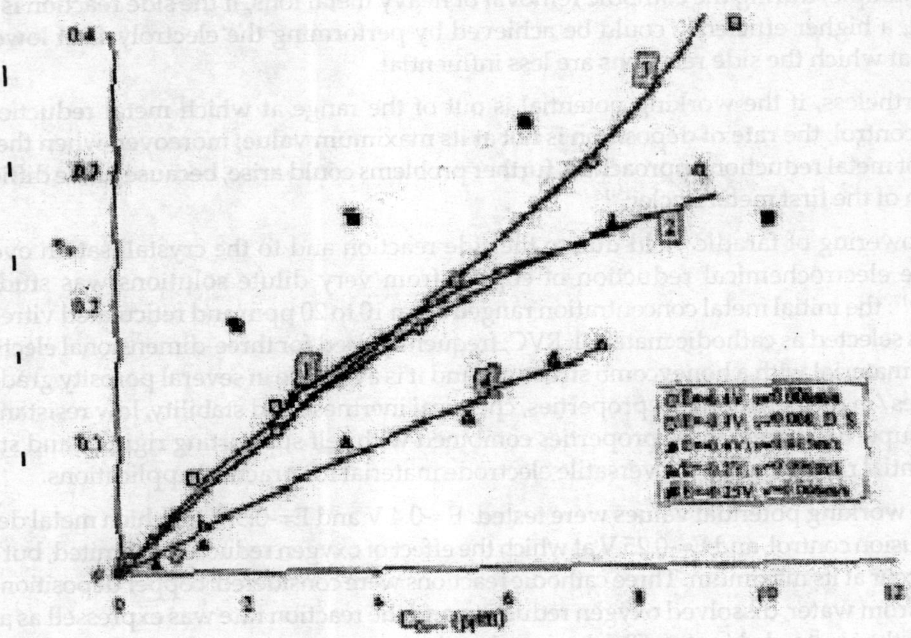

Fig. 5: Fradic Yields as a Functino of Residual Metal Ion Concentration at Different Experimental Conditions. Full lines refer to curves calculated by elaborated model

been obtained using electrodes based on metal dioxides such as PbO_2 or SnO_2[20-21], with or without dopant, or, more recently, using boron doped diamond anodes[24-25], however porous carbon electrodes, which are economical or at least readily available, could be also used.

For example, the ability of a fixed bed of carbon pellets, analogous to that commonly used as adsober in wastewater treatment, to work as anode in destroying phenolic compounds was verified in our laboratory[22].

The experimental results showed that the applied current was able to continuously regenerate the external active sites of carbon surface on which organic solutes were adsorbed: active oxygen, produced by electrolysis, physically adsorbed at carbon surface, reacted with adsorbed organic compounds, causing their oxidation. The low organic coverage degree of carbon hinders phenol diffusion to the interior of pellets.

As far as the oxidation products is concerned, cyclic intermediates, such as quinones and aliphatic acids were detected during the treatment, the distribution of which was strongly dependent on the carbon amount per unit volume of solution (m_s). In particular, as m_s increases not only is more effective removal of phenolic compound observed, but also the ratio between toxic aromatic and biodegradable aliphatic by-products is smaller, thus indicating that the higher amount of carbon enhances the oxidation. Moreover, only low corrosion effects were observed at the end of the runs and the superficial characteristics of the pellets were well maintained after several hours working. The acceptable faradic yield (H=30% for reactant and cyclic intermediates removal) and the low potential value of the anode during the runs (E=0, 98 V vs SCE after 70 h working at a current density equal to 20 mA/cm[3] of electrode) suggest that the energy consumption may be very limited using this economical three dimensional electrode.

REFERENCES

1. C.L. Lopez-Cacicedo, 1981, *J. Separ. Process Thecnol.*, 2: 34.
2. Marketed by BEWT (watr Engineers) Ltd, Tything Road, Arden Forest Industrial Estate, Alcester, Warwicks, England.
3. M.G. Konicek and G. Platek, 1983, *New material and Processes*, 2: 232.
4. Marketed by Eltech Systems Corp., 12850 Bournewood Drive, Sugar Land, Texas 77478, USA.
5. G. Kreysa and C. Reynvaan, 1982, *J. Applied Electrochem.*, 12: 241 Marketed by EnViroCell, Umveltttechnik GmbH, Gattenhoter Weg 29, D-6370, Oberurel, Germany.
7. R.C. Alkire and P.K. Ng, 1977, *J. Electrochem. Soc.*, 124: 1220.
8. F. Leroux and F. Coreuet, 1985, *Electrochim. Acta*, 30: 831.
9. G. Kreysa, 1978, *Electrochim. Acta*, 23: 1351.
10. M. Matlosz and J. Newman, 1986, *J. Electrchem. Soc*, 133: 1850.
11. A. Stork M. A. Enriquez-Granados, M. Roger and F. Coueret, 1982, *Electrochim. Acta*, 27: 293.
12. K. Kinoshita and S.C. Leach, 1982, *J. Electrochim. Soc*: 129: 1993.
13. J. Newman and W. Teideman, 1987, *Advances in Electrochem, and Electrochem. Engineering*, 11: 353.
14. B.G. Ateya, 1980, *J. Appl. Electrochem.*, 16: 627.
15. C, Ponce de Leon, D. Pletcher, 1996, *Electrochim. Acta*, 41: 533.
16. B. Scharifker G. Hills, 1983, *Electrochim. Acta*, 28: 879.
17. A.M. Polcaro, S. Palmas, M. Mascia and R.L. Correa dos Santos, 1999, *Ind. Eng. Chem. Res.* 38: 1400.
18. E. Podlaha and J.M. fenton, 1995, *J. Appl: Electrochem.*, 25: 299.
19. A. Walkerand A. Wraag, Cem., 1980, *Eng. Sci.*, 35: 405.
20. J. Feng and J. Johnson, 1991, *J. Electrochem. Soc.*, 138: 3328.
21. Ch. Comninellis and C., Pulgarin, 1991, *J, Appl. Electrochem.* 21
22. A.M.Polcaro, S.Palmas F. Renoldi and M. Mascia, 2000, *Electrochim. Acta*, 46: 389.

12

Electrochemical Studies on Indigocarmine: A Textile Dye

RAJEEV JAIN, MEENAKSHI BHARGAVA & NIDHI SHARMA

Department of Environmental Chemistry, Jiwaji University, Gwalior - 474011

ABSTRACT

Indigoid dyes are widely used for dyeing and printing protein and cellulose fibers. Indigo dyes are certified food colours and also used for colouring nylon surgical sutures. For the electrochemical study the synthetic Indigocarmine (disodium salt of 5, 5′-indigotinsulphonic acid) of Fluka Chemika with dye content of 85% is taken. Controlled potential electrolysis (CPE) was carried out at three-electrode cell system. A constant potential of-1.20 V was applied. Voltammograms were recorded prior to electrolysis and thereafter when the solution becomes colourless and peak current is significantly reduced. COD of the solutions were also recorded before and after electrolysis. End products of electrochemical reduction are studied using TLC and GC-MS. Using platinum foil working electrode cpe took 2 hours for complete colour removal and significant current decrease. COD decreased substantially from 1200 mg/l to 400 mg/l at the completion of electrolysis.

Introduction

Indigo has been known as Indian for many centuries, and the colouring power of the indigo plant has been mentioned in Sanskrit literature of remote antiquity. The dye was introduced into Europe from India after 1500 A.D and until the early years of this century when synthetic indigo achieved commercial success in competition with the natural product. Indigo was one of the major commodities in the export trade of India. Natural indigo contains besides indigotin, which is the essential tinctorial constituent, varying proportions of indirubin (as isomer of indigotin formed from indoxyl by a side reaction[1].

Indigo and Indigoid dyes are vat colours applicable to both cellulose and protein fibres, but more important for the latter. Indigo itself has considerably better fastness on wool, than it has on cotton. Halogenation normally improves the properties of indigos on cotton, particularly their affinity for the fibre. Synthetic Indigo appeared on market in 1897 and not withstanding the discovery of faster vat Blues of the anthraquinone class, indigo has continued to hold its own. The annual consumption of Indigo was estimated at 10 millions pounds, a figure exceeded by not more than four or five other dyes [2]. Moreover tedious extraction of colouring component from the raw material, low colour value and long dyeing time, push the cost of dyeing with natural dyes considerably higher than with synthetic dyes[3].

Indigoid dyes are mainly used as pigments and vat dyes. The solubilised form of the vat dye is known as a leuco compound. The leuco compound in its free form is water insoluble also; however the alkali salts of these compounds are water soluble, and are widely used for dyeing and printing protein and cellulose fibres [2]. Other than being used in textile industry indigo has been used in dry powdered drinks, candy and confectionery products that are oil and fat free, bakery products and cereals and puddings. The FDA for use in colouring nylon surgical sutures also approves it [4,5].

This dye is a dark blue powder with coppery lustre and is sensitive to light. Its solution has a blue or bluish purple colour. One gm dissolves in 100 ml water at 25 °C. It is slightly soluble in alcohol and practically insoluble in most other organic solvents [6]. The compound *"Indigocarmine"* chosen as model compound for the present study finds vast applications. Moreover in the conventional indigo-dyeing process, significant quantities of excess indigo was found in the wash water from the rinsing operations which became part of the plant's effluent [7]. Amongst different industrial wastewaters with different types of colouring agents, dye colour wastes occupy a prominent position. A considerable part of the dye consumed in industrial processes is residual liquors through incomplete exhaustion and washing process. Dyes are confirmed heavy metal carriers since some heavy metals like chromium, copper and cobalt may form an intrinsic part of the dye molecule. Textile industrial processes generate large quantities of wastewater, which are very toxic and difficult to be disposed by physical, chemical and microbiological methods. The textile wastewater is characterized by high chemical oxygen demand (COD), low biological oxygen demand (BOD), and heavy colour. In particular, the COD and colour of the wastewater are resistant to conventional wastewater treatment. In addition to textile industries, indigo has been used for colouring various materials including foodstuff. Although these chemicals are recommended under specific rules and their use are very specific yet use of these chemicals in day to day consumable products is of great human concern and warrants testing of their genotoxic effect[8-12].

Synthesis of Indigocarmine

Indigo (Indigotin) is prepared by following methods:

Scheme I

Reduction of isatin chloride (obtained by the action of phosphorus pentachloride on isatin) with zinc dust in glacial acetic acid yields indoxyl, which upon oxidation in air gives indigotin.

Scheme I

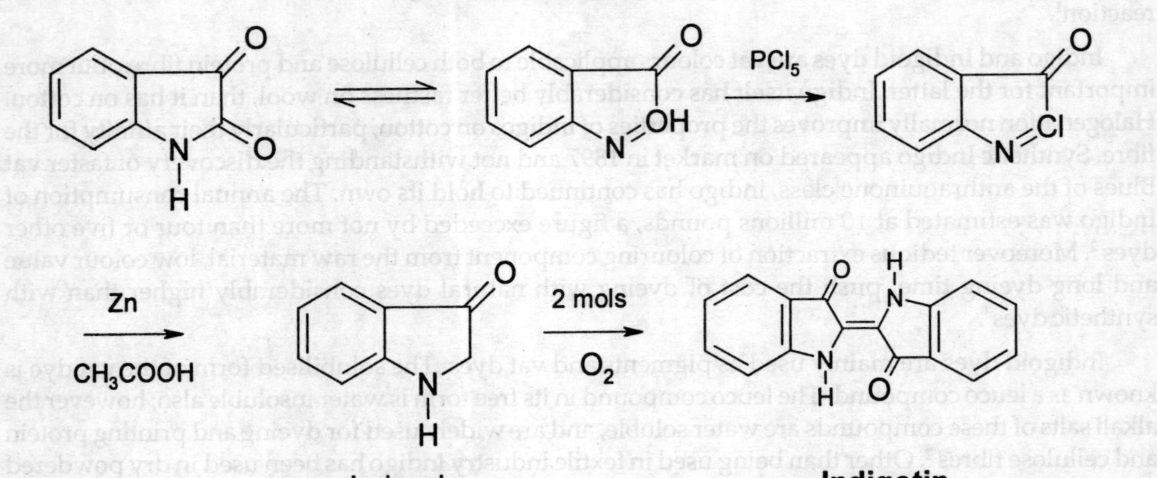

PCl_5

Zn, CH_3COOH → Indoxyl; 2 mols, O_2 → Indigotin

Scheme II

Indigo is also synthesizd from anthranilic acid. When anthranilic acid is made to react with chloroacetic acid, it gives the phenyl glycine-o-carboxylic acid, which undergoes ring closure and decarboxylation to indoxyl on fusion with a mixture of potassium hydroxide and sodamide. Atmospheric oxidation of indoxyl gives indigotin [13].

Scheme II

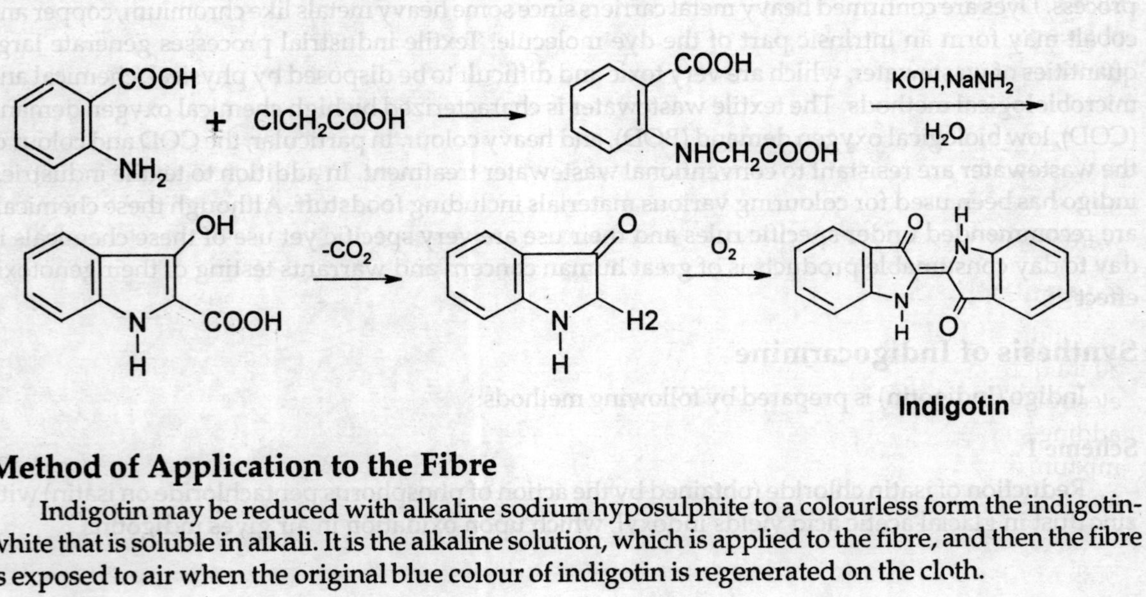

COOH, NH_2 + $ClCH_2COOH$ → COOH, $NHCH_2COOH$; $KOH, NaNH_2$, H_2O →

OH, COOH; $-CO_2$ → H2; O_2 → Indigotin

Method of Application to the Fibre

Indigotin may be reduced with alkaline sodium hyposulphite to a colourless form the indigotin-white that is soluble in alkali. It is the alkaline solution, which is applied to the fibre, and then the fibre is exposed to air when the original blue colour of indigotin is regenerated on the cloth.

Reduction

Oxidation

Indigotin **Indigotin - White (colourless)**

The principle use of large amount of synthetic indigo is in the dyeing of cotton for working clothes such as overalls. A considerable amount of indigo is also used for dyeing wool in heav shades. Indigo has only moderately good fastness to light (approx 4 on cotton or 4-5 on wool). It has great advantage over other dyes that as fading proceeds there is little or no change in the hue.

Indigocarmine is chemically known as disodium salt of 5, 5′-indigotinsulfonic acid. It is commercially known as Indigo blue or Vat blue 1 and has a C.I name CI Food Blue 1. For the electrochemical study the synthetic indigo dye (disodium salt of indigotin) of Fluka Chemika with dye content of 85% was taken.

Experimental

Instrumentation

Controlled potential electrolysis (CPE) was carried on BAS CV-27 Cyclic Voltammograph in connection with a Digital Electronic 2000 Omnigraph X-Y/t recorder. A three-electrode cell system was used for the purpose. The cell consists of a glass container with a cap having holes for introducing electrodes and nitrogen gas. The working electrodes are Platinum foil (3 * 3 cm^2) and Steel foil (4.5* 3.5 cm^2), the reference is Ag/AgCl and auxiliary electrode a platinum wire which was directly poured in the solution. Cyclic Voltammograms were taken using EG&G Princeton Applied Research VersaStat-II Potentiostat. The pH-metric measurements were made on Hach EC-40 Bench top pH/ISE meter. For identification of end products thin layer chromatography (TLC) was also employed. Various mobile phases were applied for the separation. Good resolution could be achieved in Dioxane:Water (9:1) and Acetone:Water (8:2) mobile phases. Components of the electrolysed solution on the TLC plate could be visualised under UV lamp.

The spectral studies were carried out using UV-Vis spectrophotometer (Elico SL-159). The kinetics of current and absorbance decrease was studied by recording changes in current and absorbance with time. Chemical Oxygen Demand (COD) of both coloured and decolourised/electrolysed solutions were recorded using COD Digestion Apparatus (Spectralab-2015-S).

Reagents and Solutions

Indigocarmine of Standard Fluka Chemika was used. $2*10^{-3}$ M stock solution of indigocarmine in 50 ml of distilled water was prepared. 1.0 M KCl was prepared in distilled water as a supporting electrolyte. BR (Britton-Robinson) buffers in the pH range 2.5-12.0 were prepared in distilled water by adding suitable amounts of 0.4M NaOH solution (basic solution) to a stock solution composed of a mixture of phosphoric acid, boric acid and glacial acetic acid (acidic solution). Solutions were stirred and left overnight to attain equilibrium. All the chemicals used were of Anal R grade.

For COD determination potassium dichromate (0.25 N) and ferrous ammonium sulphate (0.1 N), both of Anal R grade were prepared in distilled water. Sulphuric acid-silver sulphate reagent was prepared by dissolving 10 gm silver sulphate in 1000 ml conc. H_2SO_4 and kept for about 2 days for digestion.

Procedure

For cyclic voltammetry (CV) and controlled potential electrolysis (CPE) the solution was prepared by taking 9.0 ml of the stock solution and 1.0 ml of 1.0M KCl (as supporting electrolyte).

1.0 ml of stock solution was diluted to 10.0 ml with appropriate BR buffer solution. The solution was deareated by passing nitrogen gas for about 15 minutes and, thereafter, a blanket of nitrogen gas was maintained throughout the experiment. The cyclic voltammograms were recorded in the beginning of the experiment and then the solution was subjected to cpe at-1.20 V.

Further at the end of electrolysis *i.e.* when the solution becomes colourless and current value is significantly decreased, the voltammograms were recorded again. For measuring the rate of decrease in current as well as absorbance the, current values were recorded continuously with time and a small aliquot of the solution was taken in the cuvette at regular intervals to note the decrease in absorbance values.

For the identification of end products being formed, the electrolysed solution was subjected to TLC and GC-MS. Ready made silica gel plates (60_{254}) of E-Merck company were used. Both coloured and decolourised/electrolysed solutions were loaded on the plate and developed in the mobile phase of Dioxane:water (9:1) and acetone:water (8:2). The plates were viewed under UV lamp and a broad band not well separated into any clear spots was observed.

For COD determination, the solutions (both coloured and electrolysed) were taken in sample tubes and refluxed for 2 hrs. in COD digestion apparatus and afterwards allowed to cool to room temperature. It was then diluted with 80 ml of distilled water and titrated with ferrous ammonium sulphate using ferroin indicator.

Coulometric Studies

For coulometric determination of number of electrons 'n' consumed in the reduction, solution was prepared in the similar fashion as for cyclic voltammetric studies. The solution was deareated by passing nitrogen gas for 15 minutes and then cyclic voltammograms were recorded at a negative potential slightly higher than peak potential. With the progress of the electrolysis the colour of the solution gradually faded away and finally a colourless solution was obtained.

The current decreases exponentially with time and become practically zero after some time. Value of Q was read directly and by the application of Q = nFN, the value of 'n' could be calculated.

Results and Discussion

Cyclic Voltammetry

Cyclic Voltammograms of Indigocarmine were recorded in B.R buffers of pH 2.5-12.0 and in aqueous neutral medium at various scan rates ranging from 50-2000 mV/sec. Indigocarmine gives a single cathodic peak at –0.400 V, which shows a negative shift with increasing scan rate. A small anodic peak at –0.787 V is also observed which shows positive shift with increasing scan rate. This negative shift in peak potentials with increasing scan rate indicates the irreversible nature of the electrode process (Table 1). With both the working electrodes (*i.e.* platinum foil and steel foil) a linear plot of ip.c and $v^{1/2}$ (v = scan rate) is obtained indicating diffusion-controlled nature of the electrode process (Fig. 1).

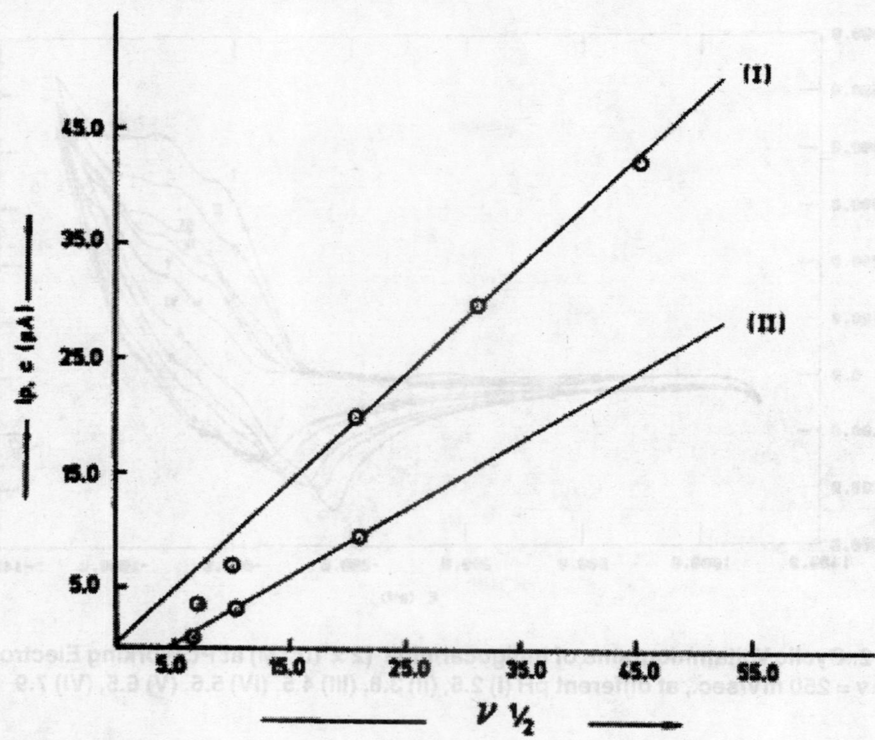

Fig. 1: Plots of Ip c (μA) vs *V*½ for Indigocarmine (2 × 10^{-4} M in Distilled Water) at (I) Platinum foil, (II) Steel foil working electrodes

With steel foil as working electrode there is slight deviation from the origin indicating electrode reaction to be mainly diffusion controlled with some adsorption contribution. The diffusion-controlled process is also supported by the fact that forward peak current is proportional to the square root of scan rate and plot of log ip versus log *v* is linear with a slope of nearly 0.50 V.

Peak current ip, c shows linear behaviour with increasing concentration but at higher concentration it tends to limit down and shows constancy. Ep, c shows negative shift with increasing concentration and linearity in behaviour (Table 2). This behaviour indicated that electrode process was mainly diffusion-controlled with some adsorption contribution. This behaviour also indicated difficult electro-reduction at higher concentration.

Electrochemical Behaviour at Different pH

Cyclic voltammograms af Indigocarmine in the pH range 2.5-12.0 shows well defined reduction peaks and corresponding oxidation peaks at the platinum electrode (Table 3, Fig. 2). In the acidic range (2.5-5.6), the voltammograms show two cathodic peaks while in the alkaline range single (not so well defined) peak is observed. Both cathodic and anodic peak potentials show negative shift with increasing pH. Plot of Ep, c vs. pH (Fig. 3) is almost linear upto pH 5.6 *i.e.* peak potential shifted towards more negative value with pH linearly till 5.6 thereafter, there was little change in Ep, c with pH. The behaviour indicated the effect of pH on the redox behaviour of Indigocarmine through

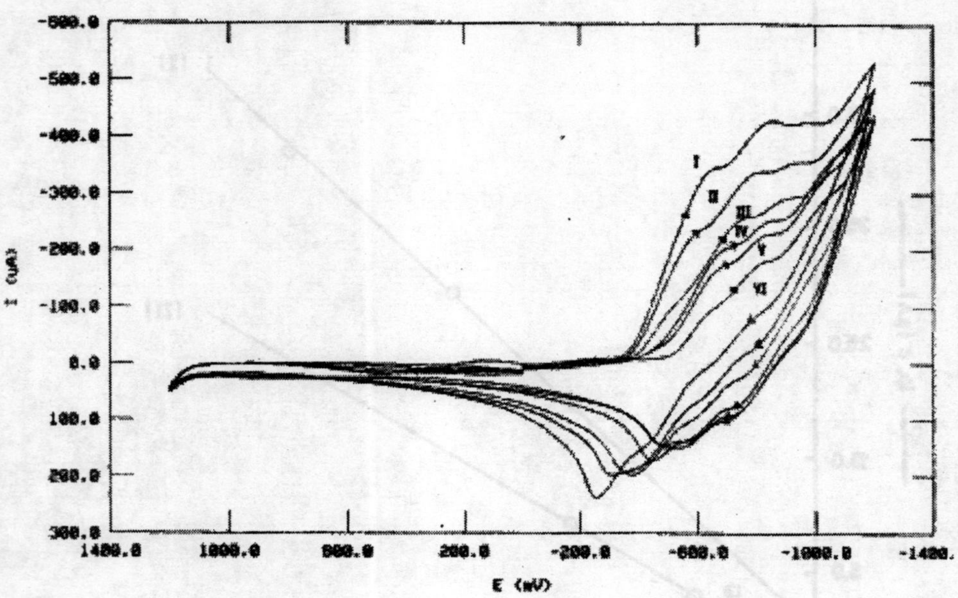

Fig. 2: Cyclic Voltammograms of Indigocarmine (2 × 10^{-4} M) at Pt. Working Electrode, v = 250 mV/sec., at different pH (I) 2.5, (II) 3.8, (III) 4.5, (IV) 5.6, (V) 6.5, (VI) 7.9

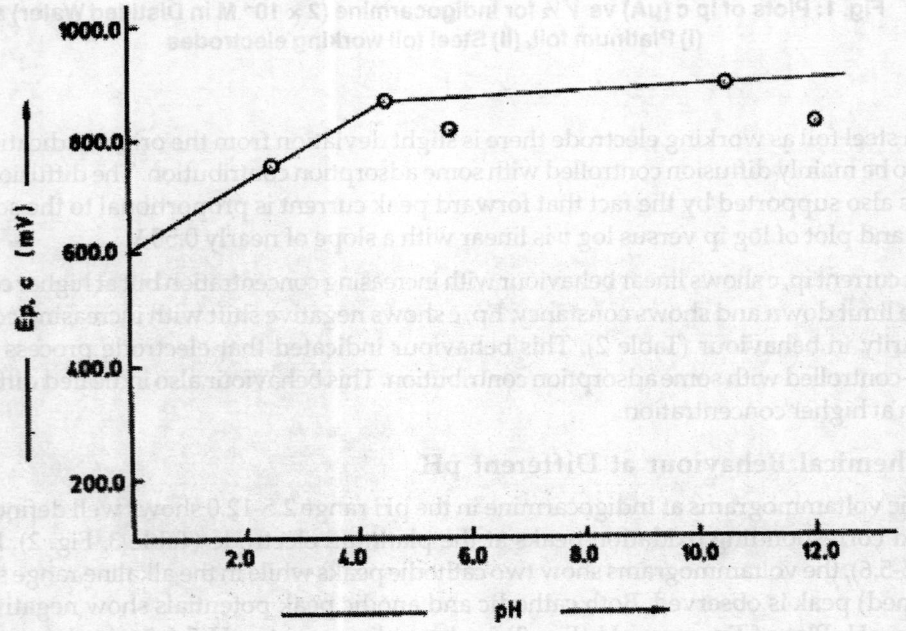

Fig. 3: Plots of Ep c (µV) vs pHfor Indigocarmine (2 × 10^{-4} M) at Platinum Foil Working Electrode, Scan Rate = 100 mV/sec.

protonation reaction. Plot of ip, c vs. pH (Fig. 4) shows a linear decreasing trend till 5.6 thereafter, showing little change in ip, c with pH. The decrease in peak current with increasing pH indicated difficult electro reduction at higher pH. The break in linearity at pH 5.6 in both plots (Ep, c vs pH and ip, c vs pH) corresponds to the pKa value [14]

Table 1: Electrochemical Characteristics of Indigocarmine in Aqueous Nedium *at Various Scan Rates* at Platinum Foil and Steel Foil Working Electrodes, conc. = 2 * 10^{-4} M

Working Electrode	*Scan Rate mV/sec*	$V^{1/2}$	*Ip, c (μA)*	*Ep, c (V)*	*ip, a (μA)*	*Ep, a (V)*
Platinum Foil	50	7.07				
	(1) A		– 3.55	– 0.400	– 0.53	– 0.787
	(2) B		—	—	—	—
	100		10.00			
	(1) A		– 7.24	– 0.403	2.43	0.793
	(2) B	—	—	—	—	—
	500		22.36			
	(1) A		– 20.79	– 0.561	3.17	0.653
	(2) B		– 10.00	– 0.561	55.00	0.750
	1000					
	(1) A	31.62	– 29.47	– 0.613	5.53	0.595
	(2) B		– 18.00	– 0.613	60.00	0.710
	2000					
	(1) A	45.00	– 42.87	– 0.729	5.439	– 0.543
	(2) B		—	—	—	—
Steel Foil	50 7.07					
	(1) A		– 46.80	– 0.267	35.54	– 0.840
	(2) B		– 0.77	– 0.263	108.70	– 0.893
	100	10.00				
	(1) A		– 59.62	– 0.310	44.08	– 0.826
	(2) B		– 3.41	– 0.310	176.20	– 0.865
	500	22.36				
	(1) A		– 158.40	– 0.391	63.61	– 0.758
	(2) B		– 9.58	– 0.391	232.00	– 0.822

A = Before electrochemical treatment; B = After electrochemical treatment

Table 2: Electrochemical Characteristics of Indigocarmine in Aqueous Medium *at Various Concentration*, *V* = 250 mV/sec.

Working Electrode	*Concentration (Conc.* 10^{-4} M)*	*Ip, c (μA)*	*Ep, c (V)*	*ip, a (μA)*	*Ep, a (V)*
Platinum Foil	2.00	– 11.66	– 0.503	6.71	– 0.518
	4.00	– 15.69	– 0.518	5.62	– 0.418
	8.00	– 19.51	– 0.515	7.39	– 0.427
	20.00	– 20.63	– 0.610	8.54	– 0.250

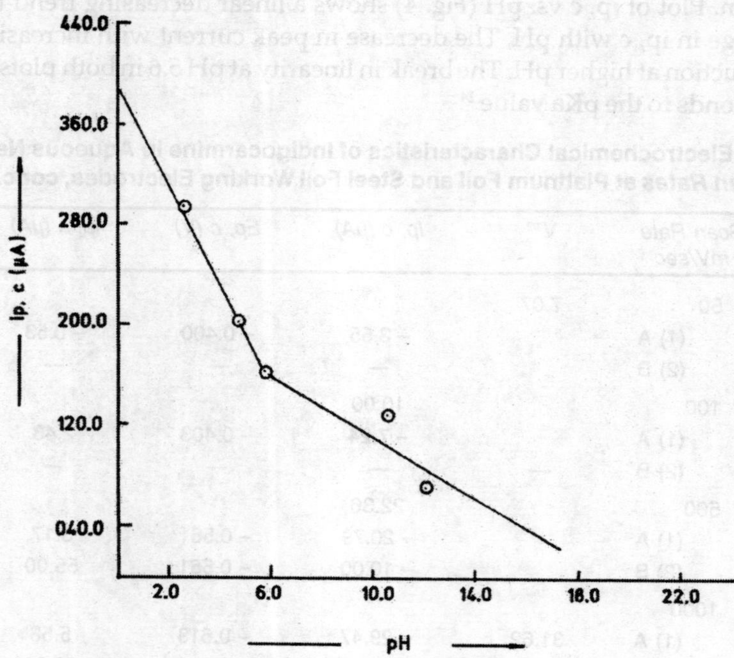

Fig. 4: Plot of Ip c (μV) vs pH for Indigocarmine (2×10^{-4} M) at Platinum Foil Working Electrode, Scan Rate = 100 mV/sec.

Controlled Potential Electrolysis

Controlled potential electrolysis (CPE) of Indigocarmine in aqueous medium (Conc. $2*10^{-4}$ M) was carried out at electrolysis potential of –1.20 V using platinum foil and steel foil as working electrode. The electrolysis reaction took about 2 hours with platinum foil electrode while with steel foil electrode the electrolysis duration was 3 hours. At both these working electrodes effective decolourisation and peak reduction was obtained (Fig. 5). The value of 'n' *i.e.* number of electrons involved in the reduction of indigocarmine were determined and was found to be six. Rate of current decrease at both the working electrodes was studied by recording change in peak current (ip, c) with time (Table 4). The peak current remains constant in the beginning thereafter showing a significant decrease.

Spectral Studies

The UV-Visible spectra of Indigocarmine were recorded in the region 350-650 nm. Indigocarmine dye solution has a Λ_{max} of 608 nm. The rate of decay of current and absorbance was monitored as a function of time at this Λ_{max} using platinum foil working electrode. Peak at 608 nm gradually decreases with time and after 2 hours of electrolysis there is significant decrease in absorbance and a colourless solution is obtained. The absorbance when monitored as function of time (Fig. 6) showed decline initially and later it becomes practically parallel to the base line. This behaviour indicates that the decrease in absorbance follows first order rate expression. The resulting Log absorbance vs time plot was found to be linear. The values of the first order rate constant are presented in Table 5.

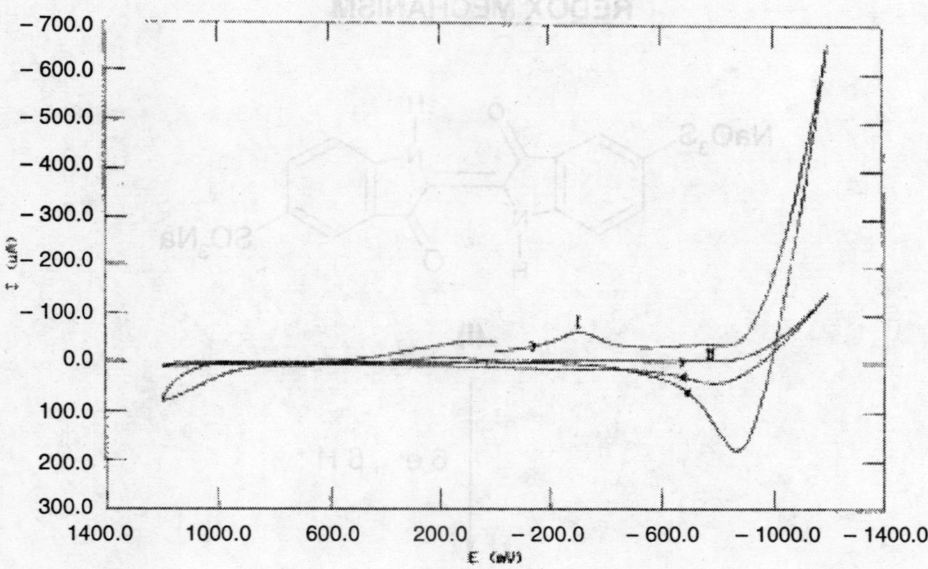

Fig. 5: Cyclic Voltammograms of Indigocarmine (2* 10^{-4} M in distilled water), v = 100 mV/sec.
(I) Electrochemical treatment using steel electrode
(II) Electrochemical treatment using Pt. electrode

Table 3: Electrochemical Characteristics for Indigocarmine *at different pH*, v = 100 mV/s, Conc. = 2 * 10^{-4} M at Pt. Foil Working Electrode

pH		*ip, c (μA)*		*Ep, c (V)*		*ip, a (μA)*	*Ep, a (V)*
		(i)	*(ii)*	*(i)*	*(ii)*		
2.5							
	(1) A	– 293.70	– 237.70	– 0.758	– 0.591	124.70	– 0.335
	(2) B	– 8.26	—	– 0.145	—	8.20	– 0.206
4.5							
	(1) A	– 173.90	– 199.60	– 0.701	– 0.872	126.40	– 0.434
	(2) B	– 8.84	—	– 0.584	—	5.38	– 0.434
5.6							
	(1) A	– 135.70	– 169.90	– 0.658	– 0.818	96.28	– 0.505
	(2) B	– 6.43	—	– 0.167	—	73.26	– 0.683
10.5							
	(1) A	– 48.88	– 131.40	– 0.715	– 0.911	70.67	– 0.890
	(2) B	– 23.09	—	– 0.231	—	90.36	– 0.772
12.0							
	(1) A	– 74.31	—	– 0.836	—	95.26	– 0.687
	(2) B	– 24.29	—	– 0.238	—	116.00	– 0.769

A = Before electrochemical treatment; B = After electrochemical treatment

REDOX MECHANISM

NaO_3S ... SO_3Na

(I)

$6e^-$, $6H^+$

(II)

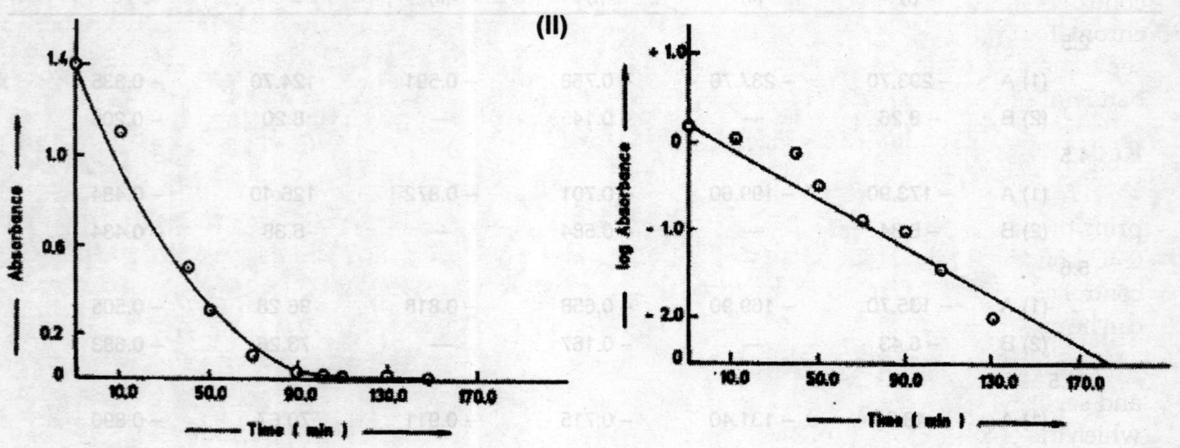

Fig. 6: Plot of Absorbance vs Time and Log Abs. vs Time Observed at 612 nm in Aqueous Medium During the Electroreduction of Indigocarmine at Platinum Foil Working Electrode

Table 4: Rate of Current Decrease with Time for Indigocarmine at Platinum and Steel Foil Working Electrodes, Conc.= 2 * 10^{-4} M

Time (min)	*Pt. Foil Working Electrode*		*Steel Foil Working Electrode*	
	ip c (mA)	*log ip, c*	*Ip, c (mA)*	*log ip, c*
0	– 3.49	0.543	Out of Range	—
10	– 3.37	0.528		
50	– 3.85	0.585	– 19.45	1.289
90	– 3.38	0.528		
100			– 18.50	1.267
105	– 3.33	0.522		
115	– 1.69	0.228		
130	– 1.02	0.008		
150			– 15.00	1.176
185			– 10.50	1.021

Table 5: First Order Rate Constants for Indigocarmine at Platinum Foil Working Electrode, Conc. = 2 * 10^{-4} M

Working Electrode	*Concentration (M)*	*Initial Applied Potential Vs SCE (V)*	*Rate Constant (k) (min^{-1})*	*Half-Life ($t_{1/2}$) (min^{-1})*	*R_{obs}/min^{-1}*
Pt. Foil	2*10^{-4}M	– 1.20	3.45* 10^{-2}	20	6.9*10^{-6}

Chromatographic Studies

Attempts to separate the electroreduction products were usually carried out following conventional controlled product electrolysis (cpe). After cpe the electrolysed solution was subjected to thin layer chromatography. Dioxane: Water (9:1) and Acetone: Water (8:2) were used as the mobile phase for separation. But these mobile phases could not resolve the mixture into clear spots. Rather a broad band not clearly separated from the mobile phase was obtained.

Redox Mechanism

For compounds such as indigocarmine having carbonyl group as well as-C=C-, the general principle is that when an electro-reducible group is conjugated with olefinic groups, the multiple-C=C-bond is certainly reduced before the carbonyl group [15-22]. On the basis of above facts, coulometry, controlled potential electrolysis (cpe), cyclic voltammetry (cv) and spectral studies following mechanism can be postulated for the reduction of indigocarmine.

Since the number of electrons that participated in this electroreduction are six and no significant and separate spots could be obtained, therefore the reaction is assumed to be stopped at stage II at which the carbonyl groups themselves gets reduced.

Chemical Oxygen Demand

Lastly the toxicity parameter was monitored for the electrochemically treated dye solution of Indigocarmine. COD of initial solution was recorded and again after electrolysis the COD was recorded.

A significant decrease in COD values from *1200 mg/l* for initial coloured dye solution ($2 * 10^{-4}$ M) to *400 mg/l* for the final electrolysed solution was observed.

REFERENCES

1. K.M Shah, 1998, *Handbook of Synthetic Dyes and Pigments, Multi-Tech Publishing Co.*
2. F.D Snell and L. S. Ettre, 1973, *Encyclopedia of Industrial Chemical Analysis*, Eds, 5, John Wiley and Sons, Inc.
3. P.S. Vankar, 2000, Chemistry of Natural Dyes, *Resonance*, 10, 73.
4. Fluka Riedel-d Haen, *Laboratory Chemicals and Analytical Reagents*, 1999-2000.
5. *Aldrich Catalog Handbook of Fine Chemicals*, 1992-1993.
6. *An Encyclopedia of Chemicals and Drugs, Merck & Co.*, N. J., USA, 1975.
7. A. Reife and H. S. Freeman, 1996, *Environmental Chemistry of Dyes and Pigments, Wiley Interscience Publication*, New York.
8. X. Hu, L. Lei, G, Chen and P. Yue, 2001, *Wat. Res.*, 35, 2078.
9. L. C. Chen, 2000, *Wat. Res.*, 34, 974.
10. Y. Al. Degs. M. A. M. Khraisheh, S. J. Allen and M. N. Ahmad, 2000, *Wat. Res.*, 34, 927.
11. R. B. Singh, 2000, *J. Environmental & Pollution*, 7, 77.
12. R. K. Trivedy, 1995, *Pollution Management in Industries, Environmental Publications*, Karad.
13. R. M. Silverstein, G. C. Bassler, T. C. Morril, 1991, *Spectrometric Identification of Organic Compounds, John Wiley & Sons*, New York.
14. B. Mabrouk, 1987, Electrochem., 3, 4.
15. M. M. Baizer, H. Lund, 1983, *Organic Electrochemistry: An Introduction and a Guide, Marcel Dekker, Inc.* New York.
16. J. Heyrovsky, 1934, *A Polarographic Study of Electrokinetic Phenomena. Actual Sci. Ind. No. 90, Hermann, Paris.*
17. R. H. Philip, Jr., T. Layloff and R. N. Adams, 1964, *J. Electrochem. Soc.*, 111, 1189.
18. S. S. Katiyar, M.Lalithambika and G. C. Joshi, 1974, *J. Electrochem. Chem.*, 53, 439.
19. S. S. Katiyar, M. Lalithambika and D. N. Dhar, 1974, *J. Electrochem. Chem.*, 53, 449.
20. L. I. Smith, I. M. Kolthoff, S. Wawzonek and P. M. Ruoff, 1941, *J. Amer. Chem. Soc.*, 63, 1018.
21. I. M. Kolthoff and J. J. Lingane, *Polarography*, 1952, 2nd *ed., Vol II. Wiley-Interscience, New York*, 699.
22. I. M. Kolthoff and J. J. Lingane, *Polarography*, 1952, 2nd *ed., Vol I, Wiley-Interscience, New York*, 153.
23. *APHA, Standard Methods for Water and Wastewater Examination, 19th Ed., Am. Public Health Assoc, Washington DC*, 1995.

13

Advanced Methods of Wastewater Treatment in Petroleum Refining Industries

UMA SHANKER

Indian Institute of Petroleum, Dehradun – 248 005, Uttaranchal

ABSTRACT

Petroleum refineries generate enormous amounts of wastes and pollutants in the form of organic and inorganic substances, wastewater and sludge. This communication intends to present an overview of an efficient methods of wastewater treatment in refining industries making it reusable or dischargeable without having an adverse impact on environment.

Introduction

The present refining capacity of India is about 113 MMTPA and is expected to grow continuously in years to come. It is thus expected that an enormous amount of waste will be generated and released to the environment from the refining sector alone. Majority of these wastes are hazardous and create significant air, water (including ground water) and land pollution having an adverse impact on public health, ecology and the surrounding environment. The most common method for wastewater effluent treatment is biological treatment. Due to more emphasis on pollution prevention and zerodischarge, some newer technologies have emerged which can treat specific wastewater streams. The treated water may be recused or discharged with the compliance of regulatory restrictions.

Refining Processes and Wastewater Generation

The basic unit processes generating wastewater can be broadly classified as:

1. Storage of crude oil
2. Fractionation and stabilization
3. Reforming
4. Catalytic and thermal cracking
5. Hydrotreatment and hydrofining
6. Solvent treatment processes
7. Utilities operations

Storage and Transportation

Wastewater resulting from storage of crude oil are generally contaminated with oil, suspended and dissolved solids. The water layer forms bottom sludge and periodically drained. The emulsified oil at the oil water interface is collected separately for oil recovery.

Fractionation

Atmospheric and vacuum distillation is practiced to distill the crude oil to different fractions such as naphtha, diesel etc. The wastewater generated can be a major source of sulfides, chlorides, mercaptans and phenols.

Reforming

The wastewater generated in the process contain free and emulsified oil and organic substances.

Thermal and Catalytic Cracking

The major source of wastewater in thermal cracking is the overhead accumulators on the fractionator. The wastewater contain oil, sulfides and phenol. The major pollutants from cat. cracking are oil, sulfides and phenol.

Hydrotreating Processes

The wastewater generated during the process contain high amounts of H_2S, mercaptans, ammonia, sulfides and phenol.

Solvent Treating Processes

The major pollutant from the process are the solvents themselves such as phenol, glycol and amines. Under ideal conditions, the solvents are continuously recirculated without draining to sewer.

Utilities Operations

The wastewater contain huge amounts of valuable hydrocarbons.

Characteristics of Refinery Wastewater

The concentration of various pollutants in the wastewater generated in a refinery depend on the sulfide and phenol content of the crude oil processed. The characteristics of Indian refinery's wastewater is presented in Tables 1 and 2.

Table 1: Wastewater Characteristics from Indian Refineries

Parameter	*Units*	*Concentration*		
		A	*B*	*C*
pH	—	7.2	7.4	7.5
Oil and Grease	mg/l	600	486	516
Phenol	mg/l	15	—	—
Sulfide	mg/l	80	47.7	—
BOD	mg/l	100	120	140

Table 2: Wastewater from Proposed Refinery in India

Parameter	*Units*	*Concentration*
pH	—	6-9
Temperature	°C	30-45
Suspended Solids	mg/l	400
BOD	mg/l	300
Free Oil	mg/l	3, 000
Emulsified Oil	mg/l	120
Sulfides	mg/l	220
Phenolic Compounds	mg/l	30

Refinery Wastewater Treatment

Stages

Three principal treatment stages of wastewater are:

1. Primary oil separation
2. Secondary oil separation with sulfide removal
3. Biological, physical, thermal, chemical and electro chemical treatment

Refinery Wastewater Treatment Technologies

Pollution prevention and zero discharge are the important considerations for the treatment of wastewater streams.

Fig. 1 summarizes the available technologies for wastewater treatment.

Biological treatment is excellent option for treating low concentration contaminant streams and large volumes. The efficiency for destruction of contaminants is however approximately 80%. Other treatment options such as electrochemical, air and steam stripping, carbon adsorption, thermal oxidation and various chemical oxidation processes are known to achieve 99% or more efficiency for destruction of organics and metal.

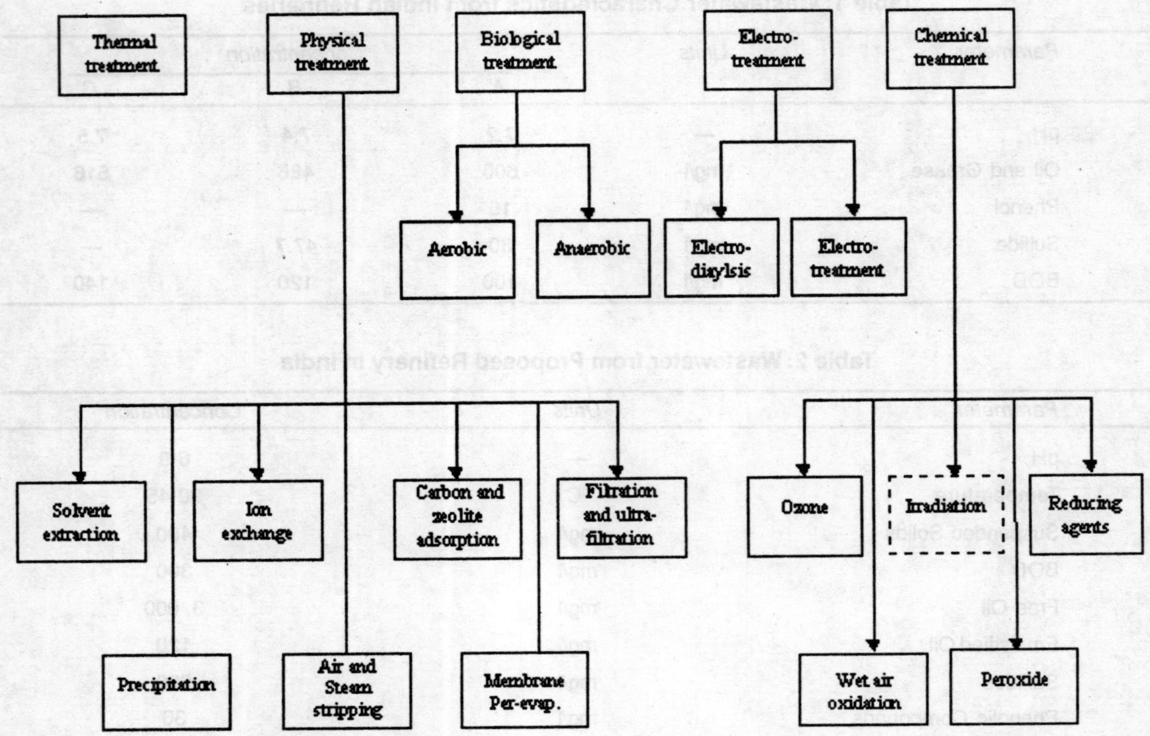

Fig. 1: Wastewater Treatment Technologies

Conclusion

The environmental regulations restrict the wastewater effluents from the petroleum refining industries. These regulations restrict the amount of chemical oxygen demand, metals, hazardous constituents and total dissolved solids discharged to the water ways. The various technological options presented in the communication may be a guiding factor for the industries to meet the regulatory requirements.

14

The Organic Effluents and Electrochemical Methods of Treatment

C. AHMED BASHA

Central Electrochemical Research Institute, Karaikudi - 630 006

Introduction

The spectacular advances in technology and quality of every day life unfortunately have come at the expense of destruction to our resource base and the environment.

The industrial activities of modern civilization are leading to an ever increasing number of new compounds, as well as increased concentrations of certain metals, inorganic, organometallic and organic substances in our environment[1].

Organic synthesis started around 1800; in 1890 some 12000 compounds were known; 1990:6-7 million compounds were synthesized and described including about 15000 organo-chlorinated hydrocarbons, 2000 fluoro-organic compounds, 2000 commercial dyes, 600 organo tin compounds, etc. 60-70 thousands compounds are in current use. The chemical industry produces about 15000 million tons of synthetic chemicals annually. According to the best estimate, 60-70 thousands inorganic (heavy metals) metalloids and organic compounds (pesticides, mineral oils, organic trace pollutants, etc.) and metabolites reach the environment via the atmosphere as exhaust gas or wastewaters and effluents. Under the best circumstances only a fraction of especially harmful 4000-6000 compounds are actually detected. For many of these compounds finding their way in the environment either during production, storage, transport, use or ultimate disposal, not much is known about their harmful effects on the environment.

In the recent years, there has been a dramatic increase in regulatory activities aimed at: safer manufacturing and safer disposal of chemicals. Regulations in themselves are not opposed by responsible industry provided they effectively serve certain needs and achieve their objective in cost efficiency.

In view of new and evermore stringent environmental regulations and rising costs, the following are a must for the (chemical) industry:

1. Avoidance/minimization of wastes, using new technologies with low or zero effluents, *e.g.*, electro synthesis
2. Utilization of wastes (recovery, recycling of wastes),
3. Ecologically acceptable waste disposal.

A Company's reputation can be severely damaged by mismanagement of its wastes.

Precipitation as hydroxides carbonates or sulfides or cementation of metals need large quantities of additional chemicals and produces a large amount of sludge, which must be contained.

Disposal through deposition after chemical or physical treatment is senseless, since these only results in a local change in the environmental problem, *e.g.* a transfer from wastewater into ground water via a leaking landfill.

The only long-term solution is recycling (recovery) and subsequent feedback into the process.

The electrochemical technologies are clean technologies and offer several promising approaches for the prevention and remedy of pollution problems (2-5). One of the strongest points in favour of electrochemical reductive removal of metal ions and metal complexes is that they are eliminated from the solution in the most useful form as pure metals, either as films or powders.

Multivalent species can be transferred by reductive or oxidative treatment to a less toxic valency, (*e.g.*Cr (VI) into Cr (III) or, when used as mediators in synthetic processes, regenerated to the state before use, *e.g.* Sn (II) /Sn (IV); Ce (III) /Ce (IV); Cr (III) /Cr (VI), etc. Finally, they can be changed to a valence state better suited for separation, *e.g.* for accumulation on an ion exchanger. Further, electrochemistry has also its part to play as treatment preceding a biological process to destroy refractory pollutants that bacteria have great difficulty in digesting.

To ensure the good operating condition of biological treatment installations, it may be practical to include in the industrial strategy an electrochemical pretreatment stage where the pollutants are degraded into smaller molecules that the bacteria can easily assimilate in purification station.

In practice, the following treatment scheme could be envisaged:

$$\text{(refractory compound)} \xrightarrow{\text{Electrical energy}} \text{(biodegradble compounds)} \rightarrow CO_2\text{+biomass}$$

The present paper summarizes the results of the investigation on electrochemical treatment of organic effluents from the following industries: Dye-house, Agro-based Pulp and paper, Tannery, Hydrogen peroxide plant and Organic Ion Exchange Resin Waste.

Electrochemical Treatment of Textile Dye-House Effluent

The textile dye industry poses a serious threat to the environment. The cotton yarns are dyed using Procion Blue reactive dye. This particular dye is obtained from the combination of 4 different dyes such as Reactive Black, Reactive Green, Reactive Navy-blue, and Reactive Yellow. 60% of reactive dyes applied on the cotton fabrics yarn is hydrolyzed and coming in the effluent. The dye house consists of two types of wastewaters via reactive dye-bath wastewater and rinse washwater. The reactive dye bath wastewater mainly consists of complex dyestuffs and various intermediate complexes. Before treatment the effluent from the dye-bath consist of COD 5000 – 6000 ppm, TDS 52, 000 ppm, SS 2, 000 ppm, pH 9.97, Chlorides 38, 995 ppm and its wash water consist of COD 400-560 ppm, TS 4000 ppm, TDS 3200 ppm, SS 800 ppm, Chlorides 1500ppm, pH 8. After dyeing the fabrics are washed by water to remove the excess dye present in the fabrics by rinsing. The wastewater generated due to this operation is commonly called as 'wash water'. The environmental pollution due to wash water is very less. But the pollution due to dye-bath wastewater is very high. The treatment of dye bath wastewater is also very tedious compared to wash water.

The present work investigates a method to remove colour and COD from the dye-house effluent by electro oxidation. The wastewater treatments involve the use of batch reactor, batch reactor with effluent recirculation and recycle reactor.

Batch Reactor

Investigations on experimental treatment of a mixed effluent (of dye-bath and washwater) directly from a textile dye-house containing 2000ppm of COD together with 35000ppm of chloride were carried out galvanostatically in a batch reactor at room temperature. Batch electrolytic undivided cell is used in the electrochemical oxidation process is shown in the Fig 1. The cell is provided with a cathode and an anode together with salt bridge with reference electrode. The cathode was a stainless steel plate used with Noble metal oxide, MO_x served as working electrode as an anode. Electrolysis was carried out at different current densities. Table 1 gives the result of the experiments, carried out.

Batch Recirculation Reactor

Recirculation reactor was nothing but the reactor that consists of plate and frame electrodes. This could also be called as filter press type cell as shown schematically in the Fig. 2. Since the electrolyte

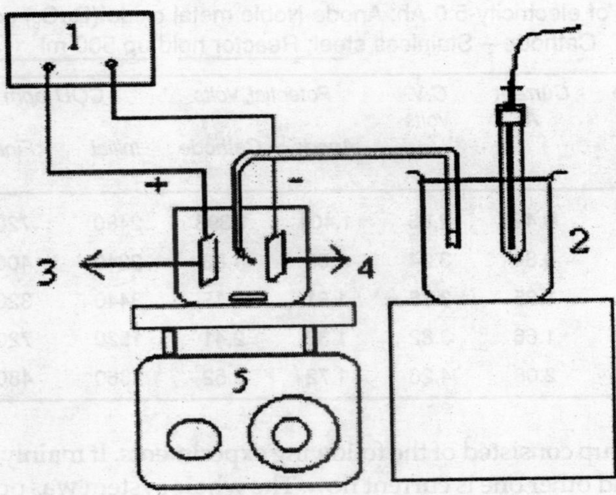

Fig. 1: Schematic Diagram of Experimental Setup Used for the Batch Electrolysis (1) Power supply, (2) Reference electrode, (3) Anode, (4) Cathode, (5) Magnetic stairrer

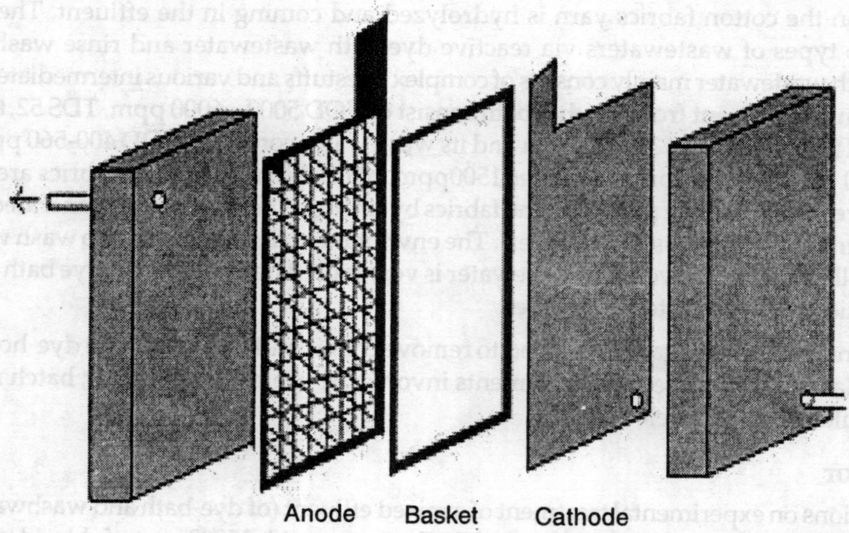

Fig. 2: Filter Press Cell

was continuously circulated through the reactor, it was named as the recirculation reactor. The reactor consisted of two-end frame made of frames made of PVC of 12.5 cm thickness. In between these two end frames were placed 1mm thick stainless steel plate act as cathode and 1mm thick expanded meshed frame, made of titanium coated with noble metal oxide (MO_x) act as anode. 1 cm thick spacing separated the electrode. Provisions for electrolyte entry had been made in PVC frames.

Table 1: Textile Dye-house Mixed Effluent-effect of Current Densities on COD Removal Experimental Batch Reactor Data on the Electrolysis

Total quantity of electricity-5.0 Ah; Anode-Noble metal oxide (RuO_2) coated on Ti; Cathode – Stainless steel; Reactor hold up 500 ml

Sl. No.	C.D A/dm²	Time hr	Current A	C.V. Volts	Potential, Volts		COD ppm		% Reduction of COD	Power Consumption kWh/kg COD
					Anode	Cathode	Initial	Final		
1.	1.0	12.0	0.42	2.85	1.404	1.398	2480	720	70.97	16.32
2.	2.0	6.00	0.84	3.24	1.39	1.83	2240	400	82.14	17.75
3.	3.0	4.00	1.25	3.75	1.61	2.15	3440	320	90.70	12.02
4.	4.0	3.00	1.66	3.82	1.33	2.41	1520	720	52.73	47.56
5.	5.0	2.40	2.08	4.26	1.72	2.52	1360	480	64.71	48.28

The experimental setup consisted of the following experiments. It mainly consisted of two parts. One is electrolyte flow and other one is current flow. The whole system was operated as recirculation reactor. This circuit consisted of the following equipments: Reservoir, Pump, Flow Meter, Reactor and Electrical or Current Circuit containing DC Power Supply together with Ammeter and Voltmeter. The experimental setup is shown in Fig. 3. Experiments were carried out under galvanostatic condition at

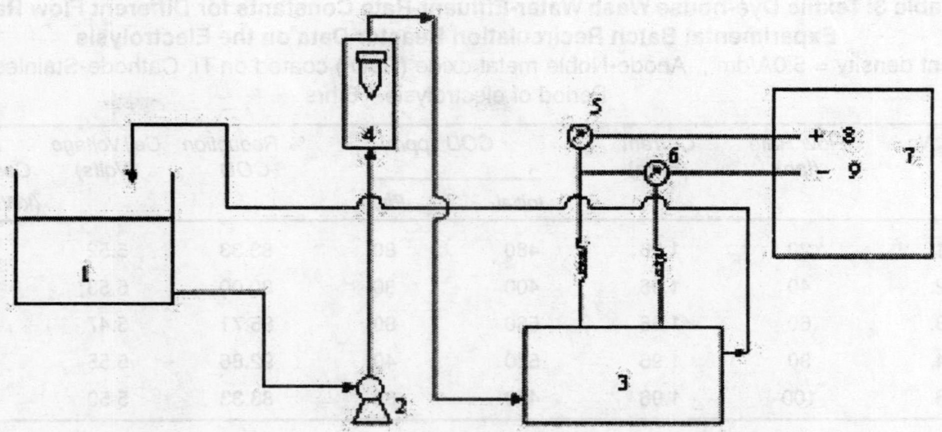

Fig. 3: Schematic Diagram of Batch Reactor with Electrolytic Recirculation (1) Reservoir, (2) Pump, (3) Reactor, (4) Flowmeter, (5) Ammeter, (6) Voltmeter, (7) Power supply, (8) Anode, (9) Cathode

specified current density and for the given volume effluent taken in the reservoir. The process of electrolysis was carried out at different flow rates of electrolyte per batch recirculation. The process of electrolysis was followed by continuous monitoring of variation of reactor voltage and COD values.

For the purpose of knowing the destruction of COD, samples were drawn at predetermined intervals. The samples were preserved with concentrated sulphuric acid COD and the analysis of COD was carried out as per ASTM standards. Tables 2-4 give the results of the experiment carried out for removal of colour and COD from mixed effluent and effluent of dye bath and wash water on the bench scale using galvanostatic electrolysis in an undivided electrochemical filter press cell in a batch recirculation system These studies envisage a method to remove the organic pollutants from the effluent by electro oxidative destruction of organic compounds.

Table 2: Textile Dye-house Mixed Effluent-Rate Constants for Different Flow Rates Experimental Batch Recirculation Reactor Data on the Electrolysis

Current density-2.0 A/dm^2; Total quantity of electricity – 5.0 Ah; Anode-Noble metal oxide (RuO_2) coated on Ti; Cathode – Stainless steel

Sl. No.	*Flow Rate lph*	*COD, ppm*		*% Reduction*	*kx10^3*
		Initial	*Final*	*of COD*	*(cm/sec)*
1.	20	2560	560	78.13	2.341
2.	40	1600	800	50.00	0.910
3.	60	1360	560	58.82	1.160
4.	80	1920	480	75.00	1.867

Table 3: Textile Dye-house Wash Water-Effluent-Rate Constants for Different Flow Rates Experimental Batch Recirculation Reactor Data on the Electrolysis

Current density = 5.0A/dm², Anode-Noble metal oxide (RuO^2) coated on Ti, Cathode-Stainless Steel; Period of electrolysis – 6 hrs

Sl. No	*Flow Rate (lph)*	*Current (Amp)*	*COD (ppm)*		*% Reduction COD*	*Cell Voltage (Volts)*	*Energy Consumption (kWh/kg COD)*
			Initial	*Final*			
1.	20	1.96	480	80	83.33	5.52	108.19
2.	40	1.96	400	80	80.00	5.53	135.48
3.	60	1.96	560	80	85.71	5.47	89.34
4.	80	1.96	560	40	92.86	5.55	83.67
5.	100	1.96	480	80	83.33	5.50	107.8

Table 4: Textile Dye-house Dye Bath Effluent-Effect of Flow Rate on COD Removal Experimental Batch Recirculation Reactor Data on the Electrolysis

Current density = 5.0A/dm²; Anode-Noble metal oxide (RuO^2) coated on Ti; Cathode –Stainless Steel; Period of electrolysis – 6hrs

Sl. No.	*Flow Rate (lph)*	*Cell Voltage (Volts)*	*COD (ppm)*		*%Reduction COD*	*Rate Constant kx10⁴ (sec⁻¹)*
			Initial	*Final*		
1.	20	3.02	5600	420	92.50	1.754
2.	40	3.04	5020	440	91.24	1.427
3.	60	3.07	5680	420	92.61	1.758
4.	80	3.13	6000	400	93.33	1.803
5.	100	3.12	5800	420	92.76	1.800

Electrochemical Treatment of Agro-Based Pulp and Paper Industry Effluent

The growth of pulp & paper industry is an index of the social, cultural, technical, industrial and economical development of a nation. The environmental pollution related with the pulp and paper industry is an appendage of its growth. Manufacture of paper required large volume of water (250 – 300m³/ton), the bulk of which is released as effluent. The present work has been an attempt to study the electro oxidative destruction as well as the decolorisation of the pulp and paper industry effluent to overcome the drawbacks of the existing treatment process.

The effluent for the experiments was taken from a pulp and paper mill. The effluent before treatment had COD of about 3800 ppm, pH of 8, chlorides of 573 mg/lit. To improve the conductivity of the effluent 4 gpl of NaCl was added to the effluent and this treatment process was done using batch reactor with electrolyte recirculation. In order to find out the optimum condition for the decolourisation as well as COD reduction, numbers of experiments have been carried out with different concentration of NaCl at constant current density and constant flow rate.

Till date, only chemical and biological treatments have been used to treat the industrial effluents. While conventional treatment methods lead to the formation of more sludge, which is a secondary pollutant, electrochemical method of treatment do not give such problems. The object of this work is to remove the COD and colour of the effluent.

The liquid waste from pulp and paper industry effluent is dark brown in colour was treated in batch reactor with electrolyte recirculation. The electrolyte (effluent) from a well-mixed reservoir is rapidly circulated through the reactor. The reactor setup was same as shown above in Fig 2 and Fig 3. The results of experiments carried out are given in the Tables 5-7.

Table 5: Agro-Based Pulp and Paper Industry Effluent: Influence of MEDIATOR (NaCl) on COD removal-Experimental Batch Recirculation Reactor Data on the Electrolysis

Reservoir volume: 1 liter; Reactor Holdup: 0.066 dm³; Anode: Lead dioxide (PbO_2) coated ox expanded Ti mesh.; Cathode: Stainless steel sheet; Flow rate of effluent: 20 LPH; Current: 2.5 A/dm²; Duration time: 8 Hrs.

Sl. No	Oxidant Amount (gpl)	Cell Voltage 'V'	COD (ppm) Initial	COD (ppm) Final	% Reduction of COD	Energy Consumption kwh/kgCOD	Average flux kgCOD/ Hrm²
1.	Nil	6.42	3840	2560	33.33	40.125	4.08×10^{-2}
2.	1.0	6.318	3680	2400	34.78	39.49	4.08×10^{-2}
3.	2.0	5.48	3520	2240	36.36	29.97	4.66×10^{-2}
4.	3.0	5.214	3200	1920	32.59	32.59	4.08×10^{-2}
5.	4.0	4.908	3584	1792	50.00	21.91	5.17×10^{-2}
6.	5.0	4.752	4640	2080	55.00	14.85	8.16×10^{-2}

Table 6: Agro-Based Pulp and Paper Industry Effluent: Influence of CURRENT DENSITY on COD Removal-Experimental Batch Recirculation Reactor Data on the Electrolysis

Reservoir volume: 1 litre; Flow rate of effluent: 20 LPH; Amount of oxidant (NaCl): 4 gpl; Duration: 8 hr

Sl. No.	C.D. A./dm²	Current A	C.V. Volts	COD (ppm) Initial	COD (ppm) Final	% Reduction COD	Power Consumption kWh/kgCOD	Average flux kgCOD/ Hr.m²
1.	1.0	0.4	3.545	3520	2080	40.9	7.88	4.59×10^{-2}
2.	2.0	0.8	4.558	3360	1760	47.6	18.232	5.1×10^{-2}
3.	3.0	1.2	5.304	3136	1568	50.0	32.476	5.0×10^{-2}
4.	4.0	1.6	5.972	3040	1280	57.9	43.43	5.61×10^{-2}
5.	5.0	2.0	6.630	3360	960	71.4	38.67	8.75×10^{-2}

Table 7: Agro-Based Pulp and Paper Industry Effluent: Influence of FLOW RATES on COD Removal-Experimental Batch Recirculation Reactor Data on the Electrolysis

Reservoir volume: 1 litre; Reactor volume: 0.066 dm³; Amount of oxidant (NaCl): 4 gpl; Current density: 2.5 A/dm²

Sl. No.	Flow Rate LPH	Duration Hr.	Cell Voltage V	COD (ppm) Initial	COD (ppm) Final	% Reduction COD	Power Consumption kwh/kgCOD	Average flux kgCOD/ Hr.m²
1.	20	7	4.93	1760	640	63.63	30.81	4.08×10^{-2}
2.	40	8	4.752	1725	800	54.0	41.0	3.0×10^{-2}
3.	60	6	4.96	1760	800	54.54	41.3	4.08×10^{-2}
4.	80	6	4.87	1824	912	50	32.09	3.8×10^{-2}
5.	100	8	4.69	1920	960	50	39.1	3.06×10^{-2}

Good results are observed for the removal of COD and colour by anodic oxidation using lead dioxide coated on expanded titanium mesh as anode and stainless steel sheet as cathode under galvanostatic process. In a batch reactor with electrolytic recirculation, the experiment was carried out for various current densities, flow rates and reservoir holdups. The results of the experiments provide strong evidence that this electrochemical treatment method found to be more efficient than that of the existing chemical treatment methods and further this electrochemical treatment does not produce any secondary pollution unlike the chemical treatment in practice.

Electrochemical Treatment of Tannery Effluent

There are nearly 3000 tanneries in operation all over India and over 60% of them are in Tamil Nadu. These tanneries generate about 1.5 billion litres/day of wastewater containing high concentrations of chlorides, ammonia, organic substances and hydrogen sulphide to be disposed on land or water bodies without proper treatment and these effluent causes severe damage to flora and fauna.

The current method of treatment by aerobic and anaerobic lagoons results in the fouling of the atmospheric due to obnoxious gases liberated during treatment and also their method has a long treatment time also due to slow reaction. The storage of effluent over long period is needed, which is considered as another disadvantage. Considering the heavy load of the pollutants in the huge volumes of effluents, there is a specific requirement of needs based technology, which will enable the proper treatment to the satisfactory level of the removal of pollutants. Unlike the conventional methods, the oxidation will be carried out electrolytically or in other words electro oxidation of the dissolved species will be carried out at anode. The basic idea in this process is any species that can be oxidized either chemically/biologically can very well be oxidized at an anode. The products of such oxidation of dissolved organics are CO_2 and water. On this count, investigation on electrochemical oxidation of tannery effluent with special emphasis on removal of COD and colour, has been carried out in a bench scale using batch reactor and batch recirculation reactor (in an undivided cell with two different anodes RuO_2/Ti and PbO_2/Ti).

The studies envisage a method to remove the pollutants from the tannery effluent by electro oxidative destruction of organic compounds from the effluent. In this study the influence of catalytic anode has been considered important since it is one of the important criteria for getting good result. The process of oxidation was followed by variation in pH, Temperature and COD. COD indicates the removal of the organic compounds from the system. The results are presented in the Tables 8-11.

Table 8a: Tannery Effluent from CETP-Effect of Current Densities on COD Removal Experimental Batch Reactor Data on the Electrolysis

Anode-Noble metal oxide (RuO_2) coated on Ti; Cathode – Stainless steel; Reactor hold up 400 ml

Sl. No.	*Current Density Adm⁻²*	*Current A*	*Time, Min*	*CV V*	*Electrode Potential*		*COD, ppm*		*%COD Removal*	*Energy Consumption Kwh/kg COD*
					Anode V	*Cathode V*	*Initial*	*Final*		
1.	1	0.248	1200	3.2	1.6	1.4	2400	80	96.66	17.1
2.	1.5	0.372	780	3.4	1.89	1.52	2000	240	88.00	23.4
3.	2.5	0.620	480	4.2	2.30	1.87	1920	240	87.50	31.0
4.	3.5	0.868	360	4.3	2.21	2.82	2000	240	88.00	32.0
5.	5	1.240	240	5.9	2.7	2.63	2240	240	89.00	36.6

Table 8.1b

Current, A	*0.248*	*0.372*	*0.620*	*0.868*	*1.240*
Time, Min			*COD ppm*		
0	2400	2000	1920	2000	2240
60	2080	1600	1600	880	1200
120	1760	1440	1280	640	1040
180	1440	1280	960	480	560
240	1280	1200	880	400	240
300	1120	1120	800	320	
360	040	1040	720	240	
420	960	480	240		
480	880	880			
540	720	800			
600	640	600			
660	560	520			
720	520	400			
780	480	240			
840	400				
900	320				
960	240				
1020	200				
1080	160				
1140	120				
1200	80				

Table 9: Tannery Effluent from CETP-Effect of Flow Rates on COD Removal-Experimental Batch Recirculation Reactor Data on the Electrolysis

Current density-2.5 A/dm^2; Volume of reservoir-2 liter;

Anode-Noble metal oxide (RuO_2) coated on Ti; Cathode – Stainless steel;

			COD ppm		
Time min.			*Flow Rate, LPH*		
	20	40	60	80	100
0	2400	2360	2320	2080	2240
60	1520	2160	2080	1824	2080
120	1200	1840	1920	1600	1680
180	1040	1600	1680	1360	1360
240	920	1360	1360	1120	880
300	800	1040	960	880	560
360	640	800	640	640	480
420	480	480	400	560	400
480	320			400	

Table 10a: Tannery Effluent-Effect of Current Densities & Electrode Material on COD Removal Experimental Batch Reactor Data on the Electrolysis

Anode-Noble metal oxide (RuO_2) coated on Ti; Cathode – Stainless steel; Reactor hold up 150ml

C.D (A/dm²)	*Current (A)*	*Duration (hrs)*	*C. V (V)*	*Electrode Potential (volts)*		*% COD Reduction*	*Power Consumption*
				Anode	*Cathode*		*Kwh/kg (COD)*
1	0.20	34.0	3.26	1.33	1.86	64.41	171.78
2	0.35	17.0	4.34	1.70	2.60	66.10	204.92
3	0.50	11.5	5.47	1.72	3.61	64.83	243.79
4	0.70	8.00	6.15	2.12	3.98	72.88	304.77
5	0.95	6.00	6.58	1.74	4.70	74.5	375.00

Table 10b: Anode-PbO_2 Coated on Ti

C.D (A/dm²)	*Current (A)*	*Duration (hrs)*	*C. V (V)*	*Electrode Potential (volts)*		*% COD Reduction*	*Power Consumption*
				Anode	*Cathode*		*Kwh/kg (COD)*
1	0.20	34.0	2.91	1.33	1.84	72.90	173.50
2	0.35	17.0	4.18	1.28	2.816	73.68	212.56
3	0.50	11.5	4.80	1.87	2.82	75.40	242.10
4	0.70	8.00	5.92	2.20	3.77	67.79	268.20
5	0.95	6.00	6.37	2.15	4.03	66.10	275.00

Table 11: Tannery Effluent from Finishing Unit-Effect of Flow Rates on COD Removal-Experimental Batch Recirculation Reactor Data on the Electrolysis

Current density-2.5 A/dm²; Volume of reservoir-2 liter; Anode-Noble metal oxide (RuO_2) coated on Ti; Cathode – Stainless steel; Duration of electrolysis-6 hrs; Conc. of NaCl added – 4 gpl

Sl. No.	*Flow Rate (!ph)*	*Cell Volt (V)*	*COD (ppm)*		*%COD Reduction*	*Power Consumption (kWh/kg COD)*	*Massflux (kgCOD/ hr.m²)*	*Rate Const.*
			Initial	*Final*				
1.	20	3.471	3840	2800	27.08	9.8122	0.0884	8.3109
2.	40	3.602	4320	3280	24.07	10.182	0.0884	6.762
3.	60	3.402	4000	3120	22	11.365	0.0748	5.6708
4.	80	3.31	4240	3440	18.86	12.164	0.0680	2.8358
5.	100	3.467	4160	3520	15.384	15.926	0.0554	4.0414

Electrochemical Treatment of Distillery Effluent

The wastewater discharged by distilleries producing industrial alcohol from sugarcane molasses posses problems of disposal to acceptable standards due to their high BOD, COD, colour, odour etc. In India, there are about 150 distilleries with an approximate total capacity of 900x10^6 liters of rectified spirit per annum. Majorities of them use molasses as the starting material in the production of alcohol. For every litre of alcohol produced about 15 to 20 liters of spent wash is discharged. The major cause of pollution is the presence of considerable amount of original matter both as suspended solids as well as dissolved solids. The allowable limit of BOD (5 days at 20°C) for disposal of effluent on land and into inland surface water is permitted up to the upper limit of 100ppm and 30ppm respectively.

The current method of treatment by aerobic and anaerobic lagoon results in the fouling of the atmospheric due to obnoxious gases liberated during treatment and also their method has long treatment time. Also due to slow reaction, the storage of effluent over along period is needed, which is considered as another disadvantage. The recent development included treatment by anaerobic digestion with a view to recover biogas for use. Use of up flow anaerobic filters has been found to improve the efficiency with a substantial reduction in pollution load. However, a process design and control criterion for full exploration of these reactors is yet to be developed. Secondly it is not possible to bring down the BOD/COD levels to acceptable standards by biological treatment alone.

The major objective is to establish the electrochemical oxidation for effluent treatment with special emphasis on removal COD and colour of distillery effluent. With this in view studies has been done. In order to find out the optimum for the decolourisation as well as COD reduction, number of experiments has been carried out with different concentration of NaCl at different current densities. The results are discussed in the present work Tables 12 and 13.

Table 12: Distillery Effluent-Effect of Concentration and Current Density Experimental Batch Reactor Data for the Electrolysis of Biologically Treated Wffluent

Anode: -RuO_2/Ti; Cathode: -Stainless Steel; Volume of the Electrolyte - 200ml

Sl. No.	Current Density (A/dm^2)	Amount of Nacl (g/l)	Current Amp	Duration (min)	C.V (V)	COD, ppm		% of Destruction	Energy consumed (KWh/Kg) COD
						Initial	Final		
1.	5	5	0.90	420	6.29	33200	2800	91.5	6.51
2.	5	4	0.90	420	6.62	32000	4000	81.5	7.44
3.	5	3	0.90	420	7.31	32800	8000	75.6	9.27
4.	5	2	0.90	420	8.45	32000	9500	70.3	11.83
5.	5	5	0.72	540	5.46	33400	6500	80.5	6.57
6.	4	5	0.72	540	6.54	34500	10000	71.0	8.64
7.	3	5	0.54	700	5.31	34000	12000	64.0	7.56
8.	2	5	0.36	1050	4.16	33800	13500	60.0	6.45

Table 13: Distillery Effluent-Effect of Concentration of Mediator Experimental Batch Reactor Data for the Electrolysis of Raw Effluent

Anode: -RuO_2/Ti; Cathode:--Stainless Steel;
Volume of the Electrolyte 200ml; Current Density: 5 (A/dm^2)

Sl. No.	Amt NaCl (g/l)	Current (Amp)	Duration (min)	C V (V)	COD, ppm		% Destruction	Energy Consumed (kWh/Kg) COD
					Initial	Final		
1.	5	0.90	840	5.45	92000	24000	73	5.04
2.	7	0.90	630	5.73	100000	20000	80	3.38
3.	9	0.90	600	4.94	120000	16000	86	2.13
4.	11	0.90	510	5.29	120000	12000	90	1.87

The results presented so far, the reaction appears to be dependent of NaCl concentration and also may consume considerable oxygen for the completion of the reaction. The increase of NaCl concentration enhances the elctrooxidation process. This is supported by the fact that the COD of the effluent is brought down considerably and the maximum % of destruction is achieved at higher NaCl concentration of 5gpl in the case of biologically treated effluent and 11 gpl in the case of raw effluent.

Further experiment was also conducted using the spent wash from distillery industry (raw effluent) which is dark brown in color was treated in batch reactor with electrolyte recirculation. The schematic diagram of experimental set up is shown in figure 2. The dimensions of the electrodes are 18cm × 18cm. The effluent was mixed with known amount of supporting electrolyte (NaCl) based on batch reactor data. The effluent of known volume for example 2 liters was taken in the reservoir. The reactor hold up is 486cm^3. The effluent treated in the reactor cell is recirculated to the reservoir. For reservoir volume of 2 liter the experiment was run at the flow rate of 20 LPH for three different NaCl concentrations. The results are sown in the Table 14.

Table 14: Distillery Effluent-Influence of Mediator Concentration on Cod Removal
Experimental Batch Recirculation Reactor Data on the Electrolysis of Raw Effluent

Reservoir Volume: 2 litres; Current Density: 5 A/dm^2; Flow Rate: 20 LPH;
Anode: -RuO_2/Ti; Cathode: -Stainless Steel

Sl. No.	*NaCl Concen-tration gpl*	*Duration Time Hr*	*Current A*	*C V Volts*	*COD (ppm)*		*% Reduc-tion COD*	*Power Consumption kWh/kg COD*	*Ave. flux kg COD/ Hr.m²*	*kx10⁶ cm/sec*
					Initial	*Final*				
1.	5	14	12.96	5.53	86756	12500	85.59	6.7561	0.5474	0.2429
2.	7	8	12.96	5.29	86756	25440	70.67	4.4720	0.5913	0.3649
3.	9	8	12.96	5.10	86756	21200	75.56	4.0329	0.6322	0.4369

Electrochemical Treatment of Effluent from Hydrogen Peroxide Plant

The present work was undertaken to investigate the problem of electrolytic effluent treatment plant of hydrogen peroxide.

Coming to the production of H_2O_2 there are several routes, one of which is the reduction oxidation process from anthraquinone in an organic solvent medium. After separation of the product the remaining work solution is returned back to the reactor for recycling. The main reaction is as follows:

$$EAQ + H_2 \xrightarrow{\text{Catalyst}} EAQH_2 + (O) \xrightarrow{\text{Air}} EAQ + H_2O_2$$

The effluent mainly contains the small carry over of the work solution, which may contain the organic reactants and degraded products of the reaction and solvent fraction. The nature of the contaminants is thus quinines, hydroquinones, esters, petroethers and highly aromatic in character. Since these are formed through reduction-oxidation cycles, it can be expected that the destruction through either reaction could be difficult, though not impossible. The main source for effluent is the wash water from the process and the composition of the effluent may vary very widely depending on even the slightest variation in the manufacturing process. Hence any process of treatment should necessarily take this point into account. In such cases the catalytic processes may be much use rather than direct oxidation/reduction processes.

The typical analysis of the effluent indicates very small amount of suspended solids, which are organic in nature and also water-soluble.

pH	2.7
Suspended solids	<0.2 mg/l
Oil and grease	<0.5 mg/l
Colour	Orange/brown
BOD	Not measured
COD	2900 mg/l
Phenolic compounds	43 mg/l
Cyanides	<0.01 mg/l
Ammoniacal Nitrogen	6.1 mg/l

Initially experiments were conducted in a static cell under galvanostatic conditions for the preliminary fixing of conditions of current density.

For any practical continuous working system dealing with large volumes of electrolyte flow, the conditions of flow, either batch recirculation or continuous flow will have to be made use of for any practical valuation. Hence further experiments were carried out in a batch-recycling mode at an average flow rate of 15-20l/hr. Both the results are sown in the Tables 15 and 16. The investigations showed that this effluent could be successfully treated through mediated electrooxidation process and the COD could be brought down from 2000 ppm to <100 ppm without andy sludge formation.

Table 15: Effluent from H_2O_2 Plant: Influence of Different Mediator COD Removal Experimental Batch Reactor Data on the Electrolysis

Current Density: 5 A/dm²; Anode: -RuO_2/Ti; Cathode: -Stainless Steel

Sl. No.	*Mediators*	*Volume (ml)*	*Q (AH)*	*CV (V)*	*COD (ppm)*		*%COD Reduction*
					Initial	*Final*	
1.	OCl	400	3.60	5.91-5.96	1020	256	74.90
2.	Redox	400	6.60	6.30-8.70	480	171	64.40
3.	OH	1000	4.80	4.98-5.40	1020	200	82.87

Table 16: Effluent from H_2O_2 Plant: Influence of Different Mediator COD Removal Experimental Batch Recirculation Reactor Data on the Electrolysis

Current Density: 5 A/dm²; Anode: -RuO_2/Ti; Cathode: -Stainless Steel

Sl. No.	*Flow Rate (l/Hr)*	*Q (AH)*	*CV (V)*	*COD (ppm)*		*%COD Reduction*
				Initial	*Final*	
1.	40	33.6	3.05-3.70	1168	530	54.62
2.	60	32.4	2.66-3.16	1168	420	64.04
3.	80	33.6	2.66-3.50	1168	350	70.03
4.	100	33.6	2.70-3.30	1168	300	74.31

Electrochemical Treatment of Organic Ion Exchange Resin Waste Effluent

Spent Organic nuclear grade resins get generated during purification of the processed streams in the main reactor and the auxiliary systems. These nuclear resins after uses are declared as organic radioactive waste. Being radioactive they need to be managed in a manner that they do not pose a long-term hazard to the environment.

Electrochemical oxidation is also a potential candidate to the achieve this mineralization and can be a adopted for solid resins as a direct feed or preheated resins in the form of "liquefied" resins using sub stoichiometric hydrogen peroxide.

In the present investigation a suitable electro-oxidative destruction processes for these organic resins (polystyrene based DVB cross linked) is identified. Mineralization is tried for synthetic cationic (Amberlite-IR 120) and anionic (Amberlite-IRA 400) resins. Batch setup with constant stirring and a continuous filter press type reactor under batch recirculation condition are tried with various current densities and flow rates using Medicated Electrochemical Oxidation (MEO) process with Fenton's mediator ($Fe^{2+}/Fe^{3+} + H_2O_2$). The electrodes used are stainless steel (cathode) and "inert anode" RuO_2/Ti. The results are given in the Tables 17–19. The results of the experiments shows that the best effect of TOC reduction is found to occur at 3.75 A/dm² in batch set up, and 10 A/dm² in flow set up with flow rate of 20 LPH

Table 17: Organic Ion Exchange Resin Waste Effluent-Effect of Current Density Experimental Batch Reactor Data on the Electrolysis

Electrolyte: Amberlite ® IRA 400 (Strong base styrenic anion exchange resin solution); Anode:RuO_2/Ti; Cathode: Stainless steel; Period of electrolysis: 7hrs.; Concentration of NaCl: 8 gpl; Volume of electrolyte: 200ml

Sl. No.	Current Density (A/dm²)	Cell Voltage (V)	pH	TOC (ppm) Initial	TOC (ppm) Final	%Reduction of TOC	Energy Concentration (kwhr/kg TOC)
1.	1.25	3.19-3.52	1.94-2.41	1579.4	1028.7	34.87	54.64
2.	2.50	4.92-5.58	2.04-8.04	1436.5	865.5	39.75	176.99
3.	3.75	5.63-6.86	2.01-8.18	1133.8	541.7	52.22	304.55
4.	5.00	7.14-8.17	1.98-8.38	1199.1	577.6	51.83	474.20

Table 18: Organic Ion Exchange Resin Waste Effluent Experimental Batch Reactor Data on the Electrolysis

Electrolyte: Amberlite ® IR 120 (Strong acid styrenic cation exchange resin solution); Anode: RuO_2/Ti; Cathode: Stainless steel; Duration of electrolysis: 7 hrs; Volume of electrolyte: 2 litres

Flow Rate (l/hr)	Current Density (A/dm²)	Concentration of NaCl (gpl)	Cell Voltage (V)	pH	TOC (ppm) Initial	TOC (ppm) Final	% Reduction of TOC
40	3.75	8	4.58-4.69	2.5-8.03	1160.7	866.9	25.31
60	3.75	8	4.44-4.52	2.49-8.16	1214.4	933.3	23.15
40	5.00	8	4.86-5.14	2.38-8.14	1387.9	1058.5	23.73
20	7.50	8	6.16-6.48	2.34-8.34	1304.4	1058.6	18.84
20	7.50	12	5.32-5.50	2.25-8.45	1227.6	984.5	19.80

The destruction of organic resin wastes by reaction with Fe (II) and NaCl in an undivided cell has been demonstrated. A series of experiments were also run to demonstrate that the organic ion-exchange resin wastes could be completely destroyed by such a process.

Experiments were ran in batch rector set up with constant stirring by varying current densities for different volumes of electrolytes investigating all the volumes of electrolytes, greater reduction of TOC was got at 3.75A/dm² current density.

Experiments were also ran by varying current densities and flow rates in filter press type continuous flow reactor and the optimum conversion (80% reduction of TOC) of ion-exchange organics to CO_2 was occurred only in 10A/dm² with flow rate of 20 LPH.

Reaction Modelling

Coming to the reaction mechanism there are two main application for the electrochemical treatment namely electrochemical conversion and electrochemical combustion. In the former, the non-biodegradable organics are transformed into biodegradable. The ideal electrode material for such reaction must have high electrochemical activity, for organic compounds. In the electrochemical combustion method the organics are completely oxidized to CO_2 and H_2O. Theoretically speaking, all organics are amenable to electrochemical oxidation either through conversion or combustion.

Table 19: Organic Ion Exchange Resin Waste Effluent-Experimental Batch Reactor Data on the Electrolysis-Results Obtained Using Fenton's as Mediator for Various Flow Rates

Electrolyte: Amberlite® IR 120 (Strong acid styrenic cation exchange resin solution);
Anode: RuO_2/Ti; Cathode: Stainless steel; Volume of electrolyte: ½ litre;
Concentration of NaCl: 5gpl; Concentration of Fenton's: 0. 37 gpl; Rate of H_2O_2 added: 1ml/hr;
H_2SO_4 added: 0.67ml (0.025N); Current density: 10 A/dm²

Sl. No.	*Flow Rate (LPH)*	*Cell Voltage (V)*	*pH*	*TOC (ppm)*		*%Reduction of TOC*	*Current Efficiency (%)*	*Energy Concentration kWhr/kg TOC*
				Inital	*Final*			
1	20	6.33-8.32	1.26-2.11	1706.5	338.4	80.17	5.31	335.81
2	40	7.35-8.78	1.29-2.03	1858.6	664.1	64.27	4.64	423.47
3	60	8.64-9.83	1.35-1.86	1951.8	785.4	59.76	4.53	434.51
4	80	5.06-6.82	1.26-2.41	1689.3	764.2	54.76	3.59	352.38
5	100	3.48-6.78	1.28-2.39	1703.4	834.7	51.76	3.48	314.33

The discharge of O_2 on anode in aqueous medium:

$$2H_2O \rightarrow O_2 + 4H^+ + 4e^- \qquad (1)$$

This was the subject to large number of studies and it is widely accepted at the 1st step in this oxidation is discharge of H_2O, which leads to hydroxyl radical at the electrode:

$$M\,(\,) + H_2O \rightarrow M\,(OH^\circ) + H^+ + e^- \qquad (2)$$

At a high over potential, this step is limiting in all cases. Further mòre, certain authors have suggested that adsorbed OH° radicals are the intermediates responsible for the transfer of O_2 in anodic oxidation of a great number of organic molecules:

$$M\,(OH^\circ) + R \rightarrow M\,(\,) + RO + H^+ + e^- \qquad (3)$$

The following competing reaction leads to oxygen evolution:

$$H_2O + M(OH^\circ) \rightarrow M(\ ) + O_2 + 3H^+ + 3e^- \quad (4)$$

where,

M ( ) represents a site that is deficient subsequent to the transfer of O_2 as in equation (3) or discharge of O_2 as per equation (4), repopulated with the oxygen (OH°) by discharge of H_2O as shown by equation (2).

The mechanism of combustion/conversion of organic on oxide anode, can generally be considered to take place through the formation of OH°; free radical, as the first step through discharge of H_2O (or) OH^- at the anode which are adsorbed on the electrode surface.

$$MO_x + H_2O \rightarrow MO_x(OH) + e^- + H^+ \quad (5)$$

The absorbed hydroxyl radicals interacts with the oxygen already present in the oxide anode with possible transition of oxygen from the adsorbed hydroxyl radicals to the lattice of the oxide anode forming the so called higher oxide.

$$MO_x(OH) \rightarrow MO_{x+1} + H^+ + e^- \quad (6)$$

Thus oxygen can be either in the physisorbed or chemisorbed state. In presence of oxidizable organic assumed that the physisorbed radicals OH° should cause predominantly complete degradation into CO_2, H_2O and H^+;

$$R + MO_x(OH^\circ)_2 \rightarrow RO_2 + 2H^+ + 2e^- + MO_x \quad (7)$$

On the other hand chemisorbed active oxygen participates in the formation of selective oxidation products.

$$R + MO_{x+1} \rightarrow MO_x + RO \quad (8)$$

For the degrading (combustion) of organic at a high concentration of OH°, radicals on the anode surface are necessary. This is the case when the rate of reaction [equation (5)] is faster than the oxygen transfer into oxide lattice equation (6), the current efficiency for the combustion of organic depends on the relative rate of the reaction [equation (7)] to the rate of the O_2 evolution.

In the absence of any oxidisable species the active O_2 reacts to produce dioxygen as follows:

$$MO_x(OH) \rightarrow \tfrac{1}{2}O_2 + H^+ + e^- + MO_x. \quad (9)$$

$$MO_{x+1} \rightarrow MO_x + \tfrac{1}{2}O_2. \quad (10)$$

It is very probable that dioxygen participate in the combustion of organic. First of all, organic radicals are formed by hydrogen abstraction, which react with the dioxygen formed at the anode according to the following reaction.

$$RH + OH^\circ \rightarrow R^\circ + H_2O \quad (11)$$

Reaction of the organic radical with dioxygen formed at the anode:

$$R^\circ + O_2 \rightarrow ROO^\circ \quad (12)$$

Further abstraction of a H_2 atom with the formation of an organic, results hydroperoxide (ROOH) and another organic radical.

$$ROO^\circ + R'H \rightarrow ROOH + R' \quad (13)$$

Since the organic hydroperoxides formed are relatively unstable, decomposition of such intermediates often leads to molecular breakdown and formation of subsequent intermediates with lower carbon members. These reactions continue rapidly until the formation of CO_2 and water. The rate of this final stage will depend on the nature of the organics and also the intermediates formed during sequence of reactions. The conversion from aromatics to aliphatics will be the most important step in the entire breaking down of complex organic so that effluents containing such organic may be treated for environmental safety.

$$(ArOH) + OH^\circ \rightarrow (\text{Aromatic Intermediates}) + OH^\circ$$

$$(\text{Aliphatic acids}) + OH^\circ \rightarrow CO_2 \quad (14)$$

Coming back to the role of OCl^-/Cl^- couple, the reaction can be through generation of the oxidized species in-situ of the anode. It is then interact with organic compounds to yield the reduced species of the couple along with formation of CO_2 and H_2O. In the situation where NaCl present in the effluent, the following reactions occur in the cell during electrolysis:

Anode

$$2Cl^- \rightarrow Cl_2 + 2e^- \quad (15)$$

$$4OH^- \rightarrow O_2 + H_2O + 4e^- \quad (16)$$

$$6OCl^- + 3H_2O \rightarrow 2ClO_3^- + 4HCl + 2H^+ + 3/2\,O_2 + 6e^- \quad (17)$$

Cathode

$$2H_2O + 2e^- \rightarrow H_2 + 2OH^- \quad (18)$$

Bulk

$$Cl_2 + H_2O \rightarrow HCl + HOCl \quad (19)$$

$$2HOCl + OCl^- \rightarrow ClO_3^- + 2Cl + 2H^+ \quad (20)$$

In the cell, anode and cathode products are not separated. The reactions are as follows:

$$Cl_2 + OH^- \rightarrow Cl^- + OCl^- + H_2O \quad (21)$$

Both the hypochlorite and the free chlorine are powerful oxidizing agents and are also chemically reactive.

Hence,

$$\text{Organic} + OCl^- \rightarrow CO_2 + N_2 + H_2O + Cl^- + \text{product.} \quad (22)$$

The variation of the composition of organic pollutants in a reactor is due to pertinent reactions. The pseudo steady state theory can be applied to each of the intermediate products taking part in the reaction. Theory is based on the concept that intermediate concentration are small, there is no significant accumulation of these components in the reactor. The rates of formation of these intermediates are practically identical to the rates of destruction of the organic pollutant present in the effluent. That is:

$$d(Cl_2)/dt = d(ClO^-)/dt = -d(E)/dt = 0 \quad (23)$$

where,

-d (E) dt is the rate of disappearance of organic pollutant or the effluent.

Assuming the TAFEL behaviour is applicable:

$$I/zFA = ka\,(Cl^-)_s \exp(bE)$$

$$= ka'\,(ClO^-) \exp(bE)$$

$$= ka''\,(E)_s \exp(bE) \quad (24)$$

It is important to note that generally waste streams contain pollutants at low concentration. This would mean that the reactive species viz the organic pollutant would have to migrate or diffuse to the anode where it would be oxidized. This diffusion process would be very likely to be slow and rate determining. The reactant Cl^- or organic pollutant of effluent is transported from bulk to electrode surface where it under goes conversion or and it may also be transported back to bulk by diffusion for reaction in the bulk. Then

$$I/zFA = k_L \{ (E)_b - (E)_s \} \quad (25)$$

Elimination of $(E)_s$ by containing eqn (A) & (B) result the eqn

$$I/zFA = k (E)_b = kC.$$

where,

$1/k = 1/k_L + 1/ka'\exp(bE)$

C: The bulk concentration of effluent

k_L: The mass transfer co-efficient

ka′: The electrochemical heterogeneous rate constant.

Reactor Modelling

The Pseudo steady state theory, which is based on the concept of no accumulation of the intermediate in the reactor, helps to obtain the rate of the reaction in the batch reactor with electrolytic circulation.

From the RTD studies and it can be assumed that the reactor is ideal CSTR and reactor is under steady state condition, and the performance equation can be written as

$$T_R = V_R/Q = (C\text{-}C') / (\text{-}r) \quad (26)$$

where,

(-r): The rate of disappearance of organic pollutants and $-r = k_2\,(C)\,S$

The rate of disappearance of organic pollutants is directly proportional to concentration of organic pollutants 'C' multiplied by the concentration of active sites (S), which is essentially a constant, S.

(-r) = kaC is a pseudo first order kinetic equation.

$$T_R = (C\text{-}C') / kaC' \quad (27)$$

where,

(-r) = kaC′

rate of disappearance of organic pollutants in reactor.

k = Electrochemical rate constant, cm/sec.

V_R = Active volume of reactor, liter.

Q = Volumetric flow rate of electrolyte, LPH.

A = Active total electrode area, dm^2.

T_R = Space time for the reactor (V_R/Q), hour.

C′ = Concentration of organic pollutants in leaving stream of reactor, ppm.

C = Concentration of organic pollutants in entering stream of reactor, ppm.

a = Specific electrode area (A/V_R), dm^2.

The equation can be written as:

$$C'/C = 1/(1 + kA/Q) \quad (28)$$

In batch recirculation reactor system, the mass balance can be written as

$$V\,dC/dt = Q(C' - C) \quad (29)$$

Substituting for C′ from the equation (28) and upon integration results

$$C_t/C_o = \exp\text{-}\{[kA/Q/(1 + kA/Q)]\,t\}/\tau \quad (30)$$

where,

C_o = initial concentration of organic pollutants in the reservoir.

C_t = the concentration of organic pollutants at time 't' in the reservoir

Q = Volumetric flow rate.

V = Volume of the reservoir.

τ = Space-time for the reservoir (V/Q).

Equation (5) can be written by taking logarithm as

$$\ln C_t/C_o = -kA/Q/(1 + kA/Q)\,t/\tau \quad (31)$$

Using the above equation the rate constant k was obtained which are given in the Tables.

Conclusions

As illustrated above, the idea here is to use an electrochemically generated redox reagent as a chemical reactant to less harmful products. The redox reagent acts then as an intermediary for shuttling electron between the pollutant substrate and the electrode, which can be electrochemically, generated either in a reversible or irreversible manner.

Oxide noble metal surfaces (Pt, Ir, Ru) are suitable for oxidation of organic species.

The results of the experiments provide strong evidence that this electrochemical treatment method found to be more efficient than that of the existing chemical treatment methods and further this electrochemical treatment does not produce any secondary pollution unlike the chemical treatment in practice.

References

(1) Pierre M. Bersier, 1995, Applied Electrochemistry for a Better Environment, *Chem. Listy*, 89: 742.

(2) R.L. Clarke, 1992, *Electrochemistry for a Cleaner Environment*, ed. J.D.Genders and N.L. Weinburg, The Electrosynthsis Co, Baffalo.

(3) *Electrochemical Engineering and the Environment*, 1992, Symposium series No.127, Institution of Chemical Engineers, UK.

(4) K. Scott, 1995, *Electrochemical Processes for Clean Technology*, The Royal Society of Chemistry, U K.

(5) K. Rajeshwar and J. Inbanez, 1997, *Environmental Electrochemistry-Fundamentals and Applications in Pollution Abatement*, Academic Press, Inc.

15

Removal of Fluoride from Fluoride Contaminated Industrial Wastewater by Electrolysis

Mrs. V.A. JOSHI & M.V. NANOTI

Scientists, NEERI, Nehru Marg, Nagpur - 440 020

ABSTRACT

Effluent discharged from fluoride process industries constitutes a potential source of fluoride contamination of surface water bodies. Wastewater containing fluoride are generally treated with lime or calcium salt supplemented with aluminium salts. Wastewater generated from different industries does not always behave in the same way due to the presence of interfering contaminants. A number of techniques have been developed and studied for the removal of excessive fluoride. Most of these are based on use of aluminium salt. In alum coagulation the sorption properties of product of hydrolysis of aluminium salts and capacity of fluoride for complex formation plays a very important role. These hydrolysis products of aluminium can be produced by passing direct current through aluminium electrode. The text presented in the paper deals with the various aspect of removal of fluoride by electrolysis using aluminium electrode from fluoride chemical based industrial wastewater. The studies had been carried out in spiked test water and industrial wastewater both. The studies on effect of current density pH and various cations and anions are carried out in spiked test water for optimisation of the technique.

***Keywords:** Fluoride in wastewater, electrolysis, aluminium electrode, current density, silica*

Introduction

Effluent discharged from fluoride process industries constitutes a potential source of fluoride contamination of surface water bodies. The industries which produce fluoride bearing wastewater are aluminium, steel, enamel, pottery, glass, bricks, welding, refrigeration, oil refinery, plastic, pharmaceutical, tooth paste, automobile, chemical, super phosphate and other fertilizers (Anon, 1972).

Wastewaters laden with fluoride are generally treated with lime or calcium salts sparingly supplemented with aluminium salt (Link & Robosky, 1978, Zabban, 1975).

Fluoride reacts with calcium to form sparingly soluble calcium fluoride precipitate. Solubility estimation reveals that, when fluoride and calcium ions are in stoichiometric amounts, 8.18 mg/L of fluoride remains in the solution. The concentration may be lower due to common ion effect. However, fluoride concentration higher than stoichiometric value is observed. Wastewaters in real life scenario are not always amenable to ideal theoretical situations, probably as a result of interfering chemical equilibrium and microbiological activities. A minimum desired fluoride level may not be consistently achieved in effluent treatment by calcium salt. The limit stipulated by Central and State Pollution Control Boards for discharge of effluent containing fluoride is 1.5 mg/L.

A number of techniques have been studied for the removal of excessive fluoride from water. Most of these technologies are based on use of aluminium salts of which alum coagulation (Nalgonda Technique) and activated alumina are extensively studied and field tested for removal of fluoride from drinking water to make it potable (Lee *et al.* 1995, Bulusu & Nawlakhe, 1990; Nawlakhe & Paramasivam, 1993). Aluminium plays a very important role in these technologies (Fatkullin, 1975). In alum coagulation, the sorption properties of product of hydrolysis of aluminium salts and capacity of fluoride for complex formation plays a very important role. These hydrolysis products of aluminium can also be produced by passing direct current through aluminium electrodes immersed in fluoride containing water. It can be successively used for removal of fluoride from water and wastewater. The studies on electrocoagulation process by using bipolar aluminium electrode have been carried out on natural water of North Africa (Mameri *et al*, 1998). The wastewater from different industries contains large number of cations and anion which affect, the process of removal.

The studies were carried out for removal of fluoride from fluoride chemical based industry wastewater using lime & alum and by passing direct current through aluminium electrodes immersed in wastewater containing fluoride. This paper deals with the laboratory studies conducted on removal of fluoride from wastewater by electrolysis using aluminium electrodes. Various aspects *viz.* effect of variation in fluoride concentration, discharge pattern of various manufacturing units, pH and interfering contaminants for achieving the residual fluoride 1-5 mg/L were studied.

Process of Electrolysis

The process of electrolysis in removal of fluoride from water proceeds in three stages. It comprise of following reaction:

Electrochemical Process

Reaction on Anode

$$Al \rightarrow Al^{3+} + 3e \; (0^{O} = -1.66V)$$

$$Al^{3+} + 6H_2O \rightarrow (Al\,(H_2O)_6)^{3+}$$

Side Reaction on the Anode

$$Cl^- + 2OH^- \rightarrow ClO^- + H_2O + 2e\ (0^o = 0.89V)$$

$$2\,Cl^- \rightarrow Cl_2 + 2e\ (0^o = 1.36V)$$

The product of side reaction is hypochlorite, which is a kind of strong oxidising agent being capable of oxidising fluoro-complex to intensify the adsorbing effect of fluoro complex.

Reaction on the Cathode

$$O_2 + 2H_2O + 4e \rightarrow 4OH^-\ (0^o = 0.401V)$$

Side reaction on the Cathode

$$2Al_3^{\ +} + 6H_2O \rightarrow 2Al\,(OH)_3 + 3H_2$$

Hydrogen generated during electrolysis comes out as minute bubbles and carries the flocs to the surface.

Coagulation, Adsorption and Formation of Fluoro-complex

During electrolysis, $(Al\,(H_2O)_6)^{3+}$ formed gets converted to $Al\,(OH)_3 \cdot (H_2O)_6$. In the process of coagulation mononuclear complexes form polynuclear complex through bridging and adsorb fluoro-complex and fluoride through Van Derwal gravitation, hydrogen bonding and physical & chemical adsorption.

$$(Al\,(H_2O)_6)^{3+} + H_2O == (Al\,(OH)\,(H_2O)_5)^{2+} + H_3O^+$$

$$2\,(Al\,(OH)\,(H_2O)_5)^{2+} == (Al_2\,(OH)_2\,(H_2O)_8)^{4+} + 2H_2O$$

Mobility of Fluoride in the Electric Field

When direct current is passed, fluoride ions present in water moves toward the anode where it reacts with aluminium oxyhydrate $(Al\,(OH)_3 \cdot (H_2O)_6)$ resulting in the formation of tri-sodium aluminium hexafluoride.

$$Al\,(OH)_3 + 6NaF \rightarrow Na_3AlF_6 + 3\,NaOH$$

According to Ostwald effect, tiny crystals first dissolves and again reseparates out forming the large crystals and finally separate out. This separation depends upon the concentration of the sodium aluminium fluoride core and the whole process is indicated by the Feike law as follows:

$$Ji = -Di\,\frac{\delta Ci}{\delta Xi}$$

where,

Ji: Dispersing Speed

Di: Dispersion Constant, and

$$\text{Concentration Grade} = \frac{\delta Ci}{\delta Xi}$$

With the increase of the concentration of sodium aluminium fluoride, the dispersing speed increases, results in quickening the separation of crystallite (Sun Li Cheng, 1985).

In view of greater efficiency of hydroxides of aluminium produced by electrolytic methods, it is feasible to utilize this method for the water treatment containing high fluoride, (Bychin & Smetankine, 1977; Novikov, GI *et al.* 1981). However, extensive adoption of this method is restricted due to the high consumption of metallic aluminium and high energy cost.

Material and Method

The studies on fluoride removal to residual fluoride 1.5 mg/L were carried out in wastewater generated during manufacture of fluoride base chemicals. The main fluoride chemicals manufactured in the industry selected for investigation, were sodium aluminium fluoride (cryolite), aluminium fluoride, hydrofluoric acid and brominated fluorocarbons.

The reactions in the manufacture of hydrofluoric acid, aluminium fluoride and cryolite are:

$$CaF_2 + H_2SO_4 \rightarrow CaSO_4 + 2\,HF$$

$$Al\,(OH)_3 + 3\,HF \rightarrow AlF_3 + 3\,H_2O$$

$$NaCl + HF \rightarrow NaF + HCl$$

$$3\,NaF + AlF_3 \rightarrow Na_3AlF_6$$

The effluent from cryolite plant contains generated dilute hydrochloric mixed along with residual hydrofluoric acid. The effluent is partially neutralised by passing over limestone pits and flows to combined effluent collection tank (CECT) which serves as equalisation chamber. The effluent from manufacturing units is collected in same CECT. The combined effluent treated with lime was subjected to conventional treatment of flocculation and sedimentation. The treated effluent contained 7.8-16.8 mg/L of soluble fluoride which was higher than the limit 1.5 mg/L stipulated by Central and State Pollution Control Board (GPCB). The flow sheet of the treatment units is depicted in Fig. 1.

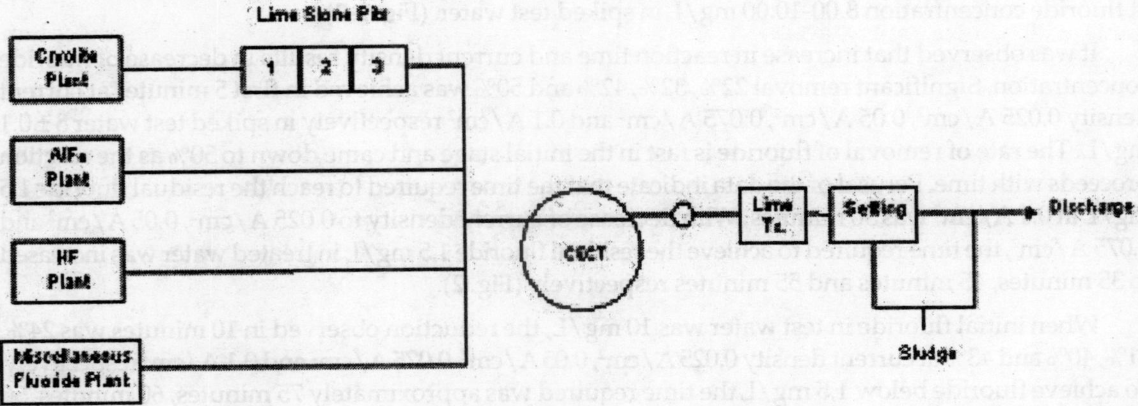

Fig. 1: Process and ETP Flowshed of Fluoride Chemical Industry

Twenty four hours flow proportional composite samples were collected from main chemical manufacturing units and each stage of wastewater treatment for physico-chemical parameters.

The laboratory experiments for removal of fluoride from spiked tap water and wastewater after neutralization from the fluoride chemical industry were carried out in batch mode with plate type aluminium electrode. Agronic-21 direct current unit was used for passing the direct current. The spiked test water of fluoride concentration 8-10 mg/L was prepared by adding sodium fluoride in tap water. The studies on spiked water were carried out at 0.025 A/cm^2-0.1 A/cm^2 current density at various time intervals to reduce the fluoride concentrated below 1.5 mg/L. The spiked test water containing different concentration of chloride, sulphate, calcium, magnesium and silica were prepared by using the salts of respective cation and anion to see the effect of various cations and anion, on removal of fluoride. The effect of pH on the formation of fluoro-aluminium complexes, its adsorption on aluminium hydroxide flocs formed during electrolysis is studied with spiked test water of fluoride 10 mg/L at pH 3.0-9.0. The pH was adjusted by using Hydrochloric acid and sodium hydrochloric The reaction period of 60 minutes with current density 0.05 A/cm^2 was constant during the studies on effect of pH. The comparative performance was studied with industrial wastewater containing the same concentration of fluoride. The fluoride analysis was done by specific ion meter (Orion) using fluoride and reference electrode. TISAB III buffer was used to maintain the pH 5.0-5.5 during analysis.

Result and Discussion

The physico-chemical characteristics of spiked tap water and fluoride chemical industry wastewaters are given in Table 1-2. Perusal of the data regarding the wastewater quality of various plants, *viz.* Cryolite, aluminium fluoride and hydrogen fluoride indicate fluoride concentration 4900-16400 mg/L, 750-3050 mg/L and 400-1100 mg/L respectively. The concentration of fluoride in combined effluent collection tank (CECT) varies 750-3350 mg/L. The variation in fluoride and other physico-chemical parameter in CECT depends upon the discharge of effluent from the individual plants but is mainly affected by the discharge from the cryolite plant. The effluent is allowed to detain in CECT for approximately 16 hours and then sent to neutralization plant for neutralization with lime. After neutralization the soluble fluoride concentration reduced to 7.8-16.8 mg/L which was above the guidelines of Pollution Control Board 1.5 mg F/L to discharge in water bodies (Table 2). The studies were carried out to see the effect of current density 0.025 A/cm^2-0.1 A/cm^2 on fluoride removal at fluoride concentration 8.00-10.00 mg/L in spiked test water. (Fig. 2-3).

It was observed that increase in reaction time and current density results in decrease of fluoride concentration. Significant removal 22%, 32%, 42% and 50% was achieved in first 5 minutes at current density 0.025 A/cm^2, 0.05 A/cm^2, 0.075 A/cm^2 and 0.1 A/cm^2 respectively in spiked test water 8 $\pm$ 0.1 mg/L. The rate of removal of fluoride is fast in the initial stage and came down to 50% as the reaction proceeds with time. Perusal of the data indicate that the time required to reach the residual fluoride 1.5 mg/L at 0.1 A/cm^2 was 30 minutes. With decrease of current density to 0.025 A/cm^2, 0.05 A/cm^2 and 0.075 A/cm^2, the time required to achieve the residual fluoride 1.5 mg/L in treated water was increased to 35 minutes, 45 minutes and 55 minutes respectively (Fig. 2).

When initial fluoride in test water was 10 mg/L, the reduction observed in 10 minutes was 24%, 31%, 40% and 43% at current density 0.025 A/cm^2, 0.05 A/cm^2, 0.075 A/cm^2 and 0.1 A/cm^2 respectively. To achieve fluoride below 1.5 mg/L the time required was approximately 75 minutes, 60 minutes, 50 minutes and 45 minutes at current density 0.025, 0.05, 0.075 and 0.1 A/cm^2 respectively (Fig. 3).

The studies on effect of various cations and anion was carried out in batch mode by keeping the reaction period 60 minutes, initial fluoride 10 mg/L and current density 0.05 A/cm^2 constant. It was also observed that the concentration of calcium, magnesium, sodium, potassium, sulphate, chloride and nitrate does not affect the removal of fluoride by electrolytic system. Silica has adverse effect on removal of fluoride. When the silica was 5 mg/L, 90% removal of fluoride was observed. The removal decreased to 47%, when silica concentration was increased 75 mg/L for same reaction period (Fig. 4). It may be because the aluminium ions generated during electrolysis may be reacting with silica to form

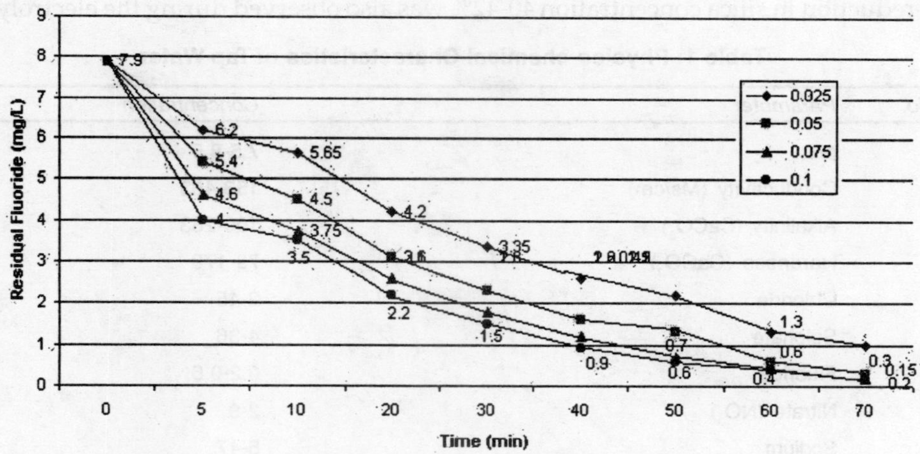

Fig. 2: Fluoride Reduction with Time at Various Current Density Spiked Test Water

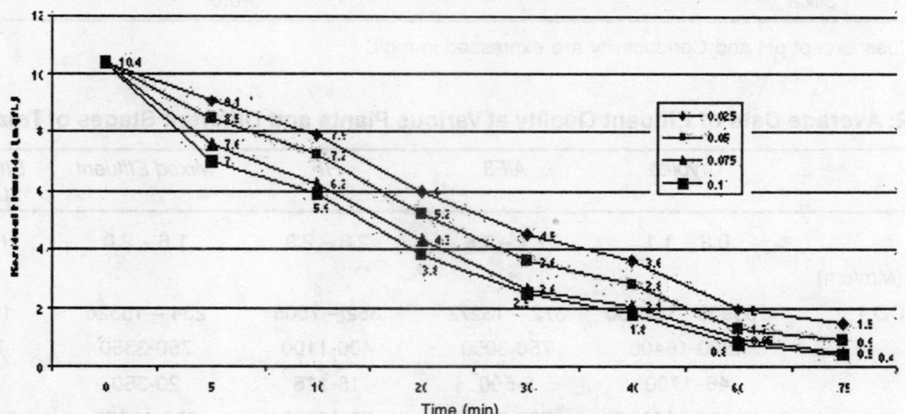

Fig. 3: Reduction in Fluoride with Time at Various Current Density Spiked Test Water

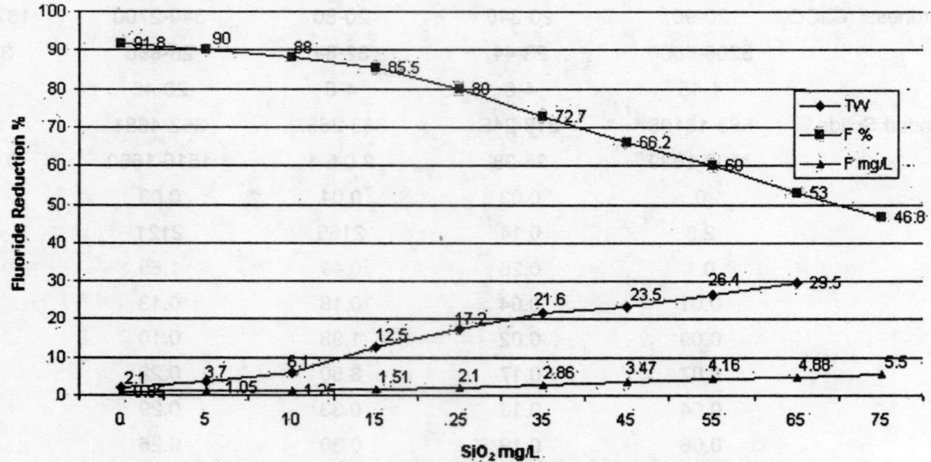

Fig. 4: Effect of SiO_2 on Fluoride Removal in Spiked Test Water

aluminium silicate results in increase consumption of aluminium whereby increasing the reaction time. The reduction in silica concentration 40-42% was also observed during the electrolysis.

Table 1: Physico-chemical Characteristics of Tap Water

Sl. No.	*Parameter*	*Concentration*
1.	pH	7.5-8.5
2.	Conductivity (Ms/cm)	153-463
3.	Alkalinity ($CaCO_3$)	100-263
4.	Hardness ($CaCO_3$)	79-170
5.	Chloride	2-45
6.	Sulphate	4-36
7.	Fluoride	0.2-0.6
8.	Nitrate (NO_3)	2-9
9.	Sodium	6-47
10	Potassium	1.3-6.0
11.	Silica.	3-8.0

All values except pH and Conductivity are expressed in mg/L

Table 2: Average Data of Effluent Quality at Various Plants and Different Stages of Treatment

Parameters	*Cryolite*	*AIF3*	*HF*	*Mixed Effluent*	*Effluent After Utilisation*
pH	0.9 – 1.1	1.1 – 2.8	2.3 – 2.9	1.6 – 2.0	6.9 – 7.5
Conductivity (Mm/cm)					
Acid/alk. ($CaCO_3$)	165900 –192760	372 – 13272	352 – 7505	234 – 15326	108 – 158
Fluoride	4900-16400	750-3050	400-1100	750-3350	7.8-16.8
Sulphate	46–1700	5-500	18-115	20-360	34-320
Chloride	70000-111000	580-5000	70-14500	580-19250	3700-6900
Total Hardness ($CaCO_3$)	2900-3600	730-3000	350-3000	730-7100	2330-8300
Calcium Hardness ($CaCO_3$)	20-90	20-340	20-80	340-2700	1670-4900
Sodium	5200-6000	23-44	32-34	26-850	30-315
Potassium	4-45	4-6	4-8	20-45	30-60
Total Suspended Solids	565-16198	218-245	363-2667	362-4681	1-183
Aluminium	1500-1689	25-38	2.0-5.1	1516-1660	0.08-1.5
Cadmium	0	0.03	0.04	0.03	ND
Iron	2.8	0.14	2166	2121	0.10
Manganese	0.1	0.26	0.49	1.69	0.08
Chromium	0.01	0.04	0.18	0.13	0.10
Copper	0.09	0.02	1.98	0.10	ND
Lead	0.07	0.17	3.50	0.22	0.15
Nickel	0.04	0.13	0.33	0.29	0.05
Cobalt	0.06	0.19	0.30	0.26	0.21
Zinc	0.02	0.03	0.26	0.13	0.06

All values except pH and Conductivity are expressed in mg/L

The studies on effect of pH indicated 90-96% removal of fluoride at pH 5.6-7.6 with maximum efficiency 96% at pH 6.5. The fluoride reduction 31-66% is seen at pH 4.4 at current density 0.025-0.1 A/cm^2 which further decrease to 13-47%, when pH was reduced to 3.0. Same trend is observed when pH was increased from 7.6 to 8.1 and 9.0 (Fig. 5).

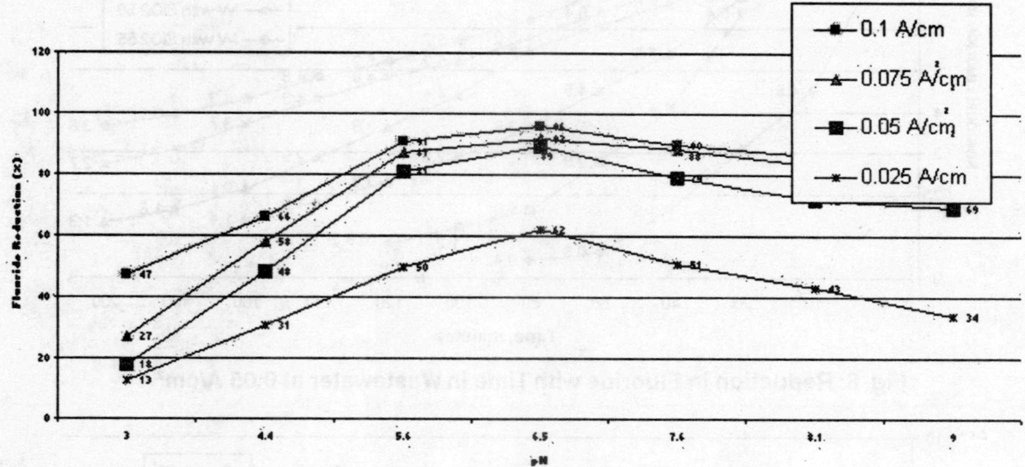

Fig. 5: Effect of pH of Fluoride Reduction

The studies carried out with the fluoride industry wastewater indicate that for same concentration of fluoride the time required to achieve the residual fluoride 1.5 mg/L is more for wastewater than spiked test water. It also varied with quality of wastewater with respect to fluoride and SiO_2 concentration.

It was observed that the time required to achieve fluoride below 1.5 mg/L at 0.05 A/cm^2 was approximately 45 and 55 minutes for spiked tap water with fluoride concentration 7.9 and 10.4 mg/L respectively whereas it was 85 minutes with wastewater of 7.7 mg/L fluoride concentration (Fig. 6). Perusal of physico-chemical characteristics of spiked test water and industry wastewater indicated silica and fluoride 5 mg/L, 7.9 mg/L and 45 mg/L, 7.7 mg/L respectively. Though the concentration of fluoride was nearly same, the time required to treat the wastewater to residual fluoride 1.5 mg/L was 40 minutes more with same experimental conditions. It may be because presence of silica has adverse effect on removal of fluoride. The concentration of silica observed in wastewater was 45 mg/L. These results were further confirmed by taking 4 more batches of wastewater with fluoride 7.7 mg/L, 8.3 mg/L, 8.5 mg/L, 9.1 mg/L and silica 55 mg/L, 52 mg/L, 60 mg/L respectively. It was observed that desired residual fluoride concentration 1.5 mg/L could be achieved in approximately 155 min. and 190 minutes in wastewater containing fluoride 7.7 mg with silica 55 mg and fluoride 8.3 mg/L with silica 52 mg/L at 0.05 A/cm^2 current density. It was also observed that fluoride reduction was 8.5 and 9.1 mg/L to 2.7 and 3.6 mg/L respectively at 0.05 A/cm^2 in 200 minutes (Fig. 6).

The studies carried out at 0.1 A/cm^2 indicated residual fluoride 1.5 mg/L in 75 minutes, 120 minutes and 160 minutes, when wastewater fluoride was 7.7 mg/L, 8.3 mg/L and 9.1 mg/L respectively (Fig. 7).

The reduction in silica was also observed during electrolysis. The initial concentration of silica was 51 mg/L, which was reduced to 8.2 mg/L after 140 minutes (Fig 8). It indicated that silica present in the wastewater was reacting with aluminium and forms insoluble complexes, which was settled with aluminium hydroxide flocs at the end of electrolysis.

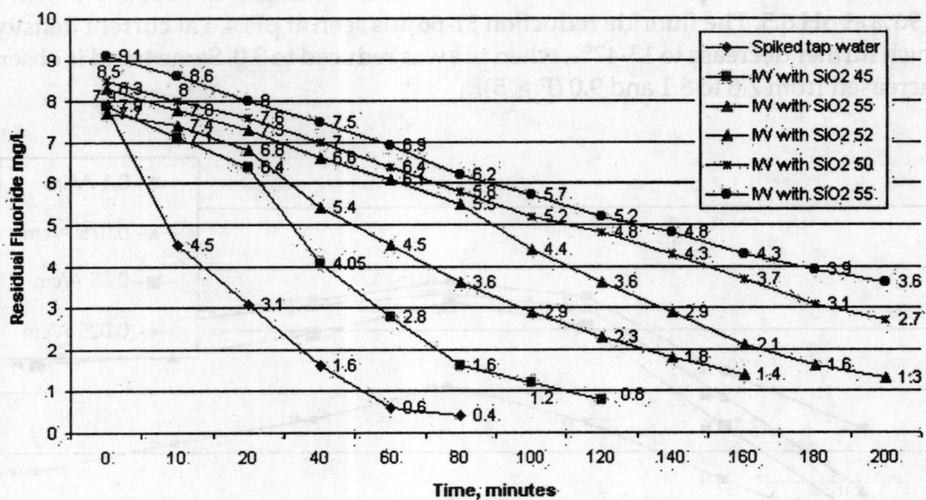

Fig. 6: Reduction in Fluoride with Time in Wastewater at 0.05 A/cm²

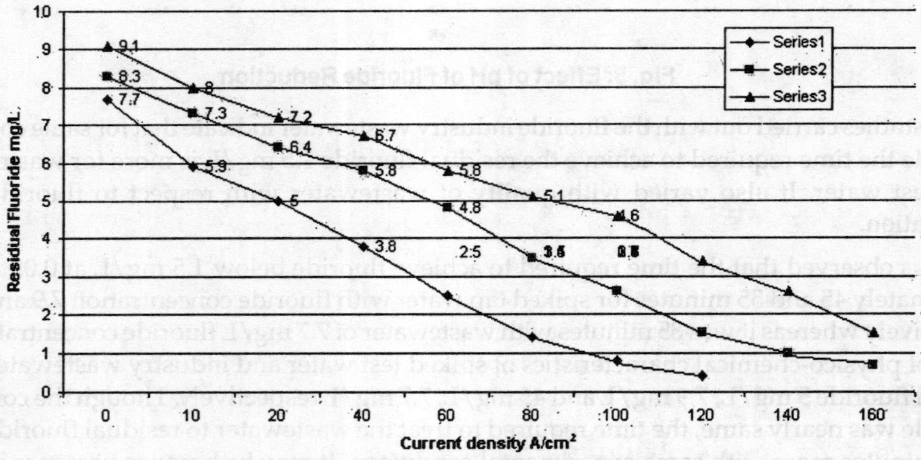

Fig. 7: Reduction in Fluoride in Wastewater at 0.01 A/cm²

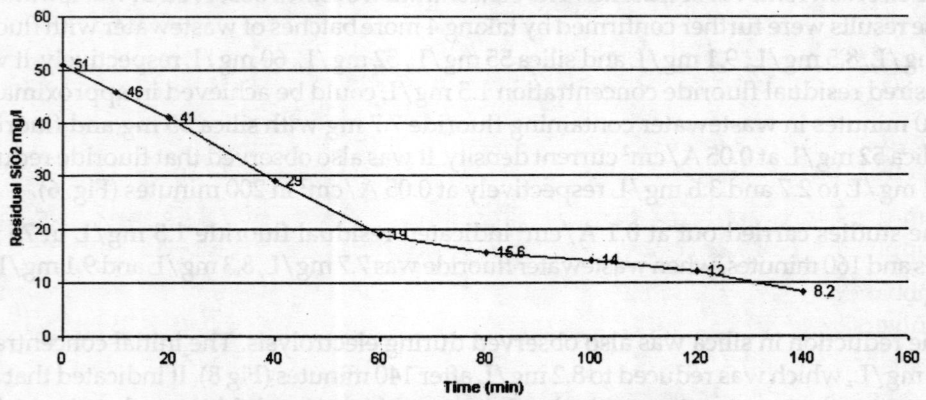

Fig. 8: Reduction in SiO_2 in Wastewater Electrolysis

Conclusion

As the quality of the waste influence the removal efficiency of the system, efforts need be made to keep the quality of wastewater nearly uniform particularly with respect to silica and fluoride. The wastewater from cryolite plant contained hydrochloric acid generated during the process along with residual hydrofluoric acid. To reduce the cost of neutralization it is passed through lime stone pit for partial neutralization where the silica present in lime stone get dissolved. It results in increase concentration of silica in wastewater. Mixing the fixed ratio of cryolite plant effluent in the effluent of other manufacturing unit and avoiding use of limestone for the neutralization may help in achieving the uniform quality of wastewater to be treated by electrolysis. As the electrolytic defluoridation is most effective in pH 6.5-7.00 the excess addition of lime need be avoided. The studies carried out indicate that the removal of fluoride is directly proportional to the current density and the time. With the increased of fluoride concentration the time and current density need be increased.

The studies carried out indicate that the electrolytic method can be used for the removal of fluoride from lime treated wastewater. Further studies are needed on use of this technique for higher concentration.

Acknowledgement

The authors express their sincere gratitude to the Director, National Environmental Engineering Research Institute (NEERI), Nagpur-440 020 (India) for according permission to publish this paper.

REFERENCES

1. Anon , 1972, Treating lead and fluoride wastes, Environmental Science & Technology, 6: 4.
2. Bulusu, K.R. and Nawlakhe, W.G., 1990, Defluridation of water with activated alumina: Continuous contacting system, Indian J. Environ. Health, 32, 197-218
3. Bychin, N.A. and Smetankiha, V.I., 1977, A study of methods of removing fluorine from water, Physical Hydrodynamics Visnch Shkola, Kiev-Donetsk
4. Fatkulin, I.G., Berezyuk, V.G. and Pushkarev, V.V., 1975, Coprecipitation of small quantity of fluoride with aluminium hydroxide precipitate, Zhurh, Pnkl, Khimii, 48 (7): 1428-31
5. Lee, D.R., Hargreaves, J.M., Badertocher, L., Rein L. and Kassir, F., 1995, Reverse Osmosis and activated alumina water treatment plant for the California State prisons located near Blythe IAHS Publ. 233, 155-161
6. Link, W.E. and Robosky, J.G., 1978, Fluoride removal from wastewater employing calcium precipitation and iron salt coagulation, Proc. Ind. Waste Conf., 485-499
7. Nawlakhe, W.G. and Paramasivam, R., 1993, Defluoridation of potable water by Nalgonda Technique, Curr. Sci. 65 (10): 743-748
8. Novikov, G.I., Romanov, A.M., Drodina, R.V. *et al.*, 1981, "Choosing the optimal parameters for electrocoagulation removal of fluoride from natural water VINITI No. 2210-81.
9. Sun Li Cheng, 1985, , Electrochemical method to remove fluorine from drinking water, Water Supply, 3, 177-186.
10. Zabban, W & Helwick, R.J., 1975, Defluoridation of Wastewater, Proc. 30th Ind. Waste Conf. Purdue University, 479-492.

16

Rate Enhancing Effect of Bromide on the Removal of Ammonia at TI/PBO$_2$ Anode

N.N. RAO*, DEEPAK J. KILLEDAR & S.N. KAUL***

**Wastewater Technology Division, National Environmental Engineering Research Institute, Nehru Marg, Nagpur – 440 020*

***Department of Civil Engineering & Applied Mechanics, S.G.S.I.T.S., Indore*

ABSTRACT

Electrochemical oxidation of ammonium (NH_4^+) at Ti/PbO$_2$ anode in chloride-free and chloride containing aqueous solutions was studied. Some experiments were carried out in the presence of bromide salt instead of NaCl. This was done with the objective of knowing whether bromide present in seawater exerts any effect on the ammonium removal kinetics. We noted that ammonium solutions were prepared in seawater or wastewater is mixed with seawater in many reported studies in this line. The experimental results have shown that oxidation of NH_4^+ is ensured only in the presence of chloride. The oxidation of NH_4^+ is found to follow pseudo first-order kinetics. Bromide in the electrolyte is found to cause substantial rate enhancing effect on the ammonium removal. The first-order rate constant (k, min^{-1}) for the ammonium oxidation in the presence of bromide is 4-7 times greater. The relative ease with which bromide is oxidised to bromine together with the greater reactivity of HOBr formed there upon appeared to account for the rate enhancing effect of Br$^-$. The dependence of k on the current density as well as halide concentration is also discussed in this study.

Introduction

Ammonia and nitrite are amongst the most commonly found pollutants in wastewater as well as in other surface water bodies. Their presence in water can cause a number of problems, but the major difficulty is their oxidation into nitrate (eq.1&2) that requires significant amount of oxygen. These reactions lead to depletion of dissolved oxygen concentration in water body. This is highly undesirable.

$$NH_4^+ + 2\,O_2 \rightarrow NO_3^- + H_2O + 2\,H^+ \qquad (1)$$

$$NO_2^- + O_2 \rightarrow 2\,NO_3 \qquad (2)$$

Further, the concentration of these nitrogenous compounds can exceed the levels that are toxic to aquatic life. Therefore, removal of nitrite and ammonia from water bodies assumes great practical importance.

Nitrification/denitrification using fixed film or fluidised bed biological reactors, adsorption by activated carbon and removal of suspended solids by sedimentation/filtration (1) are the conventional approaches for the treatment of nitrite and ammonia containing water. The study of simultaneous removal of carbon and nitrogen in a mixed culture aerobic RBC biofilm has been recently reported (2-5). Ozonation has been reported for a similar purpose notwithstanding the fact that it incurs additional high costs for ozone generation (6). A new concept based on ion exchange over zeolite and its bioregeneration has been reported for removal of ammonia present in secondary effluent (7). Another technique with great application potential is electrochemical oxidation. Electrochemical methods have been successfully applied for the treatment of wastewater (8-12), landfill leachate (13), domestic wastewater (14, 15) and tannery wastes (16-18). Many pollutants including ammonia present in water can be removed by oxidation at anode.

The electrochemical method has been applied for the treatment of domestic wastewater mixed with seawater (14, 19). This treatment resulted in a very high elimination of ammonia. Szpyrkowicz and Grandi (20) have demonstrated the feasibility to treat tannery wastewater using graphite anode and a steel cathode. In this system ammonium was fully eliminated with 19.1% current efficiency. The electrochemical treatment of tannery wastewater has been projected as an alternative to a step of biological nitrification-denitrification (16). They found that ammonium was more effectively removed at Ti/Pt/Ir electrode in comparison with Ti/Pt, but use of Ti/Pt was suggested by them because it is less expensive. Further, both direct and indirect oxidation of ammonium was reported to occur on Ti/Pt/Ir anode, while only indirect oxidation is reported on Ti/Pt. It is important to note that the tannery wastewater used in these experiments contained about 2500 mg/l chloride. Cossu *et al.* (21) have reported electrochemical treatment of MSW landfill leachate which contains high concentration of chlorides (up to 4000 mg/L) using Ti/PbO$_2$ and Ti/SnO$_2$, however only Ti/PbO$_2$ showed best electrocatalytic behaviour causing 100% ammoniacal–N removal. This was attributed to indirect oxidation effect of chlorine and hypochlorite. Lin and Wu (22, 23) have applied the electrochemical technique for the removal of nitrite and ammonia both in seawater and aquacultural water using graphite anode and TiO$_2$ cathode. These authors have added some quantity of table salt with the intention of raising the conductivity of aquacultural water but this resulted in indirect oxidation of ammonia. The authors reported 100% NO_2^--N removal in < 30 min operation time at 3000 mg/l NaCl concentration. The presence of ammonia along with NO_2^- in the electrolyte caused slow removal of NO_2^--N. In contrast to NO_2^--N removal only about 40% NH_4^+-N is removed in the same time period.

The present investigation concerns itself with the removal of ammonia at Ti/PbO$_2$ anode, both in the chloride-free and chloride-bearing electrolyte. In some experiments, a bromide salt was employed instead of NaCl. This interested us because in some of the previously reported studies (14, 19, 22, 23) the electrolyte was directly prepared in seawater or wastewater is mixed with seawater which is known to contain bromide too. It is known from the standard redox potentials (24) that $Br^- \rightarrow Br_2$ reaction requires lesser potential than does the $Cl^- \rightarrow Cl_2$ reaction. Bromine may hydrolyze into HOBr and hypobromite ion. Hypobromous acid reacts more rapidly with ammonia in water to form bromamines (25-27). It is also known qualitatively that bromamines form more quickly than do the chloramines (25). Therefore, there is a possibility that indirect electrooxidation of NH_4^+ in electrolytes prepared in or mixed with seawater or in industrial wastewater containing Br^- apart from Cl^- may be mediated through HOBr and not HOCl. Our results suggested that NH_4^+ oxidation does not occur at Ti/PbO_2 anode in halide-free electrolyte. The presence of chloride in electrolyte is necessary for observing ammonium oxidation. Further, bromide is found to be more effective in causing efficient ammonia removal in comparison with chloride. Finally the rate enhancing effect of Br^- is verified by conducting ammonium oxidation experiments using Ti/PbO_2 as well as Ti/Pt/Ir anodes in tannery industry wastewater. Some tentative chemical reactions have been presented in this paper too based on those already published in the literature merely as they help to explain the observed data.

Experimental Section

The anode was a titanium mesh over which PbO_2 was melt deposited. Two cathode plates made up of titanium were positioned on either side of the anode at a distance of 1.5 cm. The batch reactor set-up is shown in Fig. 1. The area of the mesh type anode was calculated using the formula (28) $s = 2\Pi dlb\sqrt{n}$; where, s = area of the mesh, d = thickness of mesh wire, l = length of the gauge, b = width of the gauge, n = number of wire intersections per square centimeter. Based on the values d = 0.3 cm, l = 4.0 cm, b = 5 cm and n = 4 for the present case, the effective anode area was estimated to be 75 cm^2. The dimension of the exposed portion of each cathode plate was 6x5.3 cm. Alternatively, a 10 × 10 cm titanium anode coated with platinum and iridium (60:40) according to the procedure given by Atanasoski *et al.* (29) was used in conjunction with a steel cathode of same size. A digital regulated DC Power supply (TESTRONICS-34D, India) having constant voltage (0-15 V) and constant current (0-10 A) facility was used to maintain constant current during the experimental runs. A magnetic stirrer (JSGW, India) was used to stir the contents of the reactor during runs. The test ammonium solution with 200 mg NH_4^+/l concentration was prepared by dissolving an appropriate amount of analytical reagent grade (S.D.Fine-Chem.Ltd, Mumbai, India) ammonium sulfate ($(NH_4)_2SO_4$, >99.5% purity, Cl^- = 0.0003%, NO_3^- = 0.001%) or ammonium chloride ((NH_4Cl); >99.8% purity, NO_3^- = <0.00005%) in de-ionized water. Appropriate amounts of analytical reagent grade sodium chloride or potassium bromide were added in ammonium test solutions to achieve the chloride or bromide concentrations in the range of 400-3000 and 400-2000 mg/l, respectively. Alternatively, wastewater from a tannery industry (after anaerobic treatment) was also used in some of the experiments. In a typical experiment, 500 ml of ammonium test solution was taken into the batch reactor and the electrode assembly was provided with the selected constant current. Samples for analysis were withdrawn at the chosen interval of time and changes in ammonium concentration, conductivity and pH were determined. The ammonia concentration was determined by Nesslerization and color determination (Spectronic 21, Bausch & Lomb, U.S.A.) as per the standard methods (30). Each sample was analyzed at least twice for determining ammonia concentration; the replication error was within 5%. The observations were confirmed by performing duplicate runs. The batch-to-batch error of each data point varied between 3-7 %. The dimensionless concentration ratio, $((NH_4^+)_t / (NH_4^+)_0)$, was used to follow the extent and rate of the removal of ammonia by electrooxidation process. Here $(NH_4^+)_0$ is the initial ammonium ion concentration in mg/l and $(NH_4^+)_t$ is the variable ammonium ion concentration measured at time t.

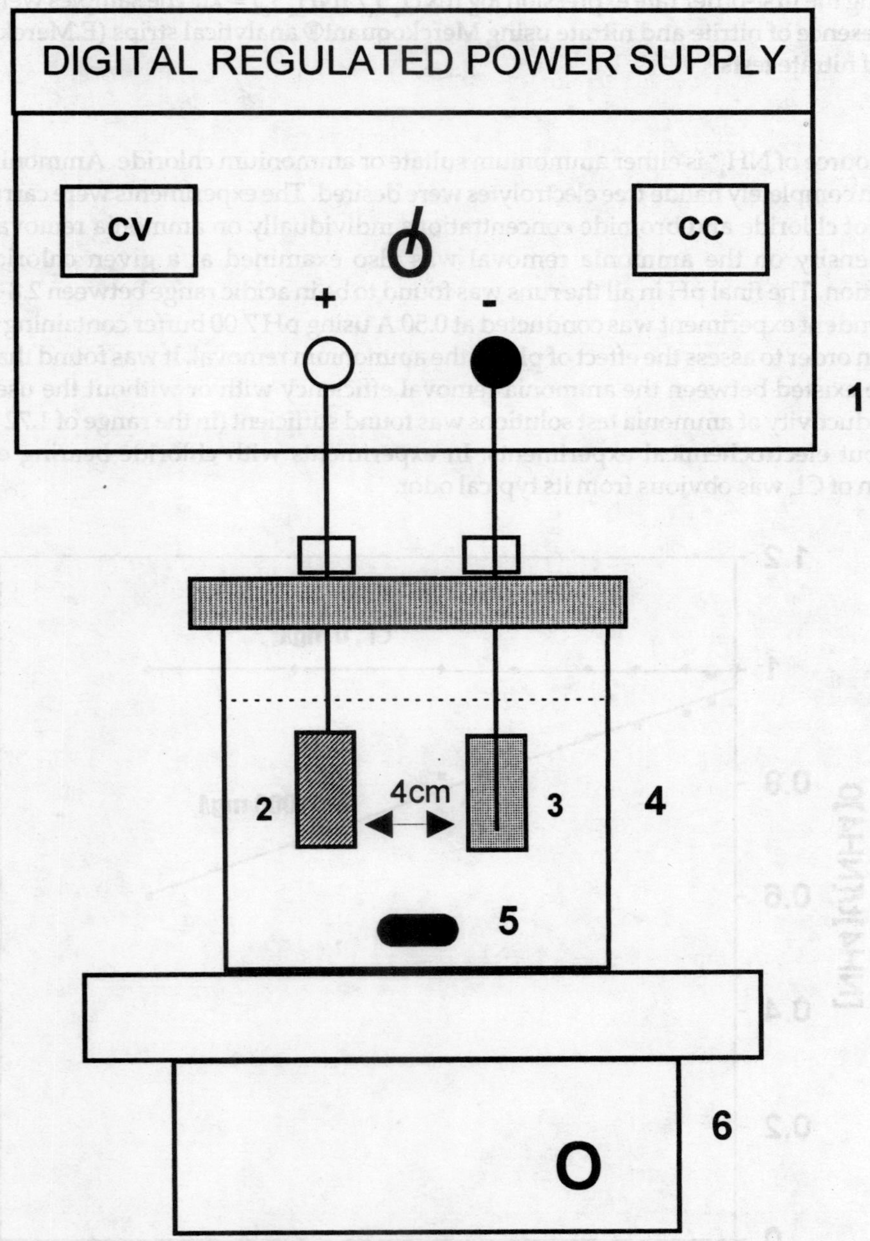

Fig. 1: Experimental Electrooxidation Setup (1) Digital regulated DC power supply, (2) Anode (3) Titanium cathode, (4) Batch reactor (electrolytic cell) (5) Magnetic bar, (6) Magnetic stirrer

The log plots of the dimensionless concentration curves were used for estimating rate constants (k, min^{-1}) using the first-order rate expression log $((NH_4^+)_t / (NH_4^+)_0) = -kt$. The samples were also analyzed for the presence of nitrite and nitrate using Merckoquant® analytical strips (E.Merck, Germany) for nitrite and nitrate tests.

Results

The source of NH_4^+ is either ammonium sulfate or ammonium chloride. Ammonium sulfate was used when completely halide free electrolytes were desired. The experiments were carried out to study the effect of chloride and bromide concentrations individually on ammonia removal. The effect of current density on the ammonia removal was also examined at a given chloride or bromide concentration. The final pH in all the runs was found to be in acidic range between 2.8-2.95. However, an independent experiment was conducted at 0.50 A using pH 7.00 buffer containing NH_4^+ and 2000 mg/l Cl^- in order to assess the effect of pH on the ammonium removal. It was found that no significant difference existed between the ammonia removal efficiency with or without the use of buffer. The initial conductivity of ammonia test solutions was found sufficient (in the range of 1.72 to 8.42 ms/cm) to carry out electrochemical experiments. In experiments with chloride-bearing electrolytes the generation of Cl_2 was obvious from its typical odor.

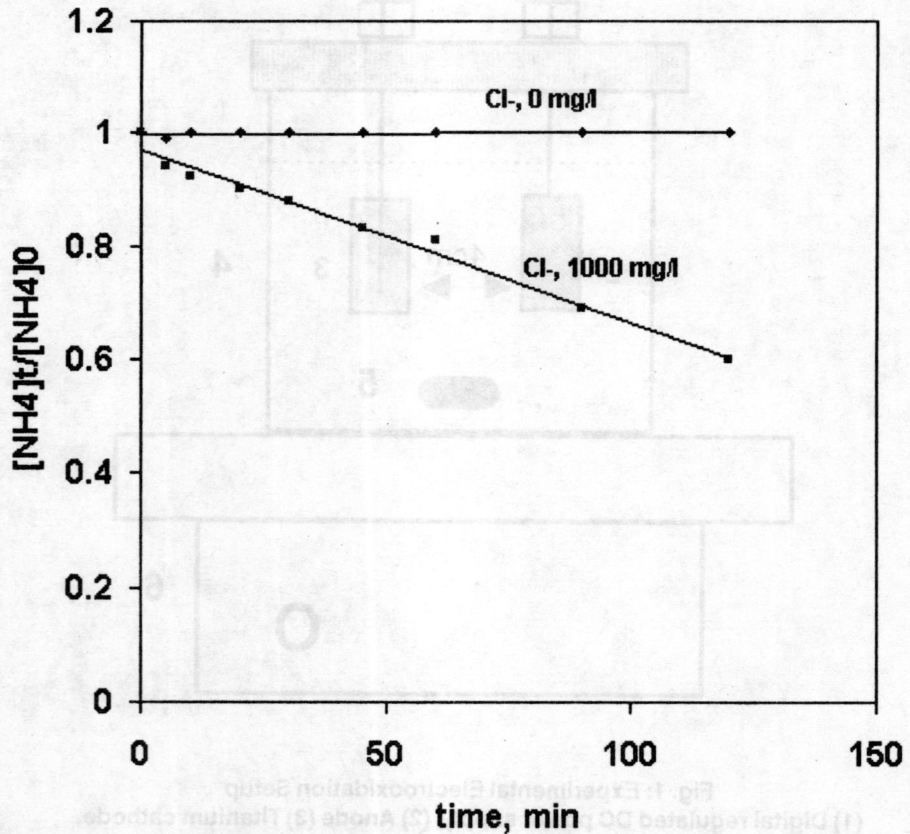

Fig. 2: Control Experiments Showing that the Presence of Chloride is Necessary for Ammonium Removal at Ti/PbO$_2$ Anode. Source of NH_4^+, $(NH_4)_2SO_4$; Current, 0.5A

Fig. 2 shows that the presence of chloride is necessary for ammonium removal in the present system. Ammonia removal is not observed at Ti/PbO$_2$ anode in an electrolyte containing ammonium sulfate without any added chloride. About 40% ammonia removal was obtained at the end of 120 min in the same electrolyte but fortified with 1000 mg/L chloride.

The effect of chloride concentration on the ammonia removal (NH_4^+source – $(NH_4)_2SO_4$) at 0.50A fixed current is depicted in Fig. 3. These log $((NH_4^+)_t / (NH_4^+)_0)$ versus time plots are approximately linear indicating that the reaction of removal of ammonia follows pseudo-first order kinetics. The R^2 values (correlation coefficient) typically > 0.90 imply that the ammonium removal data fit reasonably well into the first-order kinetic model. At the end of 120 min, the total NH_4^+ removal increased from 18% to 51% for an increase in chloride concentration from 400 to 2000 mg/l. The corresponding first-order rate constants are 8.0×10^{-4}, 1.8×10^{-3} and 2.6×10^{-3} min^{-1} for the cases of 400, 1000 and 2000 mg/l Cl^-, respectively. Similar experiments carried out at 0.50A using NH_4Cl as ammonium source and chloride addition confirmed the above results. The variation of the first-order rate constant as a function of chloride concentration is depicted in Fig. 4. The increase in the rate constant roughly saturates beyond 1000 mg/l Cl-. At this point, the concentrations of ammonium and chloride are in 1:5 ratio, and the rate constant is doubled when the chloride concentration is increased from 400 to 1000 mg/l.

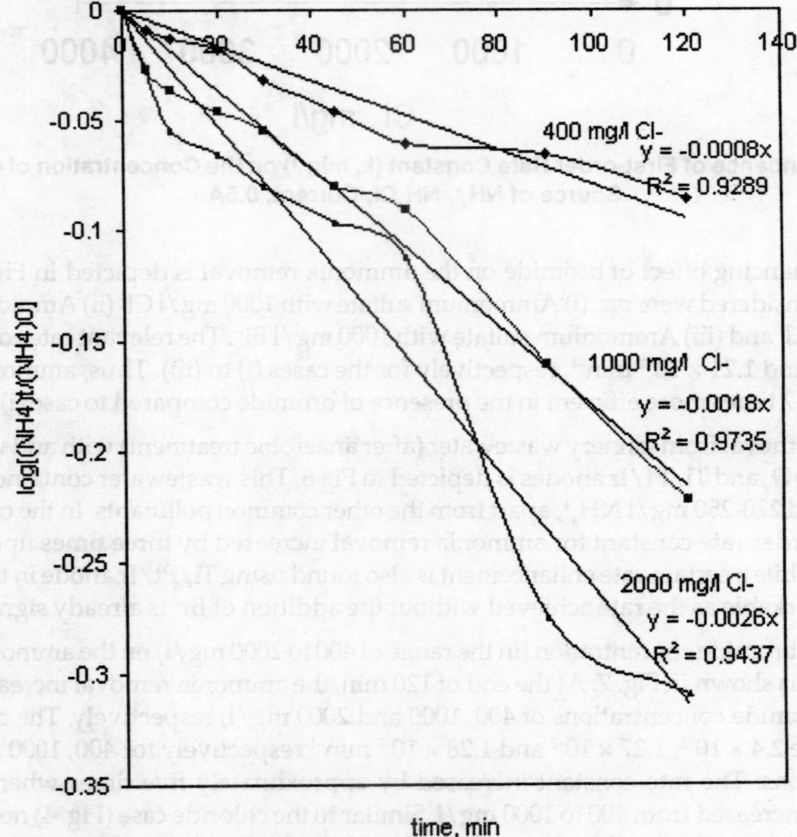

Fig. 3: Log $((NH_4^+)_t / (NH_4^+)_o)$ vs Time Plots (First-order) Depicting the Effect of Chloride Concentration on Ammonium Removal at Ti/PbO$_2$ Anode. Source of NH_4^+, $(NH_4)^2SO_4$; Current, 0.5A

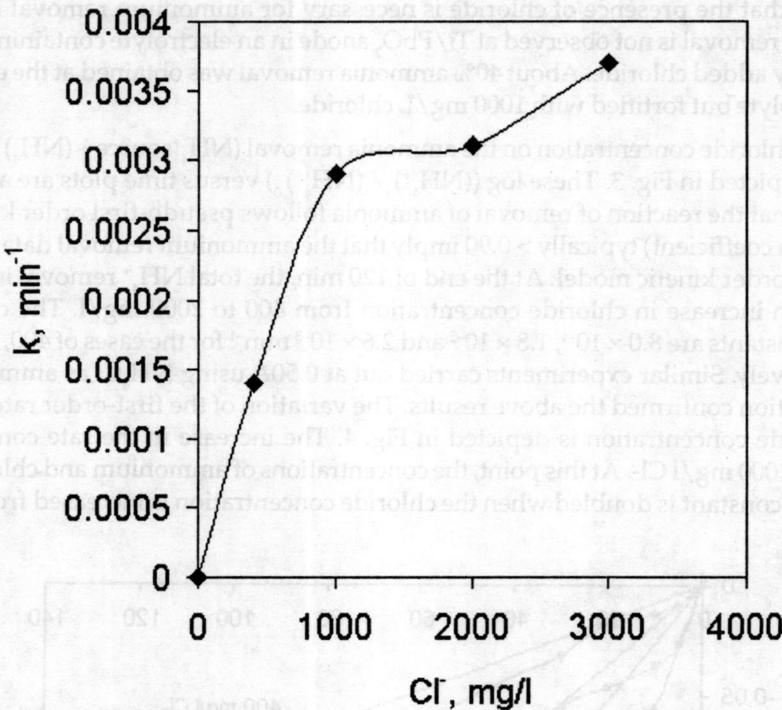

Fig. 4: Dependence of First-order Rate Constant (k, min^{-1}) on the Concentration of Chloride, Source of NH_4^+, NH_4Cl; Current, 0.5A

The rate enhancing effect of bromide on the ammonia removal is depicted in Fig. 5. The three different cases considered were *viz.* (i) Ammonium sulfate with 1000 mg/l Cl^- (ii) Ammonium chloride with 1000 mg/l Cl^- and (iii) Ammonium sulfate with 1000 mg/l Br^-. The relevant rate constants are 1.8×10^{-3}, 2.9×10^{-3} and 1.21×10^{-2} min^{-1}, respectively for the cases (i) to (iii). Thus, ammonia removal is approximately 4-7 times more efficient in the presence of bromide compared to case (i) and (ii).

Ammonia removal from tannery wastewater (after anaerobic treatment) with and without added bromide at Ti/PbO_2 and Ti/Pt/Ir anodes is depicted in Fig 6. This wastewater contained about 4500-4750 mg/l Cl^- and 220-250 mg/l NH_4^+, apart from the other common pollutants. In the case of Ti/PbO_2 anode the first-order rate constant for ammonia removal increased by three times upon addition of 1000 mg/l Br^-. While a certain rate enhancement is also found using Ti/Pt/Ir anode in the presence of Br^-, it is not remarkable as the rate achieved without the addition of Br^- is already significant.

The effect of bromide concentration (in the range of 400 to 2000 mg/l) on the ammonia removal at a current of 0.5A is shown in Fig. 7. At the end of 120 min, the ammonia removal increased to 54, 98.7 and 100% for bromide concentrations of 400, 1000 and 2000 mg/l, respectively. The corresponding rate constants are 2.4×10^{-3}, 1.27×10^{-2} and 1.28×10^{-2} min^{-1} respectively for 400, 1000 and 2000 mg/l Br^- concentrations. The rate constant increased by approximately five times when the bromide concentration is increased from 400 to 1000 mg/l. Similar to the chloride case (Fig. 4) no enhancement in the rate constant was observed for bromide concentration higher than 1000 mg/l.

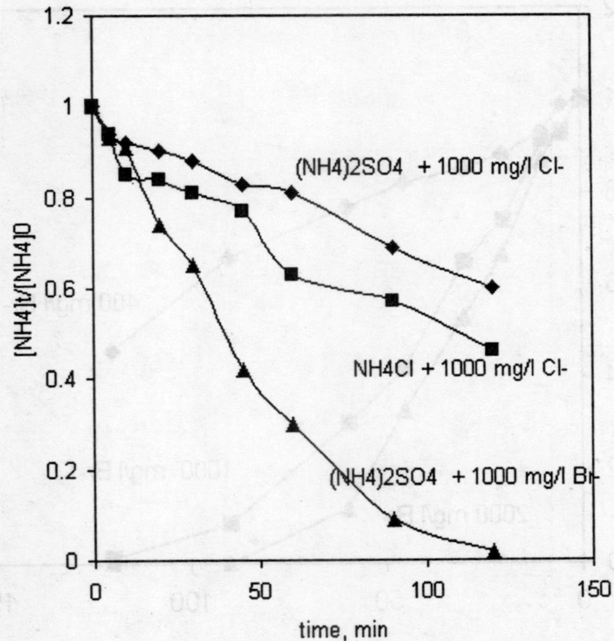

Fig. 5: Figure Depicting Rate Enhancing Effect of Bromide on Ammonia Removal at Ti/PbO$_2$ Anode. Current, 0.5A

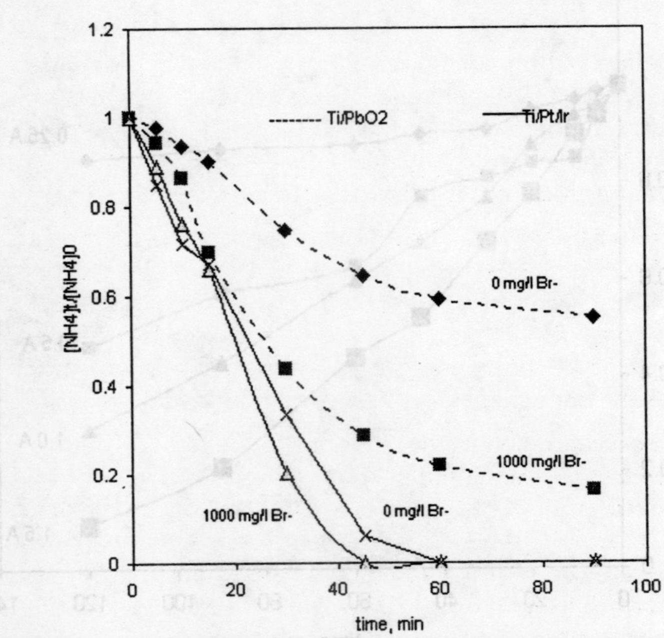

Fig. 6: Ammonia Removal from Tannery Wastewater (After Anaerobic Treatment) at Ti/PbO$_2$ with and without Added Bromide. Current, 0.5A

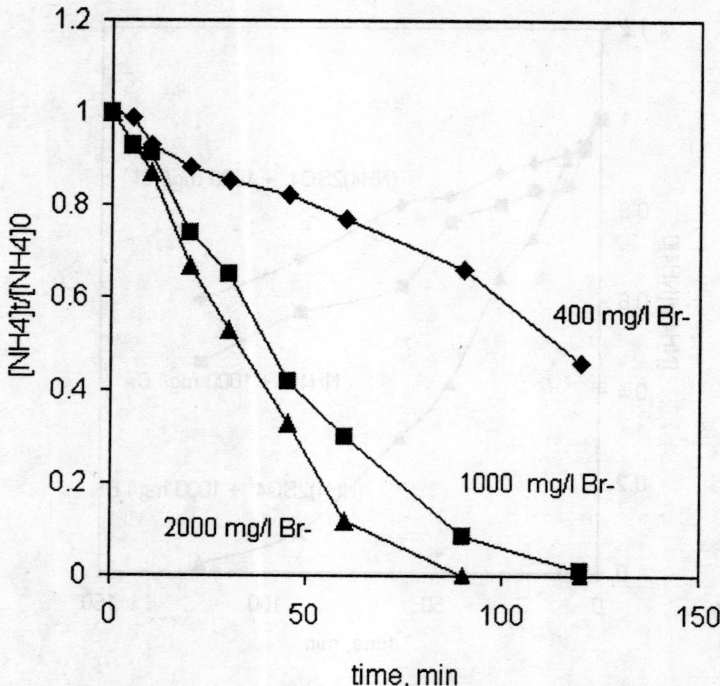

Fig. 7: Figure Depicting the Saturation Effect of Br Concentration on Ammonia Removal. Current, 0.5A

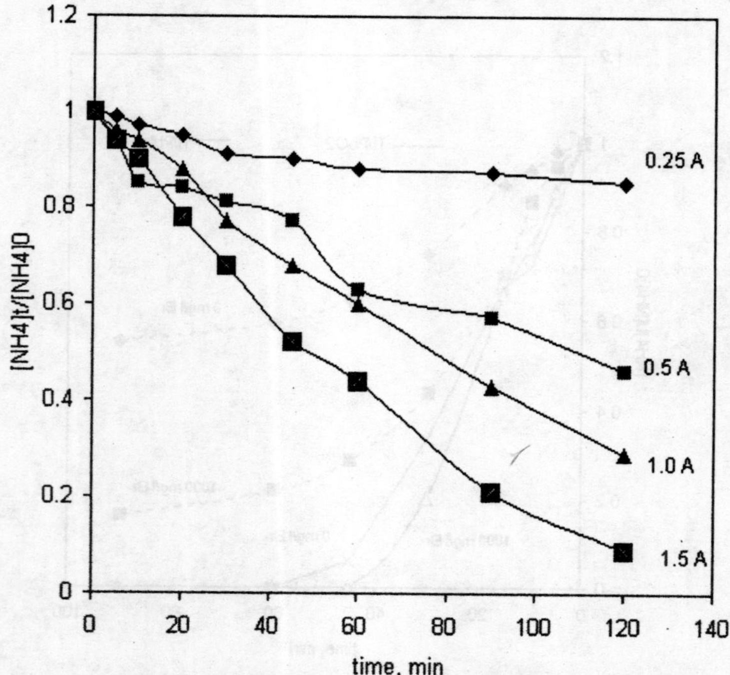

Fig. 8: Effect of Applied Current on the Ammonium Removal at Ti/PbO$_2$ Anode. Chloride Concentration: 1000 mg/L

The effect of applied current on the ammonia removal is shown in Fig. 8. The chloride concentration was fixed as 1000 mg/l. The ammonia removal obtained at different current values was 15, 54, 71 and 99% at 0.25 A, 0.50 A, 1.0A and 1.5 A respectively, based on the fraction of ammonia remained at 120 min. A plot showing current density versus rate constant shown in Fig. 9 suggests that an increase in current density generally resulted in the enhancement of rate constant.

The trend of ammonia removal at a given bromide concentration (Br^- = 1000 mg/l) for different applied current values in the range 0.25 to 1.00A is shown in Fig. 10. The relevant rate constants are 4.0×10^{-3}, 1.28×10^{-2} and 1.11×10^{-2} min^{-1} respectively for 0.25, 0.50 and 1.00A. The rate constant is found to increase by three times for twice higher current density and any further increase in current appears to be of no use as exemplified by the k values for 0.50 and 1.00A. This is in contrast to the chloride case (see Fig. 9) where increase in current density caused increase in rate constant for ammonia removal.

The nitrite/nitrate analysis of initial and final samples of the experimental runs was also carried out. Nitrite/nitrate were not detected in the initial samples. Nitrite was absent in the final samples, while 20 mg/l nitrate was present only in experiments carried out using chloride containing ammonium test solutions. Surprisingly, nitrate was not detected when bromide containing ammonia test solutions were used.

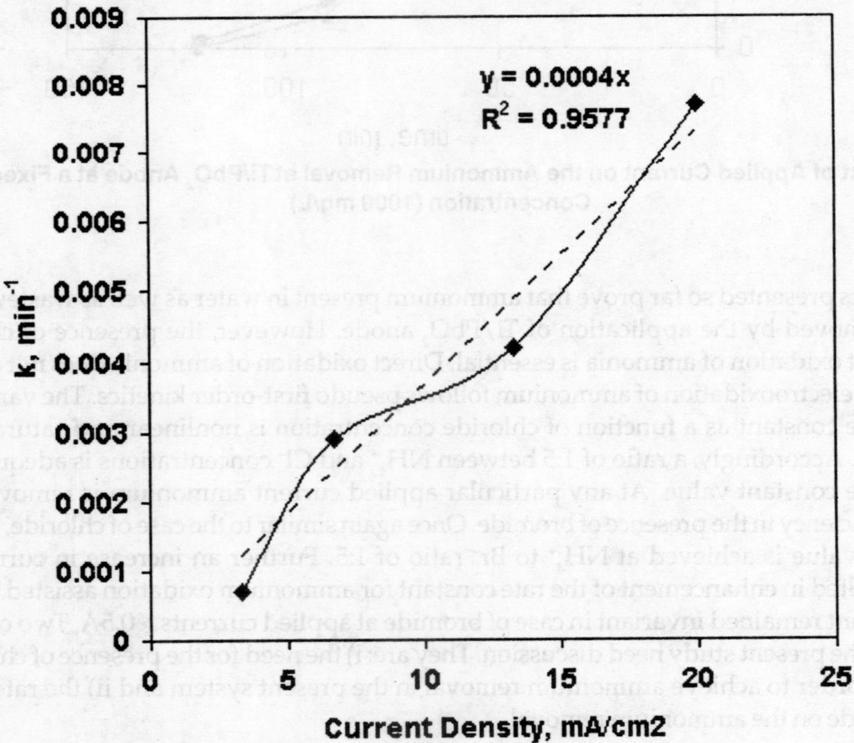

Fig. 9: First-order Rate Constant as a Function of Current Density. Chloride Concentration = 1000 mg/L

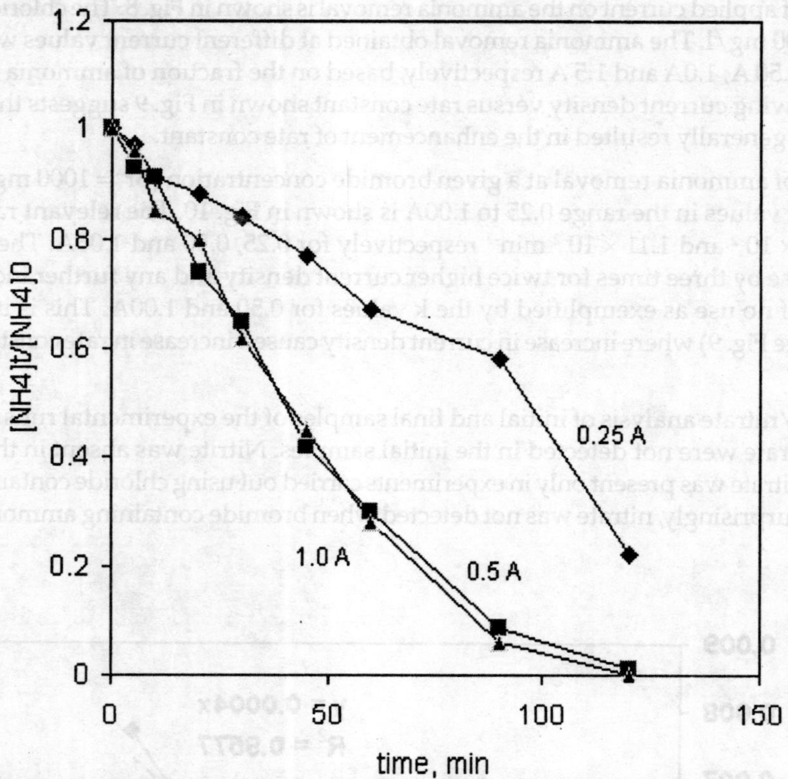

Fig. 10: Effect of Applied Current on the Ammonium Removal at Ti/PbO$_2$ Anode at a Fixed Bromide Concentration (1000 mg/L)

Discussion

The results presented so far prove that ammonium present in water as well as wastewater can be effectively removed by the application of Ti/PbO$_2$ anode. However, the presence of chloride that causes indirect oxidation of ammonia is essential. Direct oxidation of ammonia does not occur at this electrode. The electrooxidation of ammonium follows pseudo first-order kinetics. The variation of the first-order rate constant as a function of chloride concentration is nonlinear and saturates beyond 1000 mg/l Cl^-. Accordingly, a ratio of 1:5 between NH_4^+ and Cl^- concentrations is adequate to yield maximum rate constant value. At any particular applied current ammonium is removed with 3-7 times more efficiency in the presence of bromide. Once again similar to the case of chloride, a maximum rate constant value is achieved at NH_4^+ to Br^- ratio of 1:5. Further an increase in current density generally resulted in enhancement of the rate constant for ammonium oxidation assisted by chloride, but rate constant remained invariant in case of bromide at applied currents > 0.5A. Two observations made during the present study need discussion. They are: i) the need for the presence of chloride in the electrolyte in order to achieve ammonium removal in the present system and ii) the rate enhancing effect of bromide on the ammonium removal.

Role of Chloride

Investigations on direct and indirect electrooxidation of ammonia have been reported (14, 16, 19, 20, 22, 23). In the direct oxidation, ammonium ions are first adsorbed on the anode surface and then oxidised by the anodic electron transfer. A typical indirect electrooxidation is that involving chlorides.

The chloride is oxidised to gaseous chlorine at anode that may in turn form HOCl or OCl^-. Hypochlorous acid/or hypochlorite then oxidize ammonium in the bulk. Nitrogen gas is the final product as is the case with break-point chlorination (1). The possibility of direct oxidation of NH_4^+ at Ti/PbO$_2$ anode can be ignored in the present study since control experiments revealed that the presence of chloride is necessary for ammonia oxidation. Therefore, indirect oxidation of ammonium involving aqueous chlorine may be taking place. In a related study, Cossu *et al.* (21) used Ti/PbO$_2$ to remove ammonium from the MSW leachate that contained about 2500 mg/l Cl^-. This has been attributed to indirect oxidation effect of Cl_2 and HOCl formed by electrochemical oxidation of chloride. Concerning nitrite and ammonia removal from seawater and aquacultural water, Lin and Wu (22, 23) have confirmed that predominant role was played by the indirect oxidation caused by the presence of chloride in the electrolyte. However, a combined effect of direct and indirect oxidation has been reported by Szpyrkowicz *et al.* (16) on Ti/Pt/Ir anode, while indirect oxidation occurred on Ti/Pt anode.

The main electrochemical reactions during oxidation of ammonia in chloride containing electrolytes at about neutral pH are:

$$2Cl^- \rightarrow Cl_2 + 2e^- \text{ (at anode)} \quad (3)$$

$$2H_2O \rightarrow O_2 + 4H^+ + 4e^- \text{ (at anode)} \quad (4)$$

$$2H_2O + 2e^- \rightarrow H_2 + 2OH^- \text{ (at cathode)} \quad (5)$$

The generation of familiar oxidizing agents *viz.* HOCl or OCl^- becomes possible using electrogenerated Cl_2 (eq. 6&7).

$$Cl_2 + H_2O \rightarrow HOCl + HCl \text{ (in solution/near anode)} \quad (6)$$

$$2OH^- + Cl_2 \rightarrow Cl^- + OCl^- + H_2O \text{ (in solution)} \quad (7)$$

In the present case the initial pH was in the range 6.0-7.5 while the final pH was observed to be about 2.8-2.9. The lowering of the pH may be attributed to HCl produced through eq. 6. Since HOCl is the major species in water at pH < 7, the overall indirect oxidation reaction of ammonium can be written as:

$$2NH_4^+ + 3HOCl \rightarrow N_2 + 3H_2O + 2H^+ + 3HCl \quad (8)$$

The various constituent reaction steps of Eq. 8 are:

$$NH_4^+ + HOCl \rightarrow NH_2Cl + H_2O + H^+ \quad (9)$$

$$NH_2Cl + HOCl \rightarrow NHCl_2 + H_2O \quad (10)$$

$$NHCl_2 + HOCl \rightarrow NCl_3 + H_2O \quad (11)$$

$$2NH_2Cl + HOCl \rightarrow N_2 + H_2O + 3HCl \quad (12)$$

$$NH_2Cl + NHCl_2 \rightarrow N_2 + 3HCl \quad (13)$$

It was mentioned earlier that nitrate was detected in the final samples from the experiments carried out using chloride containing electrolytes. Since direct oxidation of NH_4^+ by the electrogenerated OH radicals is improbable in the present case, a side reaction producing N_2O as shown in eq.14 is possibly taking place. The sequence of reactions leading to nitrate formation may be contemplated as follows:

$$4NH_2Cl + 3Cl_2 + H_2O \rightarrow N_2O + N_2 + 10HCl \quad (14)$$

$$2\,N_2O \rightarrow 2\,NO + N_2 \quad (15)$$

$$2\,NO \rightarrow NO_2 + \tfrac{1}{2}N_2 \quad (16)$$

$$NO_2 + H_2O \rightarrow HNO_3 + H^+ \quad (17)$$

Rate Enhancing Effect of Bromide

We found that the rate constant for ammonia removal in the presence of bromide is 4-7 times higher than that in the presence of Cl^-. A similar effect is found during oxidation of ammonium in tannery wastewater using Ti/PbO_2 and Ti/Pt/Ir anodes. This effect may be understood from the following standard redox reactions (24) written in the order of ease of formation in aqueous acid solution at 25°C.

$$2Br^- \rightarrow Br_2 + 2e^- (E° = 1.065 \text{ V vs NHE}) \quad (18)$$

$$2H_2O \rightarrow O_2 + 4H^+ + 4e^- (E° = 1.23 \text{ V vs NHE}) \quad (19)$$

$$2Cl^- \rightarrow Cl_2 + 2e^- (E° = 1.36 \text{ V vs NHE}) \quad (20)$$

It is clear that bromide oxidation is relatively easy and both O_2 and Cl_2, which may form at anode, also possess the ability to oxidize bromide to bromine.

$$2Br^- + O_2 + 4H^+ \rightarrow Br_2 + 2H_2O \quad (21)$$

$$2Br^- + Cl_2 \rightarrow Br_2 + 2Cl^- \quad (22)$$

Accordingly, we observed that in the case of Ti/PbO_2 anode the first-order rate constant for ammonia removal from synthetic wastewater fortified with Br^- increased by 4-7 times (Fig. 5) while it increased by three times upon addition of 1000 mg/l Br^- in tannery wastewater (Fig 6). But no remarkable rate enhancement is found using Ti/Pt/Ir anode in the presence of Br^- (Fig. 6) This may be due to the known electrocatalytic property of Pt and Ir metals for the oxidation of chloride to Cl_2 (29) that may in turn cause effective oxidation of NH_4^+. Further, the final concentration of Cl^- is found to decrease by about 1000 mg/l even in the presence of Br^- and during this experiment using Ti/Pt/Ir anode the typical odor of Cl_2 was observed. In contrast, in experiments with Ti/PbO_2 anode in the presence of Br^- only a marginal decrease in the concentration of Cl^- is found (4660 as against the initial concentration of 4750 mg/l). However, when Ti/PbO_2 anode is used in the absence of Br^-, chloride concentration decreased to 4100 mg/l and the typical odor of Cl_2 is observed. These observations support that Br^- indeed participated in the anodic reactions even in the presence of chloride but significant rate enhancement effect may be observed at electrodes with less electrocatalytic properties for Cl^- oxidation.

Therefore in electrolytes containing bromide and chloride as well as bromide, the reactions involving bromide may predominate. Because in some of the previously reported studies (14, 19, 22, 23) the electrolyte was directly prepared in seawater or wastewater is mixed with seawater which is known to contain bromide too, this aspect becomes all the more important. Aqueous bromine, like aqueous chlorine, can form HOBr as follows:

$$Br_2 + H_2O \rightarrow HOBr + HBr \text{ (in solution/near anode)} \quad (23)$$

HOBr is a stronger oxidant than HOCl at higher pH. Part of the reason for this is that HOBr is a weaker acid and less ionized. HOBr rapidly reacts with ammonia in water to form mono-and dibromamines in reactions similar to those shown in eqs.9 & 10. It is known that at least qualitatively that bromamines form more rapidly than do the chloramines. The rate constant for the reaction of ammonia with HOBr (at pH 7 and 20°C) is about 14 times higher than with HOCl (25), the K_{obs} ($M^{-1}sec^{-1}$) being 2.9×10^5. In another context, Rook (31) has shown that the initial reaction rate of HOBr on peat extracts is much greater than the same molar concentration of HOCl. Further bromamines have been reported to be more effective disinfectants than chloramines (32, 27). Compared to chloramines, bromamines are much less stable in wastewater treatment plant effluents and receiving water and are therefore much less hazardous to aquatic life (33). The absence of nitrate formation in bromide containing electrolytes may be tentatively ascribed to the low stability of NH_2Br and $NHBr_2$ due to which the side reaction producing N_2O (as in eq.14) is absent.

Conclusions

The lead oxide anode can be used for effective removal of ammonia from aqueous solutions. Direct oxidation of ammonium does not occur at this electrode. However, the presence of chloride that causes indirect oxidation of ammonia is essential. But the use of bromide instead of chloride can result in 3-7 times more efficient removal of ammonium. The oxidation of ammonium follows pseudo first-order kinetics. The variation of the first-order rate constant as a function of chloride concentration is nonlinear and saturates beyond 1000 mg/l Cl^-. A ratio of 1:5 between NH_4^+ and Cl^- (or Br^-) concentrations is adequate to yield maximum rate constant value. Further maximum enhancement of the rate constant for ammonium oxidation is achievable in the presence of bromide at a relatively lower applied currents (0.5A).

Bromide participates in the anodic reactions even in the presence of chloride as verified in actual tannery wastewater, but significant rate enhancement effect may be observed at electrodes with less electrocatalytic properties for Cl^- oxidation. The relatively low overpotential for the oxidation of bromide at PbO_2 anode together with the high reactivity of HOBr towards ammonium appears to account for the high efficiency ammonium removal in bromide containing electrolytes.

Acknowledgments

Dr. D.J.K acknowledges Indian National Science Academy (INSA, India) for the award of Visiting Fellowship for the year 1998-99.

REFERENCES

1. Metcalf & Eddy, Inc., 1991, *Wastewater Engineering: Treatment, Disposal and Reuse*, 3rd New York, McGraw Hill
2. Henze M., 1989, Appropriate nitrogen removal from Wastewater in *Proceedings of 3rd Int. Conf. on Appropriate Waste Management Technologies for Developing Countries*, NEERI, Nagpur, India, 25-26 Feb. 1995, 2103.
3. Hellinga C., Schellen A. A. J. C., Mulder J. W., van Loosdrecht M. C. M. and Heijnen J. J., 1998, *Water Sci. Technol.*, *37 (9)*, 135-142.
4. Awadallah R.M. *et al.*, 1998, *Water Res.*, *32 (10)*, 3080-3088.
5. Gupta A.B. and Gupta S.K., 1999, *Water Res.*, *33* (2), 555-561.
6. Blogoslawski, W.J., Brown C., Rhodes E.W. and Broadhust M., 1975, Ozone disinfection of seawater supply system, *In Proc. of the 1st Int. Symp. on Ozone for Water & Wastewater Treatment*, Cleveland, Ohio,
7. Ori L. and Michal G., 1998, *Water Res.*, *32 (7)*, 2019-28.
8. Comninellis C. and Pulgarin C., 1991, *J. Appl. Electrochem.*, *21*, 415.
9. Comninellis C., 1994, *Electrochim. Acta*, *39 (11/12)*, 1857.
10. Comninellis C. and Nerini A., 1995, *J. Appl. Electrochem.*, *25*, 23.
11. Sharifman H. and Kirk D., 1985, *J. Electrochem. Soc.*, *15*, 92
12. Lin S.H. and Peng C.F., 1994, *Water Res.*, *28* (1), 277.

13. Chang L. C., and Wu T. C., 1995, *Water Res.*, *29*, 671.

14. Della Monica M., Agostizno A. and Ceglie A., 1980, *J. Appl. Electrochem.*, *10*, 527.

15. Vlyssides A. G. and Israilides C. J., 1998, *J. Environ. Sci. Health*, *A33* (5), 847-862.

16. Szpyrokowicz L, Naumczyk J, and Zilio-Grandi F., 1995, *Water Res.*, *29*, 517.

17. Naumczyk, J.; Szpyrokowicz, L.; De Faveri D. M. and Zilio-Grandi F., 1996, *Trans. IchemE.*, *74* (*B*), 59.

18. Vlyssides G. A. and Israilides J. C., 1997, *Environ. Pollution*, *97* (1-2), 147-152.

19. Poon C. P. C., Brueckner T. G., 1975, *J. Wat. Pollut. Control Fed.*, *47*, 66-78.

20. Szpyrokowicz L and Zilio-Grandi F., 1994, *Toxicol. Environ. Chem.*, *44*, 183-202.

21. Cossu R., Polcaro A. M., Lavagnolo M. C., Mascia M., Palmas S., and Renoldi F., 1998, *Environ. Sci. Technol.*, *32*, 3570-3573.

22. Lin S.H. and Wu C.L, 1996, *Water Res.*, *30* (3), 715-721.

23. Lin S. H. and Wu C. L., 1997, *J. Environ. Sci.*, *Health*, *A32* (8), 2125-2138.

24. Handbook of Electrochemical Constants (R. Parsons), Buttrworth Scientific Publications.

25. Wajon J. E., and Morris J.C., 1980, In: Water Chlorination, Environmental Impact and Health Effects (Eds. Jolley R. C., Brungs W. A. and Cumming R. B.) Ann Arbor Science Publishers Inc., The Butterworth Group, Michigan, Vol.3, p. 171.

26. Galal-Gorchev H. and Morris J. C., 1965, *Inorg. Chem.*, *4*, 899-905.

27. Johnson J. D. and Overby R., 1971, *J. Sanit. Eng. Div. ASCE.*, *97*, 617-628.

28. Alexeyev V., 1979, In: Quantitative Analysis (English Translation), MIR publishers, Moscow, pp. 477-478.

29. Atanasoski R.T., Nikolic B.Z., Jaksic M.M., and Despic A.R., 1975, *J. Appl. Electrochem.*, *5*, 155-158

30. *Standard Methods for the Examination of Water and Wastewater*, 17th ed; Clesceri, L.S.; Greenberg, A. E.; Trussel, R. R. Eds. ,1989, American Public Health Association,: Washington, D.C.

31. Rook J.J.,1978, *J. Environ. Sci. Health*, *A13* (2), 91-116.

32. Mills J.F., Bromine chloride Disinfection –State of the Art presented at the *Water Pollution Control Federation Meeting, Disinfection Workshop Proceedings*, Houston, Oct 7, 1979.

33. Mills J.F., 1978, Chemistry and Disinfectant Properties of Bromine Chloride in *Chemistry of Wastewater Technology* (Rubin A. J., Ed), Ann Arbor MI, Ann Arbor Sci. Publishers Inc.

17

Treatment of Wastewater Containing Red RB and H Acid Using Physico-Chemical Oxidation and Microbial Degradation in the Laboratory

S. SANDHYA, KANCHANA SWAMINATHAN, Y.V. SUBRAHMANYAM & S.N. KAUL

National Environmental Engineering Research Institute, CSIR Complex, Chennai - 600113

Abstract

Synthetic dyes, azo dye in particular, are widely found in the effluent from textile industries. The persistence and toxicity of these compounds cause adverse impacts in the receiving streams. Red RB is widely used azo dyes in Tiruppur and Karrur districts in Tamilnadu. H-Acid (1-amino-8-naphthol-3, 6-disulfonicacid) used in chemical industries for the synthesis of direct, acid, reactive and azoic dyes. The synthetic wastewater containing Red RB, H Acid treatment using physico-chemical and microbial degradation has been investigated. The effect of Fe^{2+}, H_2O_2, pH, initial concentration and contact time on the degradation and decolorization of the dyes were studied. It was observed that maximum color and COD removal was obtained for Red RB and H-Acid at 10 to 25 mg l^{-1} of Fe^{2+} and 400 to 500 mg l^{-1} of H_2O_2 dose. A pH range of 3 to 4 was found to be effective for all the dye samples. Release of sulfates from the Fenton's treated Red RB dye and

H-Acid indicates that the dye degradation proceeds via., substituent group cleavage. A biological treatment system using a up flow fixed bed anaerobic reactor for treatment of synthetic wastewater containing Red RB was also investigated. The effect of various hydraulic retention times on biological treatment was studied. A wastewater treatment process consisting of physico-chemical oxidation and biological oxidation in fixed bed biofilm process can be useful to treat dyestuff wastewater having colour, nonbiodegradable, recalcitrant organic concentration.

Keywords: Azo dyes, dye intermediates, Fenton's reagent, COD removal, color removal biodegradation.

Introduction

Synthetic dyestuffs are used extensively in textile, paper, printing industries and dye houses. The most commercially available synthetic dyes can be classified mainly as azo, anthraquinone, vat, phthalocyanine, indigo, polymethine, carbonium, and nitro dyes (1). Most dyes are manufactured through a number of stages and transformations involving nitration, reduction, halogenation, amination, sulfonation, diazotization and oxidation using raw materials benzene, toluene, xylene, naphthalene, anthracene etc., (2). The effluents from these industries vary widely in composition and contain organic and inorganic compounds and have high levels of color and COD (3). Colored textile effluents contain persistent dyestuffs, many of which are reported to be toxic and carcinogenic (4, 5). Direct discharge of these effluents can cause formation of toxic aromatic amines under anaerobic conditions in receiving media and are also reported to contaminate the soil and groundwater necessitating proper treatment before discharge into the environment. Color can be removed from wastewater by physical and chemical methods such as adsorption, coagulation-flocculation, oxidation, filtration and electrochemical methods (6-8). However these methods are quite expensive and have operational problems. Adsorption seems to be the most efficient method for color removal among the others tested. The most commonly used adsorbent is powdered activated carbon, which is also expensive. Recent reports indicated the possibility of using low-cost adsorbents like zeolite, bentonite fly ash, slag etc., for color removal from wastewater (9, 10).

Many of the synthetic dyestuffs are resistant to biological degradation and hence color removal by bioprocessing is difficult and not complete. Due to the large degree of aromatics present in these molecules and stability of mordant dyes, conventional biological treatment methods are not very effective for decolorization and degradation (11-13). Haugh *et al.* (14) describes a process in which a bacterial consortium could reduce mordant yellow 3 under anaerobic conditions and then mineralise the resultant amines under aerobic conditions. Most of the work reported to date appears to involve use of pure dyes in artificial media. Studies have shown that photo catalysis can be used to destroy dye compounds using TiO_2 and UV light irradiation (15-16). Recent studies indicate that advanced oxidation process (AOP's) which generate strong oxidant of hydroxyl radicals ($OH^{\cdot}$), may be an alternative for the removal of color from dye wastewater (17). Ozone treatment of textile wastewater and recycling of treated water as process water has been investigated by Koyuncu and Huseyin (18). For cotton and wool dying wastewater at an ozone flow rate of 0.2 mg/sec, decrease in COD was found to be upto 44 mg/min and for 36.6 mg/min respectively. Among the different AOP's, Fenton's reagent (which is a mixture of H_2O_2 and Fe^{2+} system) has been widely used for different treatment purposes because of its easy operation and low cost (19).

When H_2O_2 is added to an aqueous system containing organic substances and excess of ferrous ions, complex redox reactions occur. The main reactions are:

$$Fe^{2+} + H_2O_2 \rightarrow Fe^{3+} + OH^- + OH. \quad (1)$$

$$Fe^{3+} + H_2O_2 \rightarrow Fe^{2+} + H^+ + HO_2 \quad (2)$$

The hydroxyl free radical generated would attack the unsaturated dye molecules and the azo bond (N=N) in the chromophore or chromogen thus decolorizing the wastewater. It is observed that the rate of the oxidation reaction depends on the concentration of hydroxyl free radical generated from these reactions (20).

Fenton's reaction has been effectively used to degrade compounds such as aromatics (21), dyes (22, 23), and pesticides (24). Shyh Fang Kang *et al.* (25) have reported that 80-90 % color removal can be achieved by H_2O_2/UV/Fe^{2+} process. The decoloration kinetics mechanism of commercial dyes by photo Fenton's system has been studied by Tang and Chen (26). The decoloration of the dyes was attributed to be due to adsorption by iron powder and oxidation by photo-Fenton's reagent. Studies by Deng *et al.* (27) showed that the decoloration of three reactive dyes was enhanced by UV/Fe° system. It was concluded that decoloration was dependent on the molecular structure of the dyes and the reaction was of pseudo order.

The present study attempts to use Fenton's reagent to decolorize and degrade two different dyes *viz.*, Red RB, and H-acid a common dye intermediate which is used in the manufacture of a number of dyes as well as biodegradation of them under laboratory conditions and the comparison of physico-chemical and biological processes of disposal. The structures of the dyes are presented in Figure 1 and their main characteristics are given in Table 1.

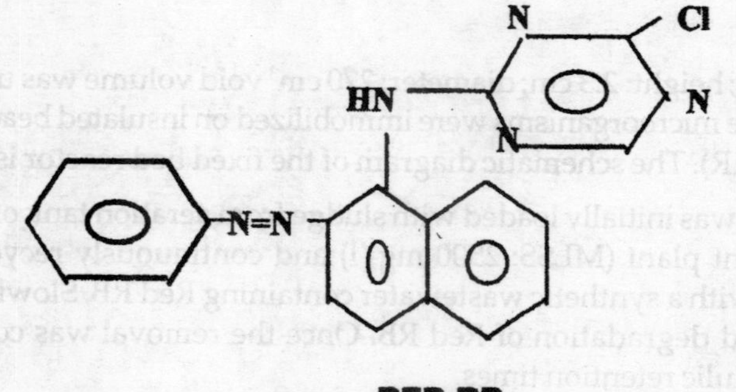

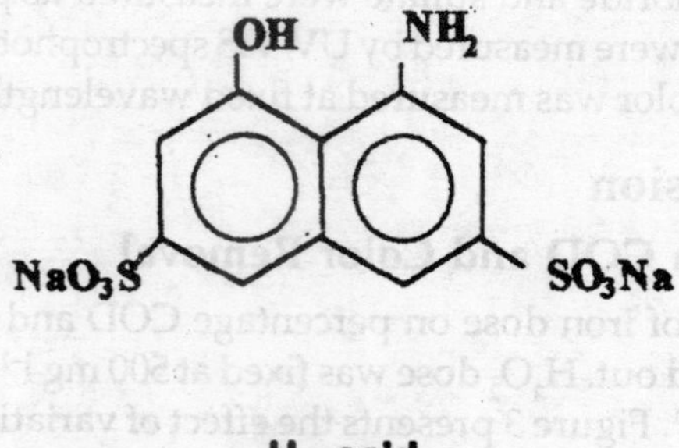

Fig. 1: Structure of Azo Dyes

Table 1: Main Characteristics of Dyes

Dye Sample	Concentration mg/l	COD mg/l	L max nm
Red RB	250	142	535.3, 235.3
H Acid	250	215	358.8, 235.3

Materials and Methods

Reagents and Stock Solutions

$FeSO_4.7H_2O$ (MERCK, purity 96%) was prepared with a predetermined concentration of 10g/L. H_2O_2 (MERCK, 30%) was used and other chemicals used in the experiment were analytical grades. The H-Acid and Red RB used were commercial grade. A known concentration of the dyes were prepared in water and used for all studies.

Fenton's Oxidation Experiment

Studies were carried out with the dye samples at original pH (acidic). The experiments were carried out in jar testing apparatus according to the following steps: (1) 100 ml dye wastewater sample was taken in a 250 ml beaker and the pH was adjusted to the desired level. (2) A definite volume of $FeSO_4$ solution was added and stirred and a known amount of 30% H_2O_2 solution was dripped into the sample. After reactions have completed samples were withdrawn and pH, COD and Ultra Violet – Visible spectra were measured.

Fixed Bed Reactor

The reactor has 49 cm; height: 2.3 cm; diameter; 270 cm^3 void volume was used in anaerobic fixed bed reactor. Sewage sludge microorganisms were immobilized on insulated beads in upflow fixed bed anaerobic reactor (UFFBAR). The schematic diagram of the fixed bed reactor is shown in Figure 2.

The fixed bed reactor was initially loaded with sludge from aeration tank of an urban wastewater activated sludge treatment plant (MLSS: 2500 mg/l) and continuously recycle. After 10 days the biomass was acclimated with a synthetic wastewater containing Red RB. Slowly the microorganisms started decolorization and degradation of Red RB. Once the removal was constant, UFFBAR was operated at various hydraulic retention times.

Analytical Methods

The pH, COD, chloride and sulfate were measured as per Standard Methods (28). The Ultra Violet scan of the sample were measured by UV-VIS spectrophotometer (Shimadzu Model UV 160A) at 200 nm to 800 nm. The color was measured at fixed wavelength for each dye sample.

Results and Discussion

Effect of Fe^{2+}Dose on COD and Color Removal

To study the effect of iron dose on percentage COD and color removal of dyes, the following experiments were carried out. H_2O_2 dose was fixed at 500 mg l^{-1} for all the dyes. $FeSO_4$ dose as $Fe^{2+}$was varied from 5 to 50 mg l^{-1}. Figure 3 presents the effect of variation of Fe^{2+} dose on the dye samples.

It is apparent from the figure that at Fe^{2+}dose of 10 mg l^{-1}, the percentage COD reduction for Red RB was 78 and above that the percentage COD removal remained unaltered. The treated dye sample showed no peak in the UV Visible range indicating that the dye has degraded completely.

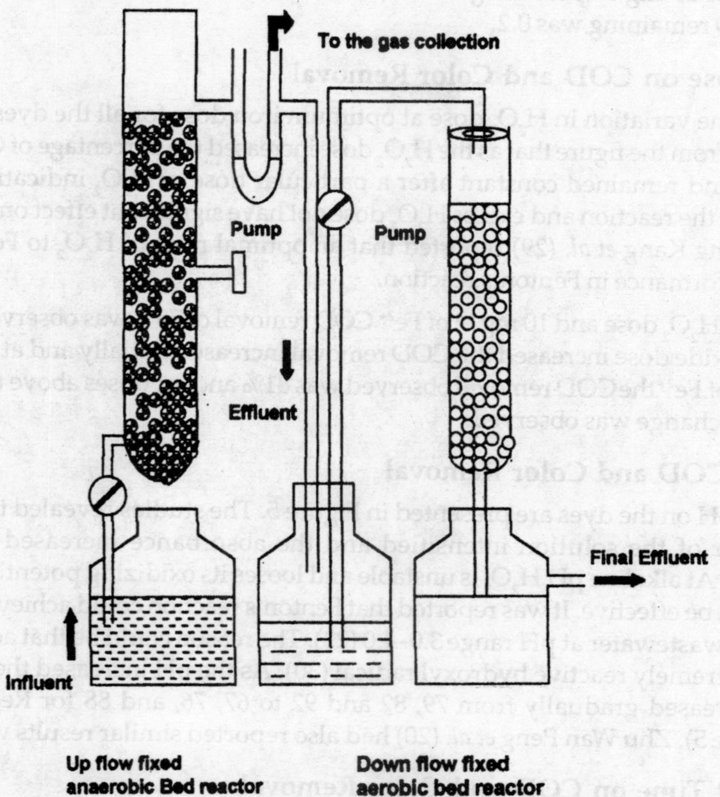

Fig. 2: Schematic Diagram of UFFBAR

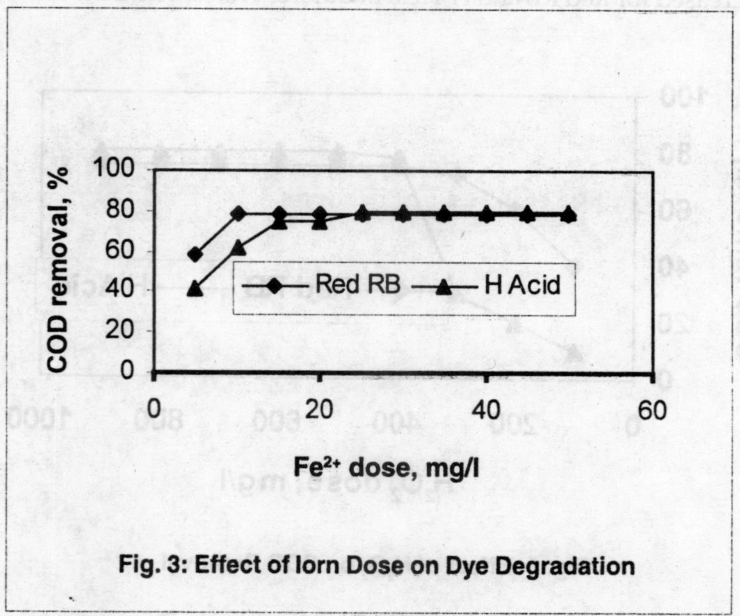

Fig. 3: Effect of Iorn Dose on Dye Degradation

At a Fe^{2+} dose of 25 mg l^{-1} percentage COD reduction for H-Acid was found to be 80 and the corresponding COD remaining was 0.2.

Effect of H_2O_2 dose on COD and Color Removal

The results of the variation in H_2O_2 dose at optimum iron dose for all the dyes are presented in Figure 4. It could be from the figure that as the H_2O_2 dose increased the percentage of COD removal also increased initially and remained constant after a particular dose of H_2O_2 indicating both iron and H_2O_2 dose influence the reaction and excess H_2O_2 dose not have significant effect on dye degradation. Studies by Shyh Fang Kang *et al.* (29) reported that an optimal ratio of H_2O_2 to Fe^{2+} is important to achieve a better performance in Fenton's reaction.

At 400 mg l^{-1} of H_2O_2 dose and 10 mg l^{-1} of Fe^{2+} COD removal of 78% was observed for Red RB. For H-Acid, as the peroxide dose increased the COD removal increased initially and at 500 mg l^{-1} of H_2O_2 dose with 25 mg l^{-1} of Fe^{2+} the COD removal observed was 81% and for doses above that COD removal remained constant. change was observed.

Effect of pH on COD and Color Removal

The effects of pH on the dyes are presented in Figure 5. The studies revealed that as the pH was increased the color of the solution intensified and the absorbance increased which may be a hyperchromic shift. At alkaline pH H_2O_2 is unstable and looses its oxidizing potential. For all the dyes pH 3.0 was found to be effective. It was reported that Fenton's reaction could achieve removal of color from synthetic dye wastewater at pH range 3.0–4.0 (23). The reason could be that acidic pH enhances the formation of extremely reactive hydroxyl radical (30). As the pH increased the percentage COD reduction also decreased gradually from 79, 82 and 92 to 67, 76, and 88 for Red RB and H-Acid respectively (Figure 5). Zhu Wan Peng *et al.* (20) had also reported similar results with H-Acid.

Effect of Contact Time on COD and Color Removal

The results presented in Figure 6 show that the COD removal increased for all the dyes when contact time was increased for Red RB and H-Acid at different rates. The COD removal at 15 min was

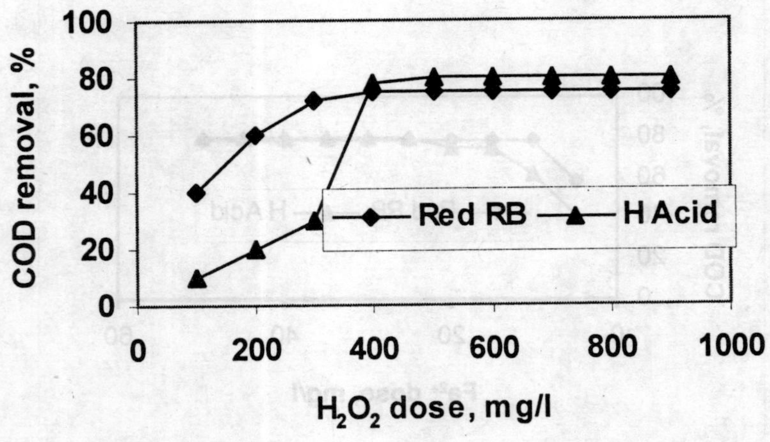

Fig. 4: Effect of H_2O_2 on COD Removal

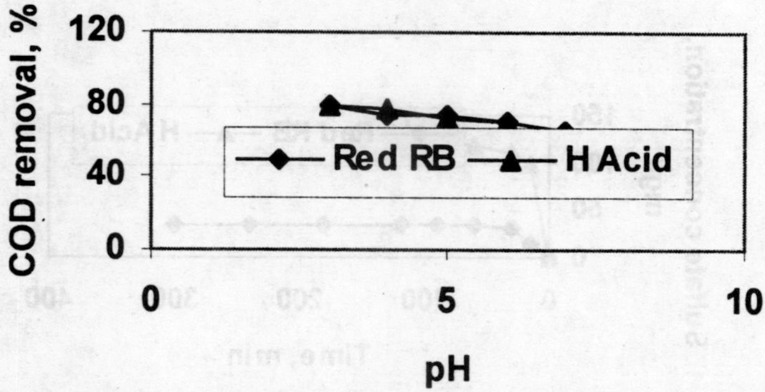

Fig. 5: Effect of pH on COD Removal

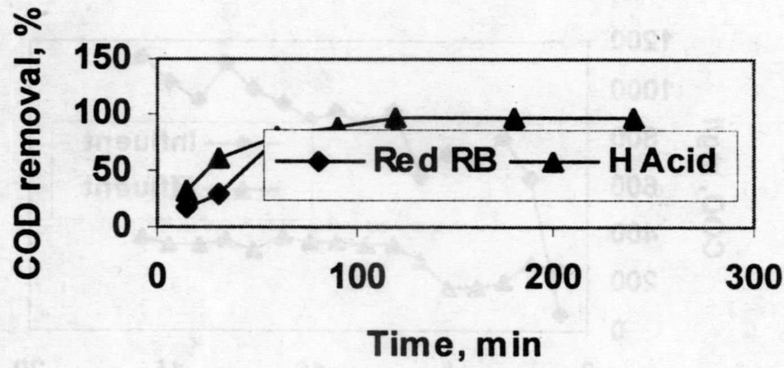

Fig. 6: Effect of Contact Time on COD Removal

18 % and 33 % respectively indicating that the H Acid dye had degraded faster. As the contact time increased the residual COD fraction obtained was very low percentage COD reduction increased to maximum 99 for all the dyes at the end of 120 min which shows that contact time is also an important factor for Fenton's reaction.

The molecular structures of the dye samples indicate that they contain different substituent groups of sulfonic acid groups in Red RB and H Acid. The release of sulfate during the reaction as presented in Figure 7 indicate the dye degradation may proceed by breaking of sulfo groups in the dye samples. Concentration of sulfate increases rapidly with increasing time for all the dyes (Figure 7) which corresponds to the trend in COD removal. This shows that during the oxidation process, sulfo group may be substituted by OH radical resulting in the release of sulfate in the solution. This observation agrees with results obtained by Zhu Wan Peng *et al.* (20) and Yizhong Wang (31). The results of the studies show that Fenton's process can be applied for the treatment of wastewaters containing dyes.

Acclimatization of Biomass in UFFBR for Red RB Biodegradation

Figure 8 shows the acclimatization results of biomass in the UFFBAR. In this acclimatizing period, the HRT and the influent COD concentration were 0.405 days and 694.51 mg/l respectively. The reactor was showing 67% COD reduction along with 72.9% decolorization.

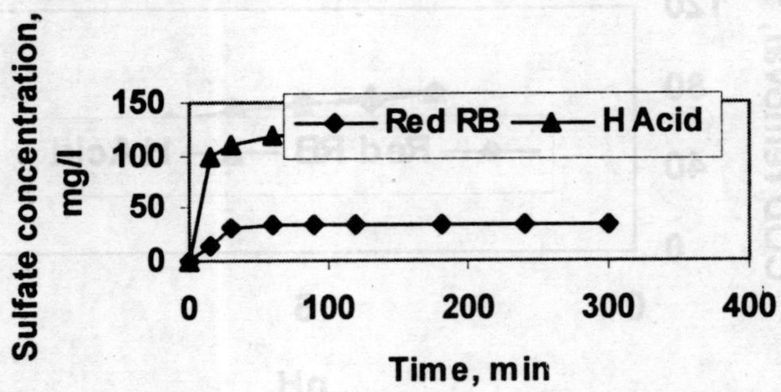

Fig. 7: Sulfate Ion Production from Fenton's Oxidation

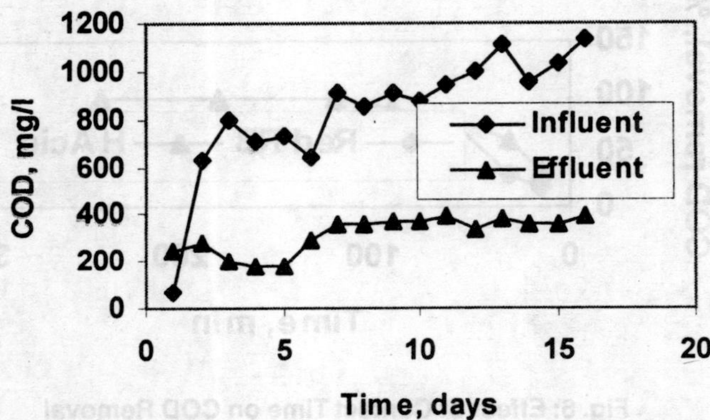

Fig. 8: Effect of Variation of Hydraulic Retention Time (HRT) on Treatment Efficiency of UFFBAR

Effect of Variation of Hydraulic Retention Time

After 48 days of operation, as the biomass of the fixed bed reactor was acclimated to the condition, the hydraulic retention time (HRT) was changed step wise from 0.405 to 0.243 days for the purpose of investigating the treatment efficiency due to HRT varation (Table 2).

Table 2: Azo dye-Red RB degradation in UFFBAR

HRT, Days	*Influent loading Kg/m³/day COD*	*COD, mg/l*			*Decolorization, %*
		Influent	*Effluent*	*Reduction, %*	
0.405	3.032	694.51	227.48	67.24	81.40
0.366	3.726	844.64	272.70	67.71	78.90
0.243	7.460	1044.58	361.67	65.37	72.91

With a HRT of 0.405 days, the effluent COD concentrations decreased continuously and a clear effluent was obtained. When HRT was decreased to 0.366 days, the effluent COD concentrations remain more or less same with marginal increase in decolorization efficiency (78.90%). Further decrease in HRT to 0.243, there is decrease in COD reduction. This may be due to the fact that decrease in HRT increases the influent COD loading. The performance of the reactor for decolorization and COD reduction is always better if influent flow is at a lower rate.

Biodegradation of H Acid Under Batch Experimentation Condition

H Acid is an important building block for large scale synthesis of azo dyes. The amino substituants are indisposable as auxochromic groups. Arylsulfonates are very rare among the natural compounds, sulfonic acid group confers a xenobiotic characteristics to this class of compounds (32). The microbial cultures isolated for azo dyes degradation were tested for H Acid degradation and results of degradation monitored by HPLC (Figure 9). Initially microbial culture were not able to degrade H Acid. But long acclimatization for over a period of six months, microorganisms could biodegrade 25 mg/l of initial H Acid within 40 hrs, but further increase in concentration was not biodegraded.

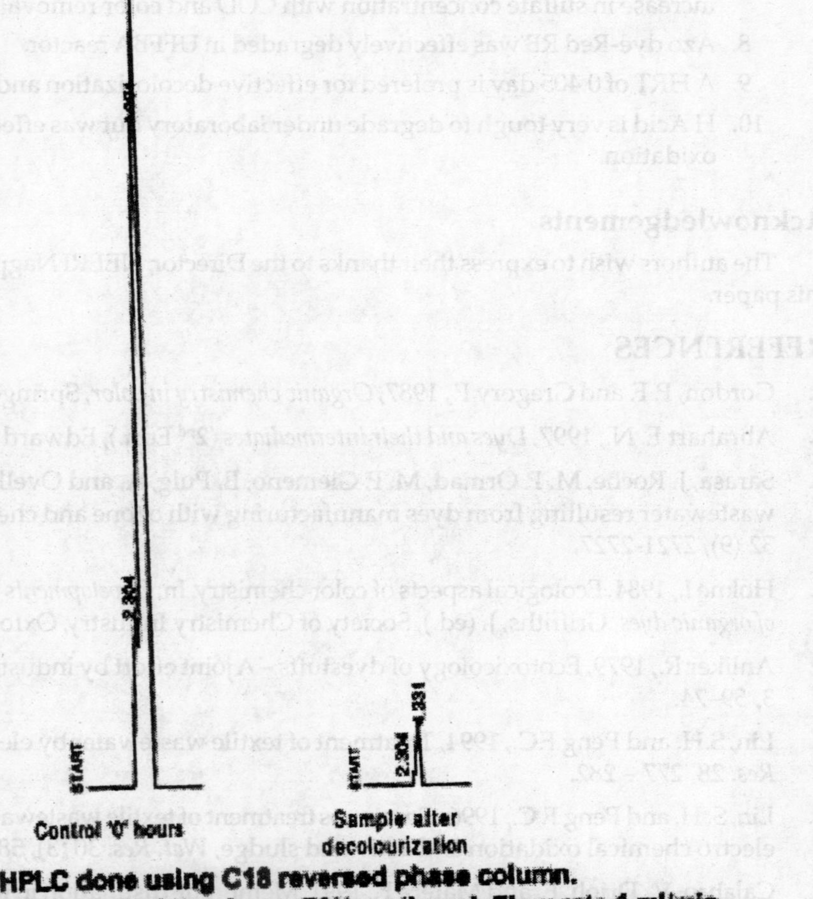

Fig. 9: Monitoring of H Acid Degradation by Using HPLC

Conclusion

The following conclusions can be drawn from the studies of oxidative degradation of Red RB and H-Acid.

1. Color and COD removal of the dye Red RB and the dye intermediate H-Acid were achieved by Fenton's oxidation process.
2. Color removal was faster than COD reduction for H-Acid and the dyes studied.
3. The doses of $Fe^{2+}$and H_2O_2 for maximum COD and color removal in Red RB and H-Acid were found to be 10 to 25 mg l^{-1} of Fe^{2+} and 400 to 500 mg l^{-1} of H_2O_2 respectively.
4. pH range of 3.0 to 4.0 was found to be effective for all the dyes.
5. As the concentration of the dye was increased the color intensified and the percentage COD removal decreased for H-Acid and Red RB. Increase in contact time and dose of Fenton's reagent may be required to treat higher concentration of dyes.
6. It was also observed from the results that the rate of degradation of the dyes were in the order H-Acid > Red RB by Fenton's oxidation.
7. The degradation of the dyes may be due to cleavage of substituent groups as observed by the increase in sulfate concentration with COD and color removal.
8. Azo dye-Red RB was effectively degraded in UFFBA reactor.
9. A HRT of 0.405 day is prefered for effective decolorization and degradation of Red RB.
10. H Acid is very tough to degrade under laboratory but was effectively removed by Fenton's oxidation.

Acknowledgements

The authors wish to express their thanks to the Director, NEERI Nagpur, for permitting to publish this paper.

REFERENCES

1. Gordon, P. F. and Gregory P., 1987, *Organic chemistry in color*, Springer – Verlag, New York, 322.
2. Abrahart E. N., 1997, *Dyes and their intermediates* (2nd Edn.), Edward Arnold Ltd., London.
3. Sarasa, J. Roche, M. P. Ormad, M. P. Giemeno, E. Puig, A. and Ovelleiro J. L., 1998, Treatment of wastewater resulting from dyes manufacturing with ozone and chemical coagulation. *Wat. Res.* 32 (9), 2721-2727.
4. Holme I., 1984, Ecological aspects of color chemistry. In: *Developments in the chemistry and technology of organic dyes*, Griffiths, J. (ed.), Society of Chemistry Industry, Oxford, pp. 111–128.
5. Anliker R., 1979, Ecotoxicology of dyestuffs – A joint effort by industry. *Ecotoxicol. Environ. Safety.* 3, 59–74.
6. Lin, S.H. and Peng F.C., 1994, Treatment of textile wastewater by electro chemical methods, *Wat. Res.* 28, 277 – 282.
7. Lin, S. H. and Peng F. C., 1996, Continuos treatment of textile wastewater by combined coagulation, electro chemical oxidation and activated sludge, *Wat. Res.* 30 (3), 587 – 593.
8. Calabro, V., Drioli, E. and Matera F., 1991, Membrane distillation in textile wastewater treatment, *Desalination*. 83, 209 – 224.

9. Ramakrishna K. and Viraraghavan T., 1997, Dye removal using low cost adsorbents, *Wat. Sci. Technol.* 36, 189 – 196.

10. Lambert, S. D. Graham, N. J. D. Sollar, C. J. and Fowler G. D., 1997, Evaluation of inorganic adsorbents for the removal of problematic textile dyes and pesticides, *Wat. Sci. Technol.*, 36 (2-3), 173 – 180

11. Souther, R. H. and Alspaugh T. A., 1957, Textile waste treatment studies, *J. Water Pollut. Control Fed.* 29, 804.

12. Hamza A. and Hamoda M. F., 1980, Multiprocess treatment of textile wastewater, In: *Proc. 35th Purdue Ind. Waste Conf.*, West Lafayette, IN.

13. Sheng, H. L. and Chi M. L., 1993, Treatment of textile waste effluents by ozonation and chemical coagulation, *Wat. Res.* 27 (12), 1743 – 1748.

14. Haug, W., Schmidt, A., Nortemann, B., Hempel, D.C., Stolz, A. and Kncckmuss, H.J., 1991, Mineralization of the sulphonated azo dyes mordant yellow 3 by a 6 amino naphthalene-2-sulphonate degrading bacterial consortum, Appli. Environ. Microbiol. 57, 3144-49.

15. Blake D. M., 1994, *Bibliography of work on the photocatalytic removal of hazardous compounds from water and air, national renewable energy laboratory*, Golden, CO.

16. Hoffmann M. R., Martin S. T., Choi W. and Bahnemann D. W., 1995, Environmental application of semiconductor photocatalysis, *Chem. Rev.* 95, 69 – 96.

17. Huang, C. P. Dong, C. D. and Tang Z. H., 1993, Advanced chemical oxidation: its present role and potential future in hazardous waste. *Waste and Management*, 13 (5/7), 361 – 377.

18. Ikbal Koyuncu and Huseyin Afsar, 1996, Decomposition of dyes in the textile wastewater with ozone, *J. Environ. Sci. Health.* A31 (5), 1035 – 1041.

19. Bigda R. J., 1995, Consider Fenton's chemistry for wastewater treatment. *Chem. Eng. Prog.* 91, 62-66.

20. Zhu Wanpeng, Yang Zhihua and Wang Li, 1996, Application of ferrous-hydrogen peroxide for the treatment of H-Acid manufacturing process wastewater, *Wat. Res.* 30 (12), 2949 – 2954.

21. Loreto Lunar, Dolores Sicilia, Soledad Rubio, Dolores Perez-Bendito and Virich Nickel, 2000, Identification of metol degradation products under Fenton's reagent treatment using liquid chromatography – mass spectrometry, *Wat. Res.* 34 (13), 3400 – 3412.

22. Kou W. G., 1992, Decolorizing dye wastewater with Fenton's reagent. *Wat. Res.* 26 (7), 881 – 886.

23. Solozhenko, E. G., Soboleva N. M. and Goncharuk V. V., 1995, Decolorization of azo dye solutions by Fenton's reagent, *Wat. Res.* 29, 2206 – 2210.

24. Patric L. Huston and Joseph J. Pignatello, 1999, Degradation of selected pesticide active ingredients and commercial formulations in water by photo assisted Fenton reaction, *Wat. Res.* 33 (5), 1238 – 1246.

25. Shyh Fang Kang, Chih-Hsiang Liao and Hung-Pin Hung, 1999, Peroxidation treatment of dye manufacturing wastewater in the presence of ultraviolet light and ferrous ions, *J. Hazard. Mater.* B65, 317 – 333.

26. Tang W.Z. and Chen, R.Z. 1996, Discolorization kinetics and mechanisms of commercial dyes by H_2O_2/iron powder system. *Chemosphere.* 32 (5), 947 – 958.

27. Deng, N. Luo, F. Xiao, M. and Wu, X., 2000, Decoloration of aqueous solution in the UV/Fe^{o} system, *Wat. Res.* 34 (8), 2408 – 2411.

28. American Public Health Association, 1995, *Standard methods for examination of water and wastewater* (19th Edition), Washington, D. C., USA.

29. Shyh Fang Kang, Teh-Hsiu Wang and Yen-Han Lin, 1999, Decolorization and degradation of 2, 4-dinitrophenol by Fenton's reagent, *J. Environ Sci. Health*. A34 (4), 935 – 950.

30. Shyh Fang Kang, Huey-Min Chang, 1997, Coagulation of textile secondary effluents with Fenton's Reagent, *Wat. Sci. Technol.* 36 (12), 215-222.

31. Yizhong Wang, 2000, Solar photocatalytic degradation of eight commercial dyes in TiO_2 suspension, *Wat. Res.*, 34 (3), 990 – 994.

18

Wet-Oxidation of Spent Caustic Wastewater from Oil Refinery-Sulphide Conversion into Sulphate

P.P. PATHE, C.V. DESHPANDE & S.N. KAUL

Scientists, National Environmental Engineering Research Institute, Nagpur – 440 020

ABSTRACT

Presently, chemical treatment involving ferrous sulphate and chlorine gas are being used to treat spent caustic wastewater from oil refinery. Toxic components from spent caustic wastewater are removed by chemical precipitation method. During chemical reaction sodium sulphide and hydrogen sulphide are produced and remain in aqueous phase which are further precipitated out as ferric sulphide and ferric hydroxide. The insoluble flocs settles at the bottom in the form of sludge. Therefore, the sludge became toxic and hazardous in nature and faces handling and disposal problems. Besides this, high cost of chemicals used is another problem in the chemical treatment of spent caustic wastewater. Considering these facts, a new methodology of wet-air oxidation is studied in details for the treatment of spent caustic wastewater and the results are presented in this paper.

Keywords: Wet air oxidation, oil refinery wastewater, spent caustic wastewater, wastewater treatment.

Introduction

Petroleum refineries contribute wide spectrum of pollutants in their wastewaters many of these are toxic and hazardous in nature. In oil refinery, spent caustic wastewater is being generated from sweetening process in which sodium hydroxide removes sulfur/sulphide from various refinery products. Spent caustic wastewater from oil refinery is alkaline in nature (pH 12-14), pungent odour, and is toxic. The wastewater contributes major inorganic toxicants *viz.*, sulfur, hydrogen sulphide etc. These toxicants are contributed either from crude oil or generated during refining processes. Removal of these components has become a major challenge for the refineries, for overall greener or cleaner performance with associated healthy and safer environment.

Presently, the treatment employed in oil refinery for removal of toxicants from spent caustic wastewater is only chemical precipitation using chlorinated copperas (Operation Manual, 1980 & NEERI Report, 1994). In this process sulphides and hydrogen sulphide from liquid phase are converted into solid insoluble sludge, which faces handiling and disposal problems. To treat spent caustic wastewater from refinery, chlorine gas is passed through ferrous sulphate solution in reaction tank. Spent caustic wastewater is introduced at a control rate to this tank where soluble and insoluble forms of sulfides, Na_2S and H_2S are converted into insoluble forms *viz.*, FeS and S. These insoluble components are removed as a sludge, on settling in settling tank. The associated chemical reactions occur during treatment in alkaline pH range are as follows:

$$2NaOH + H_2S \rightarrow Na_2S + H_2O$$

$$3FeSO_4.7H_2O + 1.5\ Cl_2 \rightarrow Fe_2\ (SO_4)_3 + FeCl_3.2H_2O$$

$$Fe_2\ (SO_4)_3 + 3Na_2S \rightarrow Fe_2S_3 + 3\ Na_2SO_4$$

$$Fe_2S_3 \text{ hydrolysis} \rightarrow 2FeS \downarrow + S \downarrow$$

$$FeCl_3 + 3\ NaOH \rightarrow Fe\ (OH)_3 \downarrow + 3\ NaCl$$

The process removes total sulfides to an extent, and the residual sulfide can further be removed in biological treatment. However, chemical precipitation process has the following disadvantages:

1. The iron salts Fe $(OH)_3$, FeS and elemental sulfur precipitates out in the form of chemical sludge, which are the main contributors to sludge management problems.
2. Sulfide precipitation has the potential of generating toxic hydrogen sulphide gas which is an operational hazard. This can be avoided by always maintaining the reaction pH above 8.0. Even at alkaline pH hydrogen sulfide may be emitted through rapid hydrolysis where a soluble sulfide salt is added to water
3. The process is ineffective in treatment of phenols, which are detrimental to the online biological systems at down streams.

In view of above disadvantages and stringent regulations imposed by statutory authorities for disposal of toxic chemical sludges, an alternative treatment technology is need to be evolved which would completely mineralize all the toxic components present in the wastewater, without leaving behind any hazardous residues.

Oxidation of spent caustic wastewater was reported to be undertaken by using hydrogen peroxide as an oxidising agent (A. Prashant Kumar, R. Ray and A. Haldar, 2000). In this treatment method, hydrogen peroxide was used along with chlorinated copper/ferric chloride ($FeCl_3$) to oxidize sulfides into insoluble sulfur. Associated chemical reaction is as follows:

In Acidic/neutral pH Conditions

$$H_2O_2 + H_2S \rightarrow 2H_2O + S\downarrow$$

Stoichiometrically, one mole of hydrogen peroxide reacts with one mole of hydrogen sulfide but in practice, two to three times hydrogen peroxide is need for completion of reaction because of the presence of reactive materials in wastewater.

In Alkaline pH Conditions

$$4H_2O_2 + Na_2S \rightarrow Na_2SO_4 + 4\ H_2O$$

Stoichimetrically 4 moles (4.25 kg) hydrogen peroxide reacts with 1 mole (1 kg) sodium sulphide to complete the reaction. The end products are sodium sulphate and water which are non-toxic and non-hazardous, and also they do not impose any oxygen demand on subsequent treatment.

Considering cost of hydrogen peroxide, transportation, storage and handling, an alternative treatment technology is need to be explored.

Accordingly, a technology based on wet-oxidation of spent caustic wastewater using air under pressure at elevated temperature was studied. Sodium sulphide from spent caustic wastewater was oxidized at higher temperature, 110±5°C. An intermediate sodium thiosulphate is formed, which finally oxidized into sodium sulphate. The end product, sodium sulphate is non-toxic and non-hazardous and also do not impose any oxygen demand on subsequent treatment. The associated chemical reactions during treatment are as follows:

$$2\ Na_2S + 2O_2 \rightarrow Na_2S_2O_3 + 2O_2 \rightarrow 2Na_2SO_4$$

Experimental details of proposed treatment and results are discussed in this paper.

Materials and Methods

A laboratory scale model of wet-air oxidation reactor with operating capacity of 5 litres was designed and operated for the treatment of spent caustic wastewater. The schematics of the reactor is in Fig. 1.

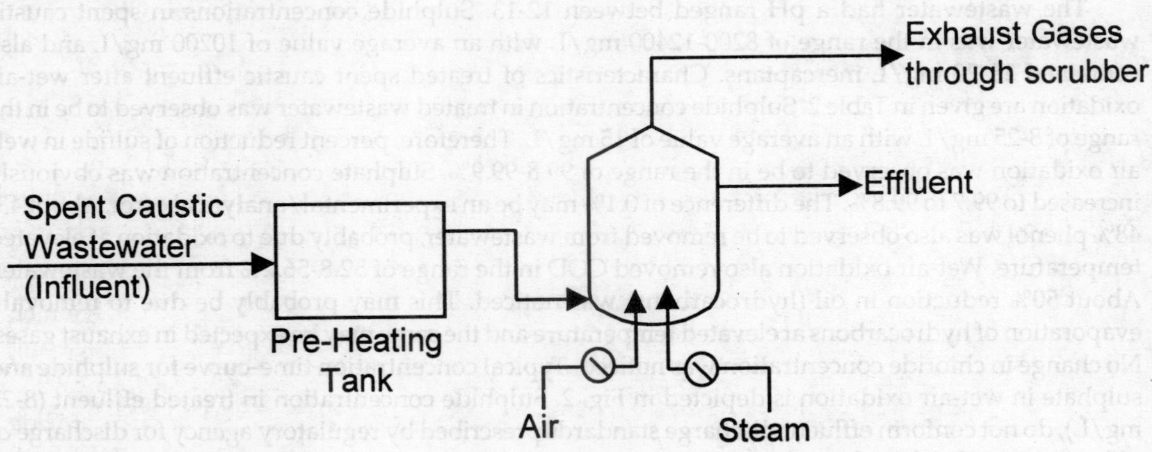

Fig. 1: Laboratory Scale Reactor for Wet-Air Oxidation in the Treatment of Spent Caustic Wastewater

A reactor made up of steel having a diameter of 210 cm and height of 250 cm was designed for wet-air oxidation. Two tubes were attached at the bottom of the reactor with non-return valves one for compressed air and other steam. A steel tank of 2.5 liter capacity was used as pre-heating chamber. Electrical heating arrangement was used for pre-heating the wastewater, to rise the temperature from ambient to 60°C. The reactor was fed by spent caustic wastewater at a flow rate of 20 ml per min. Compressed air was introduced in the reactor at a flow rate of 1.5 kg/cm² and the steam was introduced at a pressure of 1.5 kg/cm² at a temperature of 110±5°C. The exhaust gases were passed through a scrubber and cooled by a cooling systems with two spiral condesers in series.

Spent caustic wastewater before and after wet-air oxidation at various experimental conditions were analysed for various physico-chemical parameters (APPHA, AWWA, WPCF, 1989).

Results and Discussions

Characteristics of spent caustic effluent are presented in Table 1.

Table 1: Characteristics of Spent Caustic Wastewater from Oil Refinery

Sl. No.	*Parameters*	*Value*		
		Min.	*Max.*	*Avg.*
1.	pH	12	13	12-13
2.	Suspended Solids	200	400	320
3.	COD	5200	12400	8600
4.	Oil & Grease	50	54	58
5.	Phenols	300	800	540
6.	Sulphides as S	8200	12400	10200
7.	Sulphates	72	120	90
8.	Chlorides	10	30	15

All values except pH are expressed in mg/L

The wastewater had a pH ranged between 12-13. Sulphide concentrations in spent caustic wastewater was in the range of 8200-12400 mg/L with an average value of 10200 mg/L and also contained 15-20 mg/L mercaptans. Characteristics of treated spent caustic effluent after wet-air oxidation are given in Table 2. Sulphide concentration in treated wastewater was observed to be in the range of 8-25 mg/L with an average value of 15 mg/L. Therefore, percent reduction of sulfide in wet-air oxidation was observed to be in the range of 99.8-99.9%. Sulphate concentration was obviously increased to 99.7 to 99.8%. The difference of 0.1% may be an experimental/analytical error. About 43-48% phenol was also observed to be removed from wastewater, probably due to oxidation at elevated temperature. Wet-air oxidation also removed COD in the range of 52.8-56.2% from the wastewater. About 50% reduction in oil (hydrocarbons) was noticed. This may probably be due to removal/evaporation of hydrocarbons at elevated temperature and the same may be expected in exhaust gases. No change in chloride concentration was noticed. Typical concentration time-curve for sulphide and sulphate in wet-air oxidation is depicted in Fig. 2. Sulphide concentration in treated effluent (8-25 mg/L), do not conform effluent discharge standards prescribed by regulatory agency for discharge of effluent into inland surface water. This wastewater also does not conform to the standards with respect to pH, SS, COD, Oil & grease, phenols, sulphides and sulphates which needs further treatment. The quantity of spent caustic wastewater is very low and therefore, this wastewater should be mixed

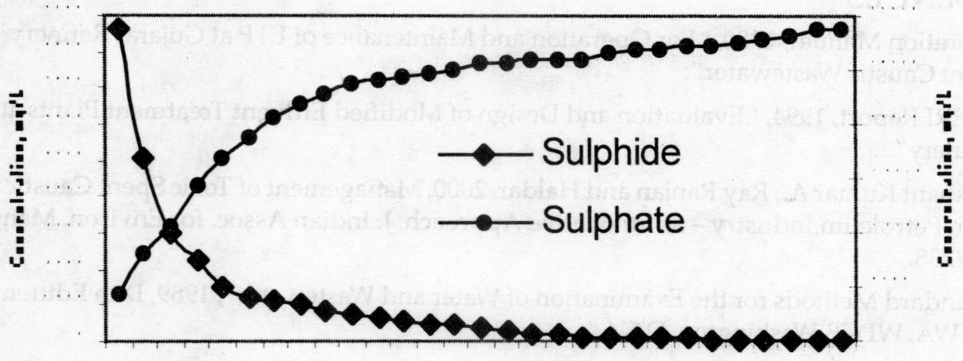

Fig. 2: Typical Concentration: Time curve for Sulphide and Sulphate in Wet-air Oxidation for Spent Caustic Wastewater

with other wastewaters at oil refinery to further reduced COD, phenol and sulphates. The combined wastewater is to be treated by usual treatment methods normally employed for treatment of wastewater from oil refinery to meet effluent discharge standards.

Table 2: Characteristics of Treated Spent Caustic Wastewater after Wet-Air Oxidation

Sl. No.	*Parameters*	*Value*		
		Min.	*Max.*	*Avg.*
1.	pH	12	13	12-13
2.	Suspended Solids	230	440	340
3.	COD	2980	5850	4400
4.	Oil & Grease	15	25	20
5.	Phenols	170	410	300
6.	Sulphides as S	8	25	15
7.	Sulphates	24050	36960	30480
8.	Chlorides	10	28	14

All values except pH are expressed in mg/L

Conclusions

Toxic and hazardous components from spent caustic wastewater can effectively be converted into non-toxic and non-hazardous chemical *viz.* sodium sulphate by using wet-air oxidation. Phenols, COD and Oils can also be partially treated during this treatment. Treated effluent from wet-air oxidation can be treated by usual technologies to further stabilized sulphide, phenols, COD and oil to conform effluent discharge standards prescribed by the regulatory authorities. As no sludge generation takes place, sludge handling and disposal problems are minimized.

REFERENCES

1. Operation Manual, 1980, "For Operation and Maintenance of ETP at Gujarat Refinery Treating Spent Caustic Wastewater".

2. NEERI Report, 1994, "Evaluation and Design of Modified Effluent Treatment Plants at Gujarat Refinery".

3. Prashant Kumar A., Ray Ranjan and Haldar, 2000, Management of Toxic Spent Caustic Effluent from Petroleum Industry – as Innovative Approach, J. Indian Assoc. for Environ. Mang. 27 (3), 233-238.

4. "Standard Methods for the Examination of Water and Wastewater", 1989, IXth Edition, APHA, AWWA, WPCF, Washington, D.C.

19

Diesel Particulate Abatement via Wall-Flow Traps Based on Perovskite Catalysts

DEBORA FINO, NUNZIO RUSSO, GUIDO SARACCO AND VITO SPECCHIA

Dipartimento di Scienza dei Materiali ed Ingegneria Chimica, Politecnico di Torino, Corso Duca degli Abruzzi 24, Italy

---

ABSTRACT

It is probably redundant to stress how extensive are nowadays the attempts to reduce the diesel particulate emissions from automotive and stationary sources. The present paper deals with a technology based on a catalytic trap based on a SiC wall-flow monolith lined with suitable catalysts for the sake of promoting a more complete and faster regeneration after particulate capture. All the major steps of the catalytic filter preparation are dealt with, including: the synthesis and choice of the proper catalyst and trap materials, the development of an *in situ* catalyst deposition technique, the bench testing of the derived catalytic wall-flow. The best catalyst selected was the perovskite $La_{0.9}K_{0.1}Cr_{0.9}O_3$. The filtration efficiency and the pressure drop of the catalytic and non-catalytic monoliths were evaluated on a diesel engine bench under various operating conditions. On the basis of the obtained results the catalysed SiC converter was found to be the most satisfactory converter to be placed on the exhaust line of the modern common-rail diesel-engine cars.

Keywords: Particulate, wall-flow monolith, $La_{0.9}K_{0.1}Cr_{0.9}O_{3-\delta}$, cromite, SiC.

Introduction

One of the most significant challenges in environmental catalysis is the removal of particulate from diesel combustion. Generally, fireplaces and cars as well as the smoke from large industrial plants are sources of particulate matter (PM). PM is present in many different sizes: the smaller the particle, the more dangerous it is, because it can travel deeper into the lungs, causing breathing problems and damage to the lungs. Besides, PM impacts the environment by decreasing visibility and harming crops and plant life. To keep on defending health and environment, the areas affected by this problem are required to develop ways to decrease PM emissions to meet future severe standards (*ė.g.* EURO IV legislation, to be enforced in 2005).

Diesel particulate filters (DPF) seem to be the primary technology for mobile sources. As high filtration efficiency as 95% can be easily reached with wall-flow traps (Heck *et al.*, 2001). Such filters (see Fig. 1) are made of high-temperature-resistant porous material such as ceramics with an as-wide-as-possible filtration area so as to decrease the pressure drop and the related fuel penalty: the monolith channels are alternatively plugged at the inlet and at the outlet; by these means the particle-laden exhaust gas is forced to pass through the wall between adjacent channels where filtration takes place according to a shallow-bed mechanism. The main hurdle for application on cars was the regeneration in all operating conditions of the trap, even though recent engine developments based on the *common rail* technology, allowed to find means to increase on purpose the exhaust gas temperature whenever needed, as later explained.

However, soot cannot be burned off effectively at temperatures (>550°C) that are likely to occur during normal driving conditions. A promising way for elimination of this pollutant are filters covered by catalytic materials, which promote the burn off of the trapped particulate (van Setten *et al.*, 2001). The role of the catalyst is to enhance the combustion rate of the soot collected on the filter by reducing its ignition temperature. As a consequence, the PM can burn off during normal driving conditions or at least at temperatures much lower than those required by non-catalytic combustion. Later on we will explain how this can turn into lower fuel penalties for the regeneration process.

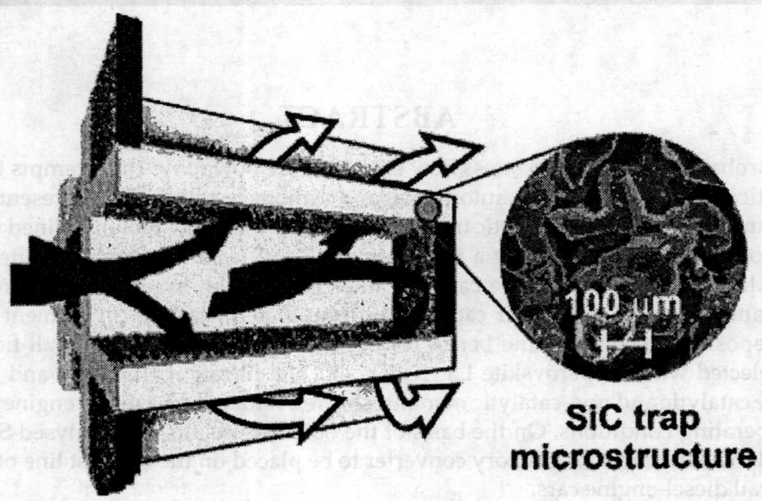

Fig. : Sketch of Wall-flow Monolith Filtration Mechanism with a SEM View of the Microstructure of a SiC Channel Wall

The focus of this study is the development of a new system based on a wall-flow SiC monoliths (shallow-bed filtration) combined with an oxidation catalyst deposited onto it (*e.g.* $LaCr_{0.9}O_3-\Delta$). As earlier introduced, a sketch concerning this last trap concept is shown in Fig. 1, which also includes micrographs of the structure of the filter.

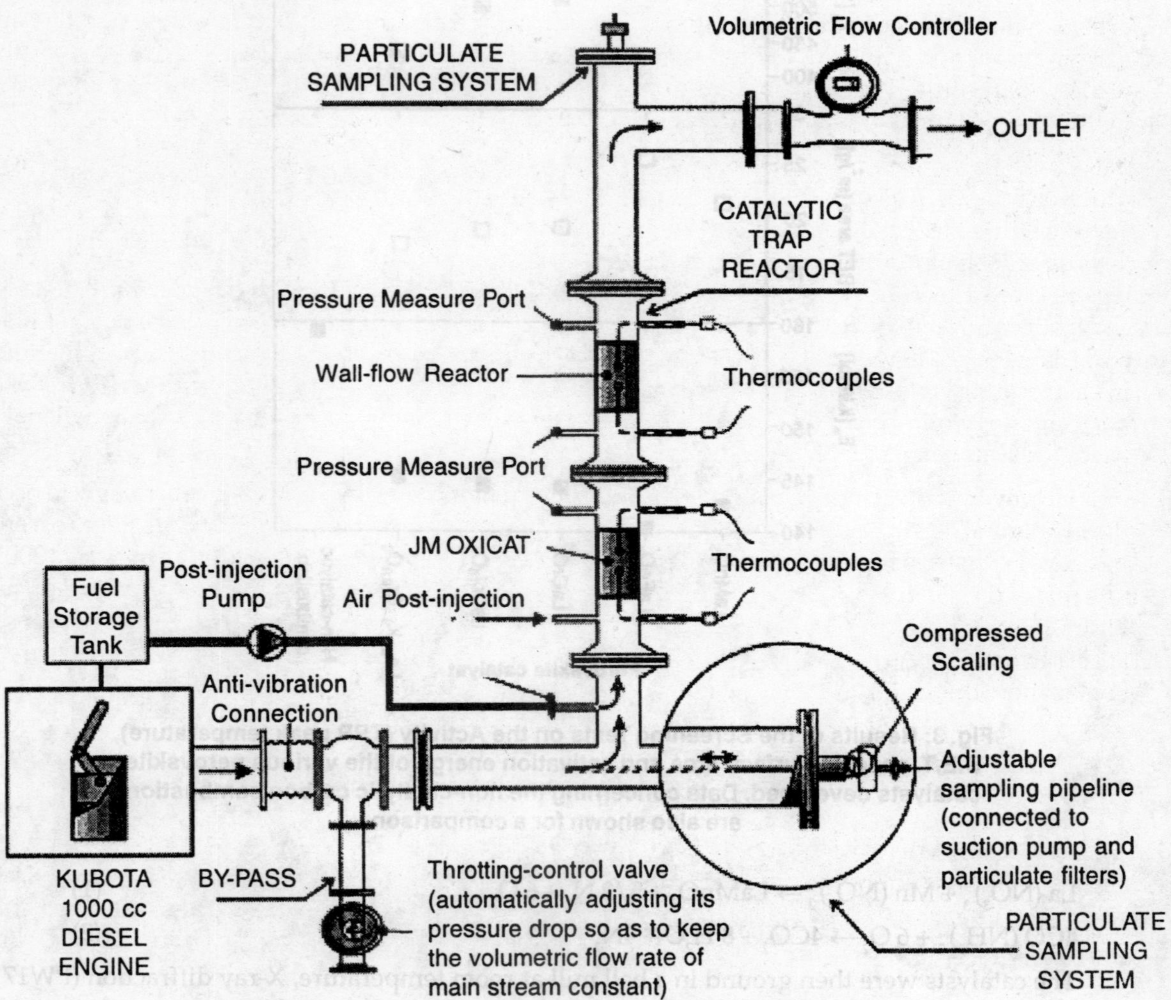

Fig. 2: Schematic View of the Engine-bench Pilot Plant

Experimental Procedure

Catalyst Preparation and Characterisation

A series of perovskite samples ($LaMnO_3$, $LaFeO_3$, $LaCrO_3$, $LaCr_{0.9}O_{3-}\delta$ and $La_{0.9}K_{0.1}Cr_{0.9}O_3-\delta$; see Fig. 3 later) were prepared via a highly exothermic and self-sustaining reaction (*combustion synthesis*; Merzahanov, 1996) to obtain nanosized particles of catalyst. The synthesis process consists of two reactions: the first concerns the real perovskite synthesis starting from the metal precursors, while the second one concerns the reaction between nitrates and urea. The entire combustion process is over within few minutes.

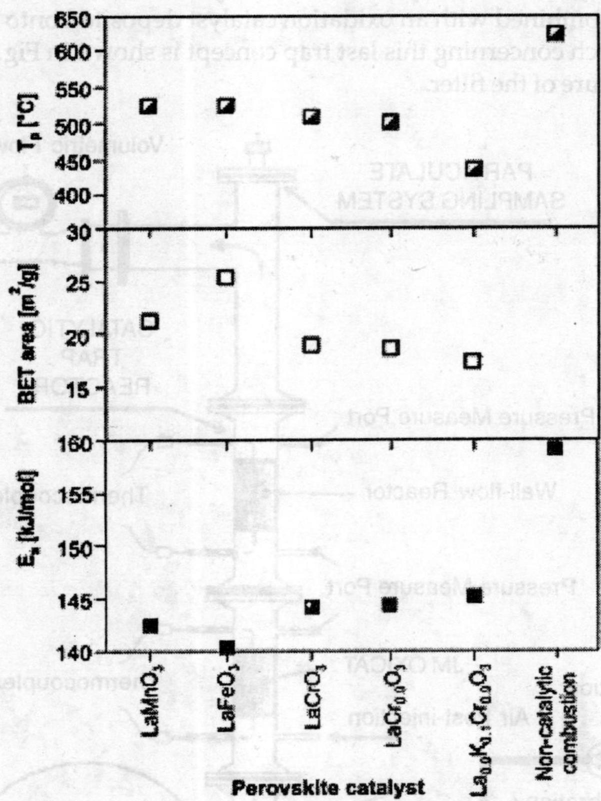

Fig. 3: Results of the Screening Tests on the Activity (TPR peak temperature), B.E.T. specific surface area and activation energy of the various perovskite catalysts developed. Data concerning the non-catalytic carbon combustion are also shown for a comparison.

$$La(NO_3)_3 + Mn(NO_3)_2 \rightarrow LaMnO_3 + 5/2\,N_2 + 6\,O_2 \qquad (1)$$

$$4CO(NH_2)_2 + 6\,O_2 \rightarrow 4CO_2 + 8\,H_2O + 4N_2 \qquad (2)$$

The catalysts were then ground in a ball mill at room temperature. X-ray diffraction (PW1710 Philips diffractometer) was used to examine whether the desired perovskite structure was actually achieved. The BET specific surface areas of the prepared catalysts were evaluated from the linear parts of the BET plot of the N_2 isotherms, using a Micromeritics ASAP 2010 analyser.

Screening Activity Tests for the Selection of the Best Catalysts

The catalytic activity of the prepared catalysts was tested in a temperature-programmed-reaction (TPR) apparatus (Fino *et al.*, 2002a, c). Air was fed at the constant rate of 50 ml·min^{-1} via a set of mass flow meters to a fixed-bed reactor enclosed in a U-shaped quartz tube. The tubular quartz reactor was loaded with of 50 mg of a 1:9 by weight mixture of carbonaceous materials and powdered catalyst and 200 mg of SiO_2 granules (0.2-0.7 mm); this inert material was adopted to reduce the specific pressure drops across the reactor and to prevent thermal runaways.

Experiments were performed by using, instead of real diesel soot, an amorphous carbon having the following properties: average particle size = 45 nm; 0.34 wt-% ashes after calcination at 800°C; 12.2 wt-% of moisture lost after drying at 110°C. This material was chosen over its real counter part so as to avoid any interfering effect due to the presence of adsorbed hydrocarbons, sulfates or fly-ashes, well-known components of real diesel soot. Furthermore, the amorphous carbon employed is more difficult to burn than real diesel soot, which renders the achieved results conservative.

A proper and reproducible degree of mixing between catalyst and carbon was obtained by grinding carefully together such substances for 15 min in a ball mill. The catalyst-carbon-SiO_2 fixed-bed was sandwiched between two quartz-wool layers, while the tip of a K-type thermocouple was located well inside the bed itself. The reaction temperature was controlled through a PID-regulation system and varied during each TPR runs from 200 to 700°C at a 5°C·min^{-1} rate.

The outlet gas composition was checked through CO_2 NDIR continuous analyzers (Hartmann & Braun URAS 10E). The signal from the analyzer was sent to a computer-controlled data acquisition system to collect CO_2 concentrations as a function of the bed temperature. The inlet and outlet flow rates were measured by means of a digital-flow meter.

Screening tests were performed in the TPR apparatus according to the standard operating procedures listed above in order to determine some critical performance parameters of each of the prepared catalysts (*e.g.* catalytic activity toward carbon combustion, selectivity to CO_2, optimal operating temperature range, etc.): the aim of these tests was to find the most promising catalysts for diesel particulate combustion. An index of their activity can be found in the peak temperature (T_p) of the TPR plot: the lower the T_p value, the higher the catalyst activity.

Differential Scanning Calorimetry (DSC)

The activation energy E_a of carbon particulate combustion over the prepared catalysts was measured via differential scanning calorimetry (Perkin Elmer DSC-Pyris) and the so-called Ozawa method (Ozawa, 1970; Ozawa, 1975). DSC experiments were performed on mixtures of catalysts with carbon to measure the heat released as an index of the evolution of catalytic combustion. Under an air flow of 100 ml·min^{-1}, a mixture of catalyst and carbon (2:1 mass basis) was placed in the sample crucible whereas an equal weight of alumina was used as a reference. Temperature was increased from 50 to 700 °C, with different heating rates (Φ = 5, 10, 20, 30, and 50 °C/min). Several plots representing exothermal combustion peaks were obtained. These patterns were processed to point out the temperature corresponding to the cumulative combustion (A) of 25, 50 and 75 wt% of the total initial carbon located in the sample holder. Some tests were also performed on an alumina/carbon mixture to evaluate the activation energy of the non-catalysed carbon combustion.

The activation energy can be evaluated through the so-called Ozawa procedure (Ozawa, 1970) by proper interpretation of thermal analysis data (DSC in the present investigation). If the oxygen partial pressure is not controlling (undiluted air is used in the DSC runs, which hinders this occurrence), the variation of the fraction A of converted carbon in a DSC test can be related to the remaining carbon fraction (1-A):

$$\frac{d\alpha}{dt} = k'(T)\,f(1-\alpha), \tag{1}$$

where

f (1-α) is a whatever function and k' is a kinetic constant depending only on temperature via an Arrhenius expression. If the DSC heating rate dT/dt=Φ is fixed, the former equation becomes:

$$\frac{d\alpha}{f(\alpha)} = \frac{Z}{\Phi}\exp\left(-\frac{E}{RT}\right)dT \tag{3}$$

which can be integrated as follows:

$$\int_0^{\alpha} \frac{d\alpha}{f(\alpha)} = \int_{T_0}^{T_\alpha} \frac{Z}{\Phi}\exp\left(-\frac{E}{RT}\right)dT \tag{4}$$

where

Tα is the temperature reached when a weight fraction α of the initial carbon is burned. The above integral cannot be solved analytically, however Ozawa provided a good approximation of its solution:

$$ln\Phi = B - 0.4567 \cdot \left(\frac{E_a}{RT_\alpha}\right) \tag{5}$$

where

B is a constant lumping A-dependent terms. If the heat released by the combustion is assumed to be proportional to the fraction α of converted carbon, once chosen a reference α value (*e.g.* 50%), the Tα value corresponding to such value can be easily derived from the DSC curves by evaluating, via a simple integration, the amount of heat released by combustion. By least square fitting of the *ln*Φ-vs.-1/Tα data series estimates of the activation energy can be derived from the slope of the best-fitting line.

In situ Catalyst Deposition Within the Traps

The most promising catalyst ($La_{0.9}K_{0.1}Cr_{0.9}O_3-\delta$, as detailed later) was deposited by *in situ* combustion synthesis directly over the wall-flow filters. The ceramic support was dipped in the aqueous solution with precursors (exactly the same used for powdered catalyst production) and then placed into an oven at 600 °C. The aqueous phase was rapidly brought to the boiling point, then the precursors mixture ignites and the main synthesis reaction takes place *in situ*.

The support selected was silicon carbide (SiC) filter, produced and industrialised by IBIDEN (cell structure = 14/200, diameter 30 mm, length = 6 – 12"; pore diameter of channel walls = 9 μm, porosity of channel walls = 42 %). A good chemical compatibility between catalysts and ceramic material was in fact assessed by means of XRD analysis after a suitable thermal treatment. Powders of both wall-flow materials were mixed with $La_{0.9}K_{0.1}Cr_{0.9}O_3-\delta$ (weight ratio of 1:1) and kept for 2 hours at 900 °C. The mixed powders were then characterised by XRD analysis in order to investigate the possible formation of new crystalline species by chemical reaction with the catalyst. Finally, this ceramic material-catalyst mixture was mixed with carbon so as to obtain samples for TPO analysis and assess the change of catalytic activity eventually occurred.

Conversely, the catalytic traps obtained underwent a microstructural and morphological characterisation by means of electron microscopy and energy dispersive analysis.

A final adhesion test was carried out by simulating the typical engine vibrations using an ultrasonic bath and weighting the samples after different periods to verify the bond of the catalytic layer to the support and eventual loss of solid materials. The adhesion properties between catalyst and ceramic surface was checked by means of ultrasonic bath analysis: a piece of the catalytic monolith was weighted before and after the ultrasonic treatment to quantify the catalyst loss. Finally, the pressure drop through the trap was measured in a specific apparatus a s a function of the superficial gas velocity and of the amount of catalyst deposited. Such amount could indeed be varied by changing the precursors concentration in the impregnating solution.

Diesel Engine Bench Tests

The developed traps were tested in the pilot plant sketched in Fig. 5 on real Diesel exhaust gases (Kubota 1000 cc IDI engine, capable of up to 23.5 hp at 3000 rpm), where the temperature and gas composition before and after the trap can be controlled and monitored as well as the filtration efficiency and the evolution of the pressure drop through the trap (a sign of soot accumulation therein). A detailed description of the plant was provided by Fino *et al.* (2002a).

The wall-flow was located along the main line of the main exhaust stream. The exhaust gas superficial velocity across the trap could be controlled at a desired value by measuring the exhaust flow rate through a volumetric flow controller connected to a throttling control valve placed on by-pass exhaust stream. The pressure drop across the trap could be measured by means of differential pressure transducers (VIKA) whereas the trap inlet and outlet temperature was measured by K-type thermocouples at axial positions. The soot concentration in the exhaust gases was measured by isokinetic sampling of a small flue-gas flow rate through a suction pump and collecting the suspended particulate using two-filters system (PALLFLEX 47 TX 40 HI 20-W). Finally, gas phase composition could be monitored through continuous analyzers by ABB (NDIR for NO, CO, CO_2, SO_2; FID for overall HC; paramagnetic for O_2).

The standard bench test procedure can be described as follows. The traps were at first loaded by letting rather cold (about 200°C) exhaust gases flow throughout it until a 110-160 mbar pressure drop is reached (corresponding to a particulate hold-up of about 10 $g \cdot L^{-1}$). Then, regeneration is induced by post injecting some fuel with a metering pump in the exhaust gases and by burning these HC with an oxidizing honeycomb catalyst (OXICAT by Johnson Matthey) placed just upstream the trap. The time needed for a complete trap regeneration (*e.g.* combustion of the soot hold-up) is an index of the catalyst performance. The higher the catalyst activity the lower the time required.

Finally, all measurements have been carried out by using as a fuel the low sulphur (< 50 ppmw) diesel produced by Haltermann named *"CEC RF 06 99 BATCH 5"* in line with the pending 2005 EEC regulations.

Results and Discussion

Figure 3 summarises the TPO peak temperature, B.E.T. specific surface area and activation energy values of all the perovskite catalysts tested. These materials show TPR peak temperatures significantly lower than 650°C (non-catalytic combustion). The superior activities of $LaCr_{0.9}O_3{-}\delta$ and $La_{0.9}K_{0.1}Cr_{0.9}O_3{-}\delta$ perovskites can be clearly noticed. $La_{0.9}K_{0.1}Cr_{0.9}O_3{-}\delta$ in particular shows a CO_2 peak temperature (T_p) at only 455°C. Owing to the nature of chromium, this transition metal can in part undergo a valence shift from Cr^{3+} to Cr^{4+} so as to re-establish electro-neutrality, despite the substoichiometric composition of the above mentioned chromite.

The presence of Cr^{4+} was demonstrated to promote an increase of catalytic activity towards carbon combustion (Saracco *et al.*, 1996). The most likely explanation is that the perovskite can act as an oxygen pump towards the soot. Oxygen can indeed be made available to PM combustion in combination with a temporary shift of Cr valence from 4+ back to 3+. Oxygen molecules coming from the gaseous atmosphere can re-oxidise the perovskite and set back Cr^{4+} species. Moreover, K-substituted lanthanum cromite is characterized by a TPR peak lower than the other chromites. As the monovalent K ion substitutes the trivalent La one a further fraction of chromium is forced to shift from the 3+ to the 4+ valence status, to the benefit of catalytic activity.

Nevertheless, it is surprising how the catalyst belonging to chromite class are more active than $LaMnO_3$ and $LaFeO_3$ catalysts, which are much better catalysts for the combustion of methane or light hydrocarbons (Saracco *et al.*, 1999b). The values of activation energy calculated for the various prepared perovskites are not very different from one another: this feature does not facilitate the comprehension of this phenomenon. Apparently, the different perovskites are capable of delivering oxygen species of similar nature to the reacting particulate, which allows for an analogous decrease of the activation energy value compared to the non-catalytic combustion case.

Shifting to the B.E.T. values, it has to be stated that despite the short duration of the combustion synthesis, it is possible to customise both the degree of crystallisation and the BET surface area, by optimising the synthesis conditions (in particular the urea-to-nitrates ratio). It is worth noticing that perovskites with quite high values of specific surface area can be obtained (compared to a lower-specific-surface-area achieving by the "citrates" method; *i.e.* a few m^2/g). Unfortunately, the value related to chromite is lower than those of the manganite and the ferrite perovskites, which does not help at all in the explanation of the superior catalytic activity of the $LaCr_{0.9}O_3{-}\delta$ pervoskite compared to its counter parts.

All things considered it has to be guessed that the surface concentration of reactive oxygen species over the surface of chromites should be much higher than that of the manganite and the ferrite. Studies are in progress to better elucidate this point by temperature-programmed desorption and reduction analysis.

Anyhow, on the grounds of its superior activity, $La_{0.9}K_{0.1}Cr_{0.9}O_3{-}\delta$ was selected for the catalytic activation of wall-flow monoliths. Shifting to the issues related to catalyst deposition, it has to be stated that the adhesion between the deposited catalyst and the channel walls of the traps was excellent, as the loss of catalyst by ultrasonic treatment was lower than 1 % (see Table 1).

Table 1: Weight Decrease of the Catalytic Wall-flow Traps After Ultrasonic Bath Treatments of Different Duration

Time (min)	*Weight (%)*
0	—
15	0.16
30	0.28
60	0.25

Different morphology of the catalytic surface was obtained using different concentration of urea, probably because of the different amount of energy and gases generated by the synthesis reaction. To these merits, Fig. 4 shows SEM pictures of catalysts lined via *in situ* deposition on the upstream channel-walls silicon carbide wall-flow traps: the catalyst is in all the cases strongly bonded to the support and the microstructure obtained appears extremely foamy. During combustion synthesis, the decomposition of reacting precursors generates a large amount of gaseous products in a very short period of time, which leads to a spongy catalyst morphology. This feature represents a great advantage, as it favours the contact between the catalyst and the soot that accumulates over it.

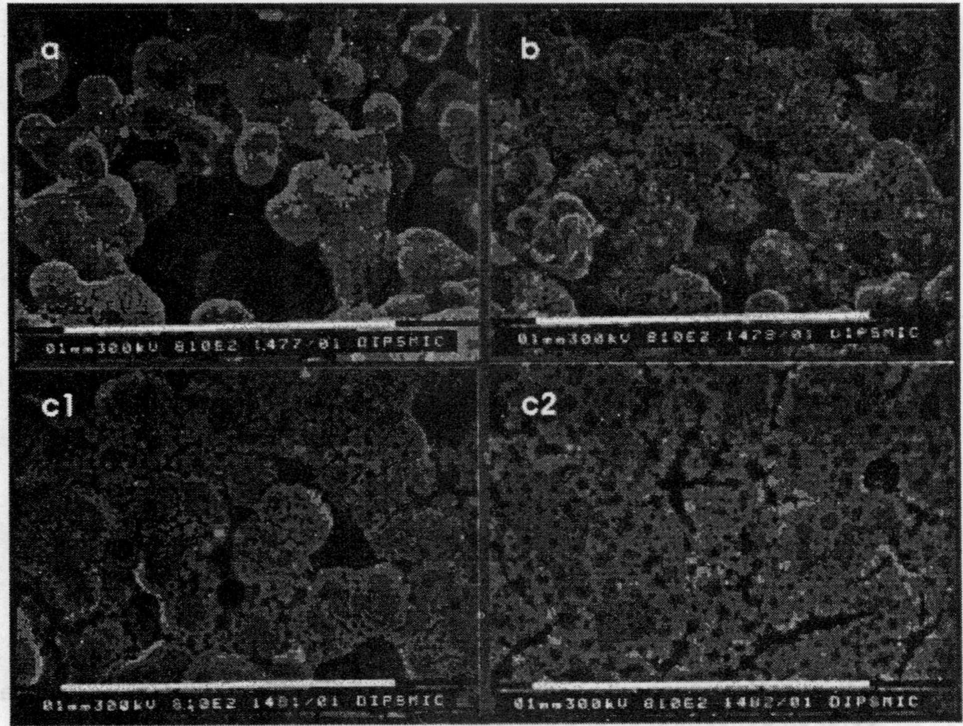

Fig. 4: Catalyst Deposition on SiC Monolith Using Different Concentrations of Precursors Solution: (a) 50 g/l, (b) 100g/l, (c1) 200 g/l-1 impregnation, (c2) 200 g/l-2 impregnations

Furthermore, in order to correlate the concentration of the solution of precursors with the load and the thickness of catalyst deposited onto the monolith, a systematic study by SEM and gravimetric analysis was carried out. Moreover the permeability of the wall flow monolith with different catalyst load was assessed at different space velocities. The results of these experiments are shown in Fig. 5.

Beyond precursors concentrations higher than 100 g/l, the pressure drop across the trap increases with a higher slope, as a likely consequence of pore plugging (see *e.g.* Fig. 4.c2). Catalyst layers with thickness ranging from 0.5 to 30 Mm were obtained using precursors solution with a concentration from 50 up to 200 g/l. A value of 140-150 g/l of mother solution concentration was selected in order to obtain a catalyst load of 10 wt% so as to maximise the catalytic wall-flow monolith performance without a remarkable permeability reduction.

Coming finally to the analysis of the performance of the catalytic trap, Fig. 6 compares the behaviour of the catalytic and the non catalytic traps during a standard loading-regeneration cycle.

After loading of the trap up to about 110 mbar pressure drops, the regeneration is achieved only for the $La_{0.9}K_{0.1}Cr_{0.9}O_3-\delta$–catalysed trap, as soon as an inlet trap temperature of about 600°C is reached by fuel-post-injection (the fuel penalty is about 20% of the entire diesel burned) and catalytic combustion over the OXICAT monolith.

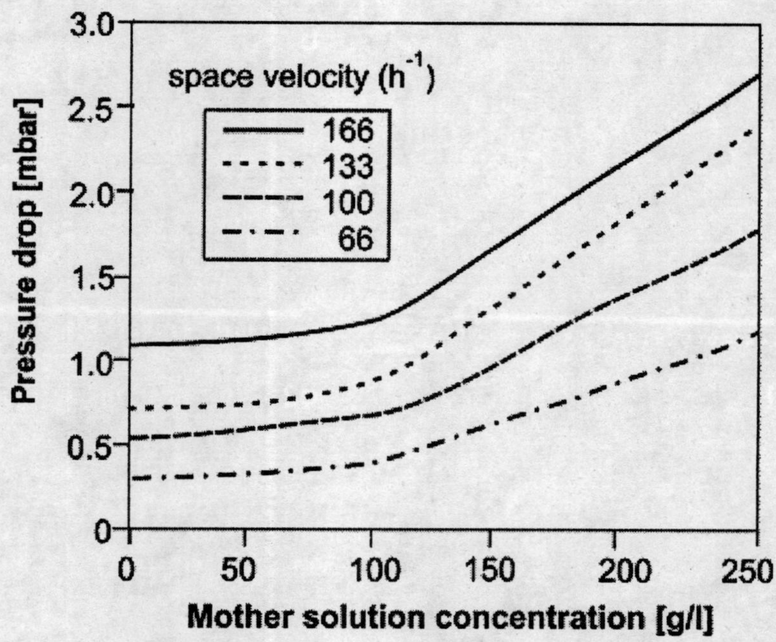

Fig. 5: Pressure Drop Throughout Catalytic SiC Wall-flow Monoliths as a Function of the Mother Solution Concentration and of the Space Velocity of Air (Room Temperature)

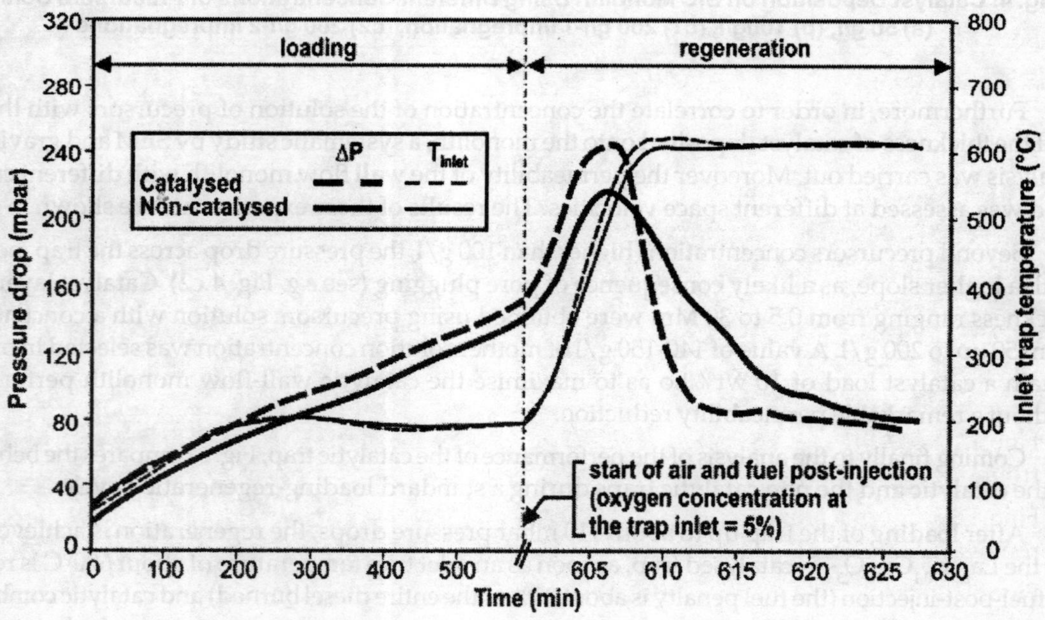

Fig. 6: Trap Loading and Regeneration for a Catalytic ($La_{0.9}K_{0.1}Cr_{0.9}O_3$) and a Non-catalytic Wall-flow Trap (SiC by Ibiden)

As shown in Fig. 6, after loading of the SiC-based traps up to a pressure drop of about 150 mbar both counter parts gets regenerated at standard test conditions; only the non-catalytic SiC wall-flow monolith needed air post-injection. This negative feature is complex to explain and is related to several factors influencing soot combustion kinetics (*e.g.* temperature, oxygen content in the flue gases). The need to burn out a significant flow rate of fuel so as to rise the exhaust gas temperature from 200 to 600°C, leads to an almost compete depletion of the oxygen concentration downstream the OXICAT converter (only 0.5 vol% remains available against the 9 vol% value emitted by the engine). While the catalyst succeeds to capitalise only this small amount of O_2, this is simply not enough for the non-catalytic combustion and, in this case only, the level of oxygen had to be raised up of 3 vol% by air post-injection.

In any case, the regeneration of the $La_{0.9}K_{0.1}Cr_{0.9}O_3-\delta$–catalysed trap is even much faster than that with the non-catalytic ceramic filter. By these means, beyond the advantage of eliminating the need of air post-injection, it becomes possible to save a significant amount of fuel at any filter regeneration, therefore reducing working costs. It has to be kept into account however that part of the fuel saving cost is compensated by a slight decrease of filter permeability owing to the presence of the catalyst.

Conclusions

Nanosized particles of several perovskite catalysts for diesel particulate combustion have been obtained by combustion synthesis. Selection of the best catalyst was operated by standardised reaction runs and a tentative explanation of the reaction mechanism has been proposed.

The best catalyst developed ($La_{0.9}K_{0.1}Cr_{0.9}O_3-\delta$) was lined via an *ad-hoc* developed technique over the channel walls of SiC wall-flow traps. The results achieved for the SiC-catalysed monoliths vs. the non-catalytic ones in terms of lower regeneration times and absence of need of air post-injection, suggest that the catalytic trap technology should provide the best tool to offset the serious problem of PM abatement down to less than 0.025 g/km, as required the 2005 EEC regulations.

Both experimental and modelling studies are currently in progress to develop new and more active catalysts, to further improve the deposition technique for the sake of reducing the entailed pressure drop increase, and to design a trap structure optimised for this promising regeneration process.

Acknowledgements

Financial support of the European Community (projects ART-DEXA and SYLOC-DEXA) is gratefully acknowledged.

REFERENCES

1. Ciambelli P., Corbo P., Gambino M., Palma V. & Vaccaro S., 1996, Catalytic combustion of carbon particulate, *Catal. Today*, 27, 99-106.

2. Fino, D., Saracco, G. and Specchia, V., 2002a, "Filtration and catalytic abatement of diesel particulate from stationary sources", *Chem. Eng. Sci.*, in press.

3. Fino, D., Saracco, G. and Specchia, V., 2002b, "Development of a catalytic trap for diesel particulate removal", *Industrial Ceramics*, in press.

4. Fino, D., Fino P., Saracco G. and Specchia V., 2002c, "Diesel Particulate traps regenerated by catalytic combustion", Korean Journal of Chemical Engineering, in press.

5. Gieshoff, J., Pfeifer, M., Schäfer-Sindlinger, A., Hackbarth, U., Teysset, O., Colignon, C., Rigaudeau, C., Salvat, O., Krieg, H. and Wenclawiak, B.W., 2001, "Regeneration of catalytic diesel particulate filters", SAE Technical Paper 2001-01-0907.

6. Heck R.M., Gulati S., Farrauto R.J., 2001, Chem. Eng. J. 82: 149-156.

7. Merzahanov, A. G., 1996, "Combustion process that synthesize materials", *J. of Mat. Proc. Techn.*, 56, 222-241.

8. Ozawa T., (1970), Kinetic analysis of derivate curves in thermal analysis, *J. Thermal Anal.*, 2, 301-324.

9. Ozawa T., (1975), Critical investigation of methods for kinetic analysis of thermo-analytical data, *J. Thermal Anal.*, 7, 601-617.

10. Saracco, G., Scibilia, G., Iannibello, A. and Baldi G., 1996, "Methane combustion on Mg-doped $LaCrO_3$ perovskite catalysts", *Appl. Catal. B: Environmental*, 8, 229-244.

11. Saracco, G., Badini, C. and Specchia, V., 1999a, "Catalytic traps for diesel particulate control", *Chem. Eng. Sci.*, 54, 3041-3047.

12. Saracco, G., Geobaldo, F. and Baldi, G., 1999b, "Methane combustion on Mg-doped $LaMnO_3$ perovskite catalysts", *Appl. Catal. B: Environmental*, 20 (4), 277-288.

13. van Setten, B.A.A.L., Makkee, M. and Moulijn, J.A., 2001, "Science and technology of catalytic diesel particulate filters", *Catal. Rev.-Sci. Eng.*, 43, 489-564.

20

Photocatalytic Degradation of Wastewater Pollutants: Titanium Dioxide Mediated Degradation of Methyl Orange and β-Naphthol Orange

P.K. SUDHADEVI ANTHARJANAM, ROBERT PHILIP & SURESH DAS

Photochemistry Research Unit, Regional Research Laboratory (CSIR), Trivandrum – 695 019

ABSTRACT

The photocatalytic degradation of azo dyes such as methyl orange and β-naphthol orange in aqueous suspensions of TiO_2 has been investigated under a variety of conditions. The kinetics of degradation was studied under different conditions such as reaction pH, substrate and catalyst concentration, types of titanium dioxide used and in the presence of electron acceptors and electron donors. The degradation rates of the dyes have been found to be strongly influenced by all the above parameters. Also the results of the photocatalyzed degradation of methyl orange in aqueous solution using titanium dioxide immobilized on glass fiber cloths are presented. Carbon dioxide yield measurements indicate that only partial mineralization occurs in the initial phase of oxidation.

Introduction

Azo dyes represent a major class of synthetic organic pigments that are manufactured worldwide and have a variety of applications such as for textiles, paper, foodstuffs and cosmetics (1). The toxicity and carcinogenic nature of these dyes and their precursors pose a threat to the environment (1-5). Moreover their degradation often leads to the formation of highly carcinogenic aromatic amines. For example, *p*-aminoazobenzene has been classified as a carcinogenic compound and there have been restrictions on the production of dyes based on this molecule. However, azo dyes continue to be a source of pollution in industrial processes, which utilize dyes to color paper, plastics as well as natural and artificial fibers. From the environment viewpoint, it is estimated that 10 %-15 % of the dye is lost during the dying processes and is released as effluent, being resistant to aerobic degradation (6). Decolourization of dye effluents has therefore received increasing attention.

The use of semiconductors for the photocatalyzed degradation of organic substrates as a method of treatment for wastewater is currently attracting considerable attention (7, 8). The initial step in the TiO_2 mediated photocatalyzed degradation is proposed to involve the generation of electron hole pair, which migrate to the photocatalyst surface to yield hydroxyl and superoxide radical anion. It has been commonly accepted that the hydroxyl and superoxide radical anions can initiate redox processes of the adsorbed substrate.

Earlier studies have shown that a wide range of organic substrates can be photomineralized in the presence of anatase titanium dioxide (7-24). However, despite its promise, the development of a practical treatment system based on the heterogeneous photocatalysis has not yet been successfully achieved because of many operational parameters that must be considered (25). In particular, the success of photocatalytic degradation in wastewater treatment will depend on how much energy costs can be brought down. This can be achieved by improving photocatalytic degradation efficiency such as by catalytic modification, optimization of process conditions and addition of specific additives to the system. The powdered TiO_2 photocatalyst has an advantage that the reaction rate is high, but it has also the disadvantage that the separation process of the powder from water is usually necessary after the reaction. Since it is difficult to separate the fine TiO_2 powder catalyst from the treated wastewater by any simple operation, immobilization of the catalyst on a support can provide a viable alternative for such processes (26-29).

Although the photoreduction of methyl orange and β-naphthol orange in the presence of colloidal TiO_2 has been reported earlier (30-32), no major efforts have been made to study the degradation kinetics of these dyes. In the present article, we present a detailed kinetic study on the photocatalytic degradation of the dyes 1 and 2 (Chart 1) sensitized by TiO_2. The effects of parameters such as pH, initial dye concentration, catalyst loading are examined. We have also presented methods to improve photocatalytic degradation of these dyes such as the addition of efficient electron acceptors and

$N(CH_3)_2$ HO N=N NaO_3S N=N NaO_3S

1 (Methyl orange) **2 (β-Naphthol orange)**

Chart I

donors. The efficiency of TiO_2 coated on a glass fiber cloth for the photocatalytic degradation of methyl orange has also been investigated here.

Experimental Details

Methyl orange and B-naphthol orange were prepared according to a reported procedure (33). Titanium isopropoxide was obtained from Aldrich and used as such without further treatment. Laboratory reagent grade hydrogen peroxide (H_2O_2), potassium bromate ($KBrO_3$) ethlylenediaminetetracetic acid (EDTA), triethylamine (Et_3N) and diethylamine (Et_2N) were used in this study. Double distilled water was used in all experiments. P25 from Degussa, Hombikat UV 100 from Sachtleben Chemie and TiO_2 obtained from Travancore Titanium Products, India (TTP) were used as photocatalysts in this study. Degussa P25 consists of 75 % anatase and 25 % rutile with a specific BET surface area of 50 m^2g^{-1} and a particle size of 30 nm (34). Hombikat UV 100 consists of 100 % pure anatase with a specific BET surface area of 250 m^2g^{-1} and a particle size of 5 nm (35). The TiO_2 obtained from Travancore Titanium Products, Trivandrum, India with a surface area of 9.82 m^2g^{-1} was used for comparative studies.

Irradiations were carried out either using a 200 W super high pressure mercury lamp on an optical bench or a 450 W medium pressure mercury lamp placed inside the well of a double walled glass photoreactor of 500 ml capacity. The output of 200 W lamp was filtered through a Pyrex filter (optical cut off 305 nm long pass filter) and focussed on to a two necked Pyrex round bottomed flask, which was magnetically stirred. The TiO_2 immobilized reactions were carried out with the 450 W lamp reactor. TiO_2 was immobilized on a glass fiber cloth by soaking the cloth in a solution of titanium isopropoxide in isopropanol. The cloth was dried in an argon chamber and calcined at 450°C for 4 h to form TiO_2 film.

Carbon dioxide measurements were carried out in aqueous solutions irradiated with the 450 W lamp. Samples were collected before and at regular intervals during the irradiation, centrifuged and taken for further analysis. The degradation of the dyes was followed by recording the absorption spectra of the irradiated samples.

Control experiments were carried out in all cases employing unirradiated blank solutions. There was no observable loss of the dyes on irradiation in the absence of TiO_2.

Results and Discussion

Effect of Oxygen on Photocatalyzed Degradation of Methyl Orange

Methyl orange undergoes slow photooxidation on irradiation with UV light. Figure 1 (curve a) shows the decrease in absorbance intensity at 464 nm on irradiation of an aqueous solution of methyl orange by pyrex filtered light from a 200 W super high pressure mercury lamp, in the absence of TiO_2.

The rate of photooxidation is however greatly enhanced when amorphous TiO_2 (Degussa P25) is added to the solution (Figure 1, curves b-d). Control experiments were carried out by keeping the TiO_2 containing solution of the dye in the dark. Only a minor loss of dye absorption, which could be attributed to adsorption of the dye on to TiO_2 could be observed. Selective excitation of the dye in the presence of TiO_2 by using 425 nm cut off filters also did not result in any significant loss in the absorbance of the dye. The rate of photodegradation of the dye in the presence of TiO_2 is very sensitive to the presence of oxygen, increasing with increasing oxygen concentration.

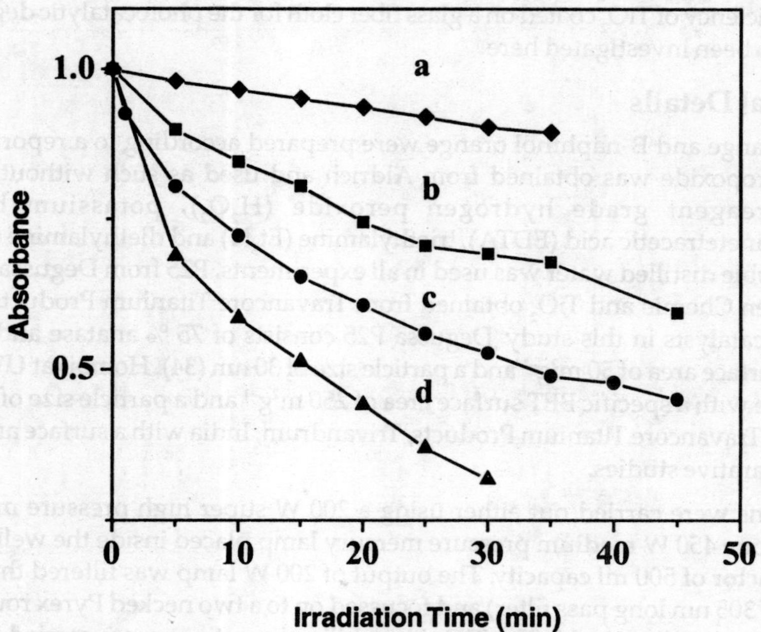

Fig. 1: Change in Intensity at 464 nm *vs* Irradiation Time for Aqueous Solution of Methyl Orange (30 μM); in absence of TiO_2 (a); in presence of 50 mg/L Degussa P 25 TiO_2 (b); nitrogen saturated (c); and oxygen saturated (d)

Figure 2 shows the effect of oxygen on the rate of photocatalyzed degradation in aqueous solution of methyl orange containing colloidal TiO_2. Colloidal TiO_2 was prepared by hydrolysis of titanium isopropoxide in isopropanol at pH 2. Under these conditions, nanosized TiO_2 particles (diameter10 nm) are formed (32). Since the particle size of TiO_2 is less than the wavelength of light such solutions are transparent to visible light. From Figure 2 it becomes evident that the effect of oxygen in the photocatalyzed oxidation in colloidal TiO_2 is the reverse of that which is observed in the presence of amorphous P25 (Degussa). In colloidal TiO_2, the rate of decolourization is the maximum for nitrogen bubbled solutions and the rate decreases with increasing oxygen concentration.

The difference in the effect of oxygen in amorphous and colloidal TiO_2 suggests that different mechanisms are involved. Increasing concentration of oxygen would lead to more efficient scavenging of the electron in the conduction band of the irradiated semiconductor. This would lead to a reduction of the electron-hole pair recombination resulting in the increased yield of hydroxyl radicals. Hydroxyl radicals thus formed can react with methyl orange both by electron transfer and adduct formation as shown in Scheme 1. The mechanism of the primary reactions of hydroxyl radicals with azo dye has been reported recently (36).

$$\text{Dye} + \dot{\text{O}}\text{H} \longrightarrow [\text{Dye}]^{\bullet+} + \overset{-}{\text{O}}\text{H}$$

$$\text{Dye} + \text{OH} \longrightarrow [\text{Dye—OH}]$$

Scheme 1

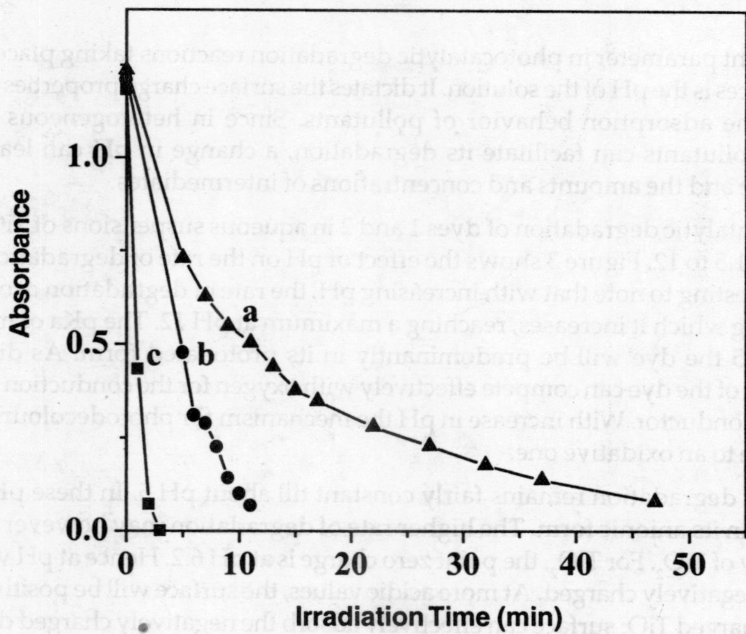

Fig. 2: Change in Intensity at 512 nm *vs* Irradiation Time for Aqueous Solution of Methyl Orange (30 µM) containing colloidal TiO_2 (6 mM); (a) oxygen saturated, (b) air saturated and (c) nitrogen saturated

Subsequent reaction of oxygen with the intermediate radicals can result in the formation of oxidized intermediates, which can undergo further TiO_2 catalyzed degradation resulting in the mineralization of the dye.

In colloidal TiO_2 solutions, the detrimental effect of oxygen on the rate of decolourization indicates that under these conditions the dye competes with oxygen for the conduction band electron. Under these conditions, decolourization occurs due to the reduction of the dye as shown in Scheme 2.

The difference in the mechanism under the two conditions may be attributed to the pH of the solutions. At the acidic pH (pH = 2) used for stabilizing colloidal TiO_2 the azo dye would be protonated, making it a better electron acceptor. The rate of pH on the photocatalyzed degradation of the dyes was investigated in more detail.

$NaO_3S-C_6H_4-N=N-C_6H_4-N(CH_3)_2 \xrightarrow{2e^-,\ 2H^+} NaO_3S-C_6H_4-NH-NH-C_6H_4-N(CH_3)_2$

$\xrightarrow{2H^+,\ 2e^-} NaO_3S-C_6H_4-NH_2 + H_2N-C_6H_4-N(CH_3)_2$

Scheme 2

Effect of pH

An important parameter in photocatalytic degradation reactions taking place on semiconductor particulate surfaces is the pH of the solution. It dictates the surface charge properties of the photocatalyst and therefore the adsorption behavior of pollutants. Since in heterogeneous photocatalysis the adsorption of pollutants can facilitate its degradation, a change in pH can lead to the change in degradation rate and the amounts and concentrations of intermediates.

The photocatalytic degradation of dyes 1 and 2 in aqueous suspensions of TiO_2 has been studied in the pH range 1.5 to 12. Figure 3 shows the effect of pH on the rate of degradation of methyl orange on pH. It is interesting to note that with increasing pH, the rate of degradation drops to a minimum at pH 7.7, following which it increases, reaching a maximum at pH 12. The pKa of methyl orange is 3.5. At pH below 3.5 the dye will be predominantly in its protonated form. As discussed earlier the protonated form of the dye can compete effectively with oxygen for the conduction band electron of the irradiated semiconductor. With increase in pH the mechanism for photodecolourization may change from a reductive to an oxidative one.

The rate of degradation remains fairly constant till about pH 5. In these pH values, the dye is predominantly in its anionic form. The higher rate of degradation may however be attributed to the surface property of TiO_2. For TiO_2, the point zero charge is at pH 6.2. Hence at pH values above 6.2, the surface will be negatively charged. At more acidic values, the surface will be positively charged. At pH 5, the positive charged TiO_2 surface can effectively adsorb the negatively charged dye anion. As the pH increases to 7, the repulsion of the dye anion by negatively charged TiO_2 surface would result in a drastic reduction in efficiency of photodegradation of the dye.

With further increase in the pH however, a large enhancement in efficiency of the photodegradation of the dye is observed. It is proposed that the enhancement in efficiency may be attributed to the

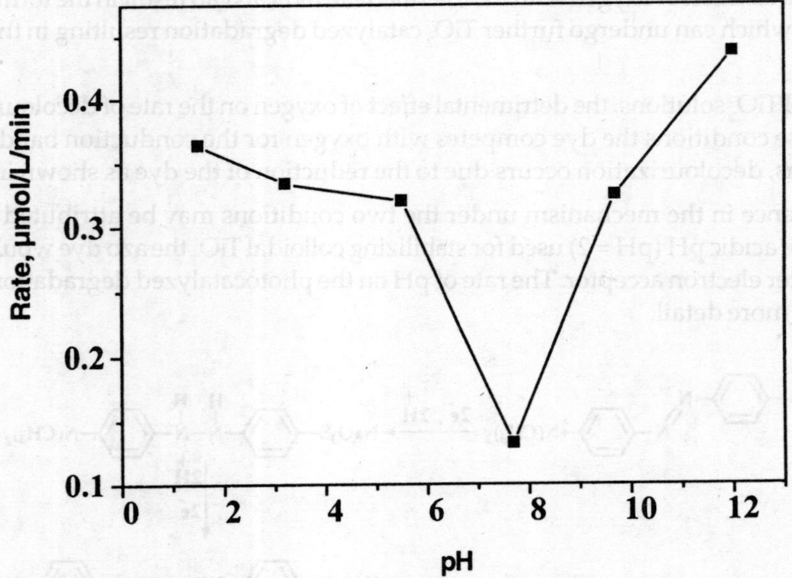

Fig. 3: Influence of pH on the Rate of Degradation of Methyl Orange (50 µM) in the Presence of Degussa P 25 type TiO_2 (1 g/L)

mechanism of the reaction of hydroxyl radical with methyl orange. As discussed earlier, hydroxyl radicals can react with the azo dyes containing amino functionalities both by electron transfer as well as by adduct formation. The pKa of the radical cation formed in the electron transfer are in the range of ~ 6 (36). Thus, at higher pH values, efficient deprotonation of the dye radical can result in more efficient degradation of the dye.

Comparison of the pH effect on the photocatalyzed degradation of β-naphthol orange confirms this observation. Since this β-naphthol orange does not possess an amino functionality the main mode of reaction of hydroxyl radicals would be *via* adduct formation. Figure 4 shows the influence of pH on the degradation rate of β-naphthol. The anomalous increase observed above pH 7 for the rate of degradation of methyl orange is not observed in this case. This confirms that electron transfer catalyzed degradation does not occur for β-naphthol orange.

Effect of Catalyst Concentration

The influence of the photocatalyst concentration on the rate of degradation of the dyes 1 and 2 has been studied using different concentration of Degussa P25 ranging from 0.5 to 5 g/L. It has been found that (Figure 5) an increase in catalyst concentration from 0.5 to 1 g/L drastically increases the degradation rate. On the other hand a further increase in loading from 1 g/L to 5 g/L did not show much improvement in the degradation rate. As the catalyst concentration increases from 0.5 g/L to 1 g/L, the number of photons absorbed by TiO_2 particles and the number of dye molecules adsorbed on TiO_2 surface also increases, thereby increasing the degradation rate. After a certain level of catalyst concentration with the same concentration of the dye, further dye molecules are not available for adsorption. The additional catalyst particles are therefore not involved in the catalytic activity. Hence the degradation rate remains constant. A similar trend has been observed in the case of Rhodamine 6

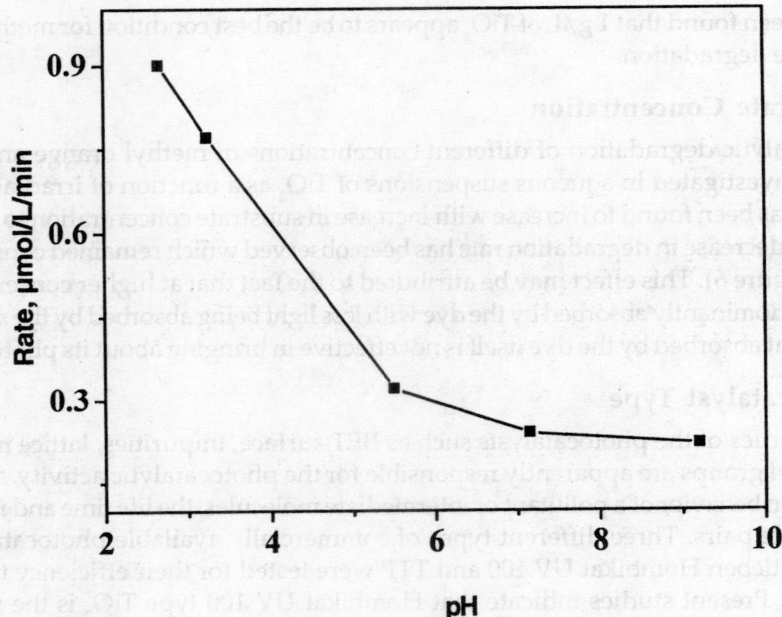

Fig. 4: Influence of pH on the Rate of Degradation of β-naphthol Orange (90 μM) in the presence of Degussa P 25 type TiO_2 (1 g/L).

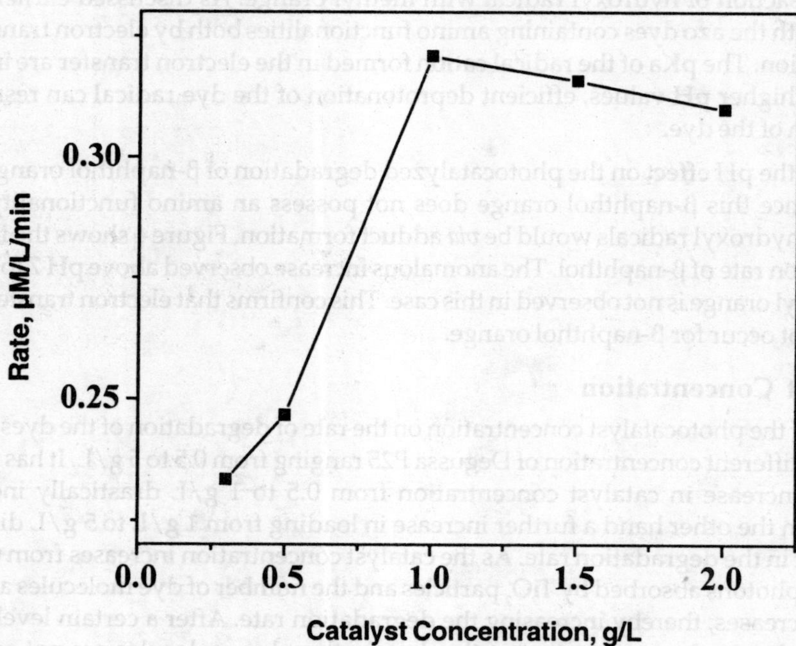

Fig. 5: Influence of the Catalyst Concentration on the Rate of Degradation of β-naphthol Orange

G dye (37). It has been found that 1 g/L of TiO_2 appears to be the best condition for methyl orange and β–naphthol orange degradation.

Effect of Substrate Concentration

The photocatalytic degradation of different concentrations of methyl orange and B-naphthol orange has been investigated in aqueous suspensions of TiO_2 as a function of irradiation time. The degradation rate has been found to increase with increase in substrate concentration to a certain level following which a decrease in degradation rate has been observed which remained constant at higher concentrations (Figure 6). This effect may be attributed to the fact that at higher concentration of the dye, the light is predominantly absorbed by the dye with less light being absorbed by the semiconductor photocatalyst. Light absorbed by the dye itself is not effective in bringing about its photodegradation.

Effect of Photocatalyst Type

Surface properties of the photocatalysts such as BET surface, impurities, lattice mismatches or density of hydroxyl groups are apparently responsible for the photocatalytic activity, since they will affect the adsorption behavior of a pollutant or intermediate molecules, the life time and recombination rate of electron hole pairs. Three different types of commercially available photocatalysts such as Degussa P25, Sachtleben Hombikat UV 100 and TTP were tested for their efficiency to bring about photodegradation. Present studies indicate that Hombikat UV 100 type TiO_2 is the most efficient photocatalyst for the degradation of methyl orange and TiO_2 from TTP and Degussa P25 was most effective for β–naphthol orange.

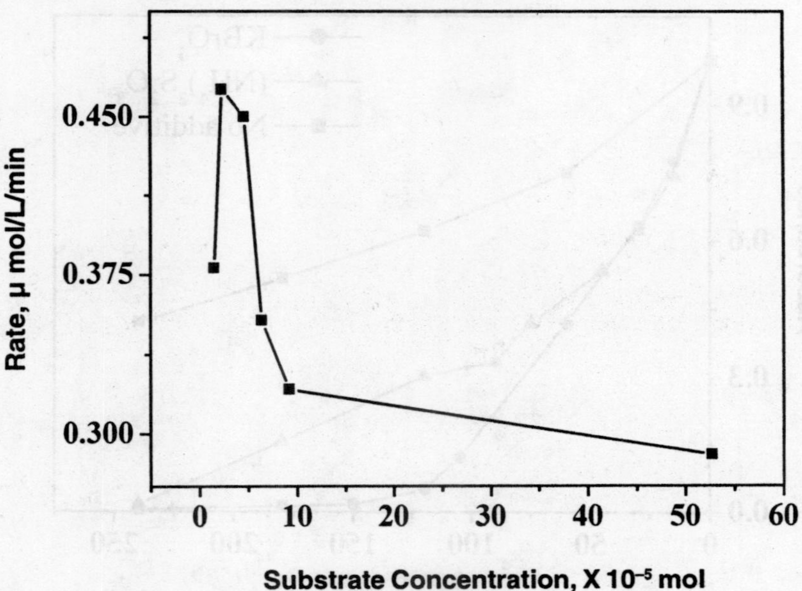

Fig. 6: Influence of Substrate Concentration on the Degradation rate of β-naphthol orange (90 MM) in the presence of Degussa P 25 type TiO_2 (1 g/L)

Effect of Electron Acceptors

One practical problem in using TiO_2 as a photocatalyst is the electron-hole recombination process. The strategy to inhibit the electron hole-recombination processes is to add efficient electron acceptors to the reaction. We have studied the effect of electron acceptors such as H_2O_2, $KBrO_3$ and $(NH_4)_2S_2O_8$ on the photocatalyzed degradation of dyes 1 and 2. The rate of degradation has been found to be markedly increased in the presence of these additives. It was found that the degradation was strongly influenced by the presence of electron acceptors and following trend has been observed. For methyl orange, the photodegradation efficiency was $KBrO_3 > H_2O_2 >$ no additive and for B-naphthol orange it was $KBrO_3 > (NH_4)_2S_2O_8 >$ no additive (Figure 7).

H_2O_2 is not only known to inhibit the electron-hole recombination process but also it generates hydroxyl radicals on abstraction of an electron from the conduction band (8, 38) (equation 1).

$$H_2O_2 + \bar{e}_{cb} \longrightarrow \dot{O}H + \bar{O}H \qquad (1)$$

Potassium bromate appears to be more effective than H_2O_2 and $(NH_4)_2S_2O_8$ in the photocatalytic degradation of methyl orange and B-naphthol orange respectively. The bromate ions are known to accept electrons and dissociate into bromide ions according to the following equation (2),

$$BrO_3^- + 6e^- + 6H^+ \longrightarrow Br^- + 3H_2O \qquad (2)$$

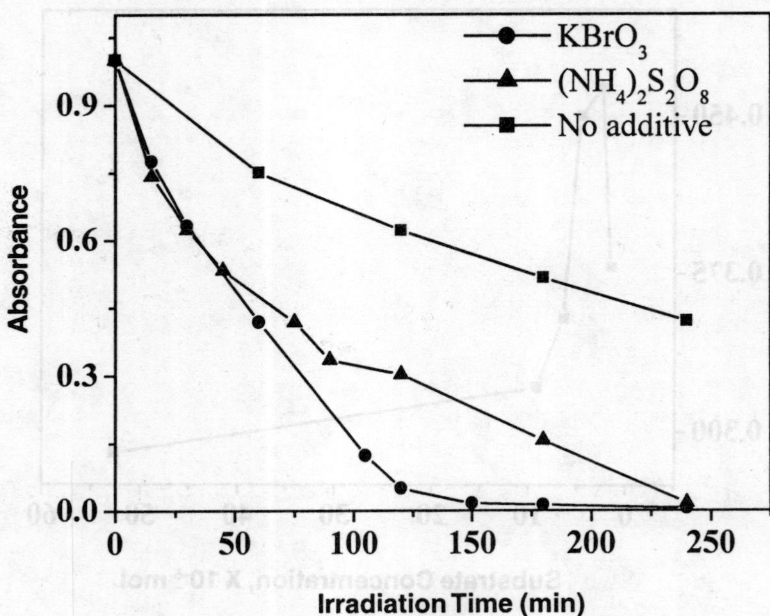

Fig. 7: Photocatalytic Degradation of β-naphthol Orange (90 μM) with Different Kinds of Electron Acceptors (3 mM) in Presence of Degussa P 25 type TiO_2 (1 g/L)

Effect of Electron Donors

The effect of different sacrificial electron donors on the rate of photobleaching of methyl orange and β-naphthol orange have been investigated. It has been observed that the rate of degradation is drastically reduced in the presence of Et_2NH and Et_3N (Figure 8). The decrease of rate of photooxidation in the presence of these additives is due to the competition for the oxidative sites of the irradiated semiconductor by the dye and the electron donor species. Similar results have been reported by Mills *et al.* (37). On the other hand in the presence of EDTA the rate of photobleaching is enhanced for both the dyes. This effect could be explained in terms of the fact that in the presence of EDTA, the solution becomes acidic resulting in the protonation of the dye. The protonated dye undergoes reduction by the conduction band electrons.

Photomineralization Studies (CO_2 Measurement)

The mineralization of the dyes 1 and 2 was studied by estimating the amount of carbon dioxide evolved on irradiation. Aqueous solution of methyl orange and β-naphthol orange containing 1 g/L of TiO_2 were irradiated with continuous purging of oxygen. The outgoing gas was bubbled through an aqueous solution of barium hydroxide to give barium carbonate, which was filtered, dried and weighed. For unirradiated blank solutions, formation of barium carbonate was not observed. Figure 9 shows the formation of CO_2 on irradiation of aqueous suspension of TiO_2 containing β-naphthol orange. The disappearance of dye absorbance measured in the same solution is also shown. The stoichiometry for complete mineralization of the dyes 1 and 2 is given by the expressions (3) and (4) respectively.

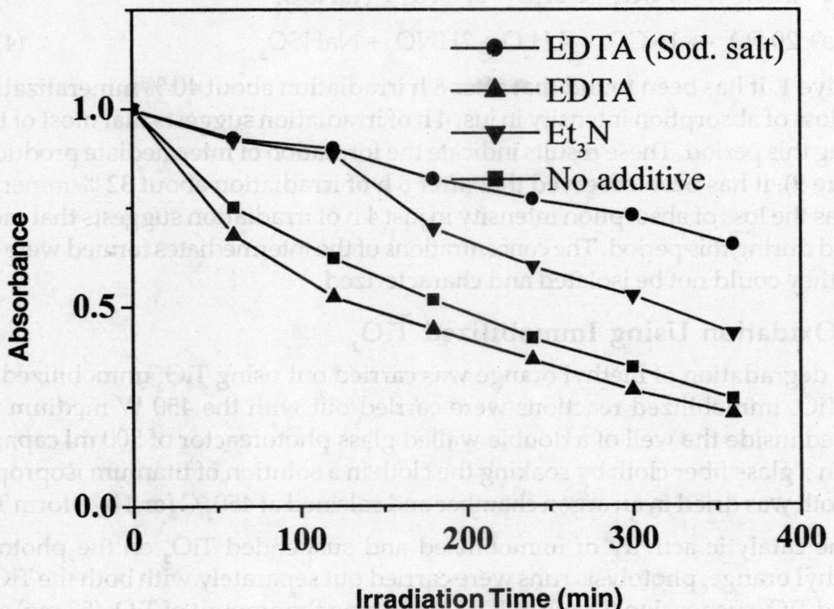

Fig. 8: Photocatalytic Degradation of β-naphthol Orange (90 μM) with Different Kinds of Electron Donors (3 mM) in Presence of Degussa P 25 type TiO_2 (1 g/L)

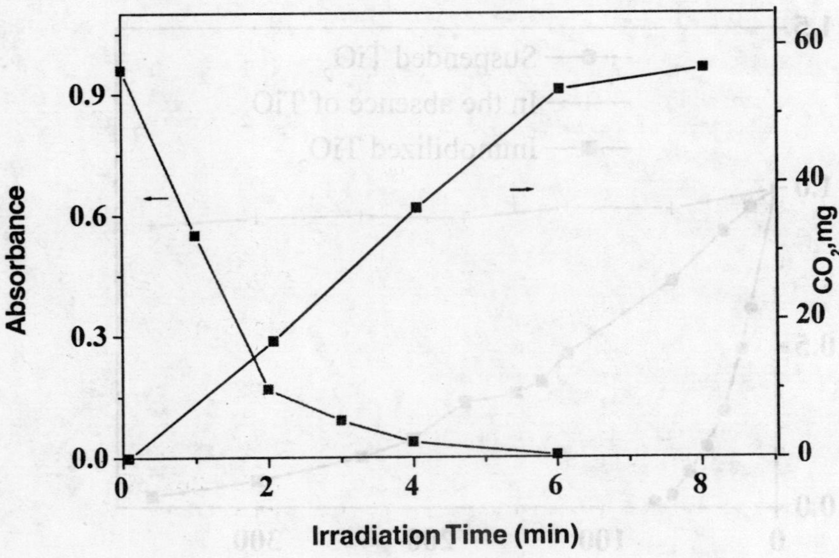

Fig. 9: Production of CO_2 and Loss of Absorbance at 484 nm on Irradiation of β-naphthol orange Solution (90 μM) containing TiO_2 (1 g/L)

$$C_{14}H_{14}N_3O_3SNa + 21\ O_2 \rightarrow 14\ CO_2 + 5\ H_2O + 3HNO_3 + NaHSO_3 \quad (3)$$

$$C_{16}H_{11}N_2O_4SNa + 20.5O_2 \rightarrow 16\ CO_2 + 4\ H_2O + 2HNO_3 + NaHSO_3 \quad (4)$$

In the case of dye 1, it has been found that after 8 h irradiation about 40 % mineralization takes place, whereas the loss of absorption intensity in just 4 h of irradiation suggests that most of the dye is photoxidized during this period. These results indicate the formation of intermediate products. In the case of dye 2 (Figure 9), it has been observed that after 8 h of irradiation about 32 % mineralization takes place, where as the loss of absorption intensity in just 4 h of irradiation suggests that most of the dye is photooxidized during this period. The concentrations of the intermediates formed were however too low and hence they could not be isolated and characterized.

Photocatalyzed Oxidation Using Immobilized TiO_2

Photocatalytic degradation of methyl orange was carried out using TiO_2 immobilized on glass fiber clothes. The TiO_2 immobilized reactions were carried out with the 450 W medium pressure mercury lamp placed inside the well of a double walled glass photoreactor of 500 ml capacity. TiO_2 was immobilized on a glass fiber cloth by soaking the cloth in a solution of titanium isopropoxide in isopropanol. The cloth was dried in an argon chamber and calcined at 450 °C for 4 h to form TiO_2 film.

To compare the catalytic activity of immobilized and suspended TiO_2 on the photocatalytic degradation of methyl orange, photolysis runs were carried out separately with both the TiO_2 coated glass fiber cloths and TiO_2 suspension (Degussa P25) containing the amount of TiO_2 (57 mg) as affixed to the glass fiber cloth using the same photoreactor set up in both cases. The photocatalytic degradation obtained in the two cases is compared in Figure 10. The comparison reveals that the activity of the suspended form of TiO_2 is greater than that of the immobilized one. However, if we consider the advantages of practical usability, the reduction in efficiency in the immobilized reactions may be

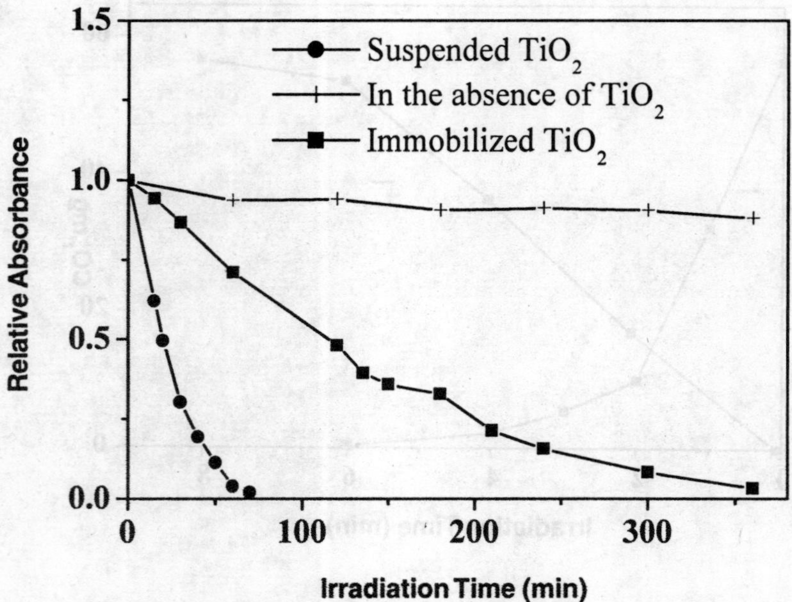

Fig. 10: Comparison of Photocatalytic Activity of Suspended TiO_2 and Immobilized TiO_2

acceptable. Similar results were reported in which TiO_2 immobilization was carried out on glass surfaces (11).

We have also carried out immobilization by varying the time of soaking of the cloth in the titanium isopropoxide solution. It was found that time of soaking has not much effect in the degradation processes. One advantage in using glass fiber cloth for degradation is that it can be used repeatedly. We have carried out degradation studies three times with the same cloth. It was found that the activity is not lost by repeated use.

Conclusions

The results of our studies clearly indicate the importance of choosing the optimum degradation parameters such as pH, nature of the catalyst, catalyst concentration and additives to obtain a high degradation rate of the pollutant. The best conditions strongly depend on the nature of the pollutant. The pH is markedly influencing the overall efficiency of degradation processes. The photocatalytic degradation of methyl orange and β–naphthol orange leads to partial mineralization as observed from CO_2 measurements. The present study also indicates the feasibility of using immobilized catalyst for degradation processes.

Acknowledgements

The authors thank the Ministry of Environment and Forests, Government of India and the Regional Research Laboratory (CSIR), Trivandrum for financial support of this work.

REFERENCES

1. H. Zollinger, 1987, In: "*Colour Chemistry*" I[st] ed., H. F. Ebel and C. D. Brenzinger (eds.), VCH, New York, Chapter 16.
2. C. E. Searle (ed.), 1976, *Chemical Carcinogenesis,* ACS Monograph, American Chemical Society, Washington, DC.
3. C. T. Helmes, C. C. Sigman, Z. A. Fung, M. K. Thompson, M. K. Voeltz, M. Mackeit, E. P. Klein and B.J. Lent, 1984, *Environ. Sci. Health, A19,* 97 –113.
4. M. Boeninger, 1980, *Carcinogenicity and Metabolism of Azo Dyes, Especially those Derived from Benzidine, DHHS* (*NiOSH*) *Publication* No. 80 –119, July.
5. J. J. Roxon, A. J. Ryan and S. E. Wright, 1967, *Food Cosmet. Toxicol., 5:* 367.
6. D. F. Ollis, E. Pelizzetti and N. Serpone, 1989, In: *Photocatalysis: Fundamentals and Applications,* M. Serpone and E. Plizzetti (eds.), Wiley Intersciences, New York, pp. 603 –637 and references cited therein.
7. R. W. Mathews, 1991, In: *Photochemical Conversion and Storage of Solar Energy,* E. Pelizzetti and M. Schiavello (eds.) Kluwer, Dordrecht, pp. 427-447 and references cited therein.
8. D. F. Ollis and H. Al-Ekabi (eds.), 1993, *Photocatalytic Purification and Treatment of Water and Air,* Elsevier, Amsterdam,
9. R. W. Mathews, 1988, *J. Catal., 111:* 264.
10. R. W. Mathews, 1991, *Water Res., 25:* 1169.
11. M. Muneer, S. Das, V. B. Manilal and A. Haridas, 1992, *J. Photochem. Photobiol. A: Chem., 63:* 107.
12. S. Das, M. Muneer and K. R. Gopidas, 1994, *J. Photochem. Photobiol. A: Chem., 77:* 83.

13. D. W. Bahnemann, J. Cunningham, M. A. Fox, E. Pelizzetti, P. Pichat and N. Serpone, 1994, In: *Aquatic and Surface Photochemistry*, D. Crosby, G. Hclz and R. Zepp (eds.), Lewis Publishers, Boca Raton, E. L. pp. 261-361.

14. P. V. Kamat, 1993, *Chem. Rev., 93*: 267.

15. M. R. Hoffmann, S. T. Martin, W. Choi and D. W. Bahnemann, 1995, *Chem. Rev., 95*: 69.

16. E. Pelizzetti and C. Minero, 1993, *Electrochem. Acta, 38*: 47.

17. M. L. Linsebigler, G. Lu and J. T. Yates Jr., 1995, *Chem. Rev., 95*: 735.

18. R. W. Mathews, 1987, *J. Phys. Chem., 91*: 3328.

19. R. W. Mathews, 1989, *J. Chem. Soc. Faraday Trans. 1*: 85.

20. M. A. Fox and M. T. Dulay, 1993, *Chem. Rev., 93*: 341.

21. H. Tahiri, N. Serpone and R. Le Van Mao, 1996, *J. Photochem. Photobiol. A: Chem., 93*: 199.

22. K. Tennakone and I. R. M. Kottegoda, 1996, *J. Photochem. Photobiol. A: Chem., 93*: 79.

23. C. Nasr, K. Vonodgopal, L. Fisher, S. Hotchandani, A. K. Chattopadhyay and P. V. Kamat, 1996, *J. Phys. Chem., 100*: 8346.

24. K. Vinodgopal and P. V. Kamat, 1996, *Chemtech., 26*: 18.

25. Subrata Naskar, S. Arumugom Pillay, and Manas Chanda, 1998, *J. Photochem. Photobiol. A: Chem.*, 113: 257-264.

26. F. F. Fan, Y. Lin, and A. J. Bard, 1985, *J. Phys. Chem.*, 89: 448.

27. H. Al-Ekabi, and N. Serpone, 1988, *J. Phys. Chem.*, 92: 5726

28. T. Tennakone, C. T. K. Tilakaratne, and I. R. M. Kottwgoda, 1995, *J. Photochem. Photobiol. A: Chem.*, 87: 1777.

29. Masayuki Murabayashi, Kiminori Itoh, Koichi Kawashima, Ryuji Masuda and Seiichi Suzuki, 1993, In: D. F. Ollis and H. Al-Ekabi (eds.) *Photocatalytic Purification and Treatment of Water and Air*, Elsevier, Amsterdam, p. 783.

30. G. T. Brown and J. R. Darwent, 1984, *J. Chem. Soc. Faraday Trans.*, 1, *80*: 1631.

31. G. T. Brown and J. R. Darwent, 1984, *J. Phys. Chem., 88*: 4955.

32. K. Vinodgopal, I. Bedja, S. Hotchandani and P. V. Kamat, 1994, *Langmuir, 10*: 1767.

33. Vogel, 1978, *"A Text Book of Practical Organic Chemistry"*, English Language Book Society and Longman Group Ltd., London, p. 718.

34. R. I. Bickley, T. Gonzalez-carreno, J. S. Lees, L. Palmissanmo and R. J. D. Tilley, 1992, *J. Solid State Chem., 92*: 178.

35. M. Linder, D. W. Bahnemann, B. Hirthe and W. D. Griebler, 1995, In: *Solar Engineering*, W. B. Stine, T. Tanaka and D. E. Claridge (eds.) Book No. H0932A, p. 399.

36. S. Das, P. V. Kamat, S. Padmaja, V. Au and S. A. Madison, 1999, *J. Chem. Soc. Perkin* 2: 1219.

37. Mills, A. Belghazi, R. H. Davies, D. Worsley and S. Morris, 1994, *J. Photochem. Photobiol. A: Chem.*, *79*: 131.

38. Kormann, D. W. Bahnemann and M. R. Hoffmann, 1991, *Environ. Sci. Technol.* 25: 494.

21

Application of Full-scale Fluidized Bed Reactor System for Recovery of Value-added Chemicals from Hazardous Waste Pickle Liquor

TAPAS NANDY, R.D. VYAS, S.N.KAUL & LIDIA SZPYRKOWICZ

National Environmental Engineering Research Institute, Nehru Marg, Nagpur - 440 020

ABSTRACT

The paper addresses application of full-scale fluidized bed reactor system with recourse to recovery of value-added chemicals from waste pickle liquor generated from steel pickling processes in the steel manufacturing industry. Optimization of the process variables of full-scale fluidized bed reactor for the recovery of hydrochloric acid (HCl) and iron oxide (Fe_2O_3) are discussed.

Keywords: Pickling; Waste pickle liquor; Hazardous; Fluidized bed; Resource recovery; Hydrochloric acid; Iron oxide.

Introduction

The process of chemically removing oxides and scales from the surface of hot-rolled steel by the action of water solutions of inorganic acid is known as pickling. Depending on the type of product being processed, number of acids, *viz.* sulfuric, hydrochloric, nitric, hydrofluoric, and phosphoric, may be utilized, either individually or in combination, in both batch and continuous processes. Sulfuric acid accounted for more than 90% of the pickling acid, however, in recent years an increasing number of steel mills have changed to hydrochloric acid. The use of hydrochloric acid offers several advantages over sulfuric acid, *viz.* it produces a more even and cleaner surface, requires lower liquor temperatures, and can be regenerated (1, 2, 3).

The pickling process comprises series of hydrochloric acid solutions at varying concentrations and temperatures, and hot rolled steel is passed through it. This process removes fine scale, and a bright and clean surface is produced. Iron oxides dissolved in acid are present in the form of iron chlorides, and spent acid is discarded, which forms a wastewater stream, commonly referred to as "Waste Pickle Liquor". The pickled steel is then rinsed with water by sequential immersion in rinse tanks to remove all acid and dissolved iron. The rinse water is also discarded as wastewater after certain washings. The following equations illustrate the chemical reactions occurring during the pickling process:

$$Fe_2O_3 + 6HCl \longrightarrow 2FeCl_3 + 3H_2O \quad (1)$$

$$FeO + 2HCl \longrightarrow FeCl_2 + H_2O \quad (2)$$

$$2FeCl_3 + H_2 \longrightarrow 2FeCl_2 + 2HCl \quad (3)$$

Waste pickle liquor from pickling bath contains free hydrochloric acid and iron chlorides, and are considered a source of environmental pollution. Because of their pollutant content and corrosiveness, these liquids appear in the list of hazardous wastes in all legislations (2). The discharge of untreated waste pickle liquors into surface water causes water pollution due to its high chloride and iron contents, besides lowering the pH of the receiving stream. Dissolved oxygen of water is also lowered due to hydrolysis of iron salts to iron hydroxides. Discharge onto land results in build-up of iron and chlorides in soil, making it unfertile, and further polluting the ground water.

To combat environmental pollution due to discharge of pickling liquors, two main options are available, *viz.* end-of-the-pipe treatment, and recovery of value-added chemicals leading to zero effluent discharge. End-of-the-pipe treatment of waste pickle liquor consists of neutralization, and/or deep well injection. Resource recovery techniques, *viz.* roasting, pressure oxidation, iron exchange and oxi-precipitation processes are also available for recovery of value-added chemicals from waste pickle liquors (2).

The iron oxide obtained in roasting processes can be considered a high quality raw material due to its high purity with chloride content of 0.05 – 0.1% and can be used in different industries, *viz.* components for ferromagnetic materials (ferrites and high power magnets), raw material for the production of iron dust (electrodes, use in chemical sector, etc.) and as an additive in the manufacture of different products (magnetic strips, colourings, abrasives, bricks, glass, cosmetics, etc.). In pressure oxidation processes, the capital and operational costs are very high, and the purity of regenerated acid is only 18-20%. Ion exchange process does not extract the ferrous iron in recoverable form, and environmental pollution problem is not fully solved. Oxi-precipitation process can only be used for liquors when the pickler agent has been recovered. This process obtains a similar iron elimination rate than other processes and yields an iron by-product for its use, without further transformation, in other industries like electronic, paintaing, etc. (2).

The paper highlights the application of full-scale fluidized bed reactor with recourse to recovery of hydrochloric acid (HCl) and iron oxide (Fe_2O_3) from the waste pickle liquor, and optimizing the operational parameters.

Materials and Methods

Application of full-scale fluidized bed reactor was carried out in a steel industry, processing 1000 tonnes/d of hot rolled steel coils, to achieve zero effluent discharge of waste pickle liquor with recourse to resource recovery of value-added chemicals. The reactor system was optimized through evaluation of the reactor performance by varying the operating parameters, *viz.* reactor temperature and waste pickle liquor feed rate, as also optimization of air velocity for fluidization of the reactor bed.

Fig. 1 shows the schematic of the full-scale fluidized bed reactor system under study, and the details are presented elsewhere (4).

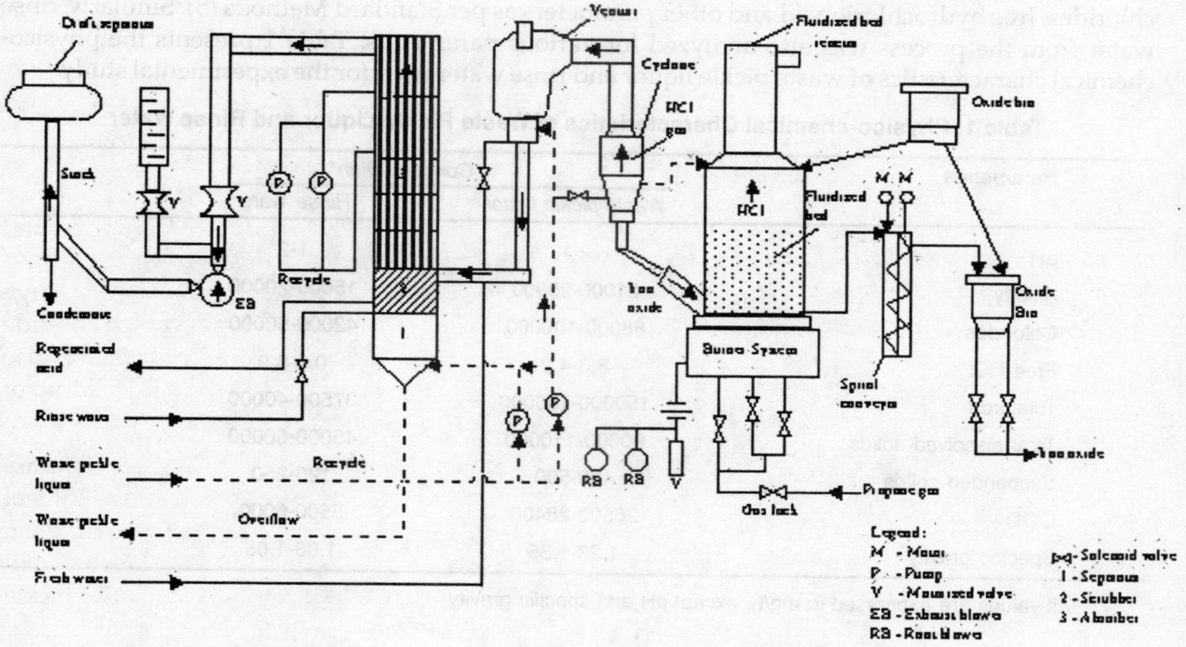

Fig. 1: Schematic of the Full-scale Fluidized Bed Reactor System

Process Reaction

The fluidized bed media consisted of iron oxide particles, which are kept in turbulent state by the combustion gas. In the reactor, the spent liquor is pumped via venturi concentration circuit. High temperature maintained in the fluidized bed reactor results in split up of $FeCl_2$ into Fe_2O_3 and HCl. Thus the reverse of the pickling reaction takes place in the reactor as follows:

$$2FeCl_2 + 2H_2O + \frac{1}{2}O_2 \longrightarrow Fe_2O_3 + 4HCl \tag{4}$$

About 95% of the splitting-up takes place on the surface of oxide particles and only 5% in the gap. While the iron oxide reacting on the surface of oxide particles contributed to the growth of oxide particles, the iron oxide formed in the gas phase generates fine iron oxide dust, which is carried out of the reactor with the combustion gases. In the cyclone, this dust is again removed from the gas and sent back to the reactor. The iron oxide formed is constantly drawn off by means of vibrating cooling trough. This facilitates maintaining the constant height of the fluidized bed in the reactor.

The hot reaction gases from the reactor are cooled in the venturi to approximately 100°C. The finest dust particles carried with the gases are captured at the same time, making use of dissipated heat to evaporate the spent liquor. A split stream is taken out of this venturi circulation system and conveyed to reactor.

The gas current passes to the absorber and the hydrochloric gas is absorbed adiabatically by water in reversible flow, achieving a solution of 16-17% hydrochloric acid.

Experimental Evaluation

Waste pickle liquor generated from steel processing was analyzed for pH, acidity, total iron, chlorides, free hydrochloric acid and other parameters as per Standard Methods (5). Similarly, rinse water from the process was also analyzed for various parameters. Table 1 presents the physico-chemical characteristics of waste pickle liquor and rinse water used for the experimental study.

Table 1: Physico-chemical Characteristics of Waste Pickle Liquor and Rinse Water

Parameters	*Concentration**	
	Waste pickle liquor	*Rinse water*
pH	< 1	1-2
Acidity	21000-25000	15000-20000
Chlorides	88000-100000	42000-50000
Free HCl	3.3-4.2	0.6-0.9
Total iron	150000-180000	37500-40000
Total dissolved solids	90000-110000	45000-60000
Suspended solids	400-500	125-250
COD	26500-28400	5500-6000
Specific gravity	1.28-1.35	1.03-1.05

*All values are expressed in mg/L, except pH and specific gravity

Propane gas was used as the fuel and air was used as the oxygen supplement and also to maintain the bed in fluidized condition. Propane gas requirement was in the range of 49.8 – 52.8 m^3/hr. In order to optimize the reactor system with respect to temperature for recovery of acid, initially studies were carried out by varying the temperature at constant pickle liquor feed rate. The temperature was varied between 400 and 1000°C, while the pickle liquor feed rate was kept at design value of 1.60 tonnes/hr. The rinse water was used to absorb the HCl gases generated in the reactor, and the feed rate of rinse water was maintained at 2.40 tonnes/hr.

Optimizing the reactor temperature, studies were carried out to assess the optimal pickle liquor feed rate at the optimum operating temperature achieved. The feed rate was varied between 0.90 and 1.70 tonnes/hr, and the temperature was maintained in the range of 850 to 900°C.

To assess the effect of airflow rate on expansion of bed and pressure gradient across the bed, studies were carried out by varying the airflow rate from 200 to 1900 m³/hr. The pressure of various heights of the reactor was measured using pressure gauges installed. Fluidized systems being extensively used because of high rate of heat transfer and uniform temperature within the bed, the heat transfer coefficient was calculated based on the expression as depicted hereunder (6):

$$\frac{hD}{k} = 0.55\left(\frac{D}{L}\right)^{0.65}\left(\frac{D}{d}\right)^{0.17}\left\{\frac{(1-e)\rho_S C_S}{e\rho C_p}\right\}^{0.25}\left(\frac{DG}{\mu A}\right)^{0.80}$$

Results and Discussions

Effect of Temperature on Recovery of HCl and Fe_2O_3

Conversion reaction of $FeCl_2$ to HCl and Fe_2O_3 occurs at temperature above 450°C, hence the reactor performance was assessed at temperatures varying from 400°C to 1000°C, and the results are presented in Table 2. The effect of temperature on recovery of acid and iron oxide is depicted in Fig. 2. Perusal of the figure indicates that hydrochloric acid and iron oxide recovery of 0.563 and 0.071 tonne per tonne of waste pickle liquor, respectively are achieved at a temperature of 500°C with waste pickle liquor feed rate of 1.6 tonnes/hr.

Increase in acid recovery is marginal (0.062 tonne/tonne) with increase in temperature from 500 to 1000°C at constant feed rate. Also in case of iron oxide, the increase in recovery is only around 0.007 tonne/tonne with increase in temperature from 500 to 1000°C. Operating the reactor system at temperature below 500°C, reduces the recovery of both acid and iron oxide.

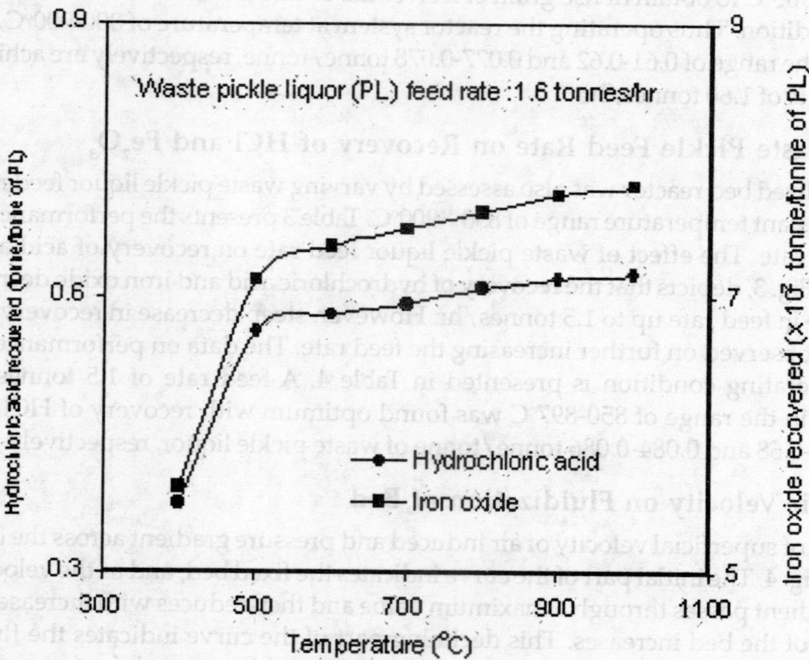

Fig. 2: Effect of Fluidized Bed Reactor Temperature on Recovery of Hydrochloric Acid and Iron Oxide

Table 2: Effect of Reactor Temperature on Recovery of Hydrochloric Acid and Iron Oxide at Constant Pickle Liquor Feed (Waste pickle liquor feed: 1.6 tonnes/hr)

Reactor Temperature, °C	*Value-added Chemicals Recovered, Tonnes/hr*	
	Hydrochloric acid	*Iron oxide (x 10^{-3})*
400	0.60	90
500	0.90	114
600	0.93	118
700	0.95	120
800	0.97	122
900	0.99	124
1000	1.00	125

Studies indicate an optimum operating reactor temperature of 500°C in terms of recovery of acid and iron oxide. However, temperature in the range of 800-900°C is required to yield dense grain of iron oxide resisting attrition. The reason being that under turbulent conditions, the granules of iron oxide developed are so small in size (in the range of 0.5-0.7 mm) that they are transported by the gas stream out of the reactor. Hence it is necessary to agglomerate these granules to such size that they are retained in the reactor in fluidized condition. The agglomeration of iron oxide presupposes a molten state in the course of reaction, and therefore temperature in the reactor needs to be maintained above the melting point of ferrous chloride, *i.e.* above 675°C. However, studies indicated temperature in the range of 800-900°C to obtain dense grain of iron oxide of size 4-5 mm to retain them in the reactor in fluidized condition. Thus operating the reactor system at temperature of 800-900°C, recovery of HCl and Fe_2O_3 in the range of 0.61-0.62 and 0.077-0.078 tonne/tonne, respectively are achieved with pickle liquor feed rate of 1.60 tonnes/hr.

Effect of Waste Pickle Feed Rate on Recovery of HCl and Fe_2O_3

The fluidized bed reactor was also assessed by varying waste pickle liquor feed rate operating the system at constant temperature range of 850 – 900°C. Table 3 presents the performance of the system at varying feed rate. The effect of waste pickle liquor feed rate on recovery of acid and iron oxide as presented in Fig. 3, depicts that the recovery of hydrochloric acid and iron oxide decreases marginally with increase in feed rate up to 1.5 tonnes/hr. However, steep decrease in recovery of both acid and iron oxide is observed on further increasing the feed rate. The data on performance of the system at optimum operating condition is presented in Table 4. A feed rate of 1.5 tonnes/hr and reactor temperature in the range of 850-897°C was found optimum with recovery of HCl and Fe_2O_3 in the range of 0.66-0.68 and 0.084-0.086 tonne/tonne of waste pickle liquor, respectively.

Effect of Air Velocity on Fluidization of Bed

The plot of superficial velocity of air induced and pressure gradient across the bed as monitored is shown in Fig. 4. The initial part of the curve indicates the fixed bed, and as the velocity increases, the pressure gradient passes through a maximum value and then reduces with increase in velocity, since the porosity of the bed increases. This declining part of the curve indicates the fluidized bed. The decrease in pressure gradient continues until the velocity is high enough for transport of the material to take place. At this point, the pressure gradient starts to increase again, which indicates the transport of bed.

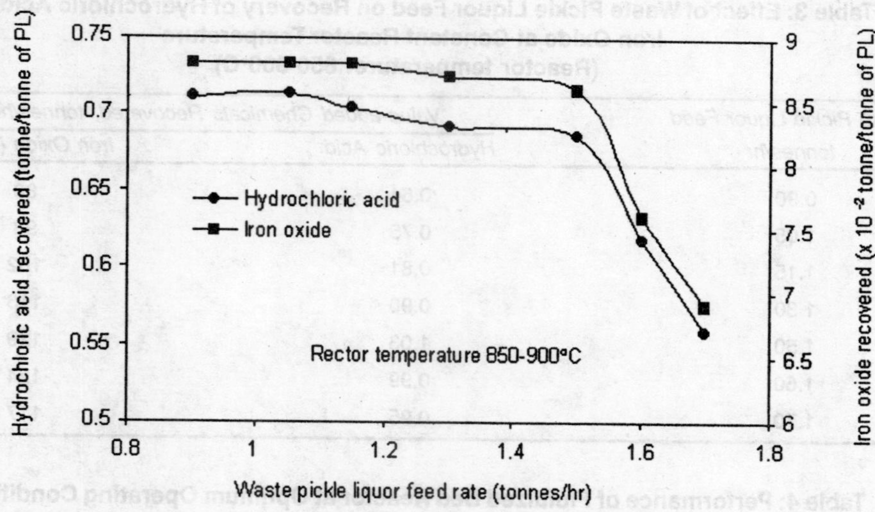

Fig. 3: Effect of Waste Pickle Liquor (PL) Feed on Recovery of Hydrochloric Acid and Iron Oxide at Constant Reactor Temperature

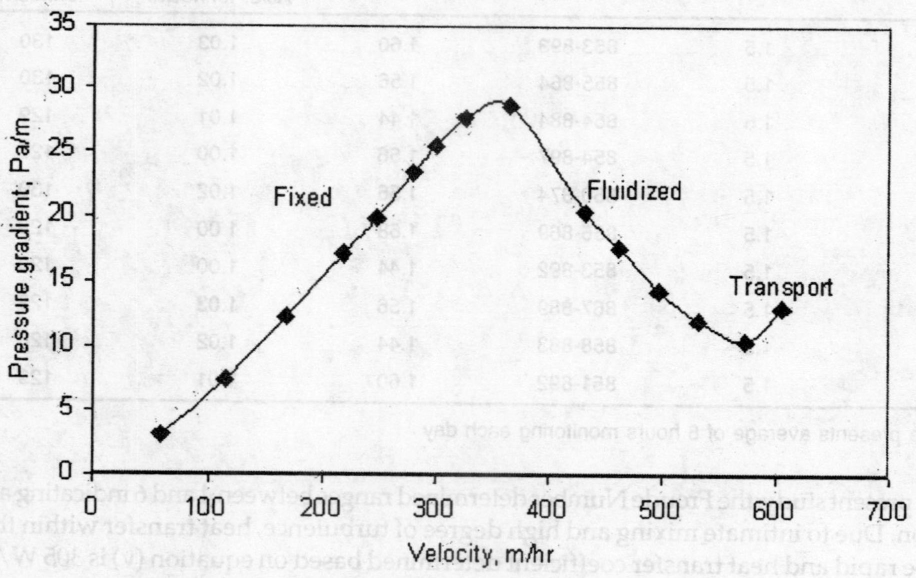

Fig. 4: Plot of Superficial Velocity of Air vs Pressure Gradient in Reactor Bed

In case of gas-solid fluidized systems, two separate phases, *viz.* dense phase and lean phase, are formed, and a condition of 'aggregative fluidization' is produced. Froude Number (u^2/gd, where u is the velocity of air flow at which fluidization takes place, d is the diameter of the particles, and g is the acceleration due to gravity) gives a criterion from which the type of fluidization taking place can be predicted. At values greater than unity, aggregative fluidization takes place (7).

Table 3: Effect of Waste Pickle Liquor Feed on Recovery of Hydrochloric Acid and Iron Oxide at Constant Reactor Temperature (Reactor temperature: 850-900°C)

Waste Pickle Liquor Feed tonnes/hr	*Value-added Chemicals Recovered, tonnes/hr*	
	Hydrochloric Acid	*Iron Oxide (x 10^{-3})*
0.90	0.64	80
1.05	0.75	93
1.15	0.81	102
1.30	0.90	113
1.50	1.03	129
1.60	0.99	124
1.70	0.95	117

Table 4: Performance of Fluidized Bed Reactor at Optimum Operating Conditions

Day	*Waste Pickle Liquor Feed, tonnes/hr*	*Reactor Temperature °C*	*Energy consumption, kWh*	*Value-added Chemicals Recovered**	
				Hydrochloric Acid, tonnes/hr	*Iron Oxide × 10^{-3} tonnes/hr*
1.	1.5	853-893	1.60	1.03	130
2.	1.5	855-864	1.56	1.02	130
3.	1.5	854-884	1.44	1.01	129
4.	1.5	854-897	1.56	1.00	127
5.	1.5	850-874	1.56	1.02	130
6.	1.5	856-869	1.68	1.00	129
7.	1.5	853-892	1.44	1.00	128
8.	1.5	867-889	1.56	1.03	129
9.	1.5	858-883	1.44	1.02	128
10.	1.5	851-892	1.60	1.01	129

*Data presents average of 8 hours monitoring each day

In the present study, the Froude Number determined ranges between 4 and 6 indicating aggregative fluidization. Due to intimate mixing and high degree of turbulence, heat transfer within the bed was found to be rapid and heat transfer coefficient determined based on equation (v) is 305 $W/m^2.°K$ and therefore, uniform temperature was quickly attained throughout the system.

Conclusions

The application of full-scale fluidized bed reactor for recovery of value-added chemicals from waste pickle liquor achieved an efficiency of 99% in terms of acid (HCl) recovery achieving Zero Effluent Discharge (ZED). In addition, iron oxide (Fe_2O_3) was also recovered with very high purity (chloride content of 0.05-0.1%). The recycling of recovered acid has resulted in reduction of HCl requirement in pickling section, achieving savings of 55.3-60.2% in a steel plant of capacity 1000 tonnes/d of hot rolled steel coils.

Nomenclature

h: heat transfer coefficient (W/m^2.°K)
k: thermal conductivity of the air (W/m.°K)
d: diameter of the particle (m)
D: diameter of the reactor (m)
L: depth of the bed (m)
e: voidage of the bed
ρ_s: density of the solid (kg/m^3)
ρ: density of the air (kg/m^3)
C_s: specific heat of the solid (J/kg.°K)
C_p: specific heat of the air at constant pressure (J/kg.°K)
μ : viscosity of the air (Pa.s)
G: mass rate of flow of air (kg/s), and
A: cross-sectional area of the bed (m^2).

REFERENCES

1. Azad, H.S., 1976, Industrial wastewater management handbook, New York: McGraw Hill.
2. Negro, C., Lattorre, R., Dufour, J., Formoso, A., Lopez, M.F., 1994, Treatment and recovery of pickling liquors, Journal of Environmental Science & Health, A29 (9): 1899-1926.
3. Rennhack, R., 1974, In: Wulfinghoff, P.E.., editor, Wastes – solids, liquids and gases, New York; Chemical publishing, 311-319.
4. NEERI report, 2000, Waste management in steel industry, Nagpur, India.
5. APHA, AWWA, WEF, 1995, Standard methods for the examination of water and wastewater, Washington, DC; 19th edn.
6. Dow, W.M., Jakob, M., 1951, Heat transfer between a vertical tube and a fluidized air-solid mixture, Chemical Engineering Progress, 47.
7. Coulson, J.M., Richardson, J.F., 1962, Chemical Engineering, London: Pergmon press.

22

Geographical Information System and Its Applications in Wastewater Management

V.K. KONDAWAR & S.M. KASHYAP

Scientist, National Environmental Engineering Research Institute, Nagpur – 20

Introduction

Geographical Information System (GIS) can be defined as organized collection of computer hardware, software, geographic data, and personnel designed to efficiently capture, store, update, manipulate, analyze and display/present all forms of geographically referenced information. In the most general sense, a GIS need not be computerized, but should be an integrated set of procedures for the input, manipulation, database management and output of geographic information. The system is concerned with spatial objects, their locations, and collateral data on specified scales of measurements referenced to some coordinate system of earth surface. The geographic data can be about natural resources, environmental status, infrastructure, demography, socio-economic parameters, etc. The geographical database can be queried for extraction of specific and meaningful information from a diverse collection of data, which is possible because the data are organized into a 'model' of the real world. The GIS can be employed for predictive modeling, planning/decision making, implementation and finally for long term monitoring of impacts of plan/decisions on the environment.

Generally information systems can be non-spatial (as Management Information System) or spatial. Spatial systems can be Non-geographic (CAD/CAM) where the data is not 'geocoded' or Geographic, where each point, line or polygon is assigned geographic co-ordinates. Geographic Information Systems are often referred based on 'what it does?' such as:

1. Land Information System
2. Land Resources Information System
3. Resources Information System
4. Natural Resource Management Information System
5. Geographically Referenced Information System
6. Spatial Information System
7. Spatial Data Management and Analysis System
8. Planning Information System
9. AM/FM (Automated Mapping and Facilities Management)

Why GIS

General objectives for the application of GIS are:

1. Spatial predictive modeling
2. Optimized resource management
3. Cost minimization
4. Ease and transparency in planning and decision making efficient data processing, communication and presentation
5. Easy availability of good quality (accuracy-position and classification) data precision, resolution, currency and completeness.

Spatial Objects: Maps

In order to represent a portion of earth's surface as map, the area must be reduced. The extent of this reduction is expressed as a ratio called Map Scale-the ratio of map distance to ground distance. Generally small scale (1: 250, 000) maps depict large ground area but they have low spatial resolution whereas large-scale maps (1:1, 200) depict small ground area with higher spatial resolution.

Maps are flat, but the earth's surface is curved (ellipsoid). A geographic coordinate system is a reference system used to locate and measure geographic features. The spherical geographic coordinates system is based on a three-dimensional sphere. Real-world locations in it are measured in degrees of latitude and longitude. The Cartesian coordinate system is based on a flat surface wherein real-world locations are measured using x-and y-coordinate values from a specified point of origin. In order lo represent three-dimensional earth's surface in two-dimensional map, a mathematical transformation called a projection is performed. Projection formulas convert data from a geographic location (latitude and longitude) on the earth to a representative location on a flat surface. The projection process will distort one or more of the following four spatial properties *viz.* Shape, Area, Distance, Direction. Because these properties are often used to make decision, anyone who uses map should know which projection distorts which properties and to what extent'.

Components

Geographical Information systems have three important components-computer hardware, sets of system and application software and a proper organizational arrangements These three components need to be in balance if the system is to function satisfactorily. The main hardware components of a GIS include central processing unit, mass storage units such as hard disk/tape drives, input and output devices/peripherals such as digitizer, scanner, printer or plotter and Visual Display Unit (VDU). The software component of GIS is related to various function and operation of GIS.

Data input, database management

Data entry, editing, verification, storage, retrieval

Data transformation, manipulation, analysis, modeling

Raster: Local, Zonal, Focal, and Global

Vector: Spatial overlay. Buffering, Map generalization. Feature extraction, Feature reduction, and theme co-ordinate transformation.

Data output, presentation and Reporting

Zooming, color assignment & change operation, view. Text writing, Annotation, Graphic User Interface (GUI).

Till recently most GIS applications were based on "map geometry" (only x, y co-ordinate) in 2-dimensional topology. Users' perception is three-dimensional. Therefore currently, more efficient GIS products (2.5 D or 3 D) are offered by vendors which integrate the third-dimension using the Digital Terrain Model (DTM). The fourth dimension is introduced through the time for dynamic systems leading to temporal GIS systems.

Data Types and Structure

GIS database includes land related data showing spatial relationship (graphical data) and attribute of the land parcels itself.

Graphical data are digital representative of spatial entities. The graphical component of land related data can be readily captured and stored in graphical data files in the form of various spatial elements. Spatial objects in the real world can be thought of as four easily identifiable objects

Map scale dictates the type of map feature used to represent real world features.

Point

Represent location feature

Lines

Represent linear feature

Polygons

Represents area features

Surface

Adding the dimension of height (elevation) to features and records surfaces as "Three dimensional object"

Points define discrete locations of geographic features, which are too small to be depicted as polygons. Lines represent the shape of geographic objects too narrow to depict as polygons such as streets and streams. Polygons are closed figures that represent the shape and location of homogeneous, real-world features such as district, parcels, soil types or land use zones. Symbols and labels area used to describe features.

Non-spatial attribute data consists of textural description on properties which may be associated with graphical entities, for example, name of location, width of road, diameter of pipeline, owner of land parcel, land use category, vegetation characteristic, etc. Many of the attributes data can be linked together through the use of a common identifiers or keys.

There are two main ways to organize the spatial data: Raster Data Structure and Vector Data Structure. The choice of a particular spatial data structure is one of the important early decisions in designing a geographic information system. Comparison of raster and vector models is given below:

Raster Model	*Vector Model*
Simple data structure	Compact data structure
Efficient overlay operation	Efficient encoding of topology
Handles high spatial variability	Easy network and geometrical analysis
Easy handling of remotely sensed images	Easy handling of line maps
Data Processing and analysis	

Data Processing and Analysis

Main data processing operations used frequently to generate a database are:

1. Digitization or scanning of data
2. Topology construction, data reduction and generalization
3. Errors detection and editing
4. Merging of points and edge matching
5. Rectification/registration
6. Format conversion

Spatial relationships between geographic features are based on the location of features. Some of the relationships can be enumerated as:

Connectivity: Describes the link (s) between different map features

Adjacency: Describes what features are adjacent to each other

Proximity: Describes what features are near each other

Intersection: Describes which features intersect (cross) each other

Containment: Describes what features are located within an area

Relative Position: Describes the comparative position between features

Elevation differences: Describe the height difference between features

GIS tools have to efficiently integrate tools of relational attribute database with spatial database characterized by above relationships. They range from simple database query system to complex analysis and decision support system. The simple analysis tools provided by GIS include:

1. User defined views involving reclassification, aggregation, overlay, logical and arithmetic operations, etc
2. Difference and connecting measurements, network analysis

3. Neighborhood characterization including buffering
4. 3 D models including estimation of point to point visibility, shadow region, slope & aspect and flow models

GIS analysis can help to develop relations between geographic features and their associated attributes drawn from different sources. Integration analysis is employed to identify and extract the useful hidden relationship in multi-thematic datasets. Integration can be for passive enquiry, probing possibilities and active planning. It involves conjunctive analysis of appropriate multi-thematic data and weighted integration to generate composite information sets. The relationships can be expressed as statement or tables. Determining weightages requires human judgment supplemented by mathematical/statistical tools.

Digital models can be linked to some layers of information with conventional GIS functionality. However full integration of physical models with GIS is presently not possible.

Some applications of GIS that can interest engineers for wastewater collection and treatment can be:

1. Number of road cuttings for particular segment of pipeline.
2. Find elevation difference between source of effluent and treatment plant site
3. Discover the best route between the source of effluent and treatment plant.
4. Determine a suitable site for a wastewater treatment plant, which cannot be located near a stream or residential area.
5. Ascertain which land use is contained within a given region
6. Overlays of landuse and ground water table

Some common GIS packages used frequently in India are ARC/INFO (ESRI, USA), PAMAP (PAMP graphics Limited, Canada), SPANS (TYDAC Technologies, USA), Geomedia, MGE (Intergraph System, The Netherlands), TNTlite (MicroImages Inc, USA), Geomatica (PCI Geomatics, USA), IDRISI (Clark University, USA), GRASS, MAPINFO, Auto-CAD, etc. Various departments of Government of India have developed GIS packages suitable for specific applications in mind such as ISROGIS, GEOSPACE, INGIS, GRAM, etc. However their utilization remained restricted.

Applications

The power of a GIS is to organize and display large land-use areas on a computer screen with zoom-in and zoom-out capabilities. Maps of selected areas can be printed on almost any scale and with variable amounts of detail. Output maps can be color-coded for viewing on-screen or printing. This capability is ideal for environmental planning when land uses or resources change with time. Another advantage of a GIS over conventional analyses is an increase in efficiency. A GIS is not just a relational database management system since in addition to the storage capabilities it includes spatial operators, which permit the manipulation of spatial data.

GIS in the past have been used primarily for the production of digital maps for resources management and management of land information systems. However, in the last 10 years with the wide availability of digital map data and the emergence of desktop mapping systems, GIS are applied in many different fields ranging from marketing to environmental planning and from city planning to water resources management.

In the initial years of GIS, the system was visualized as an extension of computer aided graphic/ design tools with very limited spatial analysis capabilities. It was essentially addressing the needs of the geographical community. Therefore, the traditional geological community was skeptical about its usage in solving serious geological problems and it still preferred specialized mining software or geostatistical packages for their applications. However, the GIS developers soon realized the need for incorporating multivariate geostatistical modules and powerful 3-D analysis and these components work often considered as advanced components and become the selling point of GIS packages.

GIS application in water resources is itself very broad. However, much of this work in water resources has been in the area of large-scale, river basin hydrology. GIS has a long history of use in the water resources field, this is due in large part to the early availability of remotely sensed data suited for this purpose. Work in natural areas tends to focus on grid, or raster-based, hydrology, whereas in urban areas-because of their complex hydraulic and hydrologic regimes-tends to use more complex models that are vector based. Raster based approaches use rectangles as their fundamental unit within which hydrologic characteristics are uniform. Vector-based models use coordinate geometry to define unique boundaries of hydrologic characteristics.

Recently, a new trend is emerging where GIS systems are combined with analytical, mathematical models to form spatial decision support toolkits. In these integrated systems the GIS is used to handle the spatial data and provide a user friendly interface to the complex analytical models used to simulate our knowledge about events in the real world. Such toolkits are transferable, permit users to operate the models without having a complete knowledge of the intricacies of the model or the data structures and could be used for guiding policy makers in evaluating the impact of alternative decisions.

GISs have been used in groundwater contamination studies to organize site-specific data, plot concentration contours, handle complex geological data, and run transport models linked to mapped output for hazardous waste sites. By using GIS applications in remote-sensing technology, the pipeline industry can benefit from computer imaging for planning, maintaining, and controlling pipeline systems. A prototype least-cost analysis was considered using remote sensing data and GIS to determine a least-cost routing for a small section of a proposed pipeline for the transport of crude oil. The proposed model was designed to include pipeline length, topography, geology, land use, and stream, wetland road, and railroad crossings. Custer County in South Dakota used GPS and GIS technology to develop a master street address guide to every structure and facility. Dobson[1] has discussed effectiveness of GIS as a tool for decision-makers. It supports decisions ranging from matters of state and environment to routine matters, such as siting facilities, expanding residential subdivisions, and routing transportation networks. The GIS developed for Brookline, Massachusetts was seen as a way of improving the efficiency of the maintenance of the city's water and wastewater infrastructure. The use of GIS for analysis and display in the temporal domain will help industry because a long-range view encourages investment in physical infrastructure and environmental protection.

A load allocation module allows multiyear distribution system studies in a GIS context to best assist utility planners in deciding when and where customers will grow and how to expand the system facilities to meet the demand growth. Officials in Clark County, Nevada, planned to include the GIS technology in its latest master plan designed to improve the design process on more than $1 billion in flood control projects. It is claimed that GIS will give the Regional Flood Control District's engineers and developers valuable new planning tools for the rapidly growing greater Las Vegas, Nevada area.

GIS has increasingly provided a foundation for integration and analyses of water environment data in context with other geographically related data. The degree of data diversity within a geographic framework employed in many water environment applications today extend to include economic, ecological, political and hydrological, to name a few examples. Technical tools, such as global positioning system (GPS) technology, satellite remote sensing and modeling integration, has allowed GIS to continue to expand beyond digital mapping to embrace a growing number of water resource protection and management domains.

Wastewater Management

Holliman[2] *et al.* described the development of an GIS to support the management, maintenance and improvement of the Long Beach reclaimed water distribution systems and wastewater collection systems. By generating databases that combine requirements from a range of organization activities, wastewater utility managers can more easily automate their operation and maintenance activities. An integrated data environment with GIS as a central component allows utilities to respond to changing user requirements cost effectively. The affordability of a GIS by municipalities was discussed by Mitschele R[3] in terms of sharing the data, costs, and benefits.

Fienberg and Uhrick[4] discuss integrating an infrastructure database in Broward County, Florida, with a GIS and water distribution and wastewater models. The HydroWorks model is used to simulate wastewater collection system, with close integration with the database of infrastructure characteristics and the GIS. Refsgaard[5] *et al.* describes the evaluation the Danish Hydraulic Institute's (DHI) land-process hydrologic model SHE (Systeme Hydrologique European) and its extensive use of GIS. Ribeiro[6] describes the use of raster-based GIS to interface the HSPF (Hydrologic Simulation Program-Fortran) to analyze the effects of basin urbanization. Hellweger[7] developed an ArcView application (ArcView popular GIS software) using the Avenue scripting language to perform calculations of an SCS (Soil Conservation Service) based model TR-55. Makropoulas[8] *et al.* demonstrate how a raster-based GIS (IDRISI) can be used to integrate theoretical concepts and site-specific spatial characteristics. The strength of GIS can be used as a common ground between specialists and non-specialists to help them communicate effectively.

GIS have been used extensively for visualizing the projected developments in metropolitan areas, in terms of changes in population, industrial patterns, infrastructural utilities, etc. One very interesting aspect in this context is the expected future load and behavior of sewer systems, which convey the municipal and industrial wastewater flow to sewage treatment plants. Large sewer systems are structurally very complex networks with many nodes, pumps and weirs, often, with overflow chambers and loops. The hydraulic performance and capacity can only be evaluated by use of comprehensive mathematical modeling tools. MOUSE[9] (DHI, Denmark) is a professional engineering software package for the simulation of surface runoff, flows, water quality and sediment transport in urban catchments and sewer systems. The program can be equally successfully applied in any other type of pipe networks with alternating free surface and low pressurized flows. It is a dynamic, user-friendly modeling tool for the analysis, design, management and operation of both simple and large pipe networks. It provides a complete and effective working environment for sophisticated urban drainage and sewer engineering. As one step towards an efficient integration between the MOUSE modeling system and GIS technologies, DHI has developed the MOUSE Model Editor application using the ArcView and Avenue from Environmental System Research Institute (ESRI). This application offers the user an interactive tool for the definition of sewer network systems to be used for simulation in the MOUSE package. The MOUSE simulation results (overflow volumes, average and extreme filling of pipes etc.) are available to the MOUSE Model Editor application for graphical display and further analyses.

By applying MOUSE it is possible to obtain answers to questions such as:

1. Which parts of existing sewer systems will be overloaded under different conditions?
2. What are the main causes of overloading-backwater or insufficient pipe capacity?
3. What are the implications of replacing critical sewers, installing new basins, weirs, etc.?

Sewer asset and management database systems stores high level of details, the level of detail required for hydraulic modeling is much simpler, including every single minor component will not improve the accuracy of the model due primarily to the uncertainty in the hydraulic input data (rainfall, foul flow, etc.). Hence there is a need to simplify the network from the database level to the model network. MOUSE, an ArcView application allows the user the possibility to extract data from the asset management system, condense it semi-automatically-while monitoring consistency in the description of the physical system-and then store the data as a model description for use.

The MOUSE Model Editor includes a wide range of facilities:

1. Display the entire sewer system network based on the external database.
2. Interactive selecting of relevant nodes, pumps, weirs, etc. to be included in the final MOUSE sewer network.
3. Editing facilities for adding or removing nodes, pumps and weirs to the sewer network.
4. Editing attributes related to the nodes, pumps, weirs or pipes through a dialogue box.
5. Facilities for viewing and analyzing the MOUSE simulation results, such as
   (a) inflow to the system
   (b) overflow volumes
   (c) average filling of pipes
   (d) extreme filling of pipes
   (e) external information-other hydraulic information

The package consist of various modules such as an advanced hydrological model for continuous time series, analyses of the potential of Real-Time Control (RTC) applications, computational models for the simulation of sediment transport and water quality in urban catchments and sewer systems, computation and statistical analysis of combined sewer overflows.

The modular structure offers great flexibility:

1. Each module can be handled separately
2. Data transfer between modules is automatic
3. Coupling of physical processes (*e.g.* sediment deposition-hydraulic conveyance) is facilitated
4. Updating or expansion of existing installations with renewed or additional modules is simple

The Pipe Flow Model performs the computations of unsteady flow in pipe networks. The computations are founded on an implicit, finite difference numerical solution of the 1D, free surface flow equations (Saint Venant). The implemented algorithm provides efficient and accurate solutions in multiply connected, branched and looped pipe networks.

The computational scheme is applicable to vertically homogeneous flow conditions, which occur in pipes ranging from small-profile collectors for detailed urban drainage, to low-lying, often-pressurized sewer mains, affected by a varying outlet water level. Both sub critical and supercritical flows are treated by means of the same numerical scheme, which is adaptive according to the local flow conditions. Naturally, flow features like backwater effects and surcharges are simulated. Pressurized flow computations are facilitated through the implementation of a narrow 'slot', as a vertical extension of a closed pipe cross section. The complete non-linear flow equations can be solved for user-specified or automatically supplied boundary conditions. In addition to the fully dynamic description, simplified flow descriptions are also available. Within the Pipe Flow Model, advanced computational formulations enable the description of a variety of pipe network elements and flow phenomena including:

1. Flexible cross-section database, including standard shapes
2. Circular manholes
3. Detention basins
4. Overflow weirs
5. Pump operations
6. Flow regulation
7. Constant or time variable outlet water level/inflows into the network
8. Non-standard headlosses at manholes and basins
9. Depth-variable friction coefficients

With such a description, the Pipe Flow Model can predict the detailed hydraulic conditions throughout the pipe network. Depending on the purpose of the study, the user may specially concentrate on the following aspects:

1. The treatment plant loads (outflow hydrographs)
2. Combined sewer overflows
3. Pump activity
4. Filling of detention basins
5. Local surface flooding, etc.

There is no question that GIS approaches will revolutionize the ability to manage and visualize large environmental and infrastructure datasets, but more importantly, they will help us understand the linkages between monitoring data and the processes being analyzed.

Future

Developments in hardware, software, manpower & networking, as also their social acceptability is increasing due to various advantages of GIS listed earlier. The increased penetration of GIS into nodal agencies, which act as infrastructure planners and information providers is likely to increase availability of digital data. If the data exchange standards are universal, data gathering would be cheaper and easier, further accelerating growth of GIS.

The diversity of analysis and simulation tools that would conceivably be needed is enormous. A more open GIS architecture or 'toolbox' approach wherein additional functions would be grafted on easily or existing functions modified, might allow unusual analysis requirements to be carried out in GIS environment easily and efficiently. Future likely developments in GIS, useful to public health engineers, can be in following areas:

1. Integration of Modeling in a GIS
2. Simulation model, Stochastic modeling, Fuzzy and probabilistic models, Optimization, Genetic Algorithm
3. Expert System, Neural Nets and Intelligent GIS
4. Dynamic Modeling & animated Visualization
5. Object Oriented GIS
6. Advanced user interface in GIS
7. Integrated use of the technologies to solve real life problems
8. Multi-dimensional GIS
9. Information packet for decision makers, mangers and citizens

REFERENCES

1. Dobson, J.E., 1997, GIS supports Decision Making Time after Time, GIS World, 10, 9, 34
2. Holliman, T.R. *et al.*, 1995, AM/FM/GIS Implementation for Water and Wastewater Systems, Proc. of the 1995 Annual Urban and Regional Info. Syst. Assoc. (1), 373.
3. Mitschele, R., 1996, Share and Share Alike: Creating A Cost-Effective GIS, Am. City County, III, 3, 24
4. Fienberg, D., and Uhrick, S.W., 1997, "Integrating GIS with Water Wastewater Hydraulic Models" Proc.., 1997 ESRI User Conf. (http:/www.esri.com/library/userconf/proc97/PROC97/TO200/PAP1999/P199.htm)
5. Refsgaard, J.C., Storm, B., and Refsgaard, A., 1995, "Recent developments of the Systeme Hydrologique European (SHE) towards the MIKE SHE". Modeling and management of sustainable basin-scale water resource systems, S.Simonovic, Z. Kundzewicz, D. Rosbjerg, and K.Takeuchi, eds., International Association of Hydrologic Sciences Pubs. No. 231, IAHS Press, Wallingford, U.K. 427-434.
6. Rebeiro, C.T., 1996, "Impact of land use on water resources: Integrating HSPF and a raster-vector GIS", Proc. HydroGIS '96: Applic. of Geographic Info. Sys. In Hydro and Water Resour. Mngmt., K. Kovar and H.P.Nachmebel, eds., International Association of Hydrologic Sciences Publ. No 235, IAHS Press, Wallingford, U.K., 349-356
7. Hellweger, R., 1996, " TABHYD-TR55 Tabular hydrograph method in Arc View " Ctr. For Research in Water Resources, University of Texas at Austin, Texas, USA (http:/www.crwr.utexas.edu/gis/gishyd98/runoff/webfiles/TABHYD/TABHYD.htm.
8. Markropoulous, C., Butler, D., and Maksimovic, C., 1998, "A GIS-based methodology for the evaluation of suitability of urban areas for source control application", UDM Proc., '98 4th Int. Conf. On Devel. In Urban Drain Modeling, D.Butler and C. Maksimovic, eds., Imperial college of Science, Technology & Medicine, London, 59-66.
9. Henrik G. Muller, Sten Lindberg, Danish Hydraulic Institute Agern Alle 5, DK-2970 Horsholm, Denmark

23

Photocatalytic Degradation of Organic Pollutants on Anatase-Rutile Titania

S.K. SAMANTARAY, P. MOHAPATRA & K.M. PARIDA

Regional Research Laboratory (CSIR), Bhubaneswar - 751 013, Orissa

ABSTRACT

Unmodified and sulfate modified TiO_2 samples of different crystal forms and crystal sizes were prepared by varying pH, source and concentration of sulfate ion and by activating at different temperatures. The XRD patterns revealed that TiO_2 sample prepared at low pH (pH 3) exhibit lower crystallite size than that of high pH (pH 9). Sulfate ion impregnation decreases the crystallite size and stabilises the anatase phase of TiO_2 up to higher activation temperature. Photocatalytic degradation of 4-nitrophenol depends on the crystal form as well as crystallite size of TiO_2. TiO_2 containing mixture of anatase and rutile phase exhibit highest percentage of degradation of 4-nitrophenol than that of other samples.

Keywords: Sulfated TiO_2, crystal forms and size, 4-nitrophenol.

Introduction

Nitrophenols are some of the most refractory pollutants, which are present in industrial wastewater. Traditional wastewater treatment includes activated carbon adsorption, chemical oxidation, biological digestion, etc. The feasibility of wastewaters containing 4-nitrophenol being subjected to an ozonisation process was investigated by Beltran *et al.* (1) whereas Pintar and Levec (2) studied catalytic oxidation in a temperature range of 423-463 K. The photocatalytic abatement of 4-nitrophenol in TiO_2 aqueous slurries have been investigated (3-11). The possibility of sensitised photocatalytic degradation at pH 8.5 was demonstrated in reference (3) and the effects of catalyst load and of oxygen and organic concentration was mainly focused in reference (4), disregarding pH. It has been reported that pore structure, pore-size, particle size and activation temperature of TiO_2 affects the photocatalytic activity of organic pollutants in wastewater (5-9).

In order to understand the effect of anions on the crystallite size of TiO_2, we have prepared hydrated TiO_2 at different pH and modified this by varying the source and concentration of anions using different techniques. The unmodified and modified TiO_2 samples are activated at different temperatures by calcination and their activities towards photochemical degradation of 4-nitrophenol were studied.

Experimental

Materials and Methods

Hydrated titania was prepared at pH 3, 5, 7 and 9 by adding aqueous ammonia to the stirred aqueous solution of titanium tetrachloride. The gel was filtered and washed repeatedly to remove Cl^- (negative $AgNO_3$ test), dried at 383 K for 10 h, powdered to 45-75 Mm mess size and kept for anion impregnation.

Modified Catalysts

One series of sulfated titania samples with varying the weight percentage of SO_4^{2-} were prepared using $(NH_4)_2SO_4$ (as the source of sulfate ions) by solid-solid kneading method and the other series of samples were prepared by aqueous wetness impregnation method using dilute H_2SO_4 (strength = 1.0M). The suspended mass was evaporated to dryness on a hot plate while stirring. The samples were then dried in an air oven at 383 K and subsequent activation (by calcination) at 573, 673, 773, 873 and 973 K at the heating rate of 283 K/min in a muffle furnace for 3 h.

Powder X-ray Diffraction

The XRD patterns of pure and modified TiO_2 samples were recorded on a Philips X-ray diffractometer having a PW 1830 generator with copper tube, a PW 1820 goniometer fitted with a post-diffracted graphite monochromator and a scintillation detector attached to a PW 1710 diffraction control in the range of $2\theta = 6\text{-}80^\circ$ at a scanning speed of 2°/min. The instrument was operated at 40kV and 20mA. The average crystallite size (L) was determined by XRD line broadening technique using the Debye-Scherrer equation:

$$L = 0.94\lambda / b\cos\theta \qquad (1)$$

where

λ is the wavelength of the X-ray used and b is the relative peak broadening, calculated as $b^2 = b^2_{exp} - b^2_{ref}$, where b_{exp} and b_{ref} are half widths at maxima observed on a given sample and on a reference material which is ideally crystalline, respectively.

Catalytic Activity

The photo-catalytic degradation of 4-nitrophenol were performed taking 0.05g/L of substrate (4-nitrphenol in water, CDH) and 0.2 g/L of catalyst. The pH of the dispersion was adjusted by addition of sulfuric acid (BDH). The solutions were exposed to sunlight in closed pyrex flasks at room temperature (no extra cooling) with constant stirring. All the irradiation was performed during the second half of March 2002 (sunny days), from 10.00 to 15.00 and was repeated three times. The corresponding dark controls were carried out simultaneously, for comparison. After irradiation taking at different times, the suspension was filtered and analysed the 4-nitrophenol quantitatively by measuring the absorption at 315 nm using Cary-1E (Varian) spectrophotometer. The detail method for measurement was available in reference (4).

Results and Discussion

It is observed that samples prepared at pH 3 possess mixture of anatase and rutile phases whereas at pH 5, 7 and 9 possess only anatase phase. This may be due to the gradual increase in rate of hydrolysis with increase pH of the solution that prevents the formation of rutile phase. It is also observed that unmodified TiO_2 sample prepared at pH 3 and activated at 673 K possesses only anatase phase. With increasing the activation temperature up to 773 K both anatase and rutile phases are found. Whereas sample activated at 873 K only rutile phases is observed. However TiO_2 samples prepared at pH 3, modified with sulfate ion and activated up to 873 K possess only anatase phase. With increasing the activation temperature up to 973 K both anatase and rutile phases are found. Whereas sample activated at 1073 K only rutile phase is observed. So, it may conclude that sulfate ion stabilises the anatase phase of TiO_2 up to 973 K activation temperature. It is also observed that sulfate modified TiO_2 samples activated at 773 K possess anatase phase, irrespective of the source and percentage of sulfate loading. There is no indication of formation of titanium sulfate in any one of the samples, even in 10wt% SO_4^{2-} impregnation. This type of observation was also reported earlier (12-14) on TiO_2 samples prepared at pH 7 and modified with phosphate and sulfate ions.

To develop a better understanding of formation and crystallisation in the presence of sulfate ion, the average crystallite size (L) perpendicular to the 210 plane was calculated from the XRD patterns of pure titania, as well as sulfated titania. It is seen (Table 1) that TiO_2 samples prepared by increasing the pH of the solution, the crystallite size gradually increases. It is observed that the crystallite size of titania sample prepared at pH 3 is lesser than the sample prepared at pH 9. It gradually increases with increasing the pH of the solution, which suppresses the growth of the particle, may be due to rapid hydrolysis. It is observed that the crystallite size of titania decreases with 2.5wt% SO_4^{2-} loading and thereafter it remains same. However, the effect is more pronounced with sulfuric acid as the source of sulfate compared to $(NH_4)_2SO_4$. This indicates that the crystalinity is more or less dependent on the presence of sulfate ion, but not on the source and percentage of sulfate loading. In addition to stabilising anatase TiO_2 crystallites, sulfate surface species inhibit TiO_2 crystallite sintering, leading to smaller crystallites than in pure TiO_2. The crystallite size decreases in the presence of sulfate ions as SO_4^{2-} species could possibly interact with TiO_2 network, and thus hinder the growth of the particle. Even a very small amount of SO_4^{2-} species is responsible for this effect. Therefore, the change in sulfate concentration did not change crystallite size further. Therefore, it is assumed that small amount of sulfate species is responsible for the lowering of crystallite size. It is also observed that crystallite size increases with increase in activation temperature from 383 to 673 K and thereafter it decreases in case of unmodified TiO_2. However, crystallite size remains almost constant up to 973 K and thereafter decreases in case of sulfate modified TiO_2. The former may be due to desorption of surface hydroxyl group from the TiO_2 network and the latter is due to the formation of rutile phase as well as inhibits sintering. This suggests that sulfate species strongly interacting with TiO_2 crystallites inhibits sintering even at higher activation temperatures. This type of effect is also observed in PO_4^{3-}modified FeO (OH) (15), WO_3 modified ZrO_2 (16) and TiO_2 (17).

Table 1: Average Crystallite Size of Unmodified and Modified TiO_2 Samples Activated at Different Temperatures

Sample Code	*Crystallite size (nm)*	*Sample Code*	*Crystallite size (nm)*
TiO_2, pH 3, 773K	13.22	TiO_2, pH 3, 573K	9.52
TiO_2, pH 5, 773K	14.54	TiO_2, pH 3, 673K	14.85
TiO_2, pH 7, 773K	16.13	TiO_2, pH 3, 773K	13.22
TiO_2, pH 9, 773K	17.25	TiO_2, pH 3, 873K	11.38
$2.5SO_4^{2-}$-TiO_2, pH 3, 773K	5.28	$2.5SO_4^{2-}$-TiO_2 (H), pH 3, 573K	3.12
$5.0SO_4^{2-}$-TiO_2, pH 3, 773K	4.70	$2.5SO_4^{2-}$-TiO_2 (H), pH 3, 673K	3.43
$7.5SO_4^{2-}$-TiO_2, pH 3, 773K	4.38	$2.5SO_4^{2-}$-TiO_2 (H), pH 3, 773K	3.55
$7.5SO_4^{2-}$-TiO_2 (H), pH 3, 773K	3.55	$2.5SO_4^{2-}$-TiO_2 (H), pH 3, 873K	3.80
$10SO_4^{2-}$-TiO_2, pH 3, 773K	4.20	$2.5SO_4^{2-}$-TiO_2 (H), pH 3, 973K	2.75

(H) indicates the sample prepared from H_2SO_4.

From the preliminary study, no degradation of nitrophenol was observed in the absence of light and/or of catalyst. The effect of pH (adjusted by H_2SO_4) of the solution on the photocatalytic degradation of 4-nitrophenol (Table 2). It is found that with increasing the pH of the solution up to 2, the percentage of 4-nitrophenol degradation remains almost constant. However, after that it decreases gradually with increasing the pH of the solution. Since no nitrobenzene was detected during the photolysis of 4-nitrophenol at pH 2, it must be concluded that the reductive pathway is favoured in acidic media. It is known that nitroaromatic compounds such as 4-nitrophenol are reduced in acidic media more easily than in alkaline solutions (18).

Table 2: Effect of Various Parameters on the Photocatalytic Degradation of Nitrophenol

Effect of pH of solution	*% of Degradation*	*Effect of Preparation Condition (pH)*	*% of Degradation*	*Effect of wt% of SO_4^{2-}*	*% of Degradation*	*Effect of Activation Temperature (K)*	*% of Degradation TiO_2*	*% of Degradation $2.5SO_4^{2-}$/TiO_2 (H)*
1	54.67	3	42.00	0	42.00	383	31.96	34.87
2	55.63	5	37.20	2.5	48.00	573	33.85	38.22
3	45.03	7	26.20	5.0	48.50	673	35.69	43.00
4	37.32	9	21.50	7.5	49.00	773	43.20	47.20
5	30.10	—	—	10.0	50.00	873	32.57	52.40
6	29.50	—	—	—	—	973	20.00	68.63
—	—	—	—	—	—	1073	16.40	25.27

The effect of TiO_2 samples prepared at different pH on the photocatalytic degradation of 4-nitrophenol is shown in Table 2. It is observed that with increasing the pH of the samples the percentage of degradation gradually decreases. This may be due to higher surface area, lower porosity and more number of hydroxyl groups present on the TiO_2 sample prepared at lower pH than that of higher pH. As reported earlier (4, 19, 20), the surface hydroxyl groups bound to titanium atoms constitute the adsorption active sites. The pH variation on the preparation procedure affects the adsorption powder of the gel because it affects the surface area as well as crystallite size and hence the number of surface hydroxyl groups. As reported earlier (21), the mixture of anatase and rutile phase of TiO_2 are found to be more active than either anatase or rutile crystallites at the same activation temperature. Here it is also observed from the XRD pattern of TiO_2 sample prepared at pH 3 and activated at 773 K shows both anatase and rutile phase. Whereas other samples prepared at pH 5, 7 and 9 exhibits only anatase phase at that activation temperature.

The effect of wt% of sulfate in TiO_2 on the photocatalytic degradation of 4-nitrophenol is shown in Table 2. It is observed that with the addition of 2.5wt% of sulfate on TiO_2, the percentage of degradation increases. However, further addition of higher sulfate content does not affect more to the percentage of degradation. This may be due to the addition of small amount of sulfate decreases the crystallite size of titania, which facilitates higher percentage of degradation. However, increasing the wt% of sulfate on TiO_2 does not affect more towards the crystallite size results almost same percentage of degradation. So, size is an important factor for degradation of 4-nitrophenol.

The activation temperature by calcination influences the catalyst activity because it affects the physical properties of the crystal such as crystal structure as well as crystal size of TiO_2. The effects of activation temperature of unmodified and sulfate modified TiO_2 on the photocatalytic degradation of 4-nitrophenol are shown in Table 2. It is observed that the percentage of degradation gradually increases with increase in activation temperature up to 773 K and thereafter decreases for unmodified TiO_2. But the percentage of degradation increases with increase in activation temperature up to 973 K and thereafter decreases for sulfate modified TiO_2. The percentage of degradation of 4-nitrophenol in case of sulfate modified TiO_2 is higher than that of unmodified TiO_2 at all activation temperature. This is due to the presence of mixture of anatase and rutile phase of TiO_2. The decrease of photocatalytic activity at higher activation temperature because of sintering due to loss of porous structure of nano-particles and the catalyst becomes denser as well as transferred to rutile from anatase phase.

The present results can be directly compared with the data published earlier (5, 6, 11). While optimising the reaction conditions, we found that 0.05g/L of 4-nitrophenol can be completely degraded by using 0.6g/L of sulfate modified TiO_2 catalyst in 3 h. These results found better than those reported earlier with respect to the concentration of nitrophenol, amount of catalyst and time.

Conclusions

1. Modification of the hydrated titania with anions like sulfate can alter the crystallinity.
2. Method as well as condition of preparation also greatly affects the above textural properties.
3. Photocatalytic degradation of 4-nitrophenol depends on phase and crystallite size of titania.

Acknowledgement

The authors are thankful to Dr. V.N. Misra, Director, Regional Research Laboratory, Bhubaneswar for his permission to present this paper.

REFERENCES

1. F.J. Beltran, V. Gomez-Serrano and A. Duran, 1992, *Wat. Res.* 26: 9.
2. A. Pintar and J. Levec, 1994, *Chem. Eng. Sci.*, 49L 4391.
3. M.S. Dieckmann and K.A. Gray, 1996, *Wat. Res.*, 30: 1169.

4. V. Augugliaro, L. Palmisano, M. Schiavello, A. Sclafani, L. Marchese, G. Martra and F. Miano, 1991, *Appl. Catal.*, 69: 323.
5. D. Chen and A.K. Ray, 1999, *Appl. Catal. B: Environmental*, 23: 143.
6. R. Andreozzi, V. Caprio, A. Insola, G. Longo and V. Tufano, 2000, *J. Chem. Tech. and Biotech.*, 75: 131.
7. N. San, A. Hatipo Lu, G. Kocturk and Z. Cnar, 2002, *J. Chem. Tech and Biotech.*, 146(3): 189.
8. C.S. Turchi and D.F. Ollis, 1990, *J. Catal.*, 122: 178.
9. G. Riegel and J.R. Bolton, 1995, *J. Phys Chem.*, 99: 4215.
10. A. Makowski, Z. Dziewiecki and W. Wardas, 2001, Presented in 17th North American Catalysis Society Meeting, June 3-8, Toronto.
11. J. Kiwi, C. Pulgarin and P. Peringer, 1994, *Appl. Catal.*, B 3: 335.
12. K.M. Parida, M. Acharya, S.K. Samantaray and T. Mishra, 1999, *J. Coll. and Inter. Sci.*, 217: 388.
13. S.K. Samantaray, T. Mishra and K.M. Parida, 2000, *J. Mol. Catal.*, 156: 267.
14. S.K. Samantaray and K.M. Parida, 2001, *J. Mol. Catal.*, 176: 151.
15. K. Kandori, S. Uchida, S. Kataoka and T. Ishikawa, 1992, *J. Mater. Sci.*, 27: 719.
16. D.G. Barton, S.L. Soled, G.D. Meitzner, G.A. Fuentes and E. Iglesia, 1999, *J. Catal.*, 181: 57.
17. L.J. Almeny, M.A. Larrubia, M.C. Jimenez, F. Delgado and J.M. Blasco, 1997, *React. Kinet. Catal. Lett.*, 60(1): 41.
18. R. Dillert, I. Fornefett, U. Siebers and D. Bahnemann, 1996, *J. Photochemistry and Photobiology* A, 94: 231.
19. H. Chen and E. Ruckenstein, 1992, *J. Coll. and Interface Sc.*, 148, 2: 382.
20. S-J. Tsai and S. Cheng, 1997, *Catal. Today*, 33: 227.
21. Q. Zhang, L. Gao and J. Guo, 2000, *Appl. Catal.* B 26: 207.

24

Adsorption Integrated Photocatalytic Degradation of Navinon and Foron Class Disperse Dyes

PRADNYA KHARE, N.N. RAO & S.N. KAUL

National Environmental Engineering Research Institute, Nehru Marg, Nagpur – 440 020

ABSTRACT

Navinon and Foron class of commercial disperse dyes have been examined for photocatalytic degradation using UV irradiated TiO_2 catalyst in aqueous dye solutions. Navinon dyes undergo complete color removal within 2-3 h. In case of Foron class dyes, the time required for complete color removal is 1-2 hrs. In all cases, gradual reduction in COD was observed and ≥ 80% COD reduction takes place within 1-3 h. The first-order rate constants were calculated from the slope of log $[(COD)_t / (COD)_0]$ Vs time plots. The rates of photocatalytic degradation of these dyes follow pseudo-first order kinetics. Thus TiO_2/UV technique can be employed for removing color and COD due to disperse dyes present in aqueous solutions.

Introduction

Large quantities of dyes are manufactured annually worldwide and are used in a variety of applications such as textiles, paper, foodstuffs and cosmetics. Textile dyes are of environmental interest because of their widespread use, their potential for formation of toxic aromatic amines & their low removal rate during waste treatment (1). The toxicity and potential carcinogenic nature of these dyes & their manufacturing precursors pose a specific threat to the environment. Some disperse dyes and solvent dyes are of interest because of their dispersal in the environment through use as colorants in military smokes and signals. The color of waste is the most apparent indication of pollution. Such colored dye effluents pose a major problem for the manufacturing plant itself as well as municipal treatment plants down the line (2). The problems are compounded by the fact that newer dyestuffs being produced are usually far more resistant to degradation, bleaching etc. Removing of color from waste is often more important than other colorless organic substances. Methods to decolorize dye effluents have therefore acquired increasing attention.

The disadvantages of commonly used treatment methods of wastewater purification are the fact that they are not destructive but only transfer the contamination from one phase to the other. New technologies of wastewater purification leading to complete mineralisation of organic pollutants are considered as more suitable solution of the problem connected with more drastic requirements concerning the contamination level in water and wastewater (3).

Information so far available from the work done in India and abroad shows that some dyes are extremely refractory and are not biodegradable. Disperse dyes are most refractory in nature. The dye carriers used along with disperse dyes make the complexity of the problem more compounded.

Recently Advanced Oxidation Processes (AOPs) have been proposed as the alternative purification methods. Among these processes, Photocatalysis can play important role in new technologies of water and wastewater purification (4-6). The case of photocatalytic oxidation of dyes is a specific type of Photocatalysis namely sensitized Photocatalysis (7-8). Some textiles azo dyes are subjected to reductive decolorisation using TiO_2 and WO_3 catalysts (9). While the sensitized oxidation of Rose Bengal at TiO_2 and ZnO, photocatalysts is retarded at higher dye concentration (10). The TiO_2/solar light degradation of methyelene blue followed first order kinetics with 99% removal efficiency (11). Reeves *et al.* (12) used highly concentrated sunlight and TiO_2 catalysts for destroying several classes of dyes and stains from wastewater, most dyes underwent degradation and reached < 1 ppm levels (from initial 30 ppm) in less than 20 min. In the present paper, TiO_2-catalysed photodegradation of Navinon and Foron class of disperse dyes has been investigated.

Materials and Methods

Powder titanium dioxide from Degussa (P-25, predominantly anatase, surface area 50±15 m^2/g, particle size 21 nm) was used as catalyst. All the other chemicals used for COD determination were high purity analytical reagents. In all, fifteen dyes were examined in this study. All these dyes were procured from a manufacturing unit in Gujarat.

The commercial names of the fifteen dyes are as follows:

1. Procion red (PR)
2. Procion green HE4BD
3. Procion Navy blue HE2R
4. Procion red M8B
5. Procion red HE7B
6. Procion yellow M3R
7. Procion blue MR
8. Ramezal Black N200

9. Ramazal Orange 3R
10. Navinon yellow GR (U/D)
11. Navinon Blue BC (U/D)
12. Navinon Red bis (U/D)
13. Foron Violet SE3RLI
14. Foron Rubine S2GFLI
15. Foron Navy S2GRLI

Methods

Photolysis-Screening of Dyes

Dye solutions (50-ppm concentration, 400-ml volume) were prepared and stirred on the magnetic stirrer. These solutions were then exposed to UV light in the absence of photocatalyst in a photoreactor using 400 W medium pressure Hg lamp (peak emission of 365nm; 5.0*1019 photons/sec.). In addition, these solutions were aerated simultaneously with the help of aquarium pump. Samples were drawn at an interval of 15 minutes and were checked for color removal during two-hour exposure.

Photocatalytic Treatment of Selected Dyes

Dye solutions (100 ppm concentration, 400 ml volume) containing 0.2g/L of P-25 TiO_2 catalyst were stirred on the magnetic stirrer to attain adsorption of dyes onto catalyst before commencing UV irradiation. These solutions were then treated in the above mentioned photoreactor in a batch wise manner for the following intervals-15, 30, 45 min, 1, 2, 3, 4, 5, 6 and 7 h. These samples were then allowed to settle down and the supernatant was used for COD determination by open reflux method as per Standard Methods (14). The COD was estimated as a function of irradiation time. The rate constants for COD removal were calculated from slopes of log $[(COD)_t / (COD)_0]$ Vs time plots. These samples were also processed for their absorbency in the wavelength range of 200-750 nm using Perkin Elmer UV-Vis spectrophotometer (Lambda 900).

Results and Discussion

Initially all fifteen commercial dyes were screened for color removal by exposing them to UV light in the absence of photocatalyst. This was done to assess the efficacy of photocatalyst as well as background contribution from simple photolysis. Out of these fifteen dyes, six dyes which are of Navinon and Foron class did not show color removal even after 2 h of exposure to UV light. All other dyes belonging to Remazol and Procion variety showed complete color removal during this illumination period. Therefore, Navinon and Foron class dyes were selected for further photocatalytic degradation studies.

The dye solutions with initial dye concentration of 100 ppm absorb strongly in the visible region. The absorption maxima & initial absorption for each dye are given in Table1. Extinction coefficients calculated by plotting standard graph for these dyes are also given. In this study, we have utilized the decrease in the intensity of the visible absorption band to monitor the decolorization process. The dye in a typical photocatalytic experiment was present in solution as well as adsorbed on TiO_2. It is observed that solution is decolorized first and then the dye-bearing TiO_2 too gradually undergoes decolorization. Thus, time required for complete color removal (both in solution and in solid phase, t_{100} in Table 1) for Navinon class of dyes is 2-3 h, whereas for Foron class of dyes, it is 1-2 h.

Table 1: Approximate Irradiation Time Required for Complete Removal of Color from Solutions of Different Disperse Dyes

Sl. No.	*Dye*	C_o, *mg/L*	l_{max} *(nm)*	I_o	ε *(mg⁻¹L)*	t_s *min*	t_{100} *(min)*
1.	Navinon Yellow GR (U/D)	100	416	0.90	0.0088	60	180
2.	Navinon Blue BC (U/D)	100	597	0.89	0.0087	60	180
3..	Navinon Red bis (U/D)	100	548	0.5187	0.0053	45	120
4.	Foron Violet SE3RLI	100	558	0.4464	0.0043	30	120
5.	Foron Rubine S2GFLI	100	474	1.1424	0.0106	45	60
6.	Foron Navy S2GRLI	100	537	1.1930	0.0117	30	120

Co: Initial dye concentration
L_{max}: Wavelength of absorption in visible region
I_o: Initial absorption
ε: Extinction coefficient, $mg^{-1}L$
t_S: time required for color removal from solution phase
t_{100}: time required for total color removal

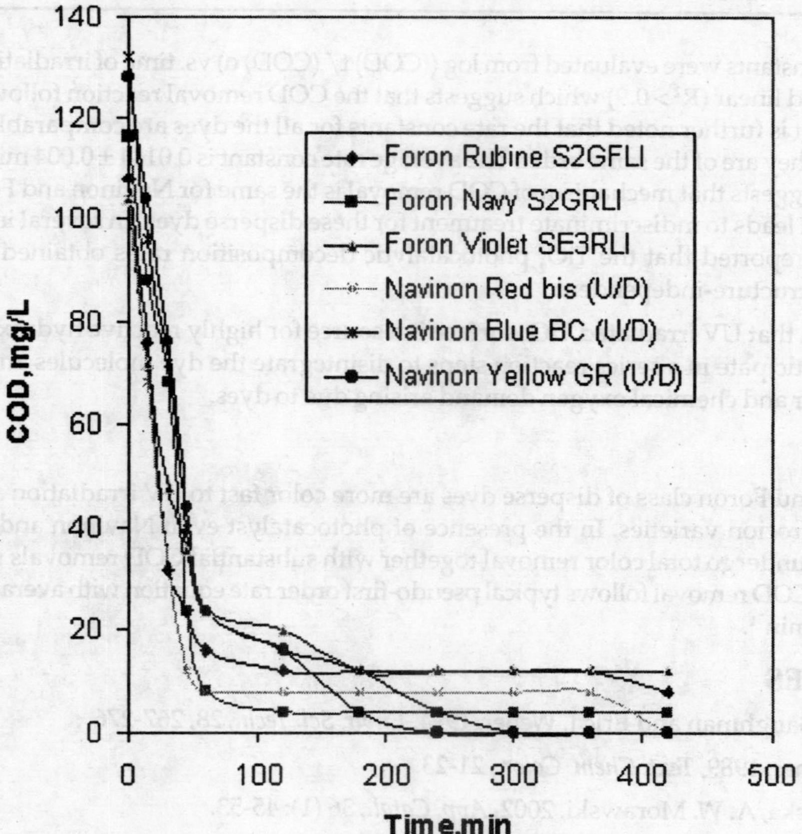

Fig. 1: Reduction of COD of Different Dyes with respect to Time

The photodegradation of Navinon and Foron class of dyes in terms of reduction in chemical oxygen demand (COD) is given in Table 2. About 80-90% reduction in COD relative to initial COD takes place in the first one hour of irradiation itself (Fig. 1). It is clear that both color and COD are removed simultaneously almost at competitive rates. This result suggests that the loss of color of these dyes upon photocatalytic treatment is due to disintegration/oxidation of the dye molecules rather than a localized chemical transformation of chromophore that may otherwise contribute color disappearance.

Table 2: Chemical Oxygen Demand Removal and Pseudo-first-order Rate Constants (k)

Dye	*Initial COD*	*Time Required for 90%COD Reduction, h*	*k. min^{-1}*
Navinon Blue BC (U/D)	132	3	0.0124
Navinon Yellow GR (U/D)	128	3	0.0122
Navinon Red bis (U/D)	136	1	0.0175
Foron Violet SE3RLI	104	3	0.0105
Foron Rubine S2GFLI	108	2	0.0144
Foron Navy S2GRLI	116	1	0.0192

The rate constants were evaluated from log ((COD) t/ (COD) o) vs. time of irradiation plots. These plots were found linear ($R^2 > 0.9$) which suggests that the COD removal reaction follows pseudo-first order kinetics. It is further noted that the rate constants for all the dyes are comparable to the second decimal place; they are of the same order. The average rate constant is 0.0144 ± 0.004 min^{-1}. The similar rate constant suggests that mechanism of COD removal is the same for Navinon and Foron class dyes and the method leads to indiscriminate treatment for these disperse dyes. In several instances earlier (15-16), it was reported that the TiO_2 photocatalytic decomposition rates obtained with different substrates are structure-independent.

It is known that UV irradiated TiO_2 surface is a source for highly reactive hydroxyl radicals (17) that in turn participate in a series reaction steps to disintegrate the dye molecules and contribute to removal of color and chemical oxygen demand arising due to dyes.

Conclusion

Navinon and Foron class of disperse dyes are more color fast to UV irradiation as compared to Remazol and Procion varieties. In the presence of photocatalyst even Navinon and Foron class of disperses dyes undergo total color removal together with substantial COD removals within 1h itself. The reaction of COD removal follows typical pseudo-first order rate equation with average rate constant 0.0144 ± 0.004 min^{-1}.

REFERENCES

1. George L Baughman and Eric J. Weber, 1994, *Envir. Sci. Tech.*, 28, 267-276.
2. W. C. Tincher, 1989, *Text. Chem. Color.*, 21-23.
3. J Grrechulska, A. W. Morawski, 2002, *App. Catal.*, 36 (1): 45-53.
4. K. Vinodgopal, D.E. Wynkoop, 1996, *Envir. Sci. Tech.* 30.
5. F. Kiriakidou, D. I. Kondaredes, K. E. Verykion, 1999, *Catalysis Today*, 54, 119-130.

6. J. M. Herrmann, 1999, *Catalysis Today*, 53, 115-129.

7. M. Schiavello (Ed.), 1988, In: *Photocatalysis & Environment*, Kluwir Academic Publishers, Dordrecht.

8. R. Andreozzi, V. Caprio, A. Insola, R. Marotta, 1999, *Cata. Today*, 53, 51-59.

9. K. Vinodgopal, Idriss Bedja, Surat Hotchandani & P. Kamat, 1994, *Langmuir*, 10.

10. K.Tennakone, W.C. Kiridena and S.Punchihewa, 1992, *J. Photo. Chem. Photobiol A Chem*, 68, 389.

11. R. F. P. Nagueira & W.F. J Jardim, 1993, *Chem. Edu.*, 70, 861.

12. P.Reeves, R.Ohlhausen., D.Sloan, K. Pamplin & T. Scoggins, 1992, *Solar Energy*, 48, 413.

13. N. N. Rao, S. Dube, 1997, *Indian J. Chem. Tech.*, 4, 1-6.

14. APHA, 1989, *Standard Methods for the examination of Water and Wastewater*. 17th Edition, American Public Health Association, Washington DC.

15. N. N. Rao. Sangeeta Dube, 1996, *J. Molecular Catalysis A: Chemical*, 104, L197-L199.

16. S. Dube, N. N. Rao, 1996, *J. Photochem Photobiol. A:Chem.*, 93, 71.

17. Craig S.Turuchi and David F.Ollis, 1990, *J. of Catalysis*, 122, 178-192.

25

Heterogenous Solar Photocatalysis of Two Commercial Reactive Azo Dyes

RITA S. DHODAPKAR, VIBHA CHATURVEDI, N.N. RAO & S.N. KAUL

National Environmental Engineering Research Institute, Nehru Marg, Nagpur – 20

ABSTRACT

Two commercial reactive azo dyes-Reactive Black 5 (RB5) and Reactive Orange 16 (RO 16) of concentration 50 mg/L have been treated by titanium dioxide (0.2 g/L) and Zinc oxide (0.2 g/L) photocatlysts separately under presence of sunlight. It is observed that solar photocatalytic treatment is effective in terms of colour and COD. The photodegradation efficiency of zinc oxide is comparable with TiO_2 at pH 5.0-6.0 for RO16. The extent of decolourization and degradation of RB5 is greater in presence of zinc oxide photocatalyst than TiO_2 at pH 5.0-6.0. Zinc oxide undergoes <1% photodisolution after 6 hours of solar irradiation at working pH.

Keywords: Solar photocatalysis, Reactive azo dyes, Zinc oxide, Titanium dioxide.

Introduction

Reactive dyes are commonly used in textile industries due to their good color fixation property by formation of covalent bond with natural or synthetic fibre. The dyehouse effluents from textile industries contain water soluble reactive azo dyes which impart intense color to wastewater. Treatment through flocculation and adsorption result in solid waste. Azo reactive dyes contain sulphonic acid groups that make the dyes very soluble and polar and difficult to adsorb onto nonpolar carbon adsorbent surface. The dyes are also recalcitrant in nature and decolourization/degradation of dye by biological means is inefficient. Recovery of unfixed dye from in wastewater is not possible due to its modified nature [1].

Photocatalysis has been studied to demonstrate decolourization and mineralization of dye compounds using TiO_2 and UV radiation [2–5]. The photocatalyst upon irradiation with UV light generates highly reactive OH° radicals which oxidize/mineralize organic pollutants. Several workers have studied homogenous and heterogeneous photocatalysis of reactive dyes using UV radiation [6–9].

Solar Photocatalysis

The UV flux near the surface of the earth is of 20-30 W/m^2 and the sun puts 0.2-0.3 mol photons/m^2h in the 300-400 nm range. The photons are suitable for destroying pollutants in photocatalytic reactors. The wavelength range UVA – 300-400 nm has corresponding diffuse energy E_{diff} (300-400 nm) = 24.3 W/m^2 and direct E_{dir} (300-400 nm) = 25.0 W/m^2. Solar radiation reaching the earth also matches with TiO_2 and ZnO band gap energy of 3.2 eV. Solar photocatalysis holds promise as a practical solution to the problem of dye removal from wastewater, however many operational parameters have to be considered.

Previous work in this laboratory has investigated the efficiency of solar photocatalysis in degradation of azo reactive dyes with Titanium dioxide (TiO_2) [10]. This paper presents solar photo-degradation studies of Reactive Black 5 and Reactive Orange 16 in presence of photocatalysts TiO_2 and ZnO.

Experimental Section

Solar Light Measurements

Solar illumination was measured hourly using Cal Zeiss Luxmeter. Although one cannot convert irradiance into rate of illuminesence accurately because of the variation in brightness of different spectral regions l gm cal/cm^2/min of sunlight multiplied by 6700 will give approximate illuminescence on a horizontal surface expressed as foot candles. The average lux reading from 11.00 AM to 12.30 PM = 31200 lux (3120 ft candle, 0.4657 gm cal/cm^2/min); average lux reading from 1.00 PM to 2.30 PM = 29806 lux (2980.6 ft. candle, 0.4449 gm cal/cm^2/min).

Solar Photocatalytic-reactor Details

All photocatalytic experiments were performed under sunlight. The duration of exposure to sun was-10.00 AM-4.00 PM and experiments were conducted in an open rooftop. Continuously stirred batch reactor experiments were conducted in 500-ml capacity Pyrex glass reactors. The flask contents were aerated using aquarium pump and air distribution manifold. The initial concentration of dyes was generally fixed at 50 mg/l while the contingent dose of photocatalyst was 0.2g TiO_2/L and 0.2 ZnO g/L.

Materials

C.I. Reactive Orange 16 and C. I. Reactive Black 5 were purchased from Colorant Pvt. Ltd. (Ahmedabad). Degussa P25 TiO_2 was a gift from Degussa (Germany). It consists of a 80:20 anatase: rutile crystal phase with a specific BET surface area of approximately 50 m^2g^{-1} and a mean particle size diameter of 30 nm. Zinc Oxide (Pure) was purchased from Merck India Ltd. This is food/pharma grade chemical and is 99% pure and surface area of 10 m^2g^{-1}. It is < 0.3% soluble in water. Deionized water was used to prepare all reagent and reaction solutions.

Dark Adsorption Studies

Adsorption studies were conducted in 250 ml Erlenmeyer flasks with stoppers containing varying concentration of dye and 0.2 g/L of photocatalyst. The suspensions were shaken in dark for 4 hours at room temperature (30°C). The photocatalyst was then removed using 0.22 um membrane filter and residual concentration determined spectro-photometrically.

Analytical Determination

Test samples were analyzed for color after filtration through 0.2μm syringe filter. Spectrophotometric measurements were carried out in the range of 200-700nm using double beam Perkin Elmer Λ 900 spectrophotometer. The reduction in the color band intensity was determined from the time-overlaid spectra. Diminution in COD was determined following standard procedure (APHA AWWA-WPCF 1989). Zinc concentration was measured using atomic absorption spectrophotometry (GRC 904 AA. Dadoneng, Australia).

Result

Photocatalytic Degradation of Dyes

Reactive Orange 16 (RO16) displays a broad colour band at $\lambda max = 493$ nm, $\varepsilon = 1.55 \times 10^4\ M^{-1}$ and many peaks in 250-330 nm range which represent aromatic structure. This is a vinyl sulphone medium intensity azo dye with a Gamma acid function (Mol. Wt-621). The Reactive Black 5 (RB5) is vinyl sulphone substituted bis azo dye with a central H-acid function (λ=597 nm, $\varepsilon = 4.0543 \times 10^4\ M^{-1}$ and Mol wt. 986). RB5 with two azo group in contrast to one in RO 16 is responsible for about 2.6 times molar absorption coefficient (ε). A 50 mg/L concentration of RB5 (5.07×10^{-5}M) has an initial absorbance of 2.1290 a.u. at λ max 597 nm. A 50 mg/L solution of RO16 (8.06×10^{-5}M) has an initial absorbance of 1.2437 a.u. at λ max 493 nm. Experiments of solar photocatalytic degradation of C. I. RB 5 and RO 16 were performed in the presence of semiconductor catalyst TiO_2 and ZnO. Results of the photolysis of 50 mg/L of RB5 in the presence of ZnO and TiO_2 are evident from time overlaid absorption spectra. (Figs. 1-2). The amount of RB5 present in clear filtrate, as estimated spectrophotometrically is plotted as a function irradiation time.

Fig. 3 shows that RB5 and RO16 solution in the presence of photocatalyst and sunlight leads to disappearance of the compound. Under these experimental conditions and in presence of 0.2 g/L TiO_2, 57% of the initial concentration of RB5 was removed after 120 mins of solar light exposure while in the case of ZnO, the decomposition of the compound after same irradiation time was 98%. The initial rate of decolorization is 1.434×10^{-2} mM/h. for TiO_2 and 1.896×10^{-2} mM/h for ZnO. After 150-175 min the degradation is more than 97% with TiO_2.

Results of the photolysis of RO16 of 8.06×10^{-2}mM concentration in the presence of ZnO and TiO_2 are given in Fig 3. The time overlaid spectra (Fig. 4-5) clearly reveals the photodegradation of RO16. In 60 min of exposure, 52% degradation is obtained with ZnO and 51% with TiO_2. After 180 mins with ZnO 95.8% and with TiO_2 97.5% removal is observed. The initial rate of disapearance of RO16/ZnO is 3.588×10^{-2} mM/h and RO 16/TiO_2 is 3.414×10^{-2} mM/h.

In the absence of photocatalyst a very small decrease in concentration was observed. After 2 h of direct radiation less than 5% of RB5 and RO16 were degraded.

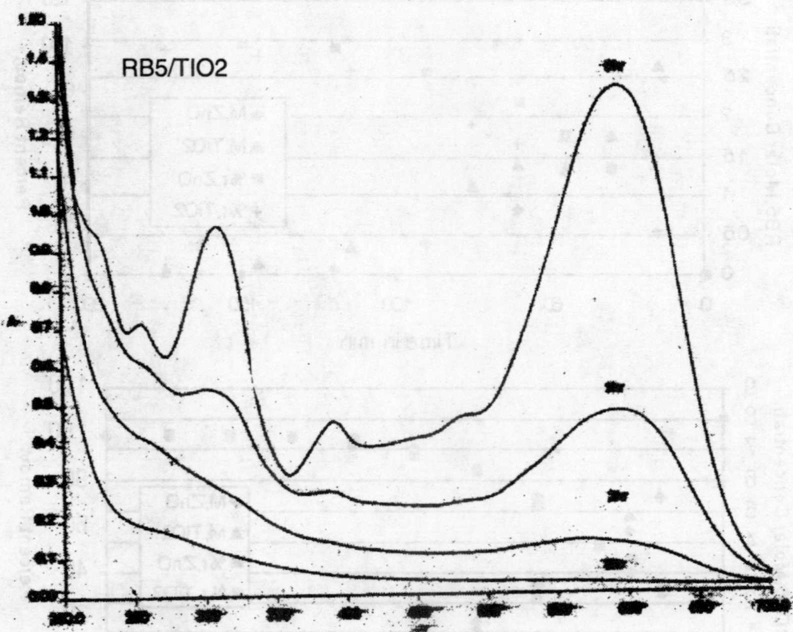

Fig. 1: UV-Vis Spectral Changes of RB5 dye (5.07 × 10^{-5}M) in aqueous TiO_2 Dispersions with Irradiation Time

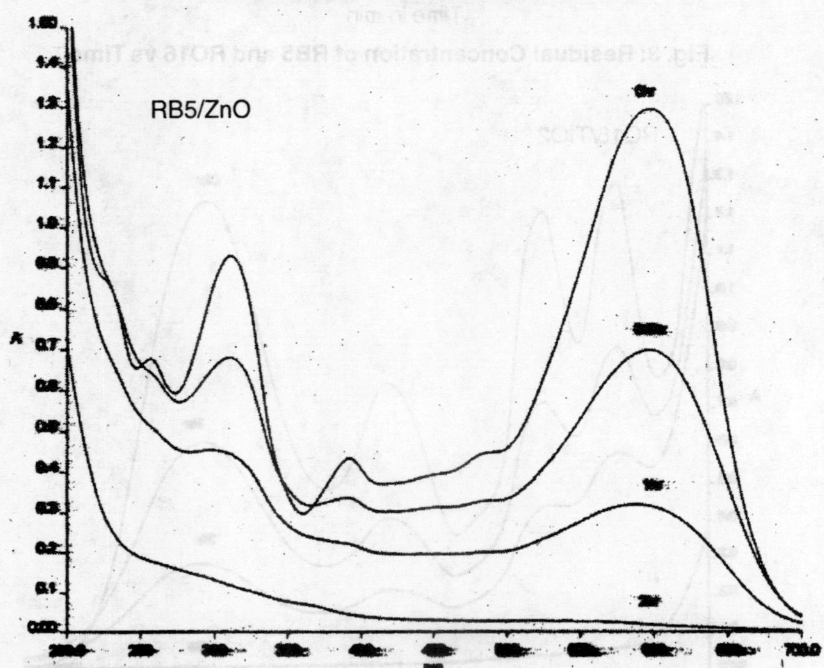

Fig. 2: UV-Vis Spectral Changes of RB5 dye (5.07 × 10^{-5}M) in Aqueous ZnO Dispersions with Irradiation Time

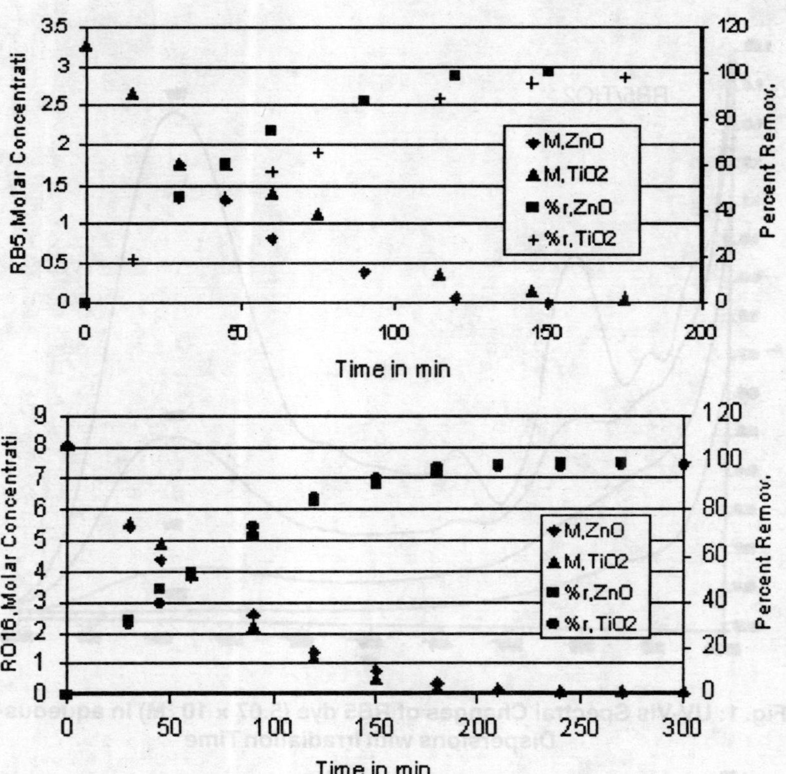

Fig. 3: Residual Concentration of RB5 and RO16 vs Time

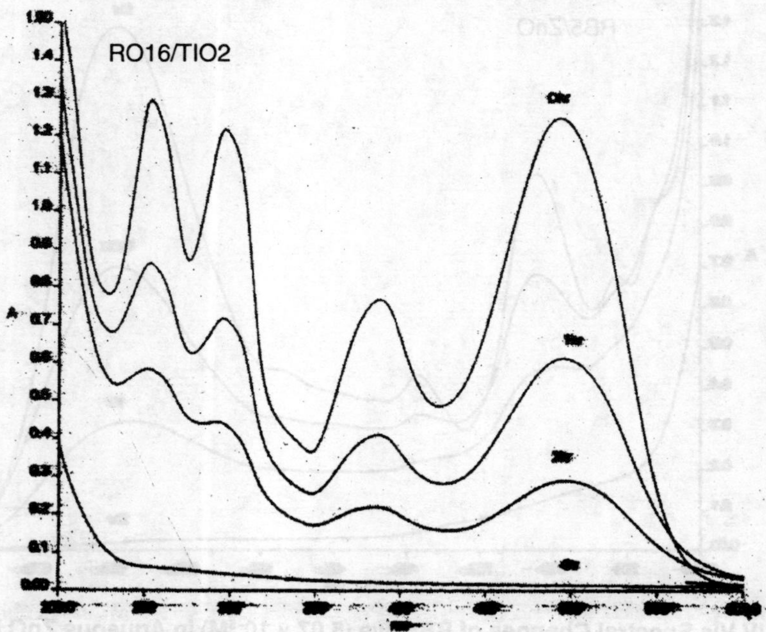

Fig. 4: UV-Vis Spectral Changes of RO16 dye (8.06×10^{-5}M) in Aqueous TiO_2 Dispersions with Irradiation Time

Simultaneous reduction in COD was also monitored by closed reflux method as shown in Fig. 6 wherein COD is plotted as a function of irradiation time. RB5 in the presence of TiO_2 50% of the initial COD was removed after 3 h while 70% was removed when ZnO was used. On extended irradiation upto 6 hours 74% reduction in COD was observed with TiO_2 and 89% with ZnO.

RO 16 after exposure for 250 min resulted in a COD reduction of 58% in presence of TiO_2 as well as ZnO. However after 6 h of extended irradiation TiO_2 gave better results than ZnO and a 65% reduction achieved with ZnO as compared to 73% with TiO_2.

Concentration of Zinc was estimated in the samples and it was observed that a maximum of 2ppm concentration was leached into solution as compared to 200 ppm concentration of ZnO powder into the reacting solution after 6 hours of exposure to sunlight.

Equilibrium Isotherm and Adsorption

The adsorption isotherms plotted between concentration of dye adsorbed per unit weight of TiO_2/ZnO and equilibrium concentration of dye suggests that Langmuir adsorption with well defined limiting adsorption beyond a certain equilibrium concentration is displayed. Fig. 7 compares Langmuir adsorption isotherm for RB5 with ZnO and TiO_2. The adsorption of RB5 on ZnO is stronger than that for TiO_2 at pH 5.5-6.0. RB5 is bound more strongly on ZnO. The X_m value for both are same indicating equivalent number of adsorption sites for RB5 molecules.

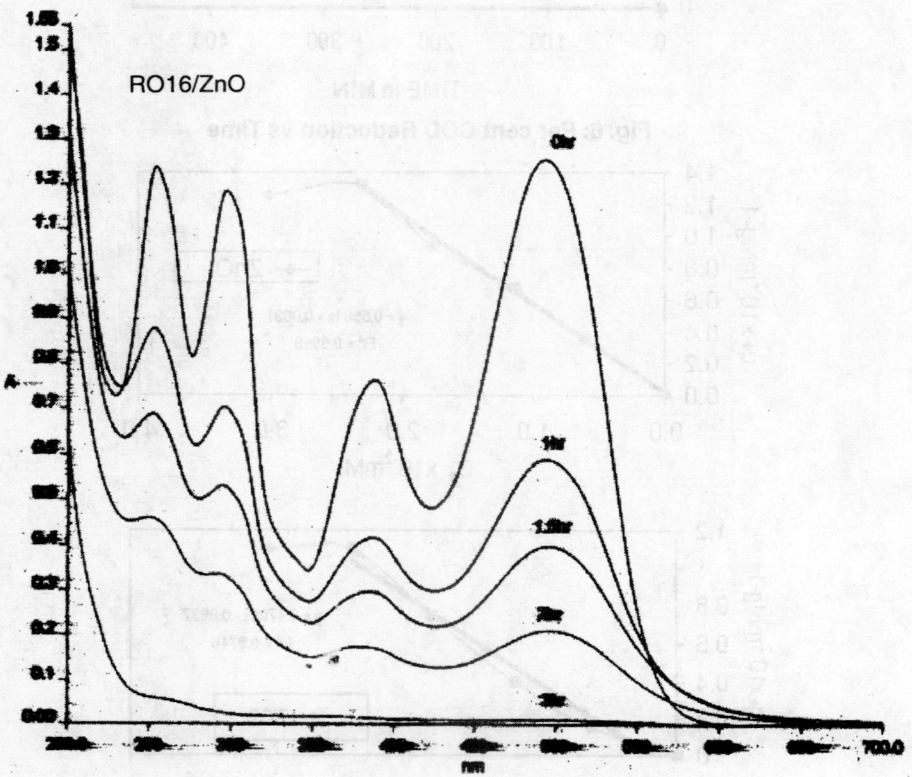

Fig. 5: UV-Vis Spectral Changes of RO16 dye (8.06 × 10^{-5}M) in Aqueous ZnO Dispersions with Irradiation Time

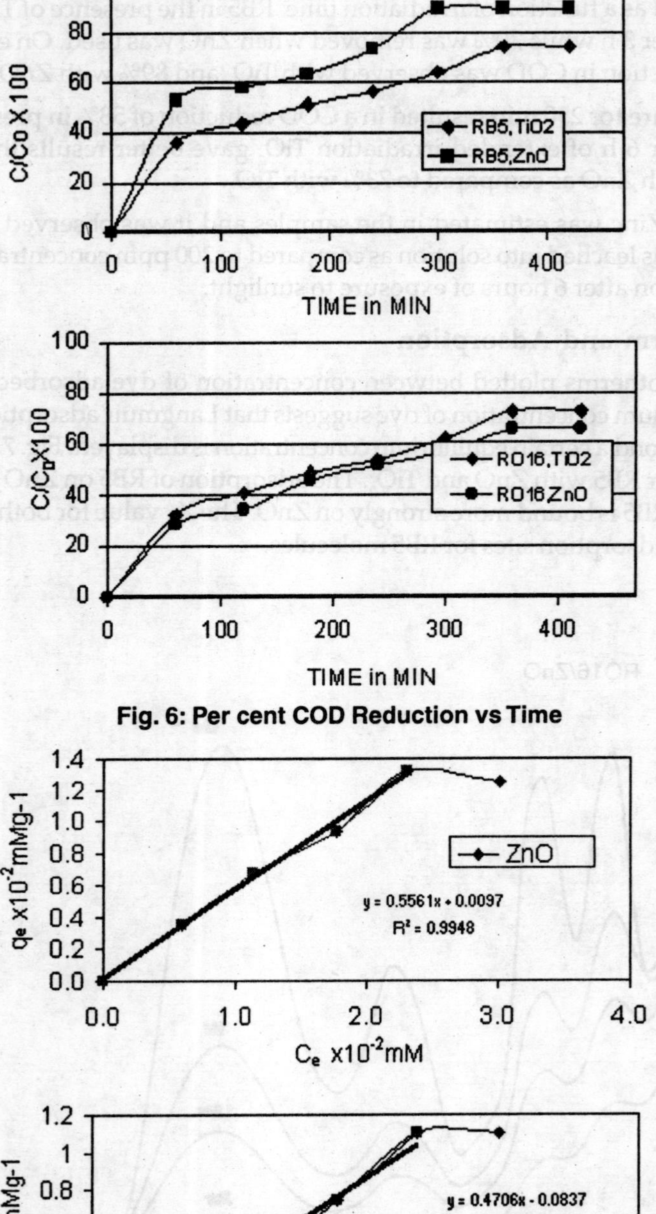

Fig. 6: Per cent COD Reduction vs Time

Fig. 7: Langmuir Adsorption Isotherm (RB5)

Similarly, Fig. 8 compares adsorption isotherms for RO16 with ZnO and TiO_2. The adsorption constant K and limiting adsorption value (X_m) are more with TiO_2 than ZnO. Both RB5 and RO16 are eqally bound to TiO_2 whereas in case of ZnO, RB5 is more strongly bound than RO16 (Table 1).

Table 1: Adsorption Isotherm Constants

Photocatalyst	*Dye*	*K*	*Xm*
TiO_2	RB5	0.4706	1.1×10^{-2}
	RO16	0.4789	2.5×10^{-2}
ZnO	RB5	0.5561	1.16×10^{-2}
	RO16	0.3687	1.9×10^{-2}

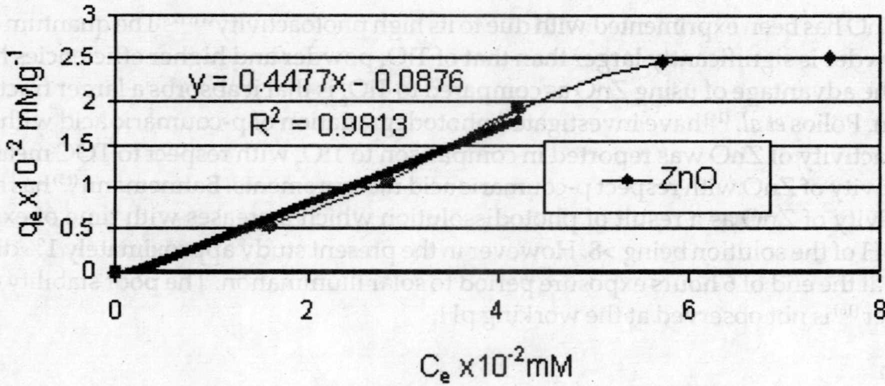

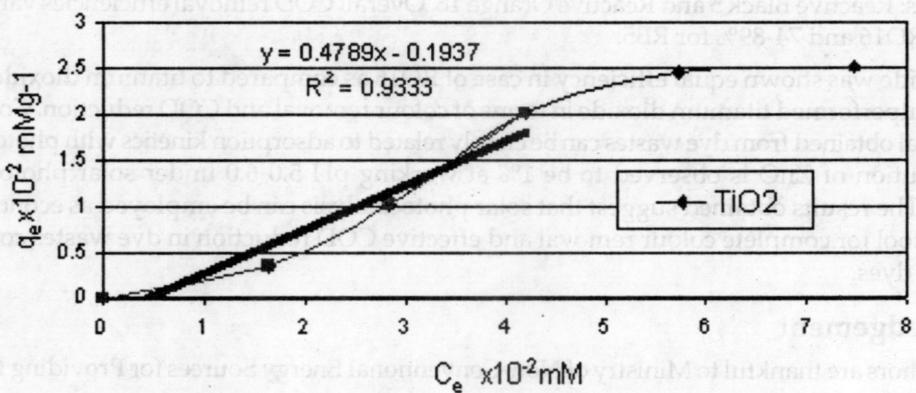

Fig. 8: Langmuir Adsorption Isother (RO16)

Discussions

Photodegradation results can be related to dark adsorption studies as the adsorption of the organic substrate determines solar photodegradation rates. It is possible that interaction of photoexcited electron or hole with absorbed dyes may depend on the affinity of adsorption of different dyes and result is different reaction rates. About 7.1% and 7.4% of initial RO16 concentration is adsorbed over ZnO and TiO_2 respectively. Further it is found that 9.7% and 7.0% RB5 is adsorbed on ZnO and TiO_2 respectively. Though the adsorption isotherm constants are not significantly different, this small difference in the case of RB5 is reflected in RB5/COD reduction. A 73% reduction in COD takes place on 360 mines of exposure with TiO_2 as compared to 240 mins with ZnO. Initial rates of reduction of RB5 is 1.3 times more with ZnO than with TiO_2. Similarly RB5 undergoes 89% COD reduction as compared to 65% for RO16 on solar illumination with ZnO after 6 h. It is also possible that after an initial breakdown of 50-58% in both above cases, more stable products/intermediates are formed on breakdown of RO16 on extended period of irradiation. Decolorization rates of RB5 are also higher with ZnO than that with TiO2.

For RO16 both ZnO and TiO_2 have almost same decolorization rates though COD reduction on extended irradiation is more with TiO_2 than ZnO.

Initially ZnO has been exprimented with due to its high photoactivity [11,12]. The quantum efficiency of the ZnO powder is significantly larger than that of TiO_2 powder and higher efficiencies have been reported [13]. The advantage of using ZnO as compared to TiO_2 is that it absorbs a larger fraction of the solar spectrum. Polios *et al.* [14] have investigated photodegradation of p-coumaric acid with ZnO and TiO_2. A lower activity of ZnO was reported in comparison to TiO_2 with respect to TOC measurement and higher activity of ZnO with respect p-coumaric acid measurements. Bahnemann [15] has attributed this lower activity of ZnO as a result of photodissolution which increases with time of exposure to sunlight, the pH of the solution being >8. However in the present study approximately 1% dissolution was observed at the end of 6 hours exposure period to solar illumination. The poor stability of ZnO as reported earlier [16] is not observed at the working pH.

Conclusion

Hetergenous solar photocatalysis using TiO_2 and ZnO is capable of completely decolourzing azo reactive dyes: Reactive Black 5 and Reactive Orange 16. Overall COD removal efficiencies varied from 65-72% for RO16 and 74-89% for RB5.

Zinc oxide was shown equal efficiency in case of RO16 as compared to titanium dioxide and for RB5 it has outperformed titanium dioxide in terms of colour removal and COD reduction. Colour and COD removal obtained from dye wastes can be closely related to adsorption kinetics with photocatalyst. Photodissolution of ZnO is observed to be 1% at working pH 5.0-6.0 under solar photocatalytic conditions. The results obtained suggest that solar photocatalysis can be employed as economic and a powerful tool for complete colour removal and effective COD reduction in dye wastes containing reactive azo dyes.

Acknowledgement

The authors are thankful to Ministry of Non-Conventional Energy Sources for Providing Financial Assistance.

REFERENCES

1. Pelegrini, R., Peralta Zamora P., De Andrade, A.R., Reyes, J. and Duran N., 1999, *Applied Catalysis B. Environmental*, 22, 83-90.

2. Dhodapkar R.S., Rao, N. N., Pande S. P. and Kaul S. N., 2000, *Environ. Conservation Journal*, (1), pp. 13-20.

3. Rao N. N. and Dube S, 1996, *J. Mol. Catal. A. Chem.*, 104 pp. 19.
4. Rao N. N. and Dube S, 1997, *Indian Journal of Chemical Technology*, 4, pp. 1-6.
5. Rao N. N. and Dube S, 1998, *Studies in Surface Science and Catalysis*. 113, pp. 1045.
6. Balciogu, I.A., Arslan I, Sacan, M.T., 2001, *Environmental Technology* 22, 7, 813-822.
7. Mohey, A., Dien, El., Libra, J.A., Wiesmann, U., 2001, *Water Science and Technology*, 44, 5, 295-302.
8. Tang, W.Z., Zhang, Z., An. H., Quinatana, M.O. and Torreo D.F., 1997, *Environmental Technology* 18, 1, 1-12.
9. Lizama, C., Yeber, M.C., Freer, J. Baeza, J., and Mansilla, H.D., 2002, *Wat. Sci. & Tech.*, 44, 5, 197-203.
10. Chaturvedi V., Dhodapkar, R.S., Rao. N.N. and Kaul, S.N., (2002 Paper presented in National Symposium on New Horizons in Heterogeneous Catalysis held on Feb. 22-24, at Banaras Hindu University, Varanasi,
11. Frank, S. N. and Bard A.J., 1977, *J. Phys. Chem.*, 81, 1484-1488.
12. Rao, M. V., Rajeshwar, V.R., Pai, V.R. and Dubow, J., 1980, *J. Phys. Chem.*, 84, 1987-1991.
13. Freund T. and Gomes, W.P., 1969, *Catal. Rev.* 3, 1-32.
14. Poulios, I., Kyriacou G., 2002, *Environmental Technology*, 23, 179-183.
15. Bahnemann, D.W., Kormann C. and Hoffman M.R., 1989, *J. Phys. Chem.*, 91, 3789.
16. Domenech J. and Prieto A., 1986, *J. Phys. Chem.* 90, 1123-1126.

26

Reduction in Pollution Load of Effluent by Replacement of Calcium Hypochlorite with Hydrogen Peroxide in Conventional Pulp Bleaching Sequence

B.P. MISHRA* R.P. MISHRA* G.G. BHARGAVA & N.K. THUSU*****

**Research Chemist **Supdt (Research) ***President (Operations)*
Orient Paper Mills, Amlai, Dist. Shahdol (M.P)

ABSTRACT

In order to achieve desired strength and optical properties for paper making the raw materials such as bamboo and/or hard woods (Eucalyptus, Casurina, Subabool, etc) are subjected to chemical cooking to remove as much Lignin as possible followed by a multi stage bleaching sequence in which Conventionally Chlorine & Chlorine compounds and/or Eco friendly chemicals Viz, Oxygen, Hydrogen Peroxide and Ozone are utilized.

In the paper being presented, bleachable grade pulp of Higher (25.3) Kappa was prepared by kraft cooking of Bamboo + mixed hard woods (65:35). The pulp was bleached under C-Ep-H-

D sequence. In order to reduce total Chlorine consumption Calcium hypochlorite was then replaced with Hydrogen peroxide to get the desired pulp brightness in C-Ep-E_1p-D bleaching sequence.The pulps bleached under C-Ep-H-D and C-Ep-E_1p-D sequences were analyzed for fibre classification and physical strength properties. The effluent at each stage of bleaching was also analyzed for different parameters to evaluate the benefit of replacement of Calcium hypochlorite with Hydrogen peroxide in terms of reduction of the pollution load of the outgoing effluent.

It was observed that replacing Calcium hypochlorite with Hydrogen peroxide in C-Ep-H-D bleaching sequence helped to reduce the total chlorine consumption by 44.3% for higher Kappa pulp bleached under C-Ep-E_1p-D sequence as compared to pulp bleached under conventional C-E-H-D sequence. Reduction in pulp shrinkage, ash content, P.C.Number and increase in pulp viscosity with respect to hydrogen peroxide utilization are the other benefits observed. What is noteworthy is that the above sequence appreciably reduces nett pollution load In terms of reduction of COD, Suspended solids, Dissolved solids and Chloride content. However since Hydrogen peroxide is a much costlier bleaching chemical as compared to Chlorine & Calcium hypochlorite, its consumption needs to be optimized which for a pulp of 25.3 Kappa is observed to be around 1.2% in order to achieve the desired pulp brightness, with out affecting the pulp yield and its strength & optical properties.

Furthermore replacement of Calcium hypochlorite with Hydrogen peroxide reduces chloride content in pulp and backwater appreciably, which are helpful in minimizing corrosion in pipelines and valves. At the same time reduction in Calcium and Calcium compounds due to utilization of Hydrogen peroxide are also observed to be beneficial in terms of improved paper machine runnability and less costly treatment of the recycled effluent.

Introduction

Lignin is the cementing material for the Cellulosic fibres in the plant kingdom and its major portion is located in the middle lamella of the wood.The remaining Lignin is distributed within secondary wall with higher concentration generally found inS_3 layer of cell wall. Since Lignin does not help in pulp and papermaking therefore its removal is essential. Complete Lignin removal at the conventional chemical cooking stage results in drastic deterioration of pulp strength properties due to action on Cellulose content, hence a small portion is allowed to remain in the cooked pulp. This residual Lignin, which makes the pulp brown in appearance, is to be eliminated before the pulp can be utilized effectively for making white grades of writing, Printing and other superior grades of papers. This is done by bleaching, which is a multistage process involving usage of various chemicals such as Chlorine, Calcium hypochlorite, Chlorine Dioxide, Oxygen, Hydrogen peroxide and ozone. Among the various bleaching agents prevalent, Chlorine and Chlorine compounds have been employed predominantly in the past, as they are highly efficient and economical. In the conventional bleaching system after direct Chlorination an alkaline stage utilizing NaOH removes Chlorolignins formed in the Chlorination process. However the presence of Chlorinated Lignin compounds in bleach effluent have detrimental effect on environment [1,2]. It may be noted that the formation of organic Chlorine compounds during bleaching, degradation and removal efficiency by microbial system differs considerably according to raw materials and the process used to manufacture [3,4].

Efforts are now being made to modify the conventional system of pulping & bleaching with the introduction of process such as White rot fungus treatment of fibrous raw materials, alkaline leaching of pulp, enzymatic pretreatment of pulp, substitution of Chlorine by ClO_2 partially or fully, Hydrogen peroxide reinforced oxidative extraction, etc.-the basic idea being to minimize the generation of pollution load[5] in the effluent.

In the paper being presented, an attempt has been made to utilize the advantage of alkaline condition as well as higher temperature at the extraction stage with the introduction of Hydrogen peroxide [6-7] in place of conventional Calcium Hypochlorite to be utilized during the dead retention time in order to obtain the final brightness of the pulp higher by one or two degrees.

The alkaline conditions in alkali extraction stage helps in the formation of per hydroxyl ions (OOH^-), which oxidize the colour and renders the pulp with increased brightness. Under the alkaline conditions only etherified phenolic nuclei or monomer unit having side chains containing Carbonyl groups appears susceptible to attack [8].

Bamboo and mixed hard woods in the 65:35 were subjected to cooking in a stationary digester at lab scale to produce pulp of 25-27 Kappa. Unbleached pulp of Higher 25-27 Kappa was then subjected to bleaching on lab scale under conventional C-Ep-H-D sequence for obtaining pulp brightness of 87±1% P.V. During the conventional bleaching process effluent characteristics were then analyzed at every stage of bleaching. Bleaching studies and effluent characteristics were also examined for the same cooked pulp bleached under modified C-Ep-E_1p-D sequence, replacing Calcium hypochlorite with Hydrogen Peroxide.

Results and Discussions

In order to bleach Higher Kappa pulp under C-Ep-H-D and C-Ep-E_1p-D sequences and to study the effect on bleached pulp quality and effluent characteristics, following studies were carried out on lab scale after cooking.

Bleaching of Higher Kappa (25.3) Pulp Under C-Ep-H-D Sequences

Use of Calcium hypochlorite has come under question due to formation of Chloroform during removal of Lignin and its adverse effect on pulp properties[9] and therefore it is being replaced gradually from the existing plants. At Hindustan News Print Ltd (H.N.L), Kottayam, bleaching of Chemical pulp under C-H stages was replaced by C-P stages with over all benefit and improvement in pulp and effluent quality[10]. At I.T.C, Bhadrachalam, successful plant trial runs have been conducted replacing C D-Ep-H-D with CD-Ep-E_1p-D bleaching sequence [11].

In our case the higher Kappa pulp (25.3) bleached under C-Ep-H-D sequence had brightness of 87.0% P.V with consumption of 9.1% Chlorine and 0.7% ClO_2. The bleached pulp yield was 36.2% on O.D chips. The Higher Kappa (25.3) pulp was also bleached under C-Ep-E_1p-D sequence with 0.8%, 1.0%, 1.2% and 1.4% Hydrogen Peroxide dosages replacing Calcium hypochlorite by Hydrogen peroxide in E_1 stage.

Process parameters maintained in above experiments are depicted in Table 1. Reaction temperature in E_1 stage was kept between 65-70°C to prevent both hornification and strength loss of the pulp in alkaline medium[12]. Total Chlorine Consumption was observed to be 5.1% and 0.7% ClO_2. The bleached pulp yield in experiment No.1 to 4 was 36.85%, 36.77%, 36.70% and 36.52% respectively. In C-Ep-E_1p-D bleaching sequence pulp yield & pulp Viscosity was higher but reverse trend was observed with P.C.NO & Ash content compared to C-Ep-H-D bleached pulp as may be seen from the data represented in Table 1.

Table 1: Summary of Process Parameters During Pulp Bleaching with C-Ep-E$_1$p-D Sequence with Variation in Hydrogen peroxide Dosages in IInd Caustic Extraction Stage (Unbleached pulp Kappa No. 25.3)

Sl. No.	Particulars	Blank Expt. C-Ep-H-D Sequence	Expt. No. 1 C-Ep-E$_1$p-D Sequence	Expt. No. 2 C-Ep-E$_1$p-D Sequence	Expt. No. 3 C-Ep-E$_1$p-D Sequence	Expt. No.4 C-Ep-E$_1$p-D Sequence
1.	**Chlorination Stage**					
	1. Cl_2 applied/consumed, %	5.1/4.97	5.1/4.97	5.1/4.97	5.1/4.97	5.1/4.97
	2. End pH	2.1	2.1	2.1	2.1	2.1
	3. Consistency, %	3.0	3.0	3.0	3.0	3.0
	4. Temperature, °C	Room	Room	Room	Room	Room
	5. Time, minutes	60	60	60	60	60
2	**Alkali Extraction Stage**					
	1. NaOH applied, %	1.53	1.53	1.53	1.53	1.53
	2. H_2O_2 applied, %	0.4	0.4	0.4	0.4	0.4
	3. End pH.	10.8	10.8	10.8	10.8	10.8
	4. Consistency, %	10.0	10.0	10.0	10.0	10.0
	5. Temperature, °C	70±1	70±1	70±1	70±1	70±1
	6. Time, minutes	60	60	60	60	60
3.	**Calcium Hypo Chlorite Stage**					
	1. Hypo chlorite applied/Consumed, %	4.0/3.96	—	—	—	—
	2. Sulphamic Acid, %	0.1	—	—	—	—
	3. Alkali added as buffer, %	0.8	—	—	—	—
	4. End pH	7.8	—	—	—	—
	5. Consistency, %	10.0	—	—	—	—
	6. Temperature, °C	40±1	—	—	—	—
	7. Time, minutes	120	—	—	—	—
4.	**Hydrogen Peroxide Stage**					
	1. H_2O_2 applied/consumed, %	—	0.8/0.75	1.0/0.86	1.2/1.0	1.4/1.1
	2. End pH	—	10.5	10.5	10.5	10.5
	3. Alkali added as buffer, %	—	0.4	0.4	0.4	0.4
	4. Consistency, %	—	10.0	10.0	10.0	10.0
	5. Temperature, °C	—	70±1	70±1	70±1	70±1
	6. Time, minutes	—	120	120	120	120
5.	Chlorine Dioxide Stage					
	1. ClO_2 applied/consumed, %	0.7/0.68	0.7/0.69	0.7/0.69	0.7/0.68	0.7/0.67
	2. End pH	5.4	4.5	4.5	5.0	5.0
	3. Consistency, %	10.0	10.0	10.0	10.0	10.0
	4. Temperature, °C	70±1	70±1	70±1	70±1	70±1
	5. Time, minutes	120	120	120	120	120

Contd...

Table 1–Contd...

Sl. No.	Particulars	Blank Expt. C-Ep-H-D Sequence	Expt. No. 1 C-Ep-E_1p-D Sequence	Expt. No. 2 C-Ep-E_1p-D Sequence	Expt. No. 3 C-Ep-E_1p-D Sequence	Expt. No.4 C-Ep-E_1p-D Sequence
6.	**Final Results**					
	1. Total Cl_2 applied/consumed, %	9.1/8.93	5.1/4.97	5.1/4.97	5.1/4.97	5.1/4.97
	2. ClO_2 applied/consumed, %	0.7/0.69	0.7/0.69	0.7/0.69	0.7/0.68	0.7/0.67
	3. Pulp Brightness, % P.V.	87.0	82.0	85.0	87.0	87.5
	4. Pulp Viscosity, (0.5% C.E.D. Cps)	6.0	9.0	8.5	8.3	7.2
	5. Pulp Srinkage, % on O.D. bleached pulp	12.8	11.2	11.4	11.6	12.0
	6. P.C. No.	2.8	1.6	1.6	1.27	1.8
	7. Ash in bleached pulp, %	0.71	0.38	0.36	0.36	0.34

Comparison of Higher Kappa Pulp Bleached Under C-Ep-H-D and C-Ep-E_1p-D Sequences

Total Chlorine demand was brought down from 9.10% in C-Ep-H-D sequence to 5.1% by replacing Calcium hypchlorite with Hydrogen peroxide in E_1 stage of C-Ep-E_1p-D sequence. For pulp of 25.3 Kappa, optimum Hydrogen peroxide dosage was observed to be 1.2% to achieve the target brightness 87±1% P.V under C-Ep-E_1p-D sequence. Theoretically AOX generation would be as follows for the two bleaching sequences:

Sl.No.	Particular	C-Ep-H-D Bleaching Sequence Higher Kappa	C-Ep-E_1p-D Bleaching Sequence Higher Kappa
1	Total Chlorine Consumption, %	8.93	4.97
2	ClO_2 consumption, %	0.69	0.68
3	AOX, Kg/Ton of Pulp	7.7	5.32

The effect of Hydrogen peroxide dosage variation on higher Kappa pulp bleached under C-Ep-E_1p-D and C-Ep-H-D sequences to achieve the final brightness is shown in Fig. 1. It may be noted that the Viscosity of C-Ep-E_1p-D sequence bleached pulp is higher compared to C-Ep-H-D sequence pulp because less damage was done to the cellulosic fibres with Hydrogen peroxide usage and the viscosity was reduced with increase in Hydrogen peroxide dosages as may be seen from Fig. 1. Replacing Calcium hypochlorite with Hydrogen Peroxide in C-Ep-H-D sequence reduces ash content in pulp. Calcium hypochlorite forms Calcium soap, which increases the ash content in pulp. Likewise pulp shrinkage and Post Colour Number (P.C.No) of final bleached pulp were observed to be on lower side as Hydrogen peroxide was used in the intermediate stage of bleaching under C-Ep-E_1p-D sequence.

Fibre Classification of Higher Kappa Pulp Bleached Under C-Ep-H-D and C-Ep-E_1p-D Sequences

Higher Kappa (25.3) pulp bleached under C-Ep-H-D sequence resulted in lowering of fibre percentage retained on 40 mesh and increase in the fines percentage passing through 140 mesh

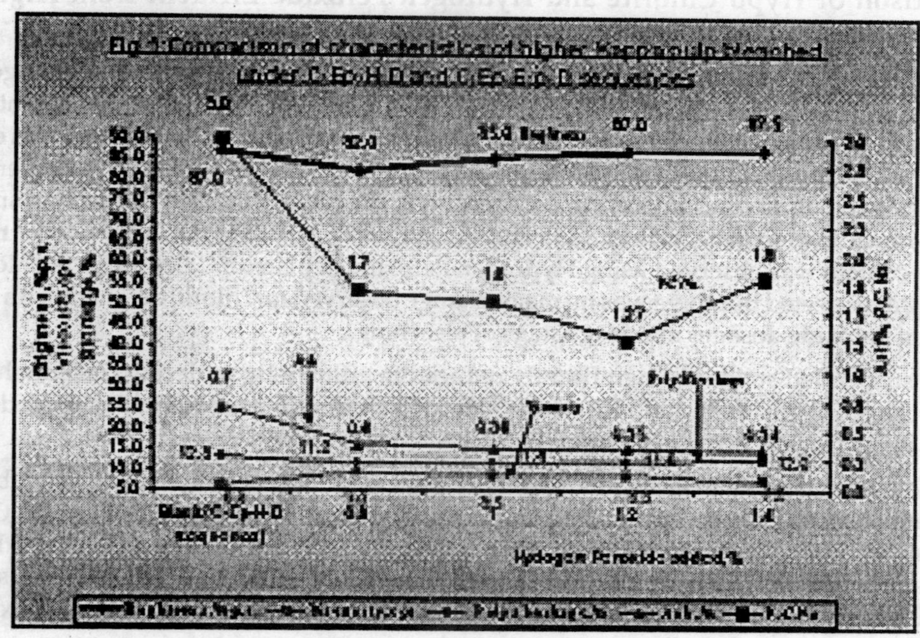

whereas under C-Ep-E_1p-D sequence fibre fraction retained on 40 mesh increases and fines passing through 140 mesh was found to be lower. (Table 2). Higher fibre fraction retained on 40 mesh & lower percentage of fines passing through 140 mesh indicates that pulp fibres are damaged to lesser extent under C-Ep-E_1p-D sequence compared to C-Ep-H-D sequence. The bleached pulp viscosity also confirms the above fact. However with increase in Hydrogen Peroxide dosages in E_1 stage fibre fraction retained on 40 mesh does not reduce appreciably and the fines fraction passing through 140 mesh increases gradually in C-Ep-E_1p-D bleached pulps.

Table 2: Fiber Classification of Unbleached Pulp of Higher Kappa (25.3) and its Bleached Pulps Under C-Ep-H-D & C-Ep-E_1p-D Sequences

Sl. No.	*Mesh Size*	*Blank Expt C-Ep-H-D Sequence bld Pulp*	*Expt. No. 1 C-Ep-E_1p-D Sequence bld Pulp (0.8% H_2O_2 Used)*	*Expt. No. 2 C-Ep-E_1p-D Sequence bld Pulp (1.0% H_2O_2 Used)*	*Expt. No.3 C-Ep-E_1p-D Sequence bld Pulp (1.2% H_2O_2 Used)*	*Expt. No.4 C-Ep-E_1p-D Sequence bld Pulp (1.4% H_2O_2 Used)*
		Fiber Retention %				
1.	+40	40.61	49.66	50.55	48.55	47.65
2.	- 40, +70	20.63	20.06	20.54	20.53	20.59
3.	- 70 +100	13.76	12.52	13.11	14.51	13.51
4.	- 100, +140	3.21	3.02	3.39	3.35	3.95
5.	- 140	22.19	13.94	12.41	13.41	14.30

Comparison of Hypo Chlorite and Hydrogen Peroxide Effluent from Higher Kappa Pulp Bleached Under C-Ep-H-D and C-Ep-E_1p-D Sequences

Calcium Hypochlorite in C-Ep-H-D bleaching sequence was replaced with Hydrogen Peroxide in E_1 stage of C-Ep-E_1P-D sequence and Comparison of Calcium hypochlorite stage effluent parameters is done with Hydrogen Peroxide stage effluent. Considerable reduction in COD value in effluent was found by replacing Calcium hypochlorite with Hydrogen Peroxide (Table 3). However increase in Hydrogen Peroxide dosages from 0.8 to 1.4% resulted in increase of the COD values in a smaller range but was much lower than Calcium hypochlorite stage effluent (Fig. 2). Appreciable reduction in Chloride content was observed using Hydrogen Peroxide in place of Calcium Hypochlorite in pulp bleaching as expected but there was no major change observed in Chloride content with variation in Hydrogen Peroxide dosages in E_1Stage stage of bleaching.

The suspended solids are significantly reduced in effluent by replacing Calcium hypochlorite with Hydrogen Peroxide in pulp bleaching although there was increase in Suspended solids with increase in Hydrogen Peroxide dosages (Fig. 2).

Ionic load increases the electrical conductivity of the effluent hence there was appreciable reduction in conductivity of the effluent by replacing Calcium hypochlorite with Hydrogen peroxide. Conductivity and Dissolved solids were appreciably reduced by replacing Calcium hypochlorite with Hydrogen peroxide dosages for bleaching of higher Kappa pulp (Fig. 2). Reduction in dissolved solids is also responsible for less pulp shrinkage and lower COD values while using Hydrogen peroxide).

Comparison of strength properties of Higher Kappa Pulps Bleached Under C-Ep-H-D & C-Ep-E_1p-D Sequences

Physical strength properties of higher kappa pulp bleached under C-Ep-H-D & C-Ep-E_1p-D sequences are represented in Table 4. Tensile Index, Burst Index, Tear Index and Double fold were

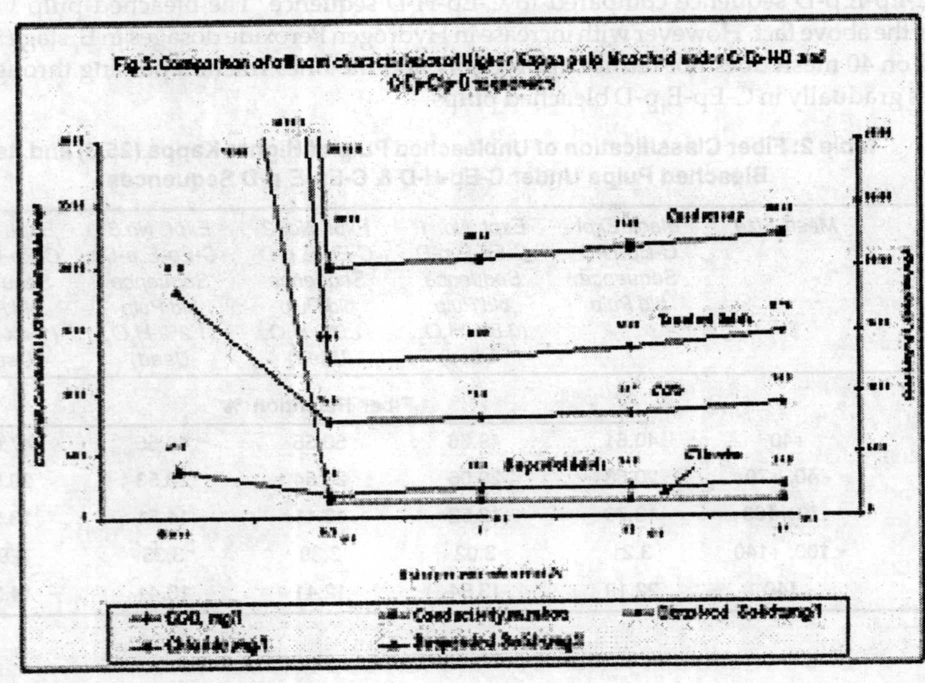

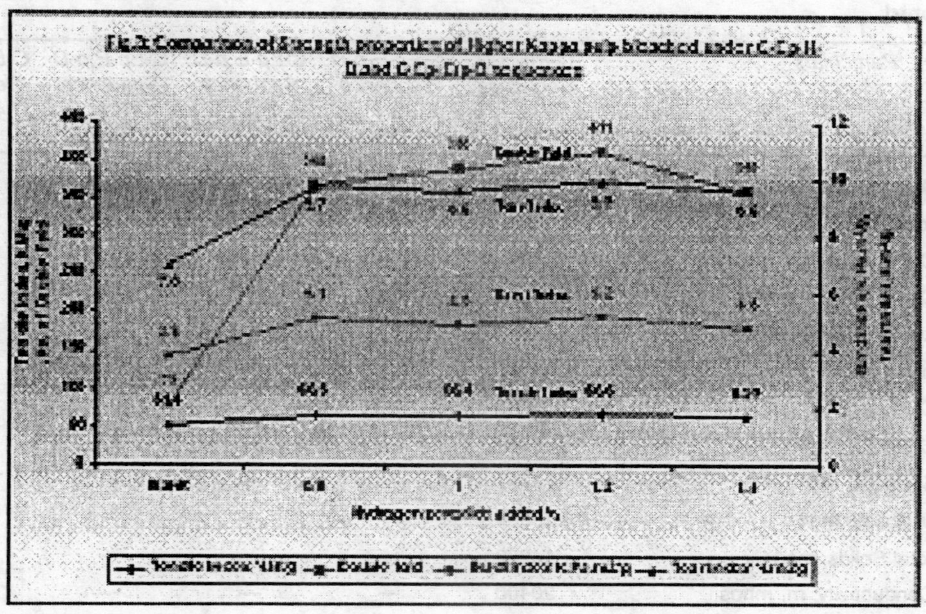

found to be higher in C-Ep-E_1p-D sequence bleached pulp than C-Ep-H-D sequence as reflected in Fig. 3 which is in accordance with higher pulp viscosity and less damage done to the fibres in utilization of Hydrogen peroxide.

Table 3: Comparison of Effluent Characteristics from Higher Kappa (25.3) Pulp Bleached Under C-Ep-H-D and C-Ep-E_1p-H-D Sequences

Sl. No.	*Particulars*	*C-Ep-H-D Sequence*	*C-Ep-E_1p-D Sequence 0.8% H_2O_2 Used*	*C-Ep-E_1p-D Sequence 1.0% H_2O_2 Used*	*C-Ep-E_1p-D Sequence 1.2% H_2O_2 Used*	*C-Ep-E_1p-D Sequence 1.4% H_2O_2 Used*
1.	**Chlorination Stage**					
	1. pH	2.1	2.1	2.1	2.1	2.1
	2. COD, mg/l	551	551	551	551	551
	3. Chloride, mg/l	2200	2200	2200	2200	2200
	4. S. Solids, mg/l	100	100	100	100	100
	5. D. Solids, mg/l	3246	3246	3246	3246	3246
	6. Total Solids, mg/l	3346	3346	3346	3346	3346
	7. Conductivity, m, mhos	5200	5200	5200	5200	5200
	8. Colour, Pt-Co unit	173	173	173	173	173
2.	**Caustic Extraction Stage**					
	1. pH	10.9	10.9	10.9	10.9	10.9
	2. COD, mg/l	2412	2412	2412	2412	2412
	3. Chloride, mg/l	770	770	770	770	770
	4. S. Solids, mg/l	212	212	212	212	212

Contd...

Table 3–Contd...

Sl. No.	Particulars	C-Ep-H-D Sequence	C-Ep-E_1p-D Sequence 0.8% H_2O_2 Used	C-Ep-E_1p-D Sequence 1.0% H_2O_2 Used	C-Ep-E_1p-D Sequence 1.2% H_2O_2 Used	C-Ep-E_1p-D Sequence 1.4% H_2O_2 Used
	5. D. Solids, mg/l	4456	4456	4456	4456	4456
	6. Total Solids, mg/l	4668	4668	4668	4668	4668
	7. Conductivity, m, mhos	7000	7000	7000	7000	7000
	8. Colour, Pt-Co unit	8700	8700	8700	8700	8700
3.	**Hypochlorite Stage**					
	1. pH	8.0	—	—	—	—
	2. COD, mg/l	1812	—	—	—	—
	3. Chloride, mg/l	7000	—	—	—	—
	4. S. Solids, mg/l	378	—	—	—	—
	5. D. Solids, mg/l	13884	—	—	—	—
	6. Total Solids, mg/l	14262	—	—	—	—
	7. Conductivity, m, mhos	20400	—	—	—	—
	8. Colour, Pt-Co unit	60	—	—	—	—
	9. Calcium, mg/l	5680	—	—	—	—
	10. Calcium hardness as $CaCO_3$, mg/l	6800	—	—	—	—
4	**Peroxide Stage**					
	1. pH		10.5	10.5	827	10.6
	2. COD, mg/l	—	760	790	120	902
	3. Chloride, mg/l	—	130	120	240	120
	4. S. Solids, mg/l	—	176	228	1358	248
	5. D. Solids, mg/l	—	1264	1282	1598	1476
	6. Total Solids, mg/l	—	1440	1510	2150	1724
	7. Conductivity, m, mhos	—	2000	2050	398	2250
	8. Colour, Pt-Co unit	—	373	376	8	373
	9. Calcium, mg/l	—	8	5	130	5
	10. Calcium hardness as $CaCO_3$, mg/l	—	150	120		110
5	**Chlorine Dioxide Stage**					
	1. pH	5.4	4.5	4.5	5.0	5.0
	2. COD, mg/l	648	478	376	361	391
	3. Chloride, mg/l	1100	420	390	440	420
	4. S. Solids, mg/l	178	196	174	164	168
	5. D. Solids, mg/l	3342	1776	1604	1800	1780
	6. Total Solids, mg/l	3520	1972	1778	1964	1948
	7. Conductivity, m, mhos	4800	2700	2500	2770	2700
	8. Colour, Pt-Co unit	33	60	71	73	53
	9. Calcium, mg/l	396	42	65	69	69
	10. Calcium hardness as $CaCO_3$, mg/l	2000	420	480	480	510

Table 4 : Comparison of Physical Strength Properties of Bleached Pulps Under C-Ep-H-D and C-Ep-E$_1$p-D Bleaching Sequences from Higher Kappa (25.2) Pulp

Sl. No.	Particulars	Blank Expt. C-Ep-H-D Bleached Pulp	Expt. No. 1 C-Ep-E$_1$p-D Bleached Pulp	Expt. No. 2 C-Ep-E$_1$p-D Bleached Pulp	Expt. No. 3 C-Ep-E$_1$p-D Bleached Pulp	Expt. No.4 C-Ep-E$_1$p-D Bleached Pulp
1.	Number of beating revolutions in a P.F.I mill	4000	5800	6000	5800	5900
2.	Final freeness, °SR of pulp	30	29	30	29	29
3.	Bulk, c.c /gram	1.43	1.4	1.39	1.40	1.38
4.	Tensile Index, N. m/g	51.39	66.5	65.36	64.63	63.95
5.	Tear Index, N.m^2/g	7.03	9.66	9.57	9.86	9.60
6.	Burst Index, K. Pa. m^2/g	3.8	5.1	4.93	5.20	4.85
7.	Double fold	79	368	389	411	359

Conclusion

It is concluded from the experimental data recorded in Table 1 to 4 that by replacing Calcium hypochlorite with Hydrogen peroxide, the bleached pulp yield and Physical strength properties in C-Ep-E$_1$p-D was higher compared to C-Ep-H-D sequence. The total Chlorine consumption could be brought down to 44.3 % by replacing Hypochlorite with Hydrogen peroxide under C-Ep-E$_1$p-D bleaching sequence. Effluent parameters mainly Dissolved solids, Chloride, COD, Calcium and Calcium hardness of the effluent are appreciably reduced with the utilization of Hydrogen peroxide in E$_1$ Stage of bleaching of pulp.

Acknowledgement

The authors are grateful to the management of Orient Paper Mills, Amlai for permitting us to present this paper in the Indo-Italian workshop at NEERI, Nagpur.

REFERENCES

1. S.Chinnaraj, *et al.*, 2000, IPPTA, Vol-12, No, P.99, Dec.
2. Corner, O.B.I, 1993, *Journal of Pulp & Paper Science*, 19 (1): 33-39.
3. Gergov, M, M, Priha, E, Talka, O.Valttila, A. Kangas and K. Kukkonen, 1998, TAPPI, 71 (12): 175-184.
4. Stromberg, L.R, Marck, F. De Sousa and O. Dahlman 1995, "Effluent of internal Process changes and external treatment of effluent chemistry Environmental fate and effects of Pulp & Paper mill effluent" P.3-19, St.Lucie Press, Delray Beach, F.L, U.S.A.
5. Makkar, M.K, and Jauhari, M.B, 2001, IPPTA, Vol-13, No. 1.
6. Tendulkar, S.R and Shinde, J.K, 1991, IPPTA, Vol-3, No. 4.
7. Andrews, D.H, singh R.P, 1991, "The Bleaching of Pulp" TAPPI Press, P216-218

8. Patricia Walsh, 1991, TAPPI, 74 (1), 81.

9. Mittal, S.K and Maheshwari, S., 2001, IPPTA Conventional Issue, P. 119.

10. Fernades, J.D, Raghavulu. V.V, and Bhargava, S.C, 1993, IPPTA, Vol-5, No.3, P. 13

11. Gupta, R.K, 1999, IPPTA, Vol-II, No.4, P.9.

12. Sridhar.P, Reddy, J.V.R., Venugopal, C.S. and Karmakar, G.R, 1998, IPPTA Conventional Issue, P.55.

27

Application of Wet Air Oxidation and Photocatalysis for Wastewater Treatment

M. SUBRAHMANYAM, V. DURGA KUMARI, A. RACHEL, M. NOORJAHAN, M. PRATAP REDDY, K.V. SUBBA RAO, & P. BOULE

Catalysis and Physical Chemistry Division, Indian Institute of Chemical Technology, Hyderabad - 500 007

ABSTRACT

The conversion of different forms of energy has been central importance in Science and Technology. The R & D developments in several aspects of environmental catalysis are required for cleaner water, air and soil. Catalysts are used to convert environmentally hazardous materials into harmless compounds. In the last twenty years photochemical processes have provided alternative routes to water purification since UV irradiation in combination with other processes can remove bacterial substances from solutions as well as dissolved organic matter. These photochemical oxidation reactions are now members of so called Advanced Oxidation Processes. Wet air oxidation is an established technology that involves the liquid-phase oxidation of dissolved organics or oxidizable inorganic compounds using a gaseous source of oxygen (usually air) at high temperatures and pressures that severely affects the economy of this technology. This presentation "focused" on the perspective of photocatalytic purification and wet air oxidation process for industrial wastewaters. It is hoped that the present overview will contribute to the improvement of the existing purification methods.

Keywords: Photocatalysis; degradation; TiO_2, immobilization, wet air oxidation.

Introduction

Industrial plants generate increasing amounts of wastewater, contaminated with toxic and hazardous organic compounds, which serve problems to the environment. The industrial wastewaters must be treated, before they meet the specifications for discharge or for recycling in the process. Especially a large amount of drug intermediates and dyestuff materials with intense color and toxicity are introduced in the aquatic systems, which are resistant to degradation. Therefore, development of efficient wastewater treatment technologies has become a major concern throughout the world. Due to the reinforcement of the regulations for quality control of effluents, the manufacturing industries have to minimize their emissions of organic and inorganic wastes using efficient routes via the so-called "atom economy" concept which consists in maximizing the atom utilization from the raw materials to the final products (1-2). In addition, it is vital to develop highly efficient processes capable of destroying hazardous xenobiotic pollutants contained non-compressible waste effluents. Among them wet air oxidation (3-19) and photocatalytic treatment (20-38) are some of the effective methods for the treatment of industrial effluents.

Wet Air Oxidation

Wet air oxidation is a liquid-phase oxidation process and is used around the world to treat industrial wastewater sludge, at moderate temperatures 200-300° C and at pressures from 75-230 atmospheres. In this presentation the development of treatment of wet air oxidation process that rely on heterogeneous catalysts will be covered to provide an environmentally attractive option to manage the growing organic sludge and toxic wastewater treatment problems. Compared to wet air oxidation process, catalytic wet air oxidation offers low energy requirements and much higher oxidation efficiencies. Further developments of this technology should include high durability/low cost catalysts. Insoluble organic matter is converted to simple soluble organic compounds which are in turn oxidized and eventually converted to CO_2 and H_2O without emissions of NO_x, SO_2, HCl, dioxins, furans, fly ash etc. More than 100 plants are in operation today, mostly to treat waste streams from petrochemical, chemical and pharmaceutical industries as well as residual sludge from waste treatment. Several plants aim at a complete decomposition of pollutants or at a partial oxidation into low-molecular weight compounds which can then be treated further by biological processes. Commercial catalytic wet oxidation processes rely either on supported precious metal and/or base metal oxide catalysts or on homogeneous catalysts such as Fe/Cu salts. The risk of hydrothermal sintering of the catalytic phase or support should also be considered carefully. The fate of the catalyst has to be studied early in the process design.

Aqueous solutions of low molecular weight carboxylic acids, such as acetic, propionic and butyric acids, were treated by catalytic wet air oxidation (CWAO) using a carbon supported platinum catalyst (3). Oxidation in the presence of the catalyst, in a stirred reactor, was carried out at 200 degrees C and 6.9 bar of oxygen partial pressure, with conversions (after 2 h) ranging from 59.4 to 75%, and selectivities to gaseous products of up to 100%.

The catalytic batch wet air oxidation of maleic acid over 5 wt.% Ru/CeO_2 catalysts was investigated at temperatures of 160 and 200°C and an oxygen partial pressure of 20 bar (4). Rivas *et al.* (5) studied over a temperature range of 415-478 K, the catalytic and non-catalytic degradation of an aqueous solution of maleic acid (0.03 M) both in the presence of oxygen and under an inert atmosphere (nitrogen). The use of 0.5 wt.% platinum on gamma-alumina catalyst significantly enhanced the degradation rate of maleic acid.

Layered double hydroxides (LDHs) with a hydrotalcite-like structure containing different amounts of $Cu^{2+}/Ni^{2+}/Al^{3+}$ cations with M^{2+}/M^{3+} ratios between 0.5 and 3.0 were prepared (6) Their catalytic behaviour for the oxidation of phenol aqueous solutions using a trickle-bed reactor and a semi-batch reactor for comparison, and air with an oxygen partial pressure of 0.9 Mpa at a reaction temperature of 413 K. Also work on to optimize the degradation yield of an activated sludge and to study the reaction mechanism on a cellulose model compound was carried out (7). This optimization is performed with a Simplex design with four parameters reaction temperature (280-350°C), injected air pressure (40-60 MPa), reaction time (10-45 minutes) and COD concentration (0.5-5 g L-1).The better yield obtained on TOC concentration is 82.5% at 330°C temperature and the yield substantially decreases.

A comparative analysis is made for a trickle-bed reactor, a packed-bubble column, a three-phase fluidized bed and a slurry-bubble column with an active and moderately deactivating catalyst for the wet oxidation at high pressure and temperature of organic-containing aqueous wastes (8). Compared to other mature industrial sectors where multiphase reactors are prevalent, the design of three-phase catalytic reactors for wet air oxidation processes is still at an emerging stage. This paper discusses, from a multiphase reactor engineering perspective, the design of such contactors by setting an exhaustive modeling framework of catalytic wet oxidation in which the molecular, particle and reactor scales are integrated. Several solid catalysts ($Co_3O_4/\gamma\text{-}Al_2O_3$, $Fe_2O_3/\gamma\text{-}Al_2O_3$, $Mn_2O_3/\gamma\text{-}Al_2O_3$, Zn-Fe-Mn-Al-O, Pt/$\Upsilon\text{-}Al_2O_3$, Ru/$CeO_2$, Ru/C) have been prepared and used to remove N-containing organic contaminants while processing toxic and hazardous industrial wastewaters using wet oxidation by air (WAO) (9)

A detailed reaction network for the catalytic wet air oxidation (CWAO) of phenol on a $CuO/\gamma\text{-}Al_2O_3$ catalyst is also reported (10). The reaction network proposed accounts for all detected intermediate products of phenol oxidation overcoming the usual lumping of compounds. The model is composed by several consecutive and parallel reactions. The parameters of the model were adjusted using experimental data obtained from a continuous trickle bed reactor using air as oxidant at different temperatures (120-160 °C) and oxygen partial pressures (0.6-1.2 MPa). Simple power law as well as Langmuir-Hinshelwood (L-H) expressions accounting for the adsorption effects were checked in the modelling of the reaction network.

Catalytic wet air oxidation of an aqueous phenol solution was carried out in a trickle bed reactor at mild conditions of temperature (120-160 ° C) and oxygen partial pressure (0.6-1.2 MPa) (11). Tests were performed at space times up to 1 h. A commercially available copper oxide supported over ϒ-alumina was used as catalyst. The results show that the phenol conversion, the chemical oxygen demand (COD) reduction and the distribution of oxidation products are considerably affected by the temperature, whereas the oxygen partial pressure only has minor influence. The use of wet air oxidation as a pretreatment step in the context of an integrated chemical/biological process was investigated for mode organic-containing wastewaters (12). It was found that the selection of suitable wastewaters and pretreatment conditions (*i.e.* temperature, residence time, use of catalysts) was significant for the effective application of an integrated process. For the systems under consideration, an integrated process was found to be advantageous for treating a polymer-containing wastewater, while mild wet air oxidation alone may suffice to treat a polyphenol-containing wastewater.

The challenging operating conditions of wet air oxidation provided a strong driving force to investigate catalysts which would allow substantial gains on temperature, pressure and residence time. Thus several types of heterogeneous catalysts were studied in the last decades, mostly based on supported base metal oxides (13-19). The design of three-phase catalytic reactors for wet air oxidation processes is still at an emerging stage. So-far, commercial-scale plants applying solid-catalyzed wet oxidation for the removal of organic pollution from industrial or domestic wastewater are rare.

Photocatalysis

Here, the importance of photocatalysis in recent years and the work carried out in our laboratory and also in collaboration with our French partner on wastewater treatment of recalcitrant organic compounds is discussed. Photocatalysis is often proposed for the elimination of biorecalcitrant organic pollutants from wastewaters. Aromatic sulfonated compounds are critical pollutants in aqueous environment. They form a large group of industrial chemicals released with dyehouse effluents. These compounds have low volatility and are highly water soluble. Consequently it is diffucult to remove them from water and they are transported from sewers to surface waters.

In recent years the photocatalytic purification of water is carried out by illumination of large band gap semiconductor such as TiO_2 in the contaminated aqueous system. The specific results on degradation of aromatic sulfonated recalcitrant compounds especially toxic and non biodegradable compounds like 4-chloro-2-methylphenoxyacetic acid (MCPA) a herbicide, 3-nitrobenzenesulfonic acid (3-NBSA), 2, 5-anilinedisulfonic acid (2, 5 ADSA), H-acid (4–amino–5-hydroxy–2, 7-naphthalene disulphonic acid monosodium salt), Orange–G (azo dye), Acid Orange–7, meta and *p*-Phenylene diamine sulphonic acid are devoted here. The strong oxidation capacities, non-toxicity, chemical inertness of TiO_2 and its low price are the attractive properties making it to use an effective photocatalyst. Also it is very simple to evaluate its photocatalytic efficiency in a device as is shown in Fig. 1. The solar wastewater de-toxicfication can be done by:

1. Suspended particles of a catalyst
2. Supported photocatalyst.

In suspended form there is a disadvantage of filtration, which can be solved by supporting catalyst on inert surfaces. Some methods are patented for instance the deposition of TiO_2 on glass for self-cleaning window panes or mirror, deposition on tiles for the elimination of bad smells and atmospheric pollutants, deposition on inorganic fibres for photocatalytic air-cleaning. The technique most commonly proposed for immobilization of TiO_2 is sol-gel, either dip-coating or spin-coating using organic Ti^{IV} derivatives, often alcoholates; but generally immobilization drastically reduces the efficiency of the photocatalyst (22-24). With some supports such as bricks and cement this deactivation may be partly attributed to the high superficial ionic concentration. Anchoring TiO_2 may also be used but on flat supports such as glass where the layer is very thin and not very stable. Major disadvantage of system containing suspended solids is the need to remove the photocatalyst from the cleaned water after the treatment. Most of the TiO_2 powder used in photocatalysis contains submicron sized particles. As a consequence the potential inexpensive separation by sedimentation or filtration of the suspension is a troublesome and treating very large volumes of water is almost impossible. Systems using a supported photocatalyst which eliminate pollutants is the need of the hour for a photocatalysis technique to use for wastewater treatment.

The photocatalytic efficiency of several TiO_2 (namely Degussa P-25 and Millennium PC50, PC 100, PC105, PC500) used as suspensions are compared for the photocatalytic degradation of 3-nitrobenzenesulfonic acid (3-NBSA) and 2, 5-anilinedisulfonic acid (2, 5 ADSA). With 3-NBSA, P-25 is clearly the most efficient and there is no apparent relationship between photocatalytic activity and specific surface area. This result is consistent with that obtained with phenol, but a contrast was noticed with 2, 5 ADSA where PC 500 is more efficient than P-25 in spite of a higher adsorption of the latter. The photocatalytic efficiency of TiO_2 immobilized on various supports (glass, cement, red brick

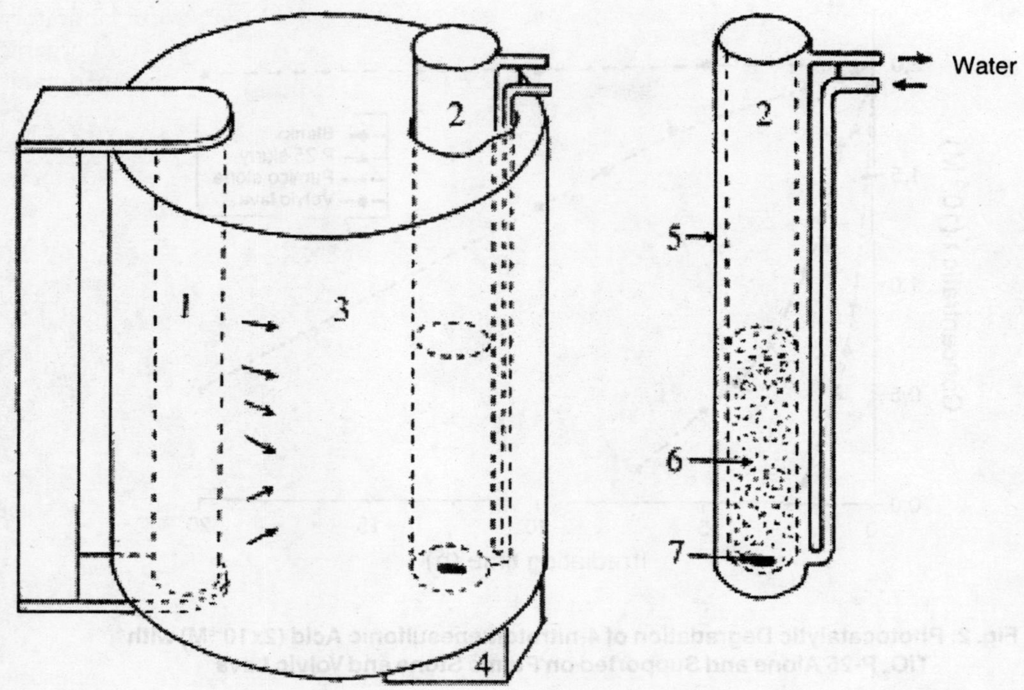

Fig. 1: A Simple Device for the Photocatalytic Evaluation of Pollutants Degradation in Slurry Mode of Catalyst
(1) Fluorescent lamp (300-450 nm), (2) Water cooled reactor (Pyrex), (3) Elliptical mirror, (4) Magnetic stirrer, (5) Cooling jacket, (6) Aqueous suspension of TiO_2, (7) Magnetic bead

and inorganic fibres) using different techniques (sputtering, sol-gel dip coating, patented method for inorganic fibres), are compared with photocatalytic efficiency of TiO_2 Degussa P25 in suspension for degradation of 3-nitrobenzenesulfonic acid and 4-nitrotoluenesulfonic acid (4-NTSA) (22-24).

Porous lavas and specially pumice stone are particularly interesting supports in photocatalysis (Fig. 2). Pumice stone has a very good retention of titanium dioxide without reducing much its photocatalytic activity after immobilization (24). Pumice stone impregnated with TiO_2 is less active than TiO_2 in suspension 2 g L^{-1} but more active than TiO_2 coated on inorganic supports such as glass and red bricks by conventional sol-gel dip-coating technique.

The photocatalytic degradation of H-acid is monitored in slurry as well as on immobilized TiO_2 system. In slurry phase we observed 90% of the H-acid degraded within one hour and at the end of the 5th hour only one major intermediate was observed. The photocatalytically degraded H-acid, with the intermediates formed during treatment are subjected to further degradation with biological treatment

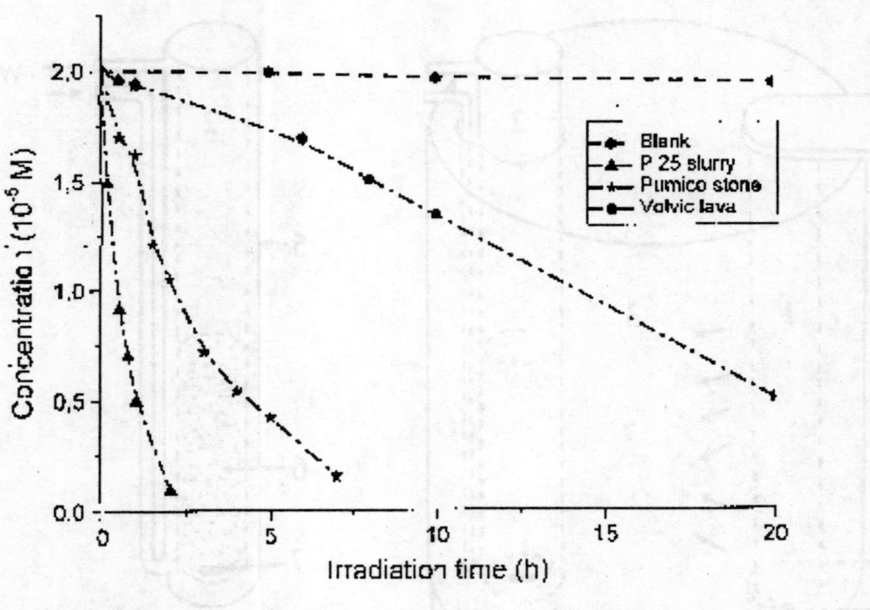

Fig. 2: Photocatalytic Degradation of 4-nitrotoluenesulfonic Acid ($2x10^{-5}$M) with TiO_2 P-25 Alone and Supported on Pumic Stone and Volvic Lava

(21). In the case of Orange-G azo dye the decolorization and degradation is monitored with UV-VIS spectroscopy and DRS, the data obtained supports the degradation pathway of the molecule.

Also several attempts have been made to improve the photoefficiency of titania by adding adsorbents like silica, alumina, zeolites, clays and active carbon. Zeolite-based photocatalysis is new and the present work reported by Durga Kumari *et al.* (25) on an easy and efficient use of TiO_2 supported and TiO_2 + HZSM-5 zeolite combinate in the photodegradation of aqueous phenol and p-chlorophenol is unique. This approach of sense and shoot of the pollutant (Fig. 3) will have a lot of scope for further development in thin film fixed bed mode of operation using photocatalysis technique. The photocatalytic oxidation of O-, m-, and p-, cresol was compared with ZnO and TiO_2. The same products were obtained with both photocatalysts (35, 36).

Stilbene derivatives correspond to a large group of compounds and they are used in textiles, detergents, paper or plastics to make products whiter and brighter by compensating for the yellowish shade of materials. Recently Rachel *et al.* (37) reported that 4, 4'-dinitrostilbene-2, 2'-disulfonate (DSD) photolysis with a low quantum yield of $2.2 – 2.4 \times 10^{-4}$ M, whereas photocatalytic transformation is found to be more efficient for elimination of DSD from wastewater.

The photocatalytic transformation of 4-choloro-2-methylphenoxyacetic acid (MCPA), a herbicide leads to 4-choloro-2-methylphenol as specific main photoproduct whereas the direct phototransformation is more complex (38).

Conclusions

The following conclusions are to summarize our views concerning wet air catalytic oxidations and photocatalytic degradations of organic recalcitrant compounds. No commercial-scale plants

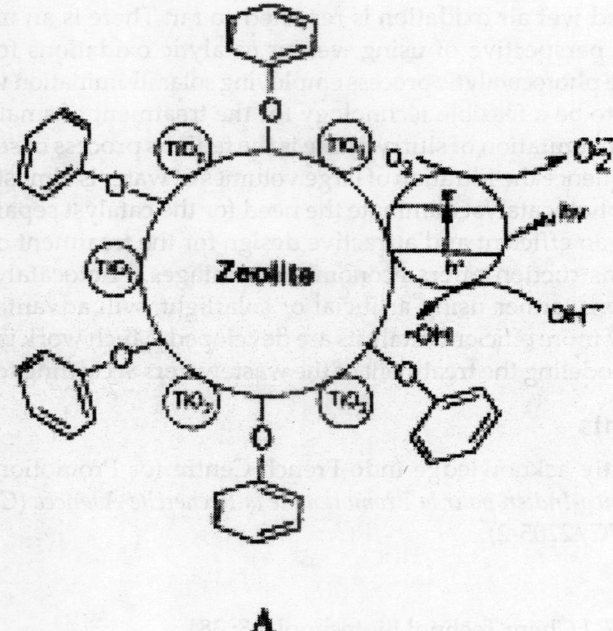

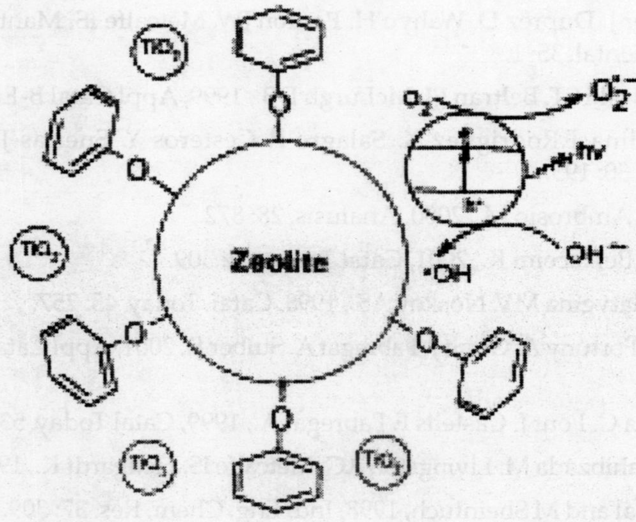

Fig. 3: Sense and Shoot Approach Using Adsorbent (Zeolite) and TiO_2 for Pollutants Degradation

applying solid-catalzed wet air oxidation is reported so far. There is an urgent need to bring new information from the perspective of using wet air catalytic oxidations for industrial wastewater treatment. Whereas the photocatalytic process employing solar illumination with a slurry type reactors operation has shown to be a feasible technology for the treatment of small to medium volumes of wastewaters. The main limitation of slurry phase is the tedious process of separation of catalyst after treatment. As a consequence the filtration of large volumes of water is almost impossible. The systems using an immobilize photocatalyst eliminate the need for the catalyst separation step. The thin film fixed bed reactors are an efficient and attractive design for the treatment of large scale wastewater systems and their construction offers economic advantages. Photocatalytic technologies for the wastewater applications, either using artificial or solarlight, will advantageously compete in the market, in particular, if more efficient catalysts are developed. Much work is needed to develop more precise methods for modeling the treatment of the wastewaters according to the type of pollutants.

Acknowledgements

The authors greatly acknowledge Indo-French Centre for Promotion of Advanced Research (IFCPAR) /*Centre Franco-Indien pour la Promotion de la Recherche Avancée* (*CEFIPRA*) for its financial support (contract n° IFC/2205-2).

REFERENCES

1. R A Sheldon, 1997, J Chem Technol Biotechnol, 68: 381.
2. A Ratnamala, V Durga Kumari and M Subrahmanyam, 2001, Proceedings of International Conference on Industrial Pollution and Control Technologies, December, Hyderabad.
3. Gomes HT. Figueiredo JL. Faria JL., 2000, Appl Catal B: Environmental. 27: L217.
4. Oliviero L. Barbier J. Duprez D. Wahyu H. Ponton JW. Metcalfe IS. Mantzavinos D. 2001, Appl Catal B:Environmental. 35: 1.
5. Rivas J. Kolaczkowski ST. Beltran FJ. McLurgh DB., 1999, Appl Catal B-Environmental. 22 279.
6. Alejandre A.Medina F.Rodriguez X. Salagre P. Cesteros Y. Sueiras JE., 2001, Appl Catal. B:Environmental. 30: 195.
7. Ricq N. Barbati S. Ambrosio M., 2000, Analusis. 28: 872.
8. Larachi F. Iliuta I. Belkacemi K., 2001, Catal Today. 64: 309.
9. Dobrynkin NM. Batygina MV. Noskov AS., 1998, Catal. Today. 45: 257.
10. Eftaxias A. Font J. Fortuny A. Giralt J. Fabregat A. Stuber F., 2001, Appl Catal B:Environmental. 33: 175.
11. Fortuny A. Bengoa C. Font J. Castells F. Fabregat A., 1999, Catal Today. 53: 107.
12. Mantzavinos D. Sahibzada M. Livingston AG. Metcalfe IS. Hellgardt K., 1999, Catal Today. 53: 93.
13. Y I Matatov-Meytal and M Sheintuch, 1998, Ind. Eng. Chem. Res. 37: 309.
14. V S Mishra, V Mahajan and J B Joshi, 1995, Ind. Eng. Chem. Res. 34: 5.
15. F Luck, 1999, Catal Today, 53: 81.
16. F Luck, 1996, Catal Today, 27: 195.
17. I Iliuta, F Larachi, 2001, Chemical Engineering and Processing 40: 175.

18. F Larachi, Iliuta K Belkacemi, 2000, Catal Today, 58: 1.

19. A Pintar, M Besson and P Gallezot, 2001, Appl Catal B: Environmental 30: 123 .

20. M Noorjahan, M Pratap Reddy, V Durga Kumari, P Boule and M Subrahmanyam, 2001, Proceedings of International Conference on Industrial Pollution and Control Technologies, December, Hyderabad.

21. V Madhuri, M Noorjahan, V Durga Kumari, M Subrahmanyam and P Boule, 2001, International Conference on Materials for Advanced Technologies, Singapore, July.

22. A Rachel, M Sarakha, M Subrahmanyam and P Boule, 2002, Appl Catal B: (Environmental) 37: 293.

23. A Rachel, M Subrahmanyam and P Boule, 2002, Appl Catal B: (Environmental) 37: 301.

24. A Rachel, B Lavedrine, M Subrahmanyam and P Boule, 2002, Catalysis Communications 3: 165.

25. V Durga Kumari, M Subrahmanyam, K V Subba Rao A Ratnamala M Noorjahan, and K Tanaka, 2002, Appl Catal A (In Press).

26. K.V.Subba Rao, D.Majumdar, M. Subrahmanyam and S.V. Ramakrishna, 1999, Recent Trends in Catalysis p 712, V Murugesan *et al*. (editors).

27. V T Supriya, D Venugopal and M Subrahmanyam, 1999, International Journal of Hydrogen Energy 24: 957.

28. Supriya V Tambwekar and M. Subrahmanyam, 1998, Int. Journal of Hydrogen Energy 23: 741.

29. Supriya V Tambwekar, G. Santosh Reddy, V. Durga Kumari and M. Subrahmanyam, 1998, Studies in Surface Science and Catalysis, 311: 329.

30. M. Subrahmanyam, V T Supriya and P Ram Reddy, 1996, International Journal of Hydrogen Energy 20: 99.

31. K V Subba Rao, D Venugopal and M Subrahmanyam, 1999, Indian Journal of Chemistry 38A: 173.

32. M Subrahmanyam, K V Subba Rao, M Rajesh Babu and B Srinivas, 1998, Indian Journal of Environmental Protection 18: 266.

33. M Subrahmanyam, 1998, National Seminar on Environment Protection-Recent trends in wastewater treatment, EPTRI, Hyderabad, January.

34. M. Subrahmanyam, 1987, Particualte and multiphase processes, Edited by T.Airman (ed) and T.N.Veziroglu, Vol 3, 391-400, Hemisphere Publishing Corporation, USA.

35. M A Malouki, T Sehili and P Boule, 1996, Toxicol. Environ. Chem. 55: 235.

36. C Richard and P Boule, N J, 1994, Chem 18: 547.

37. A Rachel, B Lavédrine, J-P Aguer and P Boule, 2002, J. Photochem. Photobiol. A: Chem. (In press).

38. A Zertal, T Sehili and P Boule, 2001, J. Photochem. Photobiol. A: Chem. 146: 37.

28

Novel Photoreactor for Heterogenous Photocatalytic Wastewater Treatment

D. PRAVEENA & T. SWAMINATHAN

Chemical Engineering Department Indian Institute of Technology, Chennai - 600 036

Introduction

With the increasingly stringent national and international regulations governing wastewater treatment in today's environment, the need to develop innovative and more efficient wastewater technologies have never been great. In the last few years, research on new methods for wastewater treatment has gone from processes involving, mere phase transfer (*e.g.* air stripping, activated carbon adsorption etc.) to complete mineralisation of the contaminant to ecologically compatible components *i.e.* carbondioxide and water.

The photocatalytic degradation of environmental pollutants, generally referred to as Advanced Oxidation Processes, has recently received much attention as an attractive method to destroy a variety of toxic, recalcitrant organic pollutants, both aliphatic and aromatic, in wastewater (Al Akabi 1988, Minero *et al.* 1991). Particular attention has been directed toward heterogeneous photocatalytic degradation which makes use of a semiconductor irradiated by a light whose energy is higher than or atleast equal to their band gap. Titaniumdioxide has been the most extensively utilised photocatalyst the reason being the high activity shown by this compound together with its chemical inertia and its non-photocorrosivity.

Photocatalytic reactions occur when charge separations are induced in a large band gap semiconductor by excitations with ultraviolet radiation. The electron/hole pairs so formed either recombine in the bulk of the catalyst solution interface or they may be trapped at defect sites. The trapped charge carriers react either by electron/hole recombination or by participation in a interface redox process with reduction occurring by the trapped holes. Photocatalytic oxidation of organic compounds is frequently supposed to occur via hydroxyl radicals formed by oxidation of surface bound hydroxide ions or water at the sites of a positive hole while predominant reductive step is the transfer of the trapped electron to adsorbed molecular oxygen to create a superoxide radical anion (Schmelling and Gray 1995). Hydroxyl radicals are very reactive neutral species; they react rapidly and non selectively with organic compounds and are the common oxidizing agent (Rideh *et al*. 1997).

Research in aqueous heterogeneous photocatalytic has mostly focused on investigation of the oxidation and reduction reactions on semiconductor photocatalysts and development of new photocatalysts. These studies examine the degradation pathways, reaction mechanisms and kinetics that influence the rate of reaction at a molecular level. There has been extensive research in this area of photocatalysis as discussed in several review articles (Ollis *et al*, 1991, Bahneman, 1991; Legrini *et al*, 1993; Hoffman *et al*, 1995).

The other area of research in photocatalysis is in reactor design. The major objective in this area is to develop practical designs to enable the economical application of photocatalysis for water purification. Reactor design and development research tries to scale up the process from laboratory pilot scale. Unfortunately, this aspect of photocatalysis has not been as intensively investigated as the mechanistic aspects.

The major problem in reactor design is inefficient light utilization which is evident from the low photo efficiencies reported for many photocatalytic reactions. Commercial photocatalytic wastewater systems must be more efficient and thus more economical to compete with relatively inexpensive but non-destructive processes such as activated carbon adsorption. The challenge is to design new photocatalytic reactors that operate with high photo efficiencies, which result in higher conversion (degradation). Such efficiencies allow the reactor to operate at higher throughputs and lower costs. The lower costs result from both better utilization of light energy and smaller reactors which significantly reduces the capital investment.

A variety of photochemical reactor configurations have been employed in photodegradation studies. Most often in laboratory experiments, well-mixed, heterogeneous batch reactors (Mills *et al*, 1993; Carraway *et al*, 1994; Legrini *et al*, 1993) and annular reactors (Augugliaro *et al*. 1992) have been used. The use of slurries in wastewater treatment systems has some disadvantages (1) the separation of photocatalyst particles from the bulk fluid phase after treatment by filtration, centrifugation or coagulation and flocculation is a slow and expensive process; (2) the depth of penetration of UV light is limited because of the strong absorption of UV light by TiO_2 and dissolved organic species. Alternative reactor configurations include either fixed or fluidized bed reactors (Braun *et al*, 1991, Legrini *et al*, 1993;). In fixed bed photocatalytic reactors the photocatalyst is immobilized on the walls of the reactor, on a solid supported matrix or around the casing of the light source. Immobilization has also been achieved either by fine particle entrapment on glass beads (Serpone at al, 1986) and glass fibers (Hofstadler and Bauer, 1999) or by compression of fine particles into ceramic membrane (Sabate *et al*, 1992). A fixed bed reactor system allows for the continuous use of the photocatalyst for wastewater treatment, eliminating the need for catalyst recovery and regeneration. However, these reactors have several limitations notably the low surface area to volume ratios and inefficiencies introduced by adsorption and scattering of light by the reaction medium.

The immobilization of a semiconductor on a support generates structural problems also. Therefore there is a need for a reactor whose design provides a high ratio or surface area to volume and also the possibility of total reactor illumination.

The objective of this study was to develop efficient reactor configurations to enhance the rate of photocatalysis. A thin film slurry photocatalytic reactor was designed based on the principle of laminar falling film reactor. This paper describes some of the results obtained with this systems. Phenol was chosen as the model pollutant for evaluation as it is an important and a common pollutant in many industrial wastewaters. The mineralization of phenol is quite difficult by the conventional methods.

Experimental

Photoreactor: Experiments were carried out in a Thin Film Slurry Photoreactor. The reactor consisted of a 4 mm thick glass sheet resting on an inclined frame over which a thin film of slurry was allowed to flow with the aid of a distributor pipe. The liquid was collected into a reservoir at the bottom of the glass plate and recirculated using a peristaltic pump. The experiments were conducted in a closed loop batch mode. Samples from reservoir were withdrawn through a valve and analyzed to determine the residual phenol and TOC concentrations. The illumination was provided by a lamp house which consisted of six 30 W UV lamps. The wavelength of the light emitted was 253.7 nm. The experimental setup is shown in Fig. 1

Reagents

Titanium dioxide (TiU_2) (EMerck,) and Phenol (Qualigen fine chemicals) were of laboratory grade. Potassium ferricyanide and 4-amino antipyrene (SRL) were of analytical grade.

Analysis

The samples were centrifuged to remove TiO_2 particles. Residual phenol was analyzed by 4-

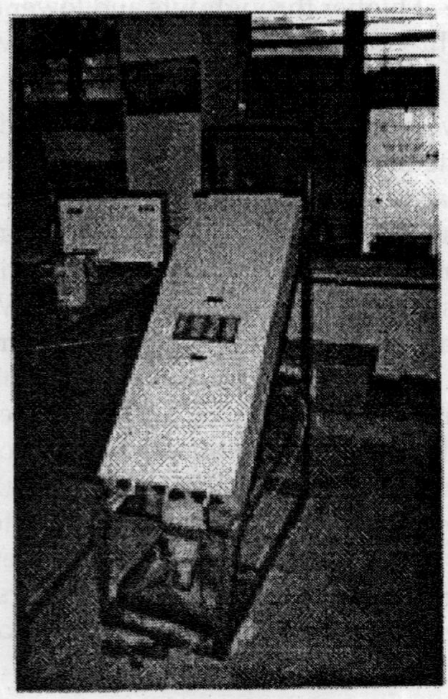

Fig. 1(a): Photograph of Thin Filk Slurry Reactor

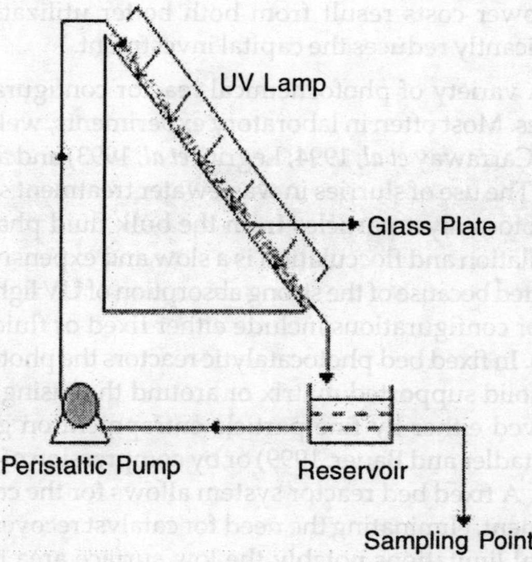

Fig. 1(b): Schematic Diagram of this Film Slurry Reactor

aminoantipyrene method using a UV-visible spectrophotometer (UV-1601 PC, Shimadzu, Japan) (APHA and AWWA 1992).

Total organic carbon (TOC) was analysed using a TOC analyzer (Euroglas, Netherlands)

Results and Discussion

For evaluating the effectiveness of the thin film slurry photoreactor, preliminary experiments were conducted at two levels of the major process variables, initial phenol concentration, catalyst concentration, light intensify and recirculation rate:

Variable	*Low*	*High*
Initial phenol concentration, mg/l	100	500
Catalyst concentration, g/l	2	8
Light intensity, no. of lamps	1	6
Recirculation rate, l/h	28.8	48.6

The effectiveness of the photoreactor in phenol removal can be seen from the time-concentration profile of the residual phenol concentration shown in Figures 2 to 7.

Near complete removal of phenol was achieved even at high concentration of 500 mg/l within 5 hours (Figs. 3 and 5). The rate of removal was high in the initial stages as seen by the steep slope but at later stages the rate of removal slowed down drastically. This may be attributed to the formation of intermediates which may also be competing for the hydroxyl ions. The absorption spectra of phenol shown in Fig. 9 indicates the change in phenol peak but the intermediates could not be identified. However, the phenol degradation mechanism via the formation of photodegradable intermediates has been proposed in the literature (Okamoto, 1985: D'Olivera, 1990).

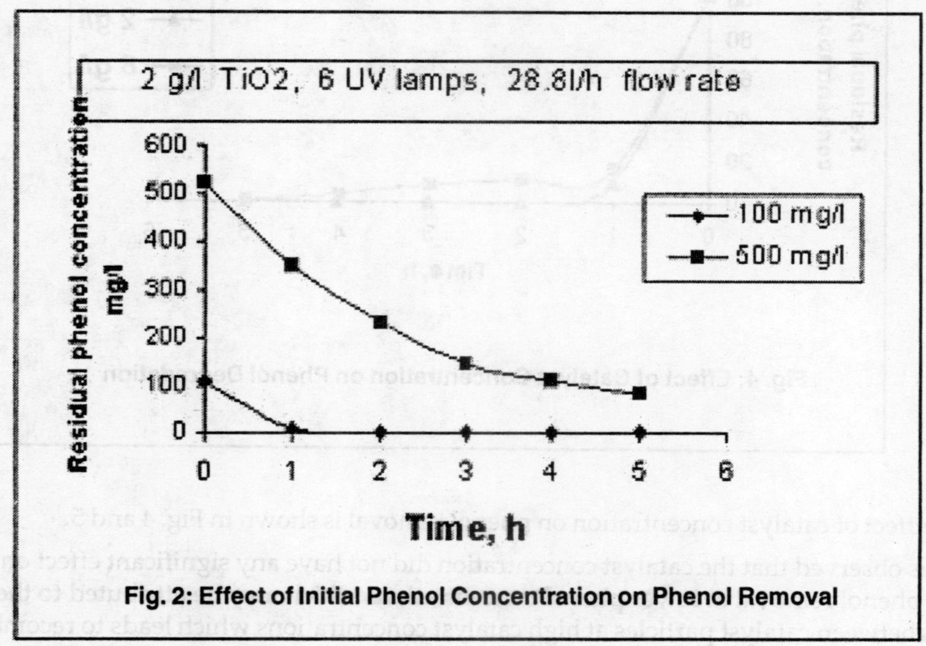

Fig. 2: Effect of Initial Phenol Concentration on Phenol Removal

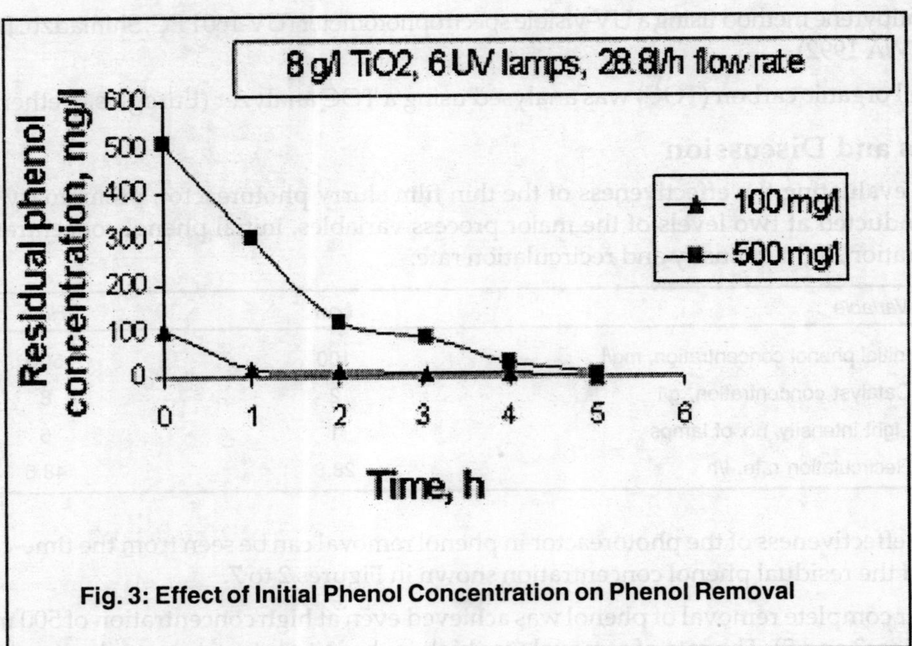

Fig. 3: Effect of Initial Phenol Concentration on Phenol Removal

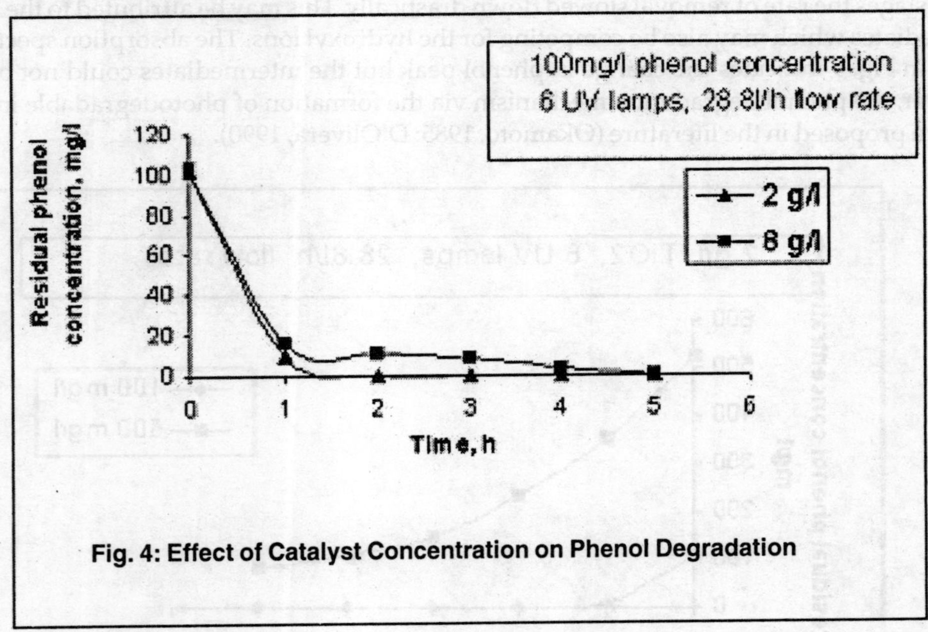

Fig. 4: Effect of Catalyst Concentration on Phenol Degradation

The effect of catalyst concentration on phenol removal is shown in Fig. 4 and 5.

It was observed that the catalyst concentration did not have any significant effect on the rate or extent of phenol removal at lower phenol concentrations which may be attributed to the increased collisions between catalyst particles at high catalyst concentrations which leads to recombination of

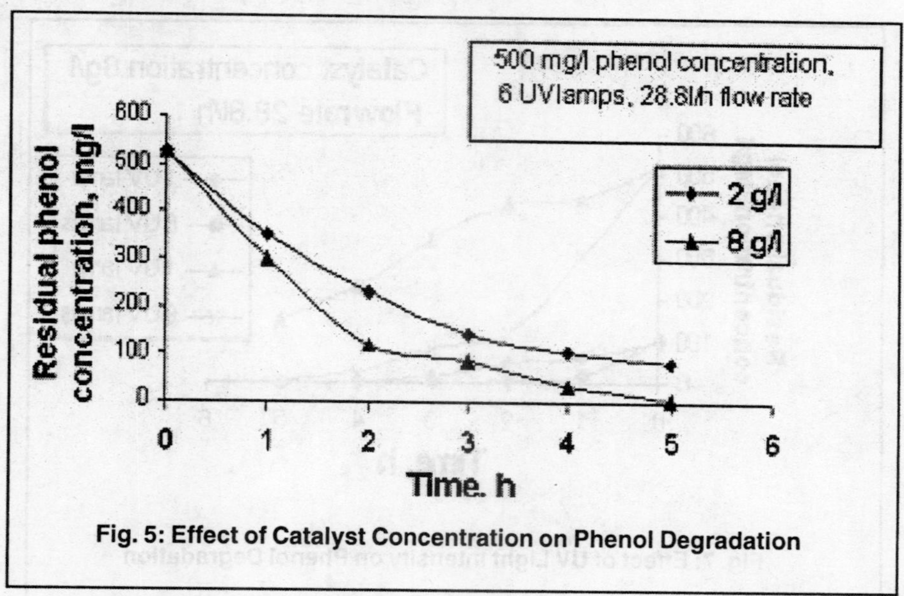

Fig. 5: Effect of Catalyst Concentration on Phenol Degradation

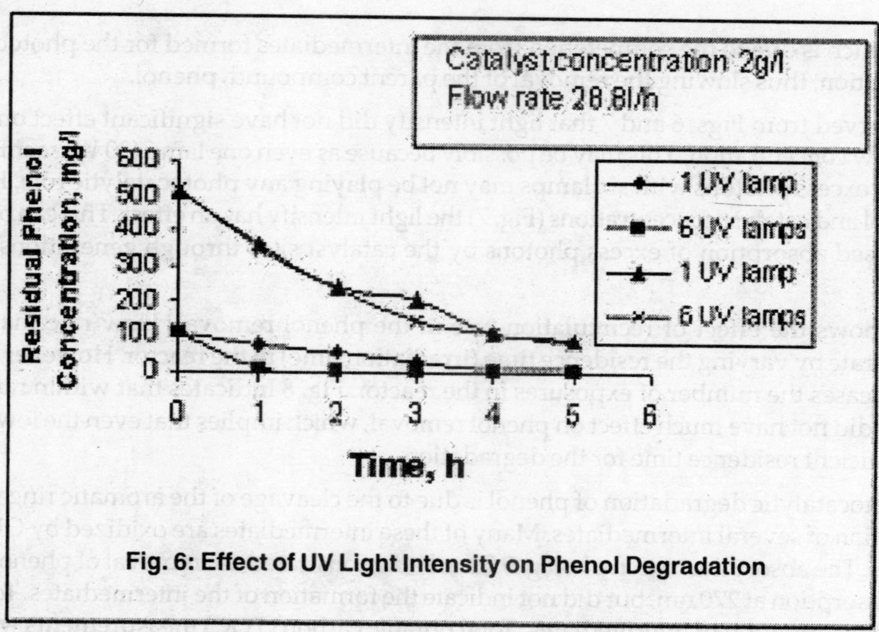

Fig. 6: Effect of UV Light Intensity on Phenol Degradation

electron-hole pairs of excited catalyst. Higher catalyst concentration may also lead to shielding effect of photon absorption. This effect was not observed at higher phenol concentration (500 mg/l). Where the removal rate was more at higher catalyst concentrations. This may be due to the availability of more active sites for adsorption of phenol and the intermediates formed. This adsorption reduces the electron hole recombination on the excited catalyst. At both concentrations, the rate of removal decreases

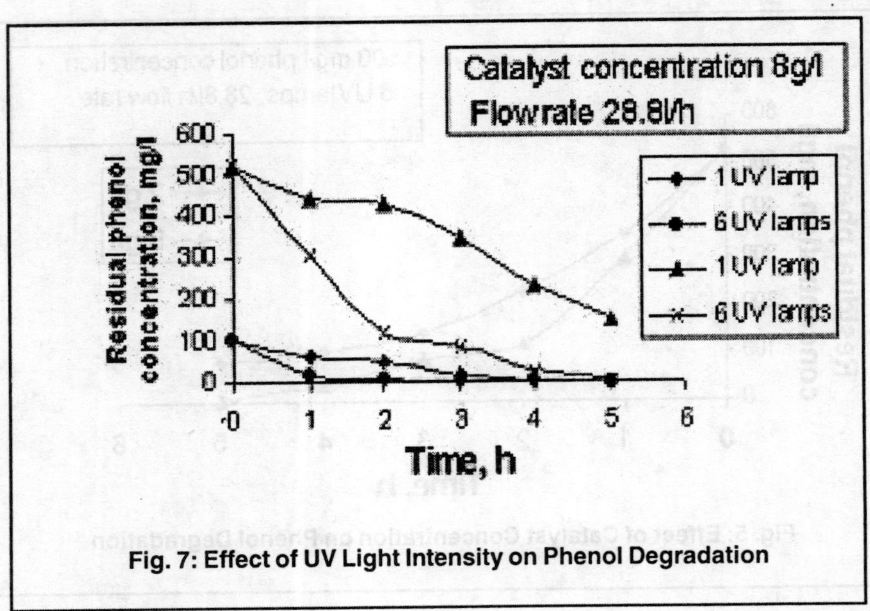

Fig. 7: Effect of UV Light Intensity on Phenol Degradation

with time, which is due to the competetion from the intermediates formed for the photocatalyst and their degradation, thus slowing the removal of the parent compound, phenol.

It is observed from Figs. 6 and 7 that light intensify did not have significant effect on the phenol removal at low concentration. This may be possibly because as even one lamp (30 W) seems to be quite effective, the excess photons with six lamps may not be playing any photocatalytic role. However, at higher phenol and catalyst concentrations (Fig. 7) the light intensify has an effect. This can be attributed to the increased absorption of excess photons by the catalyst sites through generations more OH$^{\bullet}$ radicals.

Fig. 8 Shows the effect of recirculation rate in the phenol removal. Flow rate may affect the degradation rate by varying the residence time (irradiation time) in the reactor. However higher flow rate also increases the number of exposures in the reactor. Fig. 8 Indicates that within range studied the flow rate did not have much effect on phenol removal, which implies that even the lower flow rate provides sufficient residence time for the degradation.

The photocatalytic degradation of phenol is due to the cleavage of the aromatic ring which leads to the formation of several intermediates. Many of these intermediates are oxidized by OH$^{\bullet}$ radical at varying rates. The absorption spectra in Fig. 9 Clearly demonstrated the removal of phenol, which has maximum absorption at 270 nm, but did not indicate the formation of the intermediates. To get further insight into the formation of intermediates, Total organic carbon (TOC) measurements was made for the samples at different times. The TOC the samples were compared against the TOC calculated from the residual phenol concentrations (Fig. 10). It indicates that the higher TOC of the samples is due to that contributed by the intermediates. The slowing down of TOC removal at later stages may be blue to the reduced photodegradability of some intermediates.

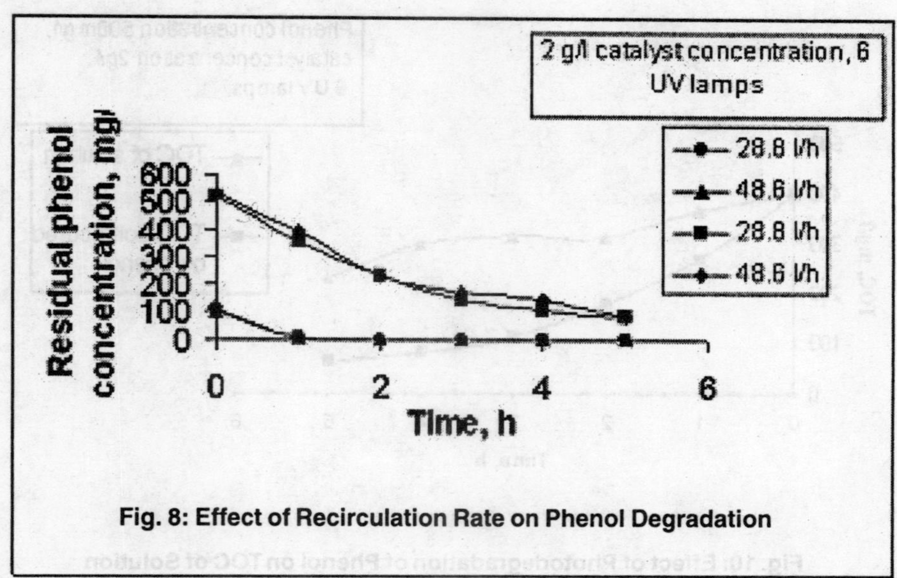

Fig. 8: Effect of Recirculation Rate on Phenol Degradation

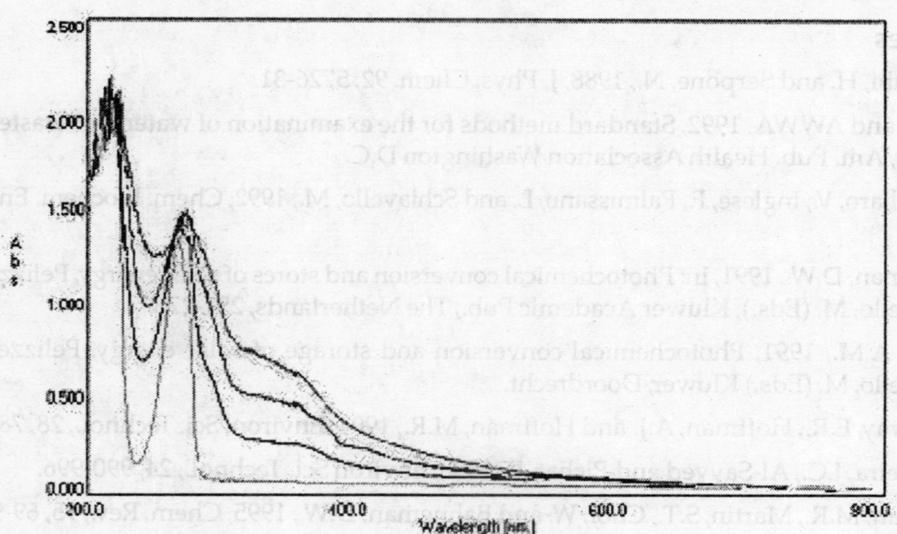

Fig. 9: Absorption Spectra of Phenol Degradation

Conclusions

The results of this study has established the effectiveness of the photocatalytic removal of phenol in a thin film slurry photoreactor, even at high phenol concentration. The process and the reactor configuration appear to be a viable alternative for the treatment of wastewaters containing toxic and hazardous contaminants.

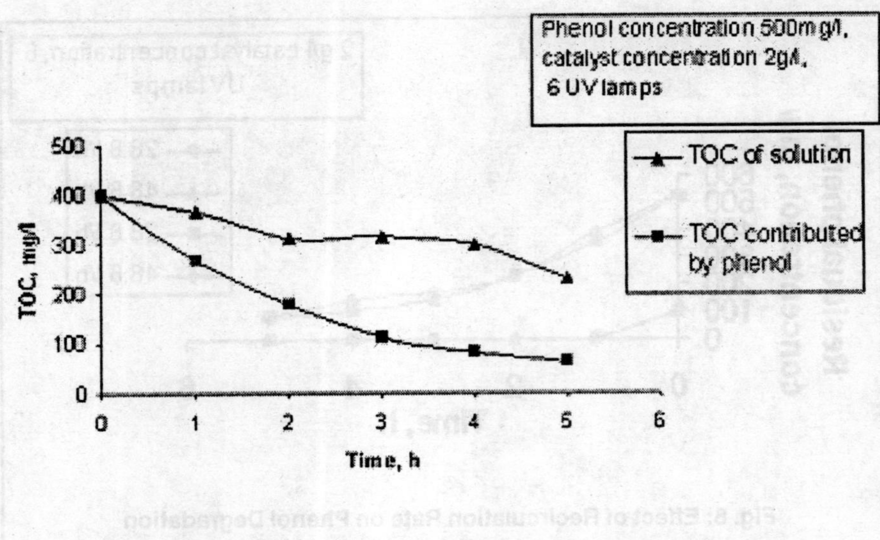

Fig. 10: Effect of Photodegradation of Phenol on TOC of Solution

References

1. Al-Akabi, H. and Serpone, N., 1988, J. Phys. Chem. 92, 5726-31.
2. APHA and AWWA, 1992, Standard methods for the examination of water and wastewater, 18th edition, Am. Pub. Health Association Washington D.C.
3. Augugliaro, V., Inglese, F., Palmissano, L. and Schiavello, M., 1992, Chem. Biochem. Eng. Q., 6 (2), 63-70.
4. Bahneman, D.W., 1991, In: Photochemical conversion and stores of solar energy, Pelizzetti, E. and Schiavello, M. (Eds.), Kluwer Academic Pub., The Netherlands, 251-276.
5. Braun, A.M., 1991, Photochemical conversion and storage of solar energy, Pelizzetti, E. and Schiavello, M. (Eds.) Kluwer, Doordrecht.
6. Carraway, E.R., Hoffman, A.J. and Hoffman, M.R., 1994, Environ. Sci. Technol., 28, 786-793.
7. D'Oliveira, J.C., Al-Sayyed and Pichat, P., 1990, Environ Sci. Technol., 24, 990-996.
8. Hoffman, M.R., Martin, S.T., Choi, W. and Bahneman, D.W., 1995, Chem. Rev., 95, 69-96.
9. Hofstadler, K., Bauer, R., Novalic, S. and Helsier, G., 1994, Environ. Sci. Technol., 28, 670-674.
10. Legrini, V., Oliveros, E. and Braun, A.M., 1993, Chem. Rev., 93, 671-698.
11. Mills, G. and Hoffman, A.J. and Hoffman, M.R., 1993, Environ. Sci. Technol., 27, 1681-1689.
12. Minero, C., Aliferti, C., Pelizzetti, E., Terzian, R. and Serpone, N., 1991, Langmuir, 7, 928-936.
13. Ollis, D.F., Pelizzetti, E. and Serpone, N., 1991, Environ. Sci. Technol., 25, 1523-1528.
14. Okamoto, K.I., Yamamoto, Y., Tanaka, A., and Tanaka, M., 1985, Bull. Chem. Soc. of Japan, 58, 2023-2028.

15. Rideh, L., Wehrer, A., Ronze, D. and Zoulalian, A., 1997, Ind. Eng. Chem. Res., 36, 4712-4718.

16. Sabate, J., Anderson, M.A., Aguado, M.A., Gimenz, J., Cerevevamarch, S. and Hill, C.G., 1992, J. Mol. Catal., 71, 57-68.

17. Schmelling, D.C., and Gray, K.A., 1995, Water Res., 2651-2662.

18. Serpone, N., Burgevello, E., Harris, R., Cahill, P., Burgevello, M. and Pelizzetti, E., 1986, Solar Energy, 14, 121-127.

30

Hydrogen Peroxide-Ultraviolet Irradiations in Water Treatment

ANJANA RUDRA, NEETA P. THACKER & SUNIL P. PANDE

Scientist, National Environmental Engineering Research Institute (NEERI)

ABSTRACT

Hydrogen Peroxide (H_2O_2) and ultraviolet (UV) irradiations in conjunction were used for the removal of trihalomethanes (THMs) in water. batch experiments were carried out in a quartz photochemical assembly using ACE Hanovia make UV lamp of 83 Watt and 40%w/w H_2O_2. Test waters at 50-200 gL^{-1} of chloroform dibromochloromethane, bromodichloromethane and bromoform were exposed to H_2O_2-UV irradiations between 2.5-10 pH range. 92-100% removal of chloroform, dibromochloromethane, bromodichloromethane and bromoform were observed with 0.1% of H_2O_2-UV irradiations in 90 mins of contact.

Keywords: Hydrogen peroxide, hydroxyl radicals, photocatalysis, trihalomethanes, ultraviolet irradiations.

Introduction

A number of physical and chemical methods of water treatments are available for organics *viz.*, airstripping, chemical oxidation, adsorption by granular activated carbon (GAC), ozonation, reverse osmosis and electron beam irradiation and photocatalysis. The photodegradation of trace volatile organics has an advantage because of the high efficiency and simplicity. Moreover the intermediate oxidation products so formed in case of trihalomethanes (THMs) are usually carbon dioxide and water (Beltran *et al.*, 1999 and Kormann *et al.*, 1991). Studies on photodegradation of trichloroethane and trichloroethylene has been reported by Beltran *et al.*, in 1997. The role of hydrogen peroxide (H_2O_2) as a photocatalytic agent was investigated by Murugan Malaiyandi *et al.* in 1980. The combine effect of H_2O_2 and UV irridations at varying exposure period and pH ranges in simulated samples were studied for the removal of THMs.

Photocatalytic Effect of Hydrogen Peroxide

H_2O_2 or ozone when used in water under UV radiation is well suited for the breakdown of organic molecules as the cleavage of the O-O bond gives OH radicals. The overall decomposition of H_2O_2 proceeds according to the reaction shown below. (Clarke *et al.* 1992)

$$H_2O_2 \rightarrow 2OH$$

$$H_2O_2 + OH. \rightarrow HO_2. + H2O$$

$$OH. + HO_2 \dashrightarrow HO_2. + OH-$$

$$2OH. \rightarrow H_2O_2$$

$$2O_2- + 2H_2O \rightarrow H_2O_2 + 2OH- + O_2$$

Hydroxyl radicals have a preference for organic chemicals as they abstract a hydrogen atom from them, which in turn results in an organic radical that initiates several chain reactions. The degradation (oxidation) of a general chlorinated hydrocarbon $C_xH_yCl_z$ can be described by the following stoichiometric equation (Kormann, C. *et al.*, 1991):

$$C_xH_yCl_z + \left(x + \frac{y-z}{4}\right)O_2 \rightarrow xCO_2 + zH^+ + zCl^- + \left(\frac{y-z}{2}\right)H_2O$$

Rates of elimination of THMs, follow pseudo first order kinetics and can be represented by the contribution of three factors, which are volatility, hydroxyl radical oxidation and direct photolysis (Beltran, *et al.*, 1999)

Experimental

Reagents and Standards

All chemicals used were of AR and residual analysis grade. The concentration of H_2O_2 (A.R. grade 40% w/w) was determined at room temperature by the potassium permanganate titration method (Vogel 1959).

Trihalomethanes Standards

Reference standards of THMs chloroform, dibromochloromethane (DBCM), bromodichloromethane (BDCM) and bromoform were procured from United States Environmental Protection Agency (USEPA).

Simulated Samples of Trihalomethanes

Simulated THMs samples were prepared separately in distilled and tap water by spiking known concentrations of individual THMs (AR grade) solutions prepared in ethanol. Distilled water of Millipore distillation assembly provided with organic free high capacity demineralizer cartridges was used. Drinking water samples were taken from the laboratory tap, which was allowed to run for 4 min before collection.

Equipments

1. ACE Hanovia model 7840 photochemical assembly (Fig. 1) having the following components was used:
    (a) UV lamp ACE Hanovia model cat no. 7825-34 'high pressure' quartz mercury vapor lamp (450 Watt) was used. It has a mercury arc length of 112.5 mm with 400.05 mm as total length of lamp, operated at 3.6 amp current and radiating 83.7 watt UV radiation out of its total energy of 450 watt. (Fig. 1)
    (b) Photochemical quartz immersion well: ACE Hanovia model 7854 double walled quartz immersion well having inlet and outlet tubes for circulating water for cooling UV radiation was selected. The length of quartz well was 76 cm and accommodated Ace Hanovia model 7825-34 UV lamp
    (c) Photochemical reaction vessel: The reaction vessel (500 ml.) for holding water samples was fabricated out of borosilicate glass. The dimension of the vessel was designed to accommodate model 7854 Quartz immersion well.

Fig. 1: Photochemical Assembly in Operation

2. *Power supply*: 6515 power supply was used to operate the ACE hanovia UV lamp and to supply extra voltage and current required to initiate the voltage across the mercury arc.
3. *Gas-chromatograph*: Perkin Elmer Sigma 3B gas chromatograph (GC) model equipped with glass lined injection port, 1.8m × 0.003m ID stainless steel column of 1% SP1000 on 80/100 mesh Chromosorb WHP and electron capture detector (ECD) (Ni63) was used for the separation and quantification of individual THMs by gas chromahography (GC).

Methodology

THMs Analysis

The THMs analysis in water samples were carried out by liquid-liquid extraction method (18th edition of standard methods, APHA/AWWA/WEF) followed by GC using ECD.

Effect of H_2O_2

The simulated samples containing 204.5, 200, 201.54 and 207.7 μgL^{-1} of chloroform, DBCM, BDCM and bromoform respectively were treated with different dilutions of 40% w/w H_2O_2 (0.1%, 2% and 5%) at neutral pH. Samples were placed on a rotatory shaker at 135 rpm for a shaking period of 60 mins and aliquots withdrawn at the end of the contact period to determine the concentration of residual THMs.

Effect of UV Irradiations

The simulated samples containing 200 μgL^{-1} of each of chloroform, DBCM, BDCM and bromoform were exposed to UV irradiations in ACE Hanovia 7840 photochemical assembly for a period of 70 mins. The samples were placed into the outer chamber of the static photo reactor. Tap water was circulated for cooling through the inner chamber. UV lamp was suspended within the photoreactor. The entire assembly was housed within a closed wooden chamber. Samples were withdrawn at the end of the exposure period to determine the concentration of residual THMs.

Effect of H_2O_2 and UV Irradiations

The simulated sample of THMs containing 200, 201.2, 204.3 and 198.4 μgL^{-1} of chloroform, DBCM, BDCM and bromoform respectively was treated with 0.1% dilution of 40% w/w H_2O_2. After vigorous shaking for 15 mins. on a rotary shaker at 135 rpm and ambient temperature, the sample was poured into the outer chamber of the above referred photoreactor. The sample was irradiated with UV light for a period of 90 mins following the above mentioned procedure. Profuse gas evolution was observed in few minutes after the onset of irradiation, which continued throughout the UV exposure period. 60 ml aliquots of the treated water sample was collected at different exposure time ranging from 15, 30, 45, 60 and 90 mins. and stored in headspace free vials till THMs analysis. The experiments were repeated by varying the percentage dilution of H_2O_2 as 1, 2 and 5% and keeping the similar initial concentrations of each THMs and UV exposure period to 90 mins.

The experiments were carried out at lower THMs concentration concentration (50 μgL^{-1}) by varying test water pH. The treated water samples were analyzed for the residual of THMs. The effect of 0.1% dilution of H_2O_2 and UV irradiations on THMs removal in field water samples were also studied.

Results and Discussion

Experiments conducted at 200μgL^{-1} of chloroform, DBCM, BDCM and bromoform with 0.1% of 40% w/w H_2O_2 removed about 10-12% of all the THMs after 60 minutes of contact. The Increase in

H_2O_2 dilutions to 1, 2 and 5% had very little effect on all THMs removal. At 50 and 100 µgL^{-1} of each THMs, the removal ranged from 11-18% and 11-14 % respectively (Fig. 2). It confirmed that H_2O_2 acts as a mild oxidizing agent in THMs removal at ambient temperature (28°C) and neutral pH.

Experiments conducted at 200 µgL^{-1} of chloroform, DBCM, BDCM and bromoform with UV irradiations for 70 mins removed 98 % of bromoform, DBCM and BDCM, and 34% of chloroform. At 50 µgL^{-1} of THMs concentrations and UV irradiations for 70 mins. 100% removal of bromoform, DBCM and BDCM and 46% of chloroform were found (Fig. 3).

Experiments at 200 µgL^{-1} of THMs with 0.1% dilution of H_2O_2 and UV irradiations for 90 mins. removed 100% of bromoform, DBCM and BDCM and 92% of chloroform [Fig. 4 (a)]. The removal efficiency decreased when the dilution of H_2O_2 increased from 0.1% to 5%. 5% of H2O2 removed 80, 79 and 79 of bromoform, DBCM and BDCM respectively, and 48% of chloroform. The decrease in THMs removal with increasing H_2O_2 dilutions can be attributed to the scavenging effect of the unreacted H_2O_2 molecules. Under same set of conditions and at 50 µgL^{-1} of the THMs the same removal trend was found. 100% removal of DBCM, BDCM and bromoform, and 92% of chloroform were recorded with 0.1% of H_2O_2 and UV exposure of 60 mins. [Fig. 4 (b)]. The test water pH had a remarkable effect on the removal of THMs. At 20 µgL^{-1} and pH 2.5 100% removal of bromoform, DBCM and BDCM, and 68% of chloroform were found with 0.1% of H_2O_2 and UV exposure of 45 mins. At 200 µgL^{-1} and pH 2.5 100% removal of bromoform, DBCM and BDCM and 86% of chloroform were found with 0.1% of H_2O_2 and UV exposure of 75 mins. At pH 10 the removal efficiency were decreased. At 20 µgL^{-1} of THMs 100% removal of bromoform, DBCM and BDCM, and 48% of chloroform were found with 0.1% H_2O_2 and UV exposure of 60 mins. At 200 µgL^{-1}, 100% removal of bromoform, DBCM and BDCM, and 57% of chloroform were found with 0.1% of H_2O_2 and UV exposure of 90 mins. The decrease in THMs removal

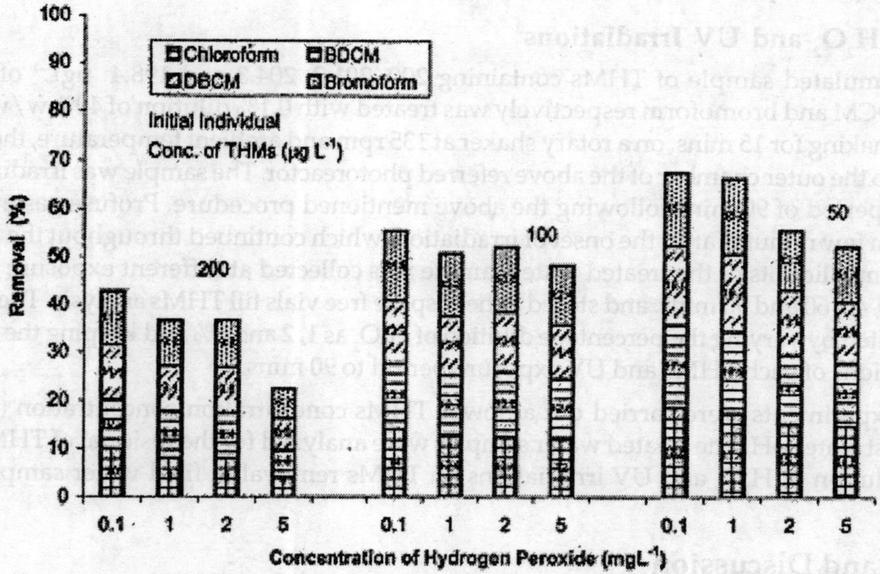

Fig. 2: Effect of Varying Concentration of Hydrogen Peroxide on Removal of Trihalomethanes

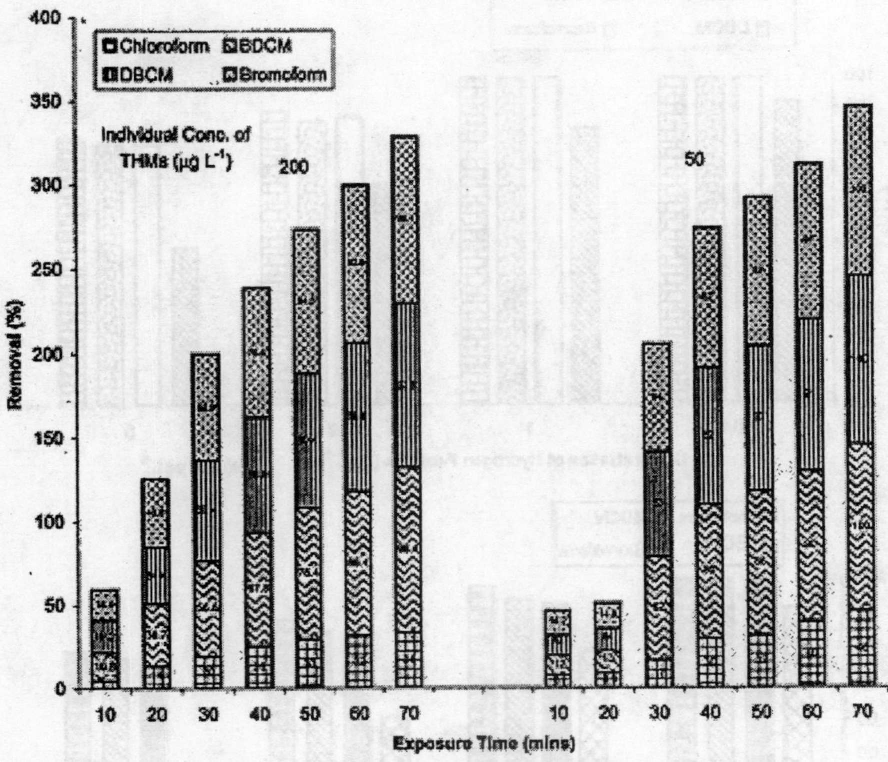

Fig. 3: Effect of UV Irradiations on Removal of Trihalomethanes

with increasing test water pH can be attributed to the H_2O_2 reactivity. Under alkaline conditions H_2O_2 reacts with the hydroxyl ion and form the perhydroxyl radical OOH^- according to the following equation:

$$H_2O_2 + OH^- \rightarrow OOH^- + H_2O$$

In the absence of other reagents the perhydroxyl anion reacts with another peroxide molecule

$$OOH^- + H_2O_2 \rightarrow H_2O + OH^- + O_2$$

In the acidic media the above equation is displaced entirely to the left thus generating more hydroxyl ions and causing greater removal of organics.

Conclusion

The H_2O_2 and UV irradiations in conjunction has higher efficiency in treating water contaminated with THMs. At neutral pH 0.1% of 40% w/w H_2O_2 and UV exposure of 90 mins. were found effective in removing 100% of bromoform, DBCM and BDCM and 92% of chloroform for the initial concentrations of 200 µgL^{-1}. The increase in H_2O_2 concentration did not effect the THMs removal. At pH 2.5 and THMs concentration of 200 µgL^{-1}, 60-100% removal was found with 0.1% H_2O_2 and UV exposure of 45 mins. The THMs removal efficiency decreased to 30-68% in alkaline pH (11.8). There has not been found any change in the general water quality on treating the water samples with H_2O_2 and UV irradiations.

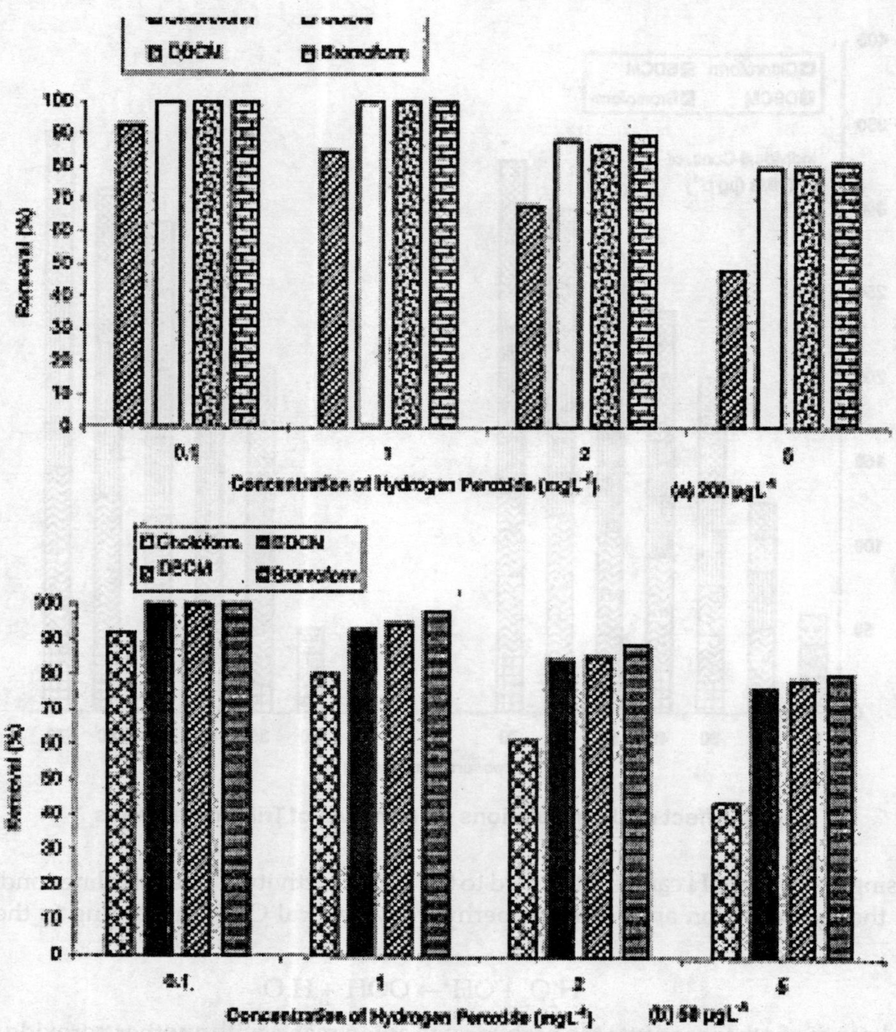

Fig. 4 (a)&(b): Effect of Varying Concentration of Hydrogen Peroxide in Conjuction with UV Irradiation of 50 mins on Removal of Trihalomethanes

Acknowledgement

Authors are grateful to the Director NEERI for granting permission to publish this work. Acknowledgements are also due to the scientists and staff of the Instrumentation Division of the Institute for their kind assistance and cooperation during sample analysis.

REFERENCES

1. Beltran, F.J., Cvejero, G, and Acedo, B., 1997, Oxidation of atrazine in water by ultraviolet radiation combined with H_2O_2. *Wat. Res.*, 26, (6) 1013-1021.

2. Beltran, F.J., Manuel Gonalez, Javier Rivas and Benito Acedo, 1999, Aqueous degradation of VOCs in the ozone combined with H_2O_2 or UV radiation process 1.Experimental results. *J. Env. Sci. Health,*. A (34), 649-671.

3. Clarke, N., and Knowles, G., 1992,High purity water using H_2O_2 and UV radiation. *Effluent and Water Treatment Journal*, 335-341.

4. Kormann C., Bahemann D.W., Hoffmann M.R., 1991, Photolysis of chloroform and other organic molecules in aqueous TiO_2 suspensions. *Environ. Sci.Technol.*, 25 494-500.

5. Murugan Malaiyandi, M. Husain Sadar, Pauline Lee and Ron O'Grady, 1980, Removal of Organics in Water using H_2O_2 in the presence of ultraviolet light. *Wat.Res.*. 14, 1131-1135.

6. Vogel, A.I., 1959, Inorganic Analysis Theory and Practice, Longman's Green and Co. 2nd edition.

31

Photocatalytic Degradation of Organic Pollutants on Anatase-Rutile Titania

S.K. SAMANTARAY, P. MOHAPATRA & K.M. PARIDA

Regional Research Laboratory (CSIR), Bhubaneswar - 751 013

ABSTRACT

Unmodified and sulfate modified TiO_2 samples of different crystal forms and crystal sizes were prepared by varying pH, source and concentration of sulfate ion and by activating at different temperatures. The XRD patterns revealed that TiO_2 sample prepared at low pH (pH 3) exhibit lower crystallite size than that of high pH (pH 9). Sulfate ion impregnation decreases the crystallite size and stabilises the anatase phase of TiO_2 up to higher activation temperature. Photocatalytic degradation of 4-nitrophenol depends on the crystal form as well as crystallite size of TiO_2. TiO_2 containing mixture of anatase and rutile phase exhibit highest percentage of degradation of 4-nitrophenol than that of other samples.

Keywords: Sulfated TiO_2, crystal forms and size, 4-nitrophenol.

Introduction

Nitrophenols are some of the most refractory pollutants, which are present in industrial wastewater. Traditional wastewater treatment includes activated carbon adsorption, chemical oxidation, biological digestion, etc. The feasibility of wastewaters containing 4-nitrophenol being subjected to an ozonisation process was investigated by Beltran *et al.* (1) whereas Pintar and Levec (2) studied catalytic oxidation in a temperature range of 423-463 K. The photocatalytic abatement of 4-nitrophenol in TiO_2 aqueous slurries have been investigated (3-11). The possibility of sensitised photocatalytic degradation at pH 8.5 was demonstrated in reference (3) and the effects of catalyst load and of oxygen and organic concentration was mainly focused in reference (4), disregarding pH. It has been reported that pore structure, pore-size, particle size and activation temperature of TiO_2 affects the photocatalytic activity of organic pollutants in wastewater (5-9).

In order to understand the effect of anions on the crystallite size of TiO_2, we have prepared hydrated TiO_2 at different pH and modified this by varying the source and concentration of anions using different techniques. The unmodified and modified TiO_2 samples are activated at different temperatures by calcination and their activities towards photochemical degradation of 4-nitrophenol were studied.

Experimental

Materials and Methods

Hydrated titania was prepared at pH 3, 5, 7 and 9 by adding aqueous ammonia to the stirred aqueous solution of titanium tetrachloride. The gel was filtered and washed repeatedly to remove Cl^- (negative $AgNO_3$ test), dried at 383 K for 10 h, powdered to 45-75 Mm mess size and kept for anion impregnation.

Modified Catalysts

One series of sulfated titania samples with varying the weight percentage of SO_4^{2-} were prepared using $(NH_4)_2SO_4$ (as the source of sulfate ions) by solid-solid kneading method and the other series of samples were prepared by aqueous wetness impregnation method using dilute H_2SO_4 (strength = 1.0M). The suspended mass was evaporated to dryness on a hot plate while stirring. The samples were then dried in an air oven at 383 K and subsequent activation (by calcination) at 573, 673, 773, 873 and 973 K at the heating rate of 283 K/min in a muffle furnace for 3 h.

Powder X-ray Diffraction

The XRD patterns of pure and modified TiO_2 samples were recorded on a Philips X-ray diffractometer having a PW 1830 generator with copper tube, a PW 1820 goniometer fitted with a post-diffracted graphite monochromator and a scintillation detector attached to a PW 1710 diffraction control in the range of $2\theta = 6\text{-}80°$ at a scanning speed of 2°/min. The instrument was operated at 40kV and 20mA. The average crystallite size (L) was determined by XRD line broadening technique using the Debye-Scherrer equation:

$$L = 0.94\lambda/b\cos\theta \quad (1)$$

where,

λ is the wavelength of the X-ray used and b is the relative peak broadening, calculated as $b^2 = b^2_{exp} - b^2_{ref}$, where b_{exp} and b_{ref} are half widths at maxima observed on a given sample and on a reference material which is ideally crystalline, respectively.

Catalytic Activity

The photo-catalytic degradation of 4-nitrophenol were performed taking 0.05g/L of substrate (4-nitrphenol in water, CDH) and 0.2 g/L of catalyst. The pH of the dispersion was adjusted by addition of sulfuric acid (BDH). The solutions were exposed to sunlight in closed pyrex flasks at room temperature (no extra cooling) with constant stirring. All the irradiation was performed during the second half of March 2002 (sunny days), from 10.00 to 15.00 and was repeated three times. The corresponding dark controls were carried out simultaneously, for comparison. After irradiation taking at different times, the suspension was filtered and analysed the 4-nitrophenol quantitatively by measuring the absorption at 315 nm using Cary-1E (Varian) spectrophotometer. The detail method for measurement was available in reference (4).

Results and Discussion

It is observed that samples prepared at pH 3 possess mixture of anatase and rutile phases whereas at pH 5, 7 and 9 possess only anatase phase. This may be due to the gradual increase in rate of hydrolysis with increase pH of the solution that prevents the formation of rutile phase. It is also observed that unmodified TiO_2 sample prepared at pH 3 and activated at 673 K possesses only anatase phase. With increasing the activation temperature up to 773 K both anatase and rutile phases are found. Whereas sample activated at 873 K only rutile phases is observed. However TiO_2 samples prepared at pH 3, modified with sulfate ion and activated up to 873 K possess only anatase phase. With increasing the activation temperature up to 973 K both anatase and rutile phases are found. Whereas sample activated at 1073 K only rutile phase is observed. So, it may conclude that sulfate ion stabilises the anatase phase of TiO_2 up to 973 K activation temperature. It is also observed that sulfate modified TiO_2 samples activated at 773 K possess anatase phase, irrespective of the source and percentage of sulfate loading. There is no indication of formation of titanium sulfate in any one of the samples, even in 10wt% SO_4^{2-} impregnation. This type of observation was also reported earlier (12-14) on TiO_2 samples prepared at pH 7 and modified with phosphate and sulfate ions.

To develop a better understanding of formation and crystallisation in the presence of sulfate ion, the average crystallite size (L) perpendicular to the 210 plane was calculated from the XRD patterns of pure titania, as well as sulfated titania. It is seen (Table 1) that TiO_2 samples prepared by increasing the pH of the solution, the crystallite size gradually increases. It is observed that the crystallite size of titania sample prepared at pH 3 is lesser than the sample prepared at pH 9. It gradually increases with increasing the pH of the solution, which suppresses the growth of the particle, may be due to rapid hydrolysis. It is observed that the crystallite size of titania decreases with 2.5wt% SO_4^{2-} loading and thereafter it remains same. However, the effect is more pronounced with sulfuric acid as the source of sulfate compared to $(NH_4)_2SO_4$. This indicates that the crystalinity is more or less dependent on the presence of sulfate ion, but not on the source and percentage of sulfate loading. In addition to stabilising anatase TiO_2 crystallites, sulfate surface species inhibit TiO_2 crystallite sintering, leading to smaller crystallites than in pure TiO_2. The crystallite size decreases in the presence of sulfate ions as SO_4^{2-} species could possibly interact with TiO_2 network, and thus hinder the growth of the particle. Even a very small amount of SO_4^{2-} species is responsible for this effect. Therefore, the change in sulfate concentration did not change crystallite size further. Therefore, it is assumed that small amount of sulfate species is responsible for the lowering of crystallite size. It is also observed that crystallite size increases with increase in activation temperature from 383 to 673 K and thereafter it decreases in case of unmodified TiO_2. However, crystallite size remains almost constant up to 973 K and thereafter decreases in case of sulfate modified TiO_2. The former may be due to desorption of surface hydroxyl group from the TiO_2 network and the latter is due to the formation of rutile phase as well as inhibits sintering. This suggests that sulfate species strongly interacting with TiO_2 crystallites inhibits sintering even at higher activation temperatures. This type of effect is also observed in PO_4^{3-}modified FeO (OH) (15), WO_3 modified ZrO_2 (16) and TiO_2 (17).

Table 1: Average Crystallite Size of Unmodified and Modified TiO_2 Samples Activated at Different Temperatures

Sample Code	*Crystallite size (nm)*	*Sample Code*	*Crystallite size (nm)*
TiO_2, pH 3, 773K	13.22	TiO_2, pH 3, 573K	9.52
TiO_2, pH 5, 773K	14.54	TiO_2, pH 3, 673K	14.85
TiO_2, pH 7, 773K	16.13	TiO_2, pH 3, 773K	13.22
TiO_2, pH 9, 773K	17.25	TiO_2, pH 3, 873K	11.38
$2.5SO_4^{2-}$-TiO_2, pH 3, 773K	5.28	$2.5SO_4^{2-}$-TiO_2 (H), pH 3, 573K	3.12
$5.0SO_4^{2-}$-TiO_2, pH 3, 773K	4.70	$2.5SO_4^{2-}$-TiO_2 (H), pH 3, 673K	3.43
$7.5SO_4^{2-}$-TiO_2, pH 3, 773K	4.38	$2.5SO_4^{2-}$-TiO_2 (H), pH 3, 773K	3.55
$7.5SO_4^{2-}$-TiO_2 (H), pH 3, 773K	3.55	$2.5SO_4^{2-}$-TiO_2 (H), pH 3, 873K	3.80
$10SO_4^{2-}$-TiO_2, pH 3, 773K	4.20	$2.5SO_4^{2-}$-TiO_2 (H), pH 3, 973K	2.75

(H) indicates the sample prepared from H_2SO_4.

From the preliminary study, no degradation of nitrophenol was observed in the absence of light and/or of catalyst. The effect of pH (adjusted by H_2SO_4) of the solution on the photocatalytic degradation of 4-nitrophenol (Table 2). It is found that with increasing the pH of the solution up to 2, the percentage of 4-nitrophenol degradation remains almost constant. However, after that it decreases gradually with increasing the pH of the solution. Since no nitrobenzene was detected during the photolysis of 4-nitrophenol at pH 2, it must be concluded that the reductive pathway is favoured in acidic media. It is known that nitroaromatic compounds such as 4-nitrophenol are reduced in acidic media more easily than in alkaline solutions (18).

Table 2: Effect of Various Parameters on the Photocatalytic Degradation of Nitrophenol

Effect of pH of Solution	*% of Degradation*	*Effect of Preparation condition (pH)*	*% of Degradation*	*Effect of wt% of SO_4^{2-}*	*% of Degradation*	*Effect of Activation Temperature (K)*	*% of Degradation* TiO_2	*% of Degradation* $2.5SO_4^{2-}$/TiO_2 (H)
1	54.67	3	42.00	0	42.00	383	31.96	34.87
2	55.63	5	37.20	2.5	48.00	573	33.85	38.22
3	45.03	7	26.20	5.0	48.50	673	35.69	43.00
4	37.32	9	21.50	7.5	49.00	773	43.20	47.20
5	30.10	-	-	10.0	50.00	873	32.57	52.40
6	29.50	-	-	-	-	973	20.00	68.63
-	-	-	-	-	-	1073	16.40	25.27

The effect of TiO_2 samples prepared at different pH on the photocatalytic degradation of 4-nitrophenol is shown in Table 2. It is observed that with increasing the pH of the samples the percentage of degradation gradually decreases. This may be due to higher surface area, lower porosity and more number of hydroxyl groups present on the TiO_2 sample prepared at lower pH than that of higher pH. As reported earlier (4, 19, 20), the surface hydroxyl groups bound to titanium atoms constitute the adsorption active sites. The pH variation on the preparation procedure affects the adsorption powder of the gel because it affects the surface area as well as crystallite size and hence the number of surface hydroxyl groups. As reported earlier (21), the mixture of anatase and rutile phase of TiO_2 are found to be more active than either anatase or rutile crystallites at the same activation temperature. Here it is also observed from the XRD pattern of TiO_2 sample prepared at pH 3 and activated at 773 K shows both anatase and rutile phase. Whereas other samples prepared at pH 5, 7 and 9 exhibits only anatase phase at that activation temperature.

The effect of wt% of sulfate in TiO_2 on the photocatalytic degradation of 4-nitrophenol is shown in Table 2. It is observed that with the addition of 2.5wt% of sulfate on TiO_2, the percentage of degradation increases. However, further addition of higher sulfate content does not affect more to the percentage of degradation. This may be due to the addition of small amount of sulfate decreases the crystallite size of titania, which facilitates higher percentage of degradation. However, increasing the wt% of sulfate on TiO_2 does not affect more towards the crystallite size results almost same percentage of degradation. So, size is an important factor for degradation of 4-nitrophenol.

The activation temperature by calcination influences the catalyst activity because it affects the physical properties of the crystal such as crystal structure as well as crystal size of TiO_2. The effects of activation temperature of unmodified and sulfate modified TiO_2 on the photocatalytic degradation of 4-nitrophenol are shown in Table 2. It is observed that the percentage of degradation gradually increases with increase in activation temperature up to 773 K and thereafter decreases for unmodified TiO_2. But the percentage of degradation increases with increase in activation temperature up to 973 K and thereafter decreases for sulfate modified TiO_2. The percentage of degradation of 4-nitrophenol in case of sulfate modified TiO_2 is higher than that of unmodified TiO_2 at all activation temperature. This is due to the presence of mixture of anatase and rutile phase of TiO_2. The decrease of photocatalytic activity at higher activation temperature because of sintering due to loss of porous structure of nano-particles and the catalyst becomes denser as well as transferred to rutile from anatase phase.

The present results can be directly compared with the data published earlier (5, 6, 11). While optimising the reaction conditions, we found that 0.05g/L of 4-nitrophenol can be completely degraded by using 0.6g/L of sulfate modified TiO_2 catalyst in 3 h. These results found better than those reported earlier with respect to the concentration of nitrophenol, amount of catalyst and time.

Conclusions

1. Modification of the hydrated titania with anions like sulfate can alter the crystallinity.
2. Method as well as condition of preparation also greatly affects the above textural properties.
3. Photocatalytic degradation of 4-nitrophenol depends on phase and crystallite size of titania.

Acknowledgement

The authors are thankful to Dr. V.N. Misra, Director, Regional Research Laboratory, Bhubaneswar for his permission to present this paper.

REFERENCES

1. F.J. Beltran, V. Gomez-Serrano and A. Duran, 1992, *Wat. Res.*, 26: 9.
2. A. Pintar and J. Levec, 1994, *Chem. Eng. Sci.*, 49: 4391.
3. M.S. Dieckmann and K.A. Gray, 1996, *Wat. Res.*, 30: 1169.
4. V. Augugliaro, L. Palmisano, M. Schiavello, A. Sclafani, L. Marchese, G. Martra and F. Miano, 1991, *Appl. Catal.*, 69: 323.

5. D. Chen and A.K. Ray, 1999, *Appl. Catal. B: Environmental*, 23: 143.

6. R. Andreozzi, V. Caprio, A. Insola, G. Longo and V. Tufano, 2000, *J. Chem. Tech. and Biotech.*, 75: 131.

7. N. San, A. Hatipo Lu, G. Kocturk and Z. C nar, 2002, *J. Chem. Tech and Biotech.*, 146(3): 189.

8. C.S. Turchi and D.F. Ollis, 1990, *J. Catal.*, 122: 178.

9. G. Riegel and J.R. Bolton, 1995, *J. Phys Chem.*, 99: 4215.

10. A. Makowski, Z. Dziewiecki and W. Wardas, 2001, Presented in 17th North American Catalysis Society Meeting, June 3-8, Toronto.

11. J. Kiwi, C. Pulgarin and P. Peringer, 1994, *Appl. Catal.*, B 3: 335.

12. K.M. Parida, M. Acharya, S.K. Samantaray and T. Mishra, 1999, *J. Coll. and Inter. Sci.*, 217: 388.

13. S.K. Samantaray, T. Mishra and K.M. Parida, 2000, *J. Mol. Catal.*, 156: 267.

14. S.K. Samantaray and K.M. Parida, 2001, *J. Mol. Catal.*, 176: 151.

15. K. Kandori, S. Uchida, S. Kataoka and T. Ishikawa, 1992, *J. Mater. Sci.*, 27: 719.

16. D.G. Barton, S.L. Soled, G.D. Meitzner, G.A. Fuentes and E. Iglesia, 1999, *J. Catal.*, 181: 57.

17. L.J. Almeny, M.A. Larrubia, M.C. Jimenez, F. Delgado and J.M. Blasco, 1997, *React. Kinet. Catal. Lett.*, 60(1): 41.

18. R. Dillert, I. Fornefett, U. Siebers and D. Bahnemann, 1996, *J. Photochemistry and Photobiology* A, 94: 231.

19. H. Chen and E. Ruckenstein, 1992, *J. Coll. and Interface Sc.*, 148(2): 382.

20. S-J. Tsai and S. Cheng, 1997, *Catal. Today*, 33: 227.

21. Q. Zhang, L. Gao and J. Guo, 2000, *Appl. Catal. B* 26: 207.

32

Treatment of Textile Wastewater Using Aqueous Oxidative Agents and their Effect on Biodegradation

R.A. PANDEY, K.V. PADOLEY, T.V. SUBBARAO, A.S. RAJVAIDYA, M.V. LAKSHANE, C.V. DESHPANDE, N.N. RAO, P.P. PATHE, T. NANDY, T. CHAKRABARTI & S.N. KAUL

National Environmental Engineering Research Institute, Nehru Marg, Nagpur - 440 020

ABSTRACT

The wastewater from a textile industry was characterized and results indicated that pH of the wastewater was alkaline (10.95) and contained average chemical oxygen demand (COD) of 1617mg/l, biochemical oxygen demand (BOD_5) of 688 mg/l, sulphide concentration of 21.6 mg/l and ammonia concentration 20.32 mg/l, alongwith intense colour. This investigation was designed to evaluate the extent of aqueous oxidative removal of organics and colour of the textile wastewater using Fenton's reagent and a combination of magnesium oxide and magnesium carbonate (magnesium peroxide) separately and also to assess the biodegradability of effluent generated after the abovesaid treatment options in comparison with the textile wastewater without any treatment, keeping same the COD concentration. The results indicate that under optimal conditions, pretreatment with Fenton's reagent could remove 69% of COD, while pretreatment with magnesium peroxide could remove 62.23% of COD. The wastewater was then subjected to biodegradation without any treatment using mixed biomass developed from bacterial cultures previously isolated on synthetic dyes and the results indicate a COD removal of 48% within a period of 72 hours. Further, the biodegradation in case of Fenton treated wastewater resulted in 96% removal of residual COD in 72 hours, while in case of magnesium peroxide treatment 93% COD could be removed in the same time period compared to only 48% COD removal for untreated wastewater within 72 hours. The chemical sludge produced during the fenton treatment was 0.6 g/l, while in case of magnesium peroxide treatment the sludge production was 3.3 g/l.

Keywords: Textile wastewater, Pretreatment, Fenton oxidation, Magnesium peroxide, Biodegradation.

Introduction

The textile wastewater poses a great challenge for treatment, since apart from being characterized with high pH, chemical oxygen demand (COD), biochemical oxygen demand (BOD), and total dissolved solids (TDS), it also contains high intensity of colour, which makes it very difficult to treat by conventional treatment methods. The untreated effluents if discharged into natural streams, can cause rapid depletion of dissolved oxygen of the receiving waterbodies, and the problem can further get aggravated due to the presence of suspended solids ultimately leading to subsequent decomposition of the sludge under anaerobic condition. The alkalinity of the wastewater and presence of chromium can severely affect the aquatic life and may also affect the biological treatment processes. Some of the dyes present in the wastewater are also reported to be toxic to aquatic life (1). The colour often renders the water unfit for use for some industrial purposes in the downstream side. It has been reported that the presence of sodium hydroxide at a concentration of 25 ppm in the textile wastewater is lethal to fish life (1). In view of the toxicity and adverse effects of the textile industry wastewater, it is essential to treat the wastewater so as to conform to the limits prescribed by the regulatory agencies

Since the wastewaters generated from textile industry are toxic and resistant to biodegradation, the physico-chemical option in combination with biological methods has been reported in the literature by several workers. A combined approach for the treatment of textile wastewater using ozonation followed by biodegradation in activated sludge process has been reported (2).

The treatment of textile wastewater using ozone/ultraviolet light/hydrogen peroxide (O_3/UV/H_2O_2) treatment as a first step followed by biodegradation in activated sludge process and the results of their study indicate that ozonation was the most suitable pretreatment option for biodegradation, since the ozone pretreated wastewater did not show any inhibitory effect on subsequent treatment with microbial flora in the activated sludge system. (2)

The textile wastewater in the present investigation was subjected to treatment using Fenton's reaction and also magnesium peroxide. Further, the pretreated wastewater was subjected to biodegradation to assess the effect of these treatment options on amenability of the wastewater to biodegradation. The results of the study are presented and discussed in this paper.

Materials and Methods

The wastewater samples were collected in 50 liter capacity containers and were preserved at source using concentrated sulphuric acid and adjusting the pH to 4.0. Two aqueous oxidative agents *viz.* Fenton, and magnesium peroxide were employed for the pretreatment of textile wastewater. The pretreated wastewater was filtered through Whatman No.42 filter paper and was analyzed for COD, and also tested for further biodegradation. All the analyses were carried out on clarified samples as per Standard Methods (3).

Both pretreated (diluted/undiluted) wastewaters were subjected to static flask biodegradation experiments. Similarly, the wastewater without pretreatment was diluted so as to get the same COD magnitude as that of pretreated wastewater and was subjected to biodegradation in order to ascertain the effect of pretreatment on biodegradability of pretreated wastewater. For biodegradation, mixed biomass was developed using bacterial cultures previously isolated in the laboratory on synthetic dyes. Wastewater was supplemented with dipotassium hydrogen phosphate and potassium dihydrogen phosphate and ammonium sulphate, the medium pH was adjusted to 7.2 ± 0.02, and the flasks were incubated at 30 ± 2°C.

Results

Evaluation of Fenton's Oxidation as a Pretreatment Option for Textile Wastewater

The textile wastewater was subjected to Fenton oxidation as one of the pretreatment options. The efficiency of the pretreatment was assessed on the basis of COD removal. Optimization of conditions for treatment of textile wastewater by Fenton's Reagent was carried out and details are given hereunder:

Optimization of pH of the Reaction

The optimization of pH was done using a constant dose of ferrous sulphate of 0.4 g/l and 2 ml/l of H_2O_2. The pH of the textile wastewater was varied in the range of 2.0, 2.5, 3.0, 3.5, and 4.0, using appropriate volume of sulphuric acid and a fixed contact time of four hours for the reactants and textile wastewater was maintained. The results on pH optimization are presented in Fig. 1, which indicate that pH of 3.0 is optimal for the treatment of textile wastewater by Fenton's reagent, since maximum COD removal of 48.96% is observed at this pH value.

Optimization of Contact Time

The optimization of contact time was carried out using a constant dose of ferrous sulphate 0.40 g/l and 2 ml/l of H_2O_2, reaction mixture at optimal of pH of 3.0. The contact time between Fenton's reagent and textile wastewater in the reaction mixture was varied from 60 minutes to 240 minutes (1 hr to 4 hrs) and the results presented in Fig. 2, indicate that Fenton's reagent acts effectively at a contact time of 3 hours (180 minutes) with a maximum COD removal of 49%.

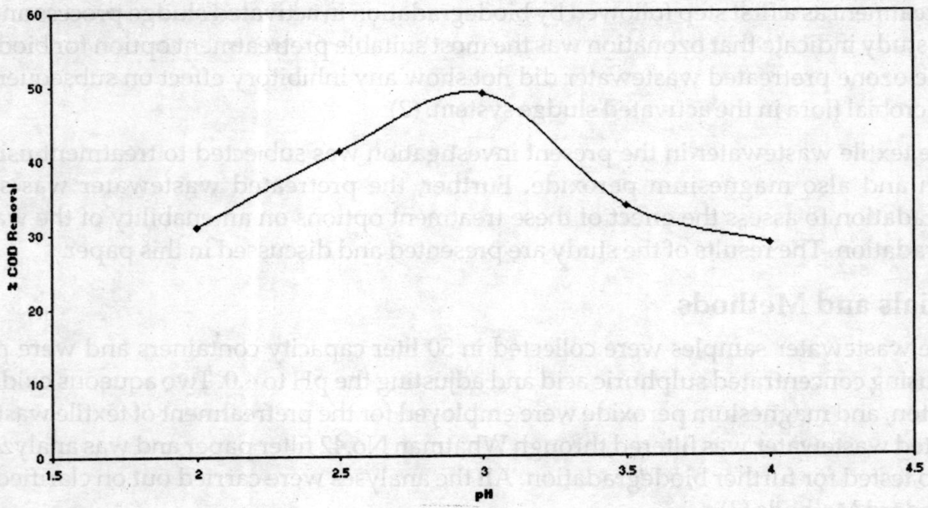

Fig. 1: Optimization of pH of Textile Wastewater for Effective Treatment

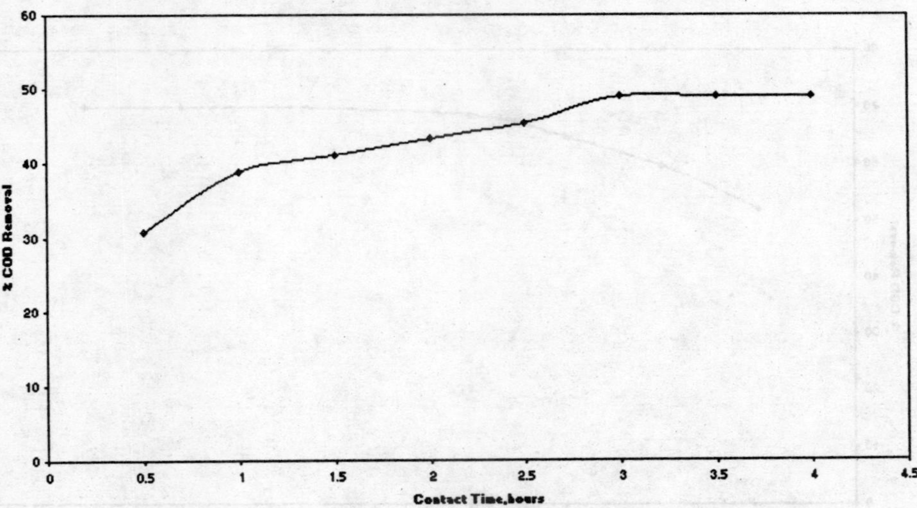

Fig. 2: Optimization of Contact Time for Fenton Treatment of Textile Wastewater

Optimization of Ferrous Sulphate Dose

Ferrous sulphate dose for the treatment of textile wastewater was varied from 0.05g/l to 0.90 g/l under optimal conditions of pH and contact time. The results of optimization of ferrous sulphate dose are presented in Fig. 3, which indicate that a dose of 0.70 g/l ferrous sulphate is optimal for treatment of textile wastewater. This resulted in 54.64% removal efficiency of COD. Further, increase in ferrous sulphate dose did not improve COD removal efficiency.

Optimization of Hydrogen Peroxide Dose

Hydrogen peroxide dose for the reaction was varied from 1 ml/l to 8 ml/l under optimal conditions of pH, contact time and ferrous sulphate dose and the results presented in Fig. 4 indicate that a

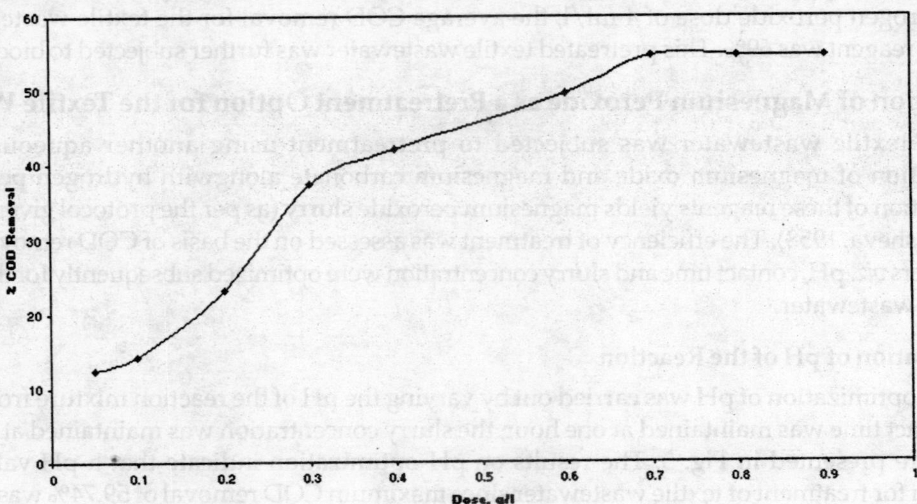

Fig. 3: Optimization of Ferros Sulphate Dose

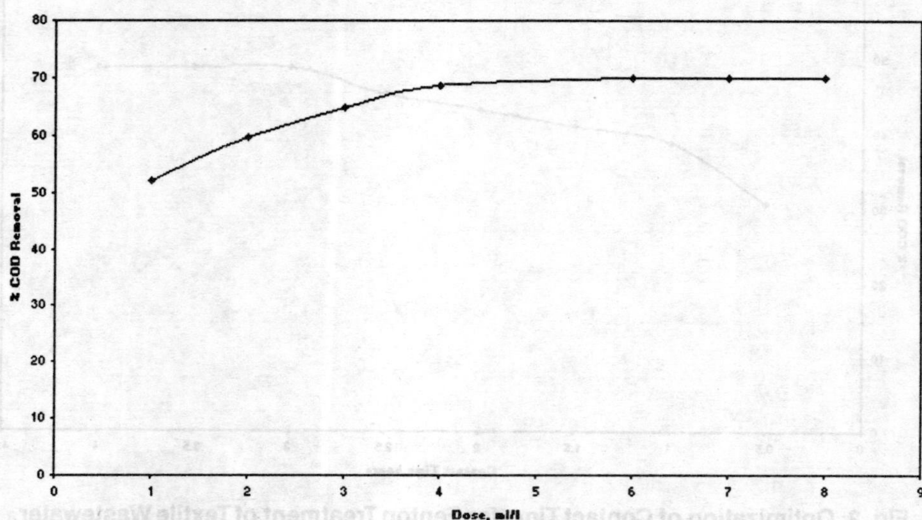

Fig. 4: Optimization of Hydrogen Perioxide Dose

hydrogen peroxide dose of 4.0 ml/l is optimal for treatment of the textile wastewater. At optimal dose of H_2O_2, 68.69% COD removal efficiency was achieved keeping other conditions constant. Further increase in the hydrogen peroxide dose did not lead to improvement in COD removal efficiency.

Treatment of Textile Wastewater Using Fenton's Reagent Under Optimal Conditions

After ascertaining all the optimal conditions *viz.* pH, contact time, and doses of ferrous sulphate and hydrogen peroxide, the textile wastewater was again subjected to treatment using Fenton's reagent. The pretreatment experiments were conducted five times under optimal conditions. The results indicate that under the optimal condition *i.e.* pH 3.0, ferrous sulphate dose of 0.70 g/l, contact time of 3 hours and hydrogen peroxide dose of 4 ml/l, the average COD removal for the textile wastewater using Fenton's reagent was 69%. This pretreated textile wastewater was further subjected to biodegradation.

Evaluation of Magnesium Peroxide as a Pretreatment Option for the Textile Wastewater

The textile wastewater was subjected to pretreatment using another aqueous oxidant-a combination of magnesium oxide and magnesium carbonate alongwith hydrogen peroxide. The combination of these reagents yields magnesium peroxide slurry (as per the protocol given by Vol'nov and Latysheva, 1958). The efficiency of treatment was assessed on the basis of COD removal and other parameters *viz.* pH, contact time and slurry concentration were optimized subsequently for the treatment of textile wastewater.

Optimization of pH of the Reaction

The optimization of pH was carried out by varying the pH of the reaction mixture from 2.0 to 8.0. The contact time was maintained at one hour, the slurry concentration was maintained at 4%, and the results are presented in Fig. 5. The results on pH optimization indicate that a pH value of 3.0 is optimum for treatment of textile wastewater since maximum COD removal of 59.74% was obtained at this pH value.

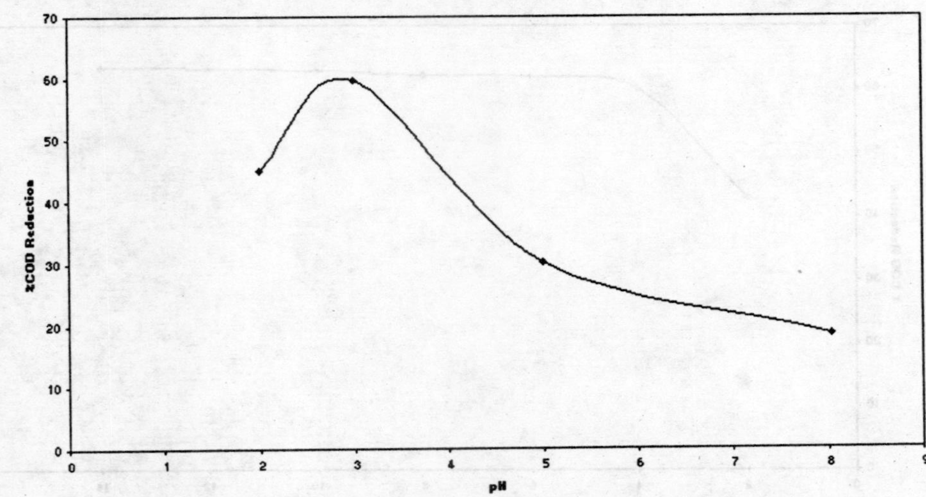

Fig. 5: Optimization of pH for Magnesium Peroxide Treatment

Optimization of Contact Time

The contact time for the treatment of textile wastewater was varied from 20 minutes to 100 minutes at pH 3.0 and with a slurry concentration of 4%. The results presented in Fig. 6 indicate that a contact time of 60 minutes (one hour) is optimum for the treatment of textile wastewater by magnesium peroxide, since maximum COD removal of 60.17% was observed at this contact time.

Optimization of Slurry Concentration

The magnesium oxide-magnesium carbonate mixture constituting the magnesium peroxide slurry was varied from 2% to 14% at a optimum pH of 3.0 and optimum contact time of 60 minutes. The results are presented in Fig. 7 indicate that 4% slurry concentration is optimal for the treatment of textile wastewater since it resulted in 60.27% COD removal.

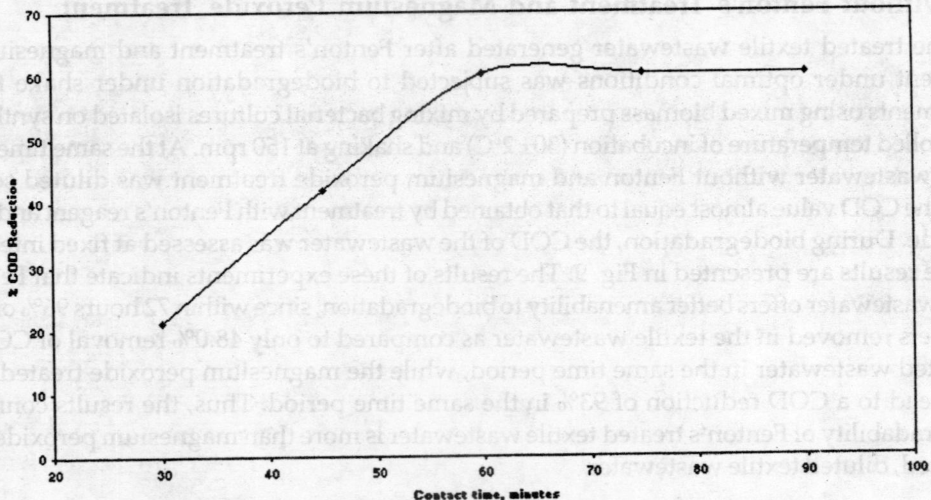

Fig. 6: Optimization of Contact Time Magnesium Peroxide Treatment

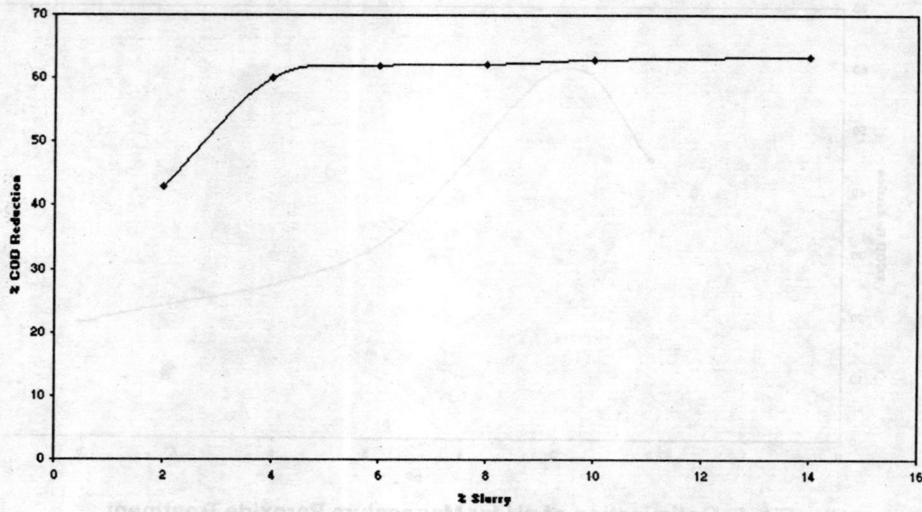

Fig. 7: Optimization of Magnesium Peroxide Slurry Concentration

Treatment of Textile Wastewater Using Magnesium Peroxide Under Optimal Conditions

After optimization of all the conditions *viz.* pH, contact time and percent slurry concentration of magnesium peroxide, the jar test experiments were conducted with the textile wastewater and the results indicate that under optimal conditions 60.27% of COD could be removed from the wastewater (average of five sets of experiments) using magnesium peroxide.

The sludge production in case of Fenton treatment system for textile wastewater was 0.6 g/l while in case of magnesium peroxide system was 3.3 g/l (Fig. 8).

Comparative Evaluation of Biodegradative Characteristics of Textile Wastewater with and without Fenton's Treatment and Magnesium Peroxide Treatment

The treated textile wastewater generated after Fenton's treatment and magnesium peroxide treatment under optimal conditions was subjected to biodegradation under shake flask culture experiments using mixed biomass prepared by mixing bacterial cultures isolated on synthetic dyes, at a controlled temperature of incubation (30± 2°C) and shaking at 150 rpm. At the same time, the original textile wastewater without Fenton and magnesium peroxide treatment was diluted so as to bring down the COD value almost equal to that obtained by treatment with Fenton's reagent and magnesium peroxide. During biodegradation, the COD of the wastewater was assessed at fixed intervals of time and the results are presented in Fig. 9. The results of these experiments indicate that Fenton treated textile wastewater offers better amenability to biodegradation, since within 72 hours 96% of the residual COD gets removed in the textile wastewater as compared to only 48.0% removal of COD in case of untreated wastewater in the same time period, while the magnesium peroxide treated wastewater could lead to a COD reduction of 93% in the same time period. Thus, the results confirm that the biodegradability of Fenton's treated textile wastewater is more than magnesium peroxide treated and untreated, diluted textile wastewater.

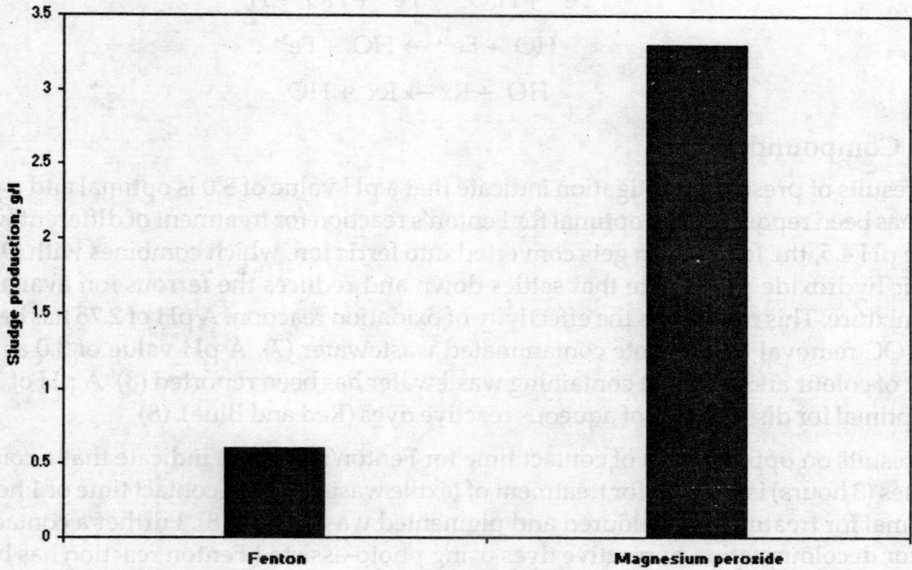

Fig. 8: Sludge Production During Pretreatment of Textile Wastewater

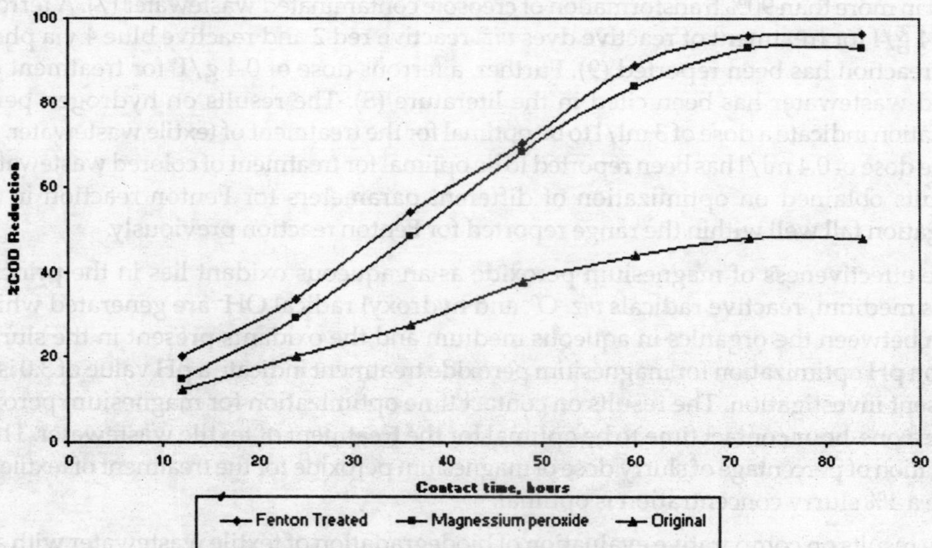

Fig. 9: Comparative Evaluation of Biodegradability of Textile Wastewater with and Without Pretreatment

Discussion

The Fenton reaction employs ferrous sulphate and hydrogen peroxide as reactants. In the mechanism of Fenton reaction, the following reactions generally are reported to occur (4, 5, 6):

$$Fe^{2+} + H_2O_2 \rightarrow Fe^{3+} + HO^- + H^+$$

$$HO^- + Fe^{2+} \rightarrow HO^- + Fe^{3+}$$

$$HO^- + Rx \rightarrow Rx^- + HO^-$$

Organic Compound (Rx)

The results of present investigation indicate that a pH value of 3.0 is optimal and a pH range of 2.0 to 4.0 has been reported to be optimal for Fenton's reaction for treatment of different wastewaters. (4) Above pH 4.5, the ferrous ion gets converted into ferric ion, which combines with OH radical to form ferric hydroxide precipitate that settles down and reduces the ferrous ion availability in the reaction mixture. This minimizes the effectivity of oxidation reaction. A pH of 2.75 has been reported for 80% TOC removal for creosote contaminated wastewater (7). A pH value of 3.0 as optimal for treatment of colour and pigment containing wastewater has been reported (8). A pH of 3.5 has been cited as optimal for dissociation of aqueous reactive dyes (Red and Blue). (8)

The results on optimization of contact time for Fenton's reaction indicate that a contact time of 180 minutes (3 hours) is optimal for treatment of textile wastewater. A contact time of 1 hour has been to be optimal for treatment of coloured and pigmented wastewater (8). Further a contact time of 40 minutes for decolourization of reactive dyes using photo-assisted Fenton reaction has been cited in the literature (9). The results on optimization of ferrous sulphate dose in the present investigation indicate that a ferrous dose of 0.70 g/l is optimal. A ferrous sulphate dose of 60.0 mg/l (0.06 g/l) resulted in more than 90% transformation of creosote contaminated wastewater (7). A ferrous sulphate dose of 4 g/l for treatment of reactive dyes *viz.* reactive red 2 and reactive blue 4 via photo-assisted Fenton reaction has been reported (9). Further, a ferrous dose of 0.4 g/L for treatment of intensely coloured wastewater has been cited in the literature (8). The results on hydrogen peroxide dose optimization indicate a dose of 3 ml/l to be optimal for the treatment of textile wastewater. A hydrogen peroxide dose of 0.4 ml/l has been reported to be optimal for treatment of colored wastewater (8). Thus, the results obtained on optimization of different parameters for Fenton reaction in the present investigation fall well within the range reported for Fenton reaction previously.

The effectiveness of magnesium peroxide as an aqueous oxidant lies in the principle that, in aqueous medium, reactive radicals *viz.* O^{2-} and hydroxyl radical OH^- are generated which lead to a reaction between the organics in aqueous medium and the oxidants present in the slurry (10). The results on pH optimization for magnesium peroxide treatment indicate a pH value of 3.0 is optimal for the present investigation. The results on contact time optimization for magnesium peroxide system indicate a one-hour contact time to be optimal for the treatment of textile wastewater. The results on optimization of percentage of slurry dose of magnesium peroxide for the treatment of textile wastewater indicate a 4% slurry concentration is optimal.

The results on comparative evaluation of biodegradation of textile wastewater with and without Fenton treatment indicate that Fenton treated wastewater is more amenable to biodegradation as compared to the wastewater treated with magnesium peroxide reagent. Inspite of diluting the original textile wastewater to the same COD magnitude as that obtained after Fenton pretreatment, the available organic pollutants were not amenable to biodegradation to the extent as observed in the Fenton treated wastewater. Similar observations have also been reported for phenolic, creosote contaminated wastewater, wherein pretreatment by Fenton has reportedly altered the constituents of the wastewater in making it more amenable to biodegradation (11, 6).

Wastewater from textile industry is very complex in nature with high COD and intense colour. This wastewater could be treated satisfactorily by applying Fenton oxidation under optimal conditions of pH, contact time and doses of ferrous sulphate and hydrogen peroxide to achieve a COD reduction of 69%. The treated effluent can further be subjected to biodegradation with efficient removal of residual COD (96%).

Further, the application of Fenton's reagent as a treatment option prior to biodegradation helps in the molecular modification of the indigenous pollutants occurring in textile wastewater which can easily be oxidized biologically, since the biodegradability of the wastewater without Fenton's treatment showed a COD removal of 48% only inspite of having same concentration of input COD. The results obtained on biodegradation of textile wastewater pretreated with magnesium peroxide indicated that 93% residual COD could be removed after treatment. However, comparative data could not be cited in the literature in context of textile wastewater. Thus, the results on the study indicate that Fenton's treatment is a better option for pretreatment of textile wastewater than magnesium peroxide treatment, since the former not only leads to a higher COD reduction but also further aids in much better COD removal during biodegradation.

Table 1: Physico-chemical and Biological Characteristics of the Textile Wastewater

Parameter	*Value/Range*	**Standard for Discharge into Inland Surface Water Body*
pH	10.95	5.5-9.0
Conductivity millisemens/cm	2.89-3.95	—
COD mg/l	1617.0	250.
BOD mg/l	688.0	30.0
Ammonia-nitrogen mg/l	20.32	50.0
Sulphide mg/l	#21.6	2.0
Heavy metal mg/l		
Cu	0.066	3.0
Cr	—	2.0
Fe	0.280	3.0
Pb	0.020	0.1
Cd	0.008	0.2
Mn	0.068	2.0
Zn	0.094	5.0

*Inputs prescribed by MoEF (Ministry of Environment and Forest)
Values more than the prescribed standard

Conclusion

The textile wastewater was characterized with high pH, and high concentration of BOD, COD, TDS, and intense colour. Studies on pretreatment of the textile industry wastewater with aqueous oxidative treatment methods like Fenton's reaction and also magnesium peroxide treatment were conducted. Biodegradation of pretreated textile wastewater was carried out using a mixed biomass developed from bacterial cultures isolated on synthetic dyes. The results obtained on biodegradation of textile wastewater without any pretreatment indicate only 48% removal of COD, while the results obtained on biodegradation of textile wastewater pretreated with Fenton's reagent indicated that 96% residual COD removal whereas treatment with magnesium peroxide resulted in 93% removal of residual COD. Aqueous oxidative pretreatment of textile industry wastewater with Fenton oxidation resulted in better biodegradation of the wastewater with minimum production of chemical sludge.

Acknowledgements

The authors are thankful to Dr. R. N. Singh, Director, NEERI, for kind permission to present this paper in Indo-Italian Workshop and also thankful to Department of Biotechnology, Government of India for providing financial assistance for the project.

REFERENCES

1. Upadhyay, R. and Pandey, G. N., 1991, Environmental pollution control in cotton textile industry, J. Ind. Poll. Cont., 7: 67-75.
2. Ledakowicz, S. and Gonera, M., 1999, Optimization of oxidants dose for combined chemical and biological treatment of textile wastewater, Water Res., 33: 2511-2516.
3. APHA, 1992, American Public Health Association. Standard Methods for the Examination of Water and Wastewater.18th Edition., Washington, DC:APHA, AWW and WEF.
4. Neodoloujko, A. and Kiwi, J., 2000, TiO_2 speciation producing mineralization of 4-Tert-Butylpyridine. Accelerated Mineralization on via Fenton photo-assisted reaction; Water Res., 3277-3284.
5. Legrini, O., Oilveros, E. and Braun, A.M., 1993, Photochemical processes for water treatment; Chem. Rev., 93: 671-698.
6. Arnold, S.M., Hickey, W.J., and Harris, R.F., 1995, Degradation of atrazine by Fentons reagent: condition optimization and product quantification, Environ.Sci.Technol., 29: 2083-2089.
7. Engwall, A.M., Pegnatello, J.J. and Grasso, D., 1999, Degradation and detoxification of the wood preservatives cresote and pentachlorophenol in water by the photo-fenton reaction; Water Res., 33: 1151-1158.
8. Park, T. J., Lee, K. H., Jung, E. J. and Kin, C. W., 1999, Removal of refractory organic and color in pigment wastewater with Fenton oxidation, Water. Sci. Technol., 39: 189-192.
9. Chang, J. S. Chou, C., Lin, Y. C., Lin, P. J. and Hu, T. L., 2001, Kinetic characteristics of bacterial azo-dyes decolorization by pseudomonas Luiteola, Wat. Res., 35: 2841-2850.
10. Kabdasali, J., Gorel, H. and Tunay O., 2000, Characterization and treatment of textile printing wastewaters, Env. Technol., 21: 1147-1155.
11. Preis S, Terentyeva Y, Rozkov A., 1997, Photocatalytic oxidation of phenolic compounds in wastewater from oil shale treatment. Water. Res., 35: 247-255.

33

Poly Aluminium Chloride (PAC): An Emerging Promising Coagulant in Water Treatment Practice

SUNIL P. PANDE, LEENA S. DESHPANDE, PRAKASH S. KELKAR AND SANTOSH N. KAUL

Scientists, National Environmental Engineering Research Institute, Nehru Marg, Nagpur – 440 020

ABSTRACT

Performance of commercial Poly Aluminium Chloride manufactured in India vis-à-vis ferric alum was studied using Jar test experiments. Research was conducted on removal of turbidity, heavy metals, cyanide, silica etc. The study showed that consumption (optimum dose) of PAC was 40 to 70% that of ferric alum and it acted as a good coagulant over a turbidity range of 5 NTU to 1000 NTU. Floc formation with PAC was larger and heavy compared to the alum flocs with low generation of sludge volume. During coagulation process, PAC and alum, both showed marginal removal of heavy metals, cyanide, silica and arsenic contaminants. Literature review on the use of PAC for coagulation in water treatment has also been presented in the paper.

Keywords: *Poly Aluminium Chloride, Coagulation, Heavy metals removal, Turbidity removal.*

Introduction

The present trend in India under water sector is to construct more number of piped water supply schemes, even for rural installations. Regional Piped Water Supply Schemes (RPWS) cover a number of villages with a central water resource having treatment facility to provide safe potable water. In Ninth Five Year Plan, the Government of India, has allocated more than 206, 000 million rupees for the water sector. The operation and maintenance cost of these water supply schemes range from Rs. 8 per capita per year for a handpump to Rs. 50 per capita per year for a regional PWS (1). The comprehensive action plan of the State Governments indicate that the total requirement of funds for providing safe water within five years period would be around Rs. 440 billion (2).

Invariably in India, commercial alum is in use for coagulation purpose in all the technologies for water treatment. An in-depth survey of water treatment plants was carried out by NEERI (3) to bring out the 'state-of-art' of operation and maintenance of water treatment plants. The evaluation study had revealed that the addition of the chemicals like alum, lime and chlorine is indiscriminate and is not supported by proper laboratory testing of water quality. The stirring mechanism was also defunct at many places leading to improper doses of alum. Maintenance of alum solutionizing facilities and dosing equipment invariably leaves much to be desired and is the weakest link in proper O&M of the water treatment facility.

In view of the above, replacement of commercial alum by another chemical was sought which is comparatively easier to handle, properly dosed and have a good efficiency for coagulation. Poly Aluminium Chloride (PAC) in one of such chemicals, now manufactured in India on commercial scale. Performance of such commercial PAC products was studied in-depth in the laboratory and plant scale for coagulation and removal of turbidity, heavy metals, cyanide and coliform organisms and was compared with that of alum.

PAC: Manufacture

Poly Aluminum Chloride (PAC) is the generic term for a group of coagulants produced by partial neutralization of aluminium chloride solution by some form of basic solution. This neutralization leads to the formulation of hydroxy aluminium species which may be monomeric, dimeric or polymeric depending on the degree of neutralization (4). The process for preparation of polyaluminium hydroxy chloride complexes having general formula $Al\,(OH)_x Cl_y$ in solution and intended to be used as flocculants and coagulants in water treatment have been patented (5) in Europe. Another process patented for presumed formula of $Al_a\,(OH)_b\,(SO_4)_c\,Cl_{3a\text{-}b\text{-}2c}$ where a, b and c may have the values 13, 34 and 1 respectively has been used in water purification technology (6). There are a number of patented formula for Poly Aluminium Chloride formulation particularly from Japan, Europe and U.S. The polymeric species $Al_{13}O_4\,(OH)_{24}^{7+}$ is thought to be most effective species (7). At the international workshop (8) on 'Aluminium in drinking water' held at Hong Kong, January 15-17, 1992 important findings were presented on use of PAC in water treatment. PAC is now manufactured on commercial basis in India and products from two industries were studied in detail for their efficacy in coagulation & flocculation.

PAC in Water Treatment

Using aluminium based compounds, water of widely different chemical characteristics and biological quality can be successfully treated. Upon addition of Al coagulant in water treatment, multiple reaction pathways are possible. Aluminium monomers, Al-polymers and $Al\,(OH)_3$ may take place depending on pH (9). Alum reacts with alkalinity in the water in form of calcium and magnesium bicarbonate producing $Al\,(OH)_3$

$$Al_2(SO_4)_3\,18\,H_2O + 3\,Ca\,(HCO_3)_2 = 3\,CaSO_4 + 2\,Al\,(OH)_3 + 6\,CO_2 + 18\,H_2O$$

$$Al_2(SO_4)_3\,18\,H_2O + 3\,Mg\,(HCO_3)_2 = 3\,MgSO_4 + 2\,Al\,(OH)_3 + 6\,CO_2 + 18\,H_2O$$

Coagulation occurs by interaction of aluminium hydrolysis products with contaminants such as dissolved organic matter and colloidal inorganic or organic polymers. Coagulation of colloidal particles in the presence of Al $(OH)_3$ is termed as enmeshment or sweep floc. Dissolved organics can be removed by adsorption on Al-precipitate. Chemical aspects of hydrolysis reactions during coagulation using alum and PAC studied by Benschoten *et al.*, (10) showed that alum did not form polymers in the water. On the other hand PAC consists of stable polymers below pH 6 and over the time period present in water treatment. Polyaluminium chloride consists of cationic hydrolysis polymers of aluminium. The polymeric structure remains intact within the PAC precipitate and particles are mere positively charged and produce low turbidity than alum floc. PAC is much less affected by temperature than alum.

Studies on coagulation kinetics of PAC and alum using kaolin clay suspensions indicated that increased alum doses reduced the time required for particle destabilization. PAC worked faster than alum. Increasing the Al doses accelerated the flocculation rate because it increased collision-attachment efficiency. PAC was less sensitive to change in pH and effectively coagulated fine particles than alum. Modelling studies on coagulation of organic contaminants like Fulvic Acid (FA) by alum and PAC showed dose requirements are nearly identical because Al binding by FA is similar for both Alum and PAC (11). Dempsey *et al.*, (12) demonstrated that PAC is more efficient than alum or $AlCl_3$ for removal of humic substances (FA) at low pH. When a lower concentration of PAC than alum is effective in forming a precipitate with FA, the resulting sludge volume is also low compared to the use of alum. A literature review (13) on use of PAC for removal of organic contaminants show that coagulation is generally found to preferentially remove colour, THM and TOC precursors. At low to moderate concentration of contaminants PAC is likely to be a better coagulant than alum especially at pH values (< 5.5 or 7.0), may be due to greater charge neutralizing capacity of PAC for the stated conditions (14). The laboratory experiments showed that PAC produced fastest settling flocs when rapid mix conditions were correctly controlled, however, these flocs were very sensitive to shear. Sludge volume per mass of contaminant removed was also low compared to alum sludge. Polymeric aluminium hydroxide solutions remain stable for few months without undergoing modifications in the properties. Thus, it is possible to store polymers for at least few months without significant loss in reactivity.

In another full scale plant study (15) demonstrated that at the Buffalo Pound Water Treatment Plant PAC coagulation could replace alum using approximately one third the dose normally required with alum. PAC consistently removed 20 percent more turbidity than alum over a wide range of water temperatures at one third the dose. The financial considerations were not attractive at the prevailing high cost of the PAC.

Experimental Method

The coagulation studies using alum and PAC were confined to the laboratory experiments using Jar tests. The performance of PAC may vary with different brand name and even batch to batch manufacturing process.

PAC manufactured by two different firms using the latest generation state-of-art technology imported from Italy were tested. These are described as PAC (I) and PAC (II). As mentioned earlier the coagulation is influenced by a large number of interrelated factors such as pH, alkalinity, turbidity, sequence of addition of coagulant, the degree of agitation during flocculation etc. Since alum and commercial PAC were to be compared, the approach in the present study was to determine the most effective doses of these chemicals for specific test conditions.

A series of Jar tests were conducted using different turbid waters for evaluating the performance of PAC vis-à-vis ferric alum. Natural turbid water was collected from the river Kanhan which flows 15 km away from Nagpur city. Dosing of these natural and synthetic coagulants was done by (i) preparing 1% solution in distilled water and using graduated pipette and (ii) neat with the help of microsyringe. It was noted that when the dosing was done with 1% solution of PAC then the treated water showed slight haziness.

Measured volumes (1000 ± 10 mL) of samples were flocculated using the Jar test apparatus. After addition of coagulant, a rapid mix was maintained at 120 (10 rpm for 30 sec followed by slow mix at 25 ± 3 rpm for 10 minutes and the contents were allowed to settle for 20 minutes. The floc size and its settleability were observed visually with illuminating device at the base of Jar test apparatus. The test waters used for turbidity removal were either natural or synthetic waters prepared with dried powdered fine (below 100 Mm sieve) black cotton soil. All the analytical estimations were done as per the procedures described in the Standard Methods (17).

Results and Discussion

Studies were conducted on natural and synthetic waters upto 1000 NTU turbidity and the results are presented in Table 1. The optimum dose of alum and PAC to achieve comparable settled water treatability was found to vary with different test waters. In general the consumption of PAC was 40%-70% that of ferric alum (Fig. 1) and reduction in pH was less as compared to alum.

Sludge volume generated by alum and PAC was compared by studying their performance on naturally available turbid waters from Kanhan river as well as synthetic turbid waters prepared in the laboratory. Optimal Jar test doses were applied so that the settled water turbidity is around 5 NTU. The results of the settling test is given in Table 2. It was observed that with 3 hrs settling period, the reduction in sludge volume with PAC (I) as compared to that obtained with alum was 16% in case of the river water and 11% in case of synthetic turbid water. The PAC coagulated water produced longer and readily settleable flocs than alum.

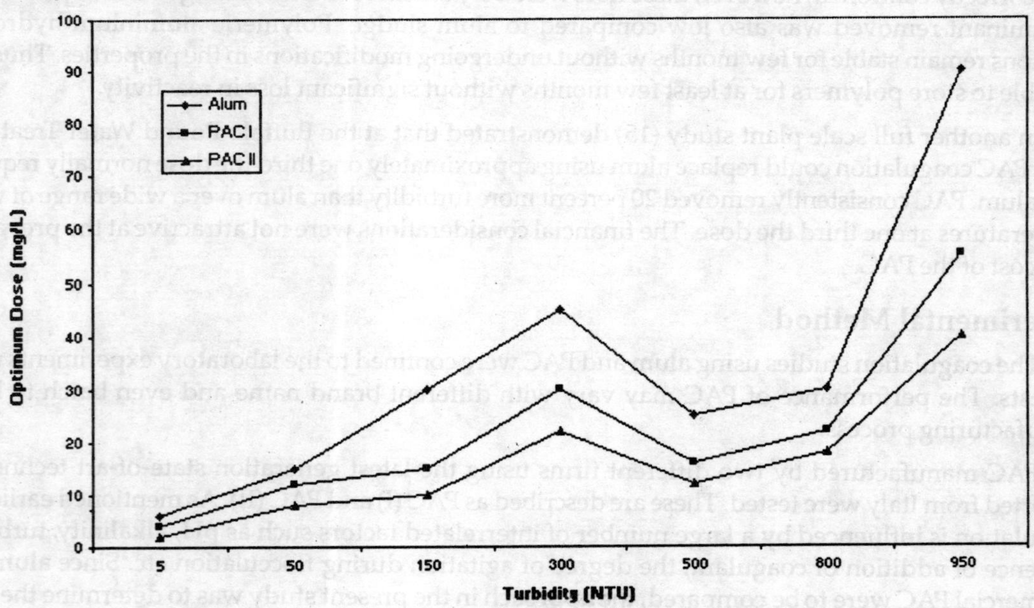

Fig. 1: Performance of PAC vis-à-vis Ferric Alum as a Coagulant

Table 1: Performance of Coagulant for Different Turbid Waters

Ferric Alum			PAC (I)			PAC (II)		
Dose mg/L	Settled Water		Dose mg/L	Settled Water		Dose mg/L	Settled Water	
	Turbidity	pH		Turbidity	pH		Turbidity	pH
Raw Water Turbidity: 5 NTU, pH: 8.0								
6	4.0	7.9	4.0	3.0	8.0	2	2.8	8.0
7	3.8	7.9	5.0	3.0	8.0	3	2.6	7.9
8	3.2	7.8	6.0	2.5	7.9	4	2.5	7.9
10	2.7	7.8	7.0	2.3	7.9	5	2.2	7.9
Raw Water Turbidity: 50 NTU, pH 7.9; (Artificial Colloidal Turbidity)								
10	25	7.6	8	25	7.8	6	22	7.8
15	15	7.5	12	15	7.7	8	15	7.7
20	7	7.4	14	7	7.6	10	6.5	7.7
25	7.1	7.4	16	6.5	7.6	12	6.2	7.6
Natural Water Turbidity: 150 NTU, pH: 8.0								
20	25	7.0	15	20	7.6	10	15	7.7
30	12	6.8	20	10	7.6	15	10	7.6
40	5.5	6.5	28	5	7.5	20	4	7.6
50	6.0	6.3	35	4	7.5	25	3	7.5
Natural Water Turbidity 300 NTU, pH 8.0								
40	22	7.5	25	24	7.9	20	18	7.8
45	15	7.3	30	20	7.8	22	15	7.8
50	10	7.0	35	12	7.7	25	12	7.7
55	6	6.7	39	5.5	7.6	28	5	7.7
Artificial Raw Water Turbidity: 500 NTU, pH: 7.8								
25	15	7.4	16	20	7.8	12	15	7.7
30	12	7.1	20	16	7.8	15	10	7.7
35	6.5	7.0	22	12	7.7	17.5	7.0	7.6
40	8.0	6.8	25	7.5	7.6	20	6.0	7.6
Artificial Raw Water Turbidity: 800 NTU, pH: 7.8								
30	18	7.4	22	16	7.7	18	15	7.7
40	9.0	7.0	25	12	7.7	20	10	7.6
45	6.5	6.8	32	5.5	7.6	23	5	7.5
50	7.5	6.7	40	6.0	7.6	30	4	7.4
Natural (Kanhan River) Water Turbidity: 950 NTU, pH 8.0								
60	58	7.5	40	38	7.9	35	25	7.8
70	40	7.3	50	30	7.8	40	20	7.7
80	25	6.5	55	25	7.8	45	5.0	7.6
90	7.5	6.3	65	6.5	7.7	50	5.0	7.5

Table 2: Sludge Volume Produced by the Coagulant

Sl. No	*Time hour*	*River Water Turbidity: 2010 NTU*		*Synthetic Water Turbidity: 1250 NTU*	
		Sludge Volume (mL/Litre of suspension)			
		Alum Dose 90 mg/L	*PAC (I) Dose 36 mg/L*	*Alum Dose 50 mg/L*	*PAC (I) Dose 20 mg/L*
1.	0	0	0	0	0
2.	0.5	19	15	35	31
3.	1.0	14	12	31	27
4.	1.5	14	12	28	24
5.	2.0	13	11	26	21
6.	2.5	13	12.5	25	20
7.	3.0	12.5	10	23	20
8.	3.5	11	9.5	20	18
9.	4.0	11	9.5	20	18
Sl. No	*Time hour*	*River Water Turbidity: 2010 NTU*		*Synthetic Water Turbidity: 1250 NTU*	
		Sludge Volume (mL/Litre of suspension)			
		Alum Dose 90 mg/L	*PAC (II) Dose 36 mg/L*	*Alum Dose 50 mg/L*	*PAC (II) Dose 20 mg/L*
1.	0	0	0	0	0
2.	0.5	24	20	46	36
3.	1.0	18	16	36	32
4.	1.5	18	16	33	29
5.	2.0	17	15	30	25
6.	2.5	17	14.5	29	24
7.	3.0	15.5	13	27	24
8.	3.5	15	13	24	22
9.	4.0	14	12.5	23	22

Jar test experiments were conducted on synthetic waters of different alkalinities prepared in distilled water. Synthetic water were spiked with 1 mg/L of Fe and Mn and 0.55 mg/L of Cd, Cr, Pb and Hg. Settled water was filtered through Whatman 42 filter paper and filtrate was analysed for heavy metals by Jobin Yvon ICP Spectrometer and Coleman Mercury Analyser. There was considerable reduction in the heavy metal levels in the raw water itself due to settling. The actual removal of heavy metals exclusively due to the coagulants was not significant as shown in the Table 3.

Alum and PAC doses of 50 to 300 mg/L showed a marginal removal of arsenic from the test water containing 0.5 mg/L of arsenic. Similarly cyanide removal from the test water contaning 0.7 mg/L cyanide was also marginal during coagulation with alum and PAC (Doses ranged from 20-120 mg/L).

Synthetic water (Alkalinity: 250 mg/L) spiked with 15 mg/L of silica was used in Jar test experiments. Alum and PAC dose were varied between 0 and 80 mg/L. In case of alum maximum removal of silica was 1.3% at 80 mg/L dose whereas for PAC I and PAC II, the removal of silica was to the extent of 2.7% (at 80 mg/L dose) and 1.3% for 80 mg/L respectively.

Table 3: Heavy Metal Removal Studies Synthetic Water Alkalinity (as $CaCO_3$): 600 mg/L

Sl. No.	Alum Dose mg/L	pH	Alkalinity mg/L as $CaCO_3$	Fe mg/L	Mn mg/L	Cd mg/L	Cr mg/L	Pb mg/L	Hg mg/L
1.	0	8.3	625	0.30	0.55	0.25	0.35	0.25	0.50
2.	50	7.7	590	0.25	0.50	0.20	0.32	0.20	0.48
3.	100	7.4	585	0.20	0.45	0.20	0.30	0.18	0.48
4.	300	6.8	510	0.15	0.42	0.20	0.28	0.18	0.48
Sl. No.	**PAC (i) Dose mg/L**	**pH**	**Alkalinity mg/L as $CaCO_3$**	**Fe mg/L**	**Mn mg/L**	**Cd mg/L**	**Cr mg/L**	**Pb mg/L**	**Hg mg/L**
1.	0	8.3	625	0.30	0.55	0.25	0.35	0.25	0.50
2.	50	7.9	600	0.26	0.52	0.22	0.32	0.22	0.50
3.	100	7.7	570	0.22	0.48	0.22	0.32	0.20	0.48
4.	300	7.4	562	0.18	0.42	0.19	0.30	0.18	0.48
Sl. No.	**PAC (II) Dose mg/L**	**pH**	**Alkalinity mg/L as $CaCO_3$**	**Fe mg/L**	**Mn mg/L**	**Cd mg/L**	**Cr mg/L**	**Pb mg/L**	**Hg mg/L**
1.	0	8.3	625	0.30	0.55	0.25	0.35	0.25	0.50
2.	50	7.9	610	0.28	0.52	0.22	0.32	0.22	0.50
3.	100	7.8	580	0.24	0.50	0.20	0.32	0.18	0.48
4.	300	7.3	570	0.20	0.46	0.18	0.30	0.18	0.47

Conclusions

Studies were carried out to evaluate effectiveness of PAC manufactured in India vis-à-vis use of alum as a coagulant in water treatment. Both natural and synthetic turbid water were used in the Jar test experiments. Comparative performance of these coagulants was studied for removal of turbidity, heavy metals and cyanide. The studies concluded that:

1. PAC acted as a good coagulant over a wide range of turbidities (5 to 1000 NTU) both in synthetically prepared and in natural waters.
2. The average dose required for PAC is 50-70% that of alum when it is prepared in 1% aqueous solution and consumed less alkalinity than alum for equitable performance.
3. Flocs produced by PAC were larger and readily settleable compared to the alum flocs.
4. Sludge volume generated by PAC was less than that generated by alum for obtaining the same quality of settled water.
5. PAC and alum both showed marginal remove of heavy metals for waters having different turbidities.
6. Similarly cyanide, silica and arsenic elimination exclusively due to coagulation was equally comparable for both coagulants and had marginal removal efficiency for these contaminants.

Acknowledgements

Authors thank Ms Jayashri Jagtap and Ms. K. Sri. Bala Kameswari, Project Fellows for conducting a Jar test experiments and Instrumentation Division, NEERI, for their help in heavy metals analysis.

Authors are also thankful to Dr. R.N. Singh, Director, NEERI for granting permission to publish these findings.

REFERENCES

1. Mazumdar K., 2001, Rural drinking water supply in India-A critical review. *Journal of Indian Water Works Association*, Vol. XXXIII No. 3: 211.
2. Annual reports published by Rajiv Gandhi National Drinking Water Mission, Ministry of Rural Development.
3. Report on "Evaluation of Water Treatment Plants in India', NEERI, Nehru Marg, Nagpur-440 020 (India); 1989: 32.
4. Bottero JY., Poirier JE. and Fiessinger F, 1980, Prog. Study of partially neutralized aqueous aluminium chloride solutions: Identification of aluminium species and relations between the composition of solutions and their efficiency as a coagulant. *Prog.Wat.Tech*, 12: 601-612.
5. Kemira Kemi Aktiebolag Heisingborg (SE), European Patent, application number 90850050.7 date of filing: 06.02.1990.
6. Carlesson C., Sweden United States Patent, application number 797, 887 date of filing 14.11.1985.
7. O'Melia, CR., Gray KA, and Yao, C., 1989, Polymeric inorganic coagulants *AWWA Research Foundation* Denver.
8. Kaeding UW, Drikas M., Dellaverde P.J., Martin D., 1992, A direct comparison between aluminium sulphate and polyaluminium chloride as coagulant in a water treatment plant, *Proceedings of International Workshop on Aluminium in Drinking Water* held at Hong Kong, 117 – 131.
9. Hundt TR, and O'Melia CR, 1988, Aluminium-Fulvic Acid interaction: mechanism and application, *J.A.W.W.A*, 80 (4): 176-186.
10. Benschoten JE, and Edzwald JK, 1990, Chemical aspects of coagulant using aluminium salt-I Hydrolytic reaction of alum and polyaluminium chloride. *Water Research*, 24 (12):1519-1526.
11. Benschoten JE, Edzwald JK, 1990, Chemical aspects of coagulation using aluminum salt-II: Coagulation of Fulvic acid using alum and polyaluminium chloride, *Water Research*, 24 (12): 1527-1535.
12. Dempsey BA, Ganho RM, O'Melia CR, 1984, The coagulation of humic substances by means of aluminium salts, *J.A.W.W.A.*, 76 (4): 141-150.
13. Randtke SJ, 1988, Organic contaminant removal by coagulation and related process combinations, *J.A.W.W.A.*, 80 (5): 40-56.
14. Dempsey BA, Shea H, Ahmed TM and Mertink, J, 1985, Polyaluminium chloride and alum coagulation of clay-fulvic acid suspensions, *J.A.W.W.A*, (3): 24-80.
15. Parthasarathy N. and Buffle J., 1985, Study of polymeric aluminium III hydroxide solutions for applications in wastewater treatment. Properties of the polymer and optimal conditions of preparation. *Water Research*, 19 (1):25-36.
16. Standard Methods for the Examination of Water and Wastewater, 1998 20th Edn. *APHA, AWWA & WEF.*

34

Studies on Immobilized TIO_2 and TIO_2 + HZSM-5 Films for an Efficient Photodegradation of Aqueous Phenol

M. NOORJAHAN, V. DURGA KUMARI, M. SUBRAHMANYAM, AND PIERRE BOULE

Catalysis and Physical Chemistry Division, Indian Institute of Chemical Technology, Hyderabad - 500 007

ABSTRACT

Advanced Oxidation Processess (AOP) using large band gap semiconductor particle TiO_2 have emerged as a promising new technology for the degradation of organic recalcitrant pollutants. Our earlier studies on suspension of TiO_2 and TiO_2 + HZSM-5 zeolite combinate in aqueous phenol and p-chlorophenol show that addition of zeolite along with TiO_2 enhances the photodegradation activity. As the major disadvantage of the system containing suspended solids is to remove the photocatalyst from water after the treatment. Thus attempts have been made to immobilize the catalyst on inert supports, while preserving the photocatalytic activity. The aim of the present study is to immobilize the TiO_2 and TiO_2+ HZSM-5 photocatalysts using acrylic emulsion on rigid ceramic tiles and the photocatalytic efficiency were compared using phenol degradation as a model system. The immobilized TiO_2 +HZSM-5 (10+40 wt%) is more active than immobilized TiO_2 (10wt%) system. The addition of HZSM-5 zeolite have enhanced the photocatalytic degradation of phenol compared to TiO_2 system. Thus TiO_2+ HZSM-5 system leads to development of an effective photocatalytic system for the treatment of wastewater.

Keywords: Photocatalysis; TiO_2; immobilization; zeolite; wastewater.

Introduction

Heterogeneous photocatalysis mediated by TiO_2 continues to attract considerable interest for the wastewater treatment[1, 2]. Many semiconductors like TiO_2, ZnO, CdS etc., have been employed as photocatalysts. As generally observed TiO_2 is an efficient photocatalyst with maximum quantum yields. TiO_2 when illuminated with UV (Λ<400nm) an electron is promoted to conduction band and a hole is created in valence band and these charge carriers are trapped at or near the surface of the particle and undergoes e^- transfer reactions across the interface with the organic pollutants from the solutions adsorbed on the TiO_2 surface. The valence band (VB) potential is + ve enough to generate $^{\bullet}OH$ radicals at the surface and CB potential is-ve enough to reduce molecular oxygen to super oxide ion. The $^{\bullet}OH$ radicals are powerful oxidizing agents and attack the recalcitrant present at or near the surface of TiO_2. Most of the studies related to photodegradation have been carried out using the suspension of power TiO_2 in aqueous solutions, however the use of aqueous suspension beside its large surface area available for reaction, it is limited for practical application because of filtration problems due to the small size of the TiO_2 from wastewater. Alternatively the catalyst may be immobilized on to a suitable solid inert support matrix which would illuminate the need for the removal of catalyst. As the catalyst is immobilized there is decrease in the surface area available for the reaction besides, the catalyst must adhere to the solid support and the reactor design is limited by the optical constraints.

Many techniques were proposed for the immobilization of TiO_2 on solid supports like glass[3], quartz[4], silica, activated carbon[5, 6], clays, zeolites[7, 8] and stainless steel etc. Brezova *et al*[9] reported immobilization of TiO_2 (P-25) on fiberglass and reported that the immobilized and slurry forms show comparable activities for the photocatalytic transformation of phenol. Supporting TiO_2 is commonly reported to be less photoactive due to the interaction of TiO_2 with support during the thermal treatment[8]. Hermann *et al*[10] have reported that TiO_2 mixture of activated carbon shows synergistic effects for the photocatalytic degradation of phenol. Zeolite based photocatalysis is new and the work reported involves the use of high TiO_2 loading[11]. In our earlier reports we have studied the preparation, characterization, and photocatalytic evaluation of TiO_2. Supported zeolites prepared byTiO_2 (SSD), Ti-isop, Ti-sol and mechanical mixtures[8] highlights that supporting TiO_2 on zeolite without interaction is enhancing the photodegradation of phenol and *p*-chlorophenol. With this background we made an attempt to immobilize TiO_2+ zeolite (HZSM-5) with the help of an emulsion on rigid ceramic tile while retaining photocatalytic activity. The photocatalytic efficiency of these immobilized systems were compared using the phenol degradation as a model system.

Experimental

Titanium dioxide Degussa P-25 (70% anatase and 30% rutile) of particle size 30nm and surface area $50m^2g^{-1}$ and HZSM-5 zeolite (PQ corporation Netherlands) were used. Phenol (analytical grade). The support used is a commercially available ceramic tile. Methanol employed was of HPLC grade.

Preparation of TiO_2+ HZSM-5 Immobilized Films

TiO_2 supported were immobilized on rigid ceramic tile. TiO_2+ zeolite of known amount were taken and to this mixture 1ml of emulsion is added while stirring. The solution is than sprayed on a previously cleaned ceramic tile with the help of laboratory spray gun and the films are dried at 100°C of oven temperature.

Analysis

The irradiated solutions were analyzed by HPLC using column C-18 and UV detector (Λ=220nm) methanol/water of ratio 50/50 v/v% as the eluent with a flow of 1ml/min.

Photocatalytic Experiments

The photocatalytic experiments were carried on TiO_2 (Ti-EF), HZSM-5 (Ze-EF) and TiO_2+HZSM-5 (Ti-ZeEF) immobilized films. The ceramic plates which are coated with the catalyst is immersed in aqueous solution of phenol of concentration (10^{-4} M) in a dish where aeration is supplied with the help of air pump and it is in continuously in movement by arranging a shaking unit in order to avoid concentration gradient and UV light of 250W is illuminated. Samples are collected at regular intervals for analysis.

Results and Discussion

The relative photocatalytic efficiency of various immobilized Ze-EF, Ti-EF, Ti-ZeEF were determined using phenol (10^{-4}M). Prior to the start of photocatalytic experiments, adsorption studies (dark) were monitored by keeping the films in phenol solution in dark for 1h. The adsorption experiments were extensively studied on Ze-EF, Ti-EF and Ti-ZeEF. The Ze-EF shows about 45% of phenol adsorption and on illumination there is a photolysis of 5% phenol degradation. The Ti-EF shows only 10% of phenol adsorption. In order to check the role of emulsion in adsorption we have also monitored adsorption of phenol on emulsion film (EF), the results shows that there is 22% of phenol adsorption on emulsion film as shown in the Fig. 1. As the TiO_2 is added to the emulsion the adsorption properties are not much significant.

The effect of TiO_2 loading on the immobilization and degradation were studied. For this we have prepared Ti-EF of different wt% like 10wt%, 20wt% and 30wt% are immobilized on ceramic tiles. Prior to the light experiment adsorption studies show that with increase of TiO_2 loading adsorption are not much increased. Whereas the degradation experiments shows that for 20wt% TiO_2 the rate of phenol disappearance is more than the 10wt% and 30wt% TiO_2 as shown in the Fig. 2. It indicates that a monolayer of catalyst is sufficient for the degradation of phenol in an immobilized system as it is known that the particles are aggregated during the immobilization and leads to the increase in the layer thickness and thus the activity is reduced[12]. In our earlier reports in aqueous suspensions of TiO_2+Zeolite the phenol disappeared within 30min on UV illumination. Whereas for immobilized system the rate is slower as the surface area is less and at the same time the transfer of charge carriers to the surface is difficult if the thickness is more for the effective degradation of organic pollutants.

The TiO_2 + HZSM-5 mixtures were also studied. Before this, the optimization of zeolite is studied by varying different wt% of HZSM-5 and from the adsorption studies of phenol it was concluded that a minimum amount 40% (400mg/gm of film) is sufficient for the effective adsorption and above which it is difficult to monitor the degradation as the percent adsorption is 100% as shown in Table 1.

Table 1. Adsorption of Phenol (10^{-4} M) on Zeolite (HZSM-5-EF)

Zeolite (mg)	*% Adsorption*
1000	100
750	100
500	100
400	45
200	27

Thus in further experiments the zeolite percent is kept constant and TiO_2 loading is increased in order to study the effect of zeolite on TiO_2 photocatalytic degradation of phenol. TiO_2+HZSM-5 mixtures

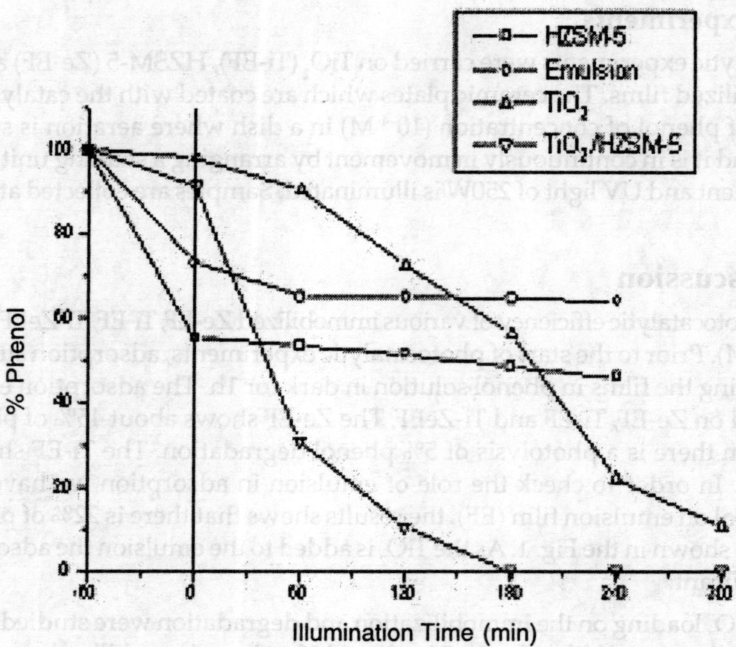

Fig. 1: Photodegradation of Phenol (10^{-4} M) on □ = ZeEF, O = EF; Δ = Ti-EF, Λ = Ti-ZeEF

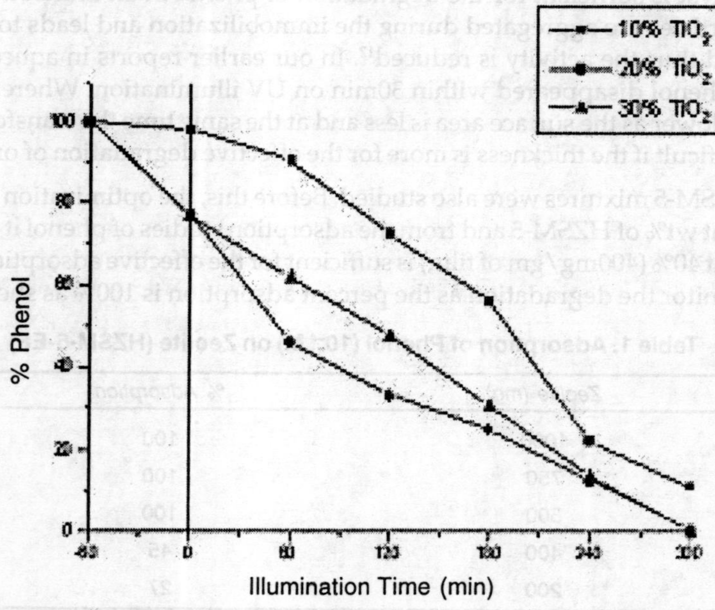

Fig. 2: Influence of the Mass of Titania on the Photocatalytic Degradation of Phenol (10^{-4} M) ■ = -10% Ti-EF, ● = -20% Ti-EF; Δ = -30% Ti-EF

of 10/40wt%, 20/40wt%, and 30/40wt% are immobilized on the ceramic tiles and dark experiments are monitored. The results shows that with the addition of TiO_2 the adsorption capacities are reduced when compared to only HZSM-5 zeolite may be due to the dispersion of TiO_2 which blocks the zeolite pores, but on illumination of these films there is an enhancement of degradation rate of phenol when compared to the Ti-EF. The 10/40wt% Ti-ZeEF shows faster degradation than the other films as shown in the Fig. 3. Thus it indicates that a minimum amount of TiO_2 is sufficient when a porous adsorbent like zeolite is used for the effective degradation of phenol.

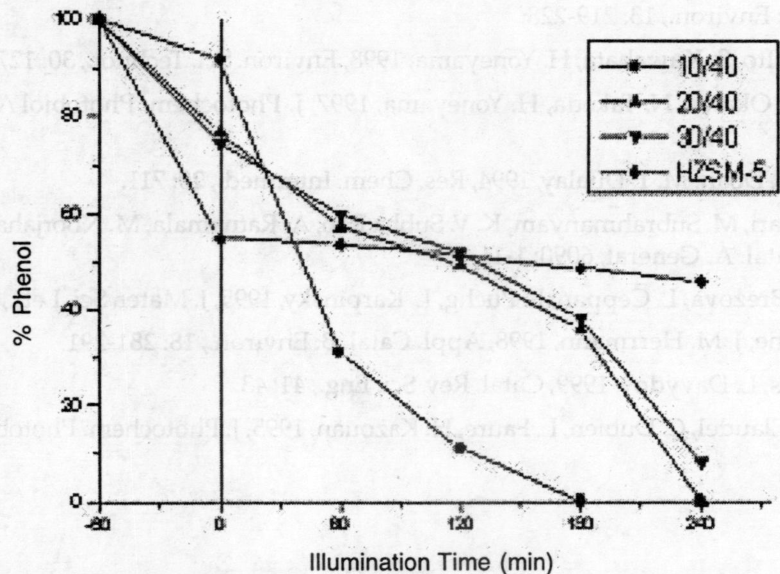

Fig. 3: Photocatalytic Degradation of Phenol (10^{-4} M) on TiO_2 + HZSM-5 Immobilized Catalysis

■ - 10/40 Ti-ZeEF ; ▲ - 20/40 Ti-ZeEF ;

▼ - 30/40 Ti-ZeEF; ◆ - ZeEF.

Conclusions

The following conclusions are to summarize our work concerning immobilization of TiO_2+ HZSM-5 for photocatalytic degradations of phenol:

1. The system using a immobilization of photocatalyst eliminates the need for filtration after treatment of wastewaters
2. The Ti-ZeEF is an efficient photocatalyst over Ti-EF.
3. The development of an immobilization of Ti-ZeEF on an inexpensive support with high photoactivity.

Acknowledgements

The authors greatly acknowledge Indo-French Center for Promotion of Advanced Research (IFCPAR) / Centre Franco-Indien Pour la Promotion de la Recherché Advance (CEFIPRA) for its financial support (Contract no IFC/2205-2).

REFERENCES

1. J. A. Byrne, B. R. Eggins, N. M. D. Brown, B. Mc. Kinney, M. Rouse, 1998, Appl. Catal. B: Environ 17: 25-36.
2. Anita Rachel, M. Subrahmanyam, Pierre Boule, 2002, Appl. Catal. B: Environ., 990: 1-8.
3. R. L. Pozo, M. A. Batanas, A. E. Cassano, 1997, Catal Today, 39: 219-231.
4. J. M. Herrmann, H. Tahiri, Y. Ait-Ichou, G. Lassaletta, A. R. Gonzalez-Elipe, A. Fernandez, 1997, Appl. Catal. B: Environ., 13: 219-228.
5. T. Torimoto, S. Ito, S. Kuwabata, H. Yoneyama, 1998, Environ. Sci. Technol., 30: 1275.
6. T. Torimoto, Y. Okawa, N. Takeda, H. Yoneyama, 1997, J. Photochem. Photobiol A: Chem., 103: 153-157.
7. M. A. Fox, K. E. Doan, M. T. Dualay, 1994, Res. Chem. Intermed., 20: 711.
8. V. Durga Kumari, M. Subrahmanyam, K. V. Subba Rao, A. Ratnamala, M. Noorjahan, K. Tanaka, 2002, Appl. Catal. A. General, 6090: 1-11.
9. M. Mikula, V. Brezova, L. Ceppan, L. Puchg, L. Karpinsky, 1995, J. Mater. Sci. Lett., 14: 615-616.
10. J. Matos, J. Laine, J. M. Herrmann, 1998, Appl. Catal. B: Environ., 18: 281-291.
11. P. G. Smirniotis, L. Davydov, 1999, Catal. Rev. Sci. Eng., 41: 43.
12. M. Bideau, B. Claudel, C. Dubien, L. Faure, H. Kazouan, 1995, J. Photochem. Photobiol A: Chem., 91: 137-144.

35

Physico-chemical Degradation (Ozonation Chlorination, UV/H_2O_2) of Toxic Pollutants Occurring in Groundwater

ANTONIO LOPEZ

National Research Council, Water Research Institute, Department of Water Chemistry and Technology, Italy

Introduction

In the last decade, the production of synthetic chemicals has continuously grown all over the world. Referring to organic chemicals, about 100,000 different compounds are used worldwide and around 1,000 new chemicals are synthesized every year. Presently, only in USA, over $150 \cdot 10^6$ ton/y are produced.

Such a massive production and use of chemicals is the main cause of environmental pollution including groundwater contamination. In fact, once reached subsoil environment, many synthetic chemicals often behave conservatively becoming a serious threat to public health particularly in countries, like Italy, where groundwater represent the primary water resource for drinking and irrigation purposes.

Referring to organic contaminants, pesticides, herbicides, chlorinated solvents, pharmaceutical drugs and dyes are the pollutants more frequently detected in groundwater.

In Europe, concern about groundwater contamination is continuously growing as proved by the recent "EU Water Framework Directive" (*EEC, 2000*) that confirms for toxic pollutants the same limits fixed by the previous Directive "Quality of Water Intended for Human Consumption" (*EEC, 1980*). For most of the above classes of pollutants, these Directives fix very stringent limits (*e.g.*, pesticides and related residues in drinking water should not exceed 0.1 μg/l for individual pesticides and 0.5 μg/l for total pesticides).

From the technological stand point, limits so stringent can be met only by specific treatments such as activated carbon adsorption, membrane filtration and advanced oxidation processes (*Hyde, 1980; Baier et al., 1987; Glaze et al., 1987; Foster et al., 1992; Lopez et al., 2001*). A few information, however, are available about the actual fate of such pollutants during conventional drinking water treatment steps such as clarification, sand-filtration, and disinfection (*El-Dib and Aly, 1977*). Chemically speaking, the latter step is particularly interesting because the commonly used disinfectants (*i.e.*, Cl_2, ClO_2, O_3, UV/H_2O_2) are all oxidizing substances that, potentially, might react with organic pollutants.

The identification of the resulting oxidation byproducts is a key step as such derivatives might result even more toxic than their parent compounds and could modify the effectiveness of successive treatments (*Lykins et al., 1986*).

On the basis of the above considerations, a long term research has been planned and carried out at the Water Research Institute of the Italian National Research Council to study the reactions of representative groundwater organic pollutants with commonly used disinfectants.

The lecture will essentially deal with the main results obtained studying the reactions between commonly used disinfectants (chlorine, chlorine dioxide, ozone, UV/H_2O_2) and selected pollutants, namely: ametryn (herbicide), isoproturon (herbicide), MMTD (pharmaceutical drug), Uniblu-A (dye), representative of toxic and recalcitrant contaminants actually found in groundwater. Of course, the technological implications of the results will be also discussed

Experimental

All the experimental details concerning the reported results can be find in the provided references.

Results

In the following, the chemical structures of the investigated pollutants, their decays during the reactions with selected disinfectants, the byproducts evolution recorded during such reactions and the degradations pathways proposed after HPLC-MS by-products identification are schematically reported. The discussion of the results will be done during the lecture.

HERBICIDES: methyl,thio-triazines

CH_3 – S; N, N, N; HN – CH_2CH_3; NH – $CH(CH_3)_2$

Chemical Structure of the Herbicide Ametryn

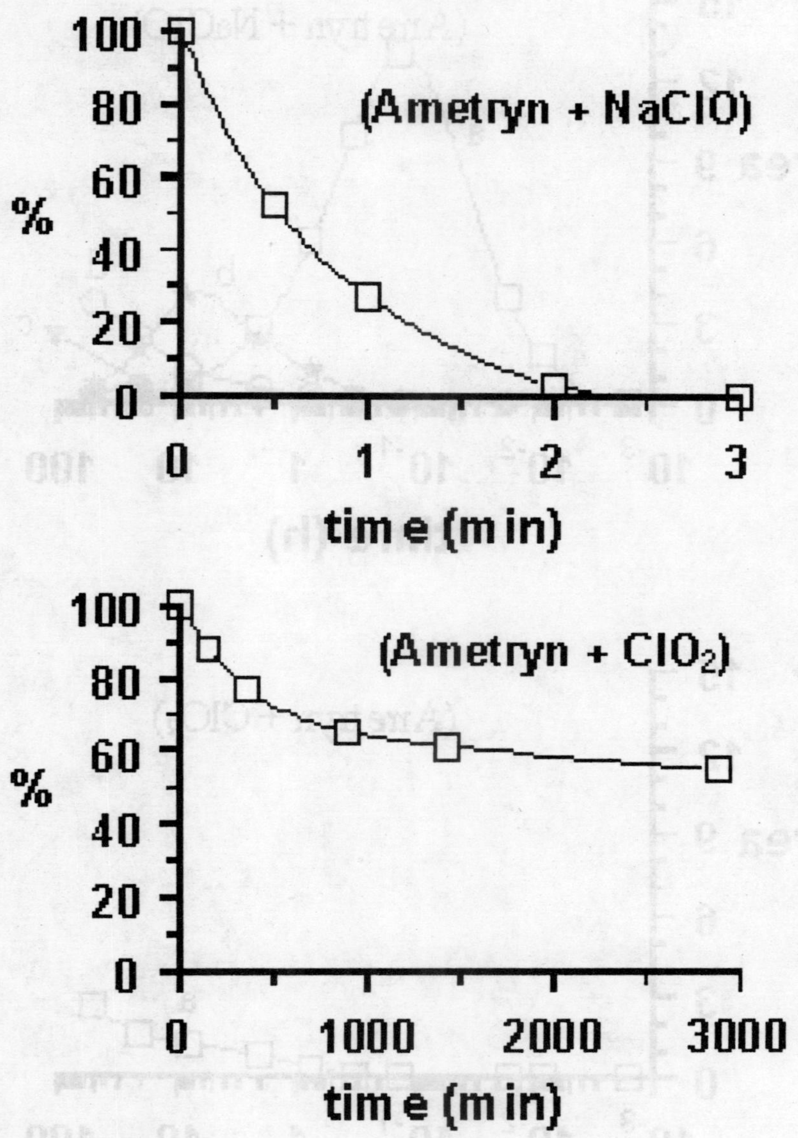

Ametryn Decays During the Reactions with NaClO and ClO_2. $(Ametryn)_o$ = 9 mg/l, $(NaClO)_o = (ClO_2)_o$ = 30 mg/l as Cl, T = 20°C, pH = 7 (Details and Experimental in *Lopez et al., 1997*)

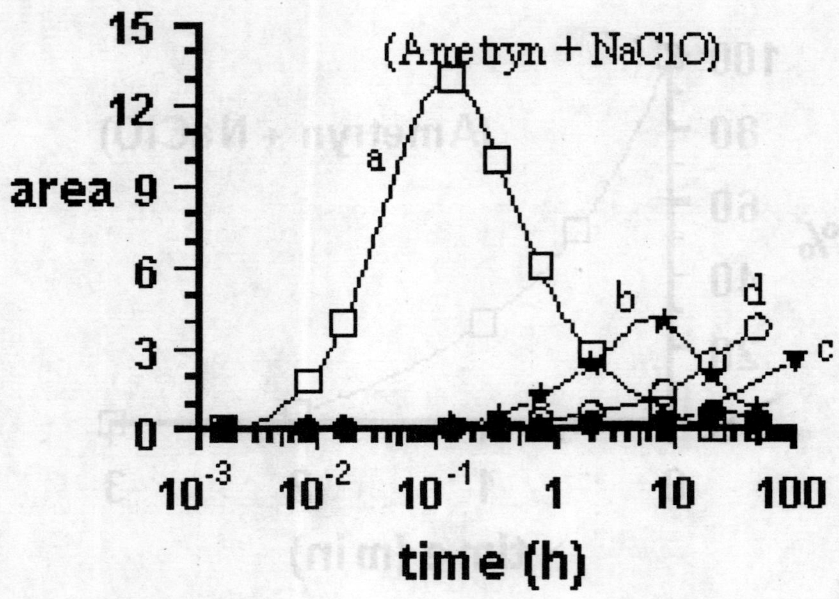

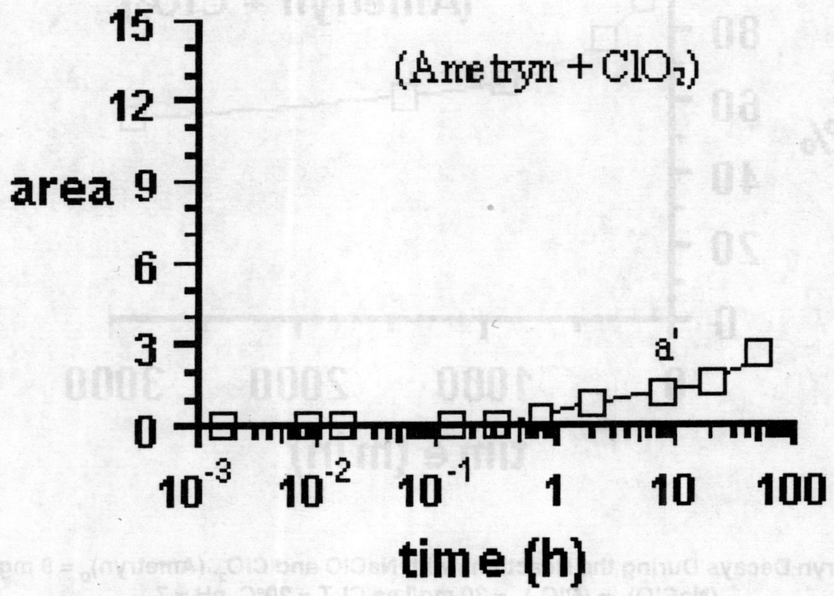

Ametryn Byproducts Trends vs Time
(Details and Experimental in *Lopez et al.*, 1997)

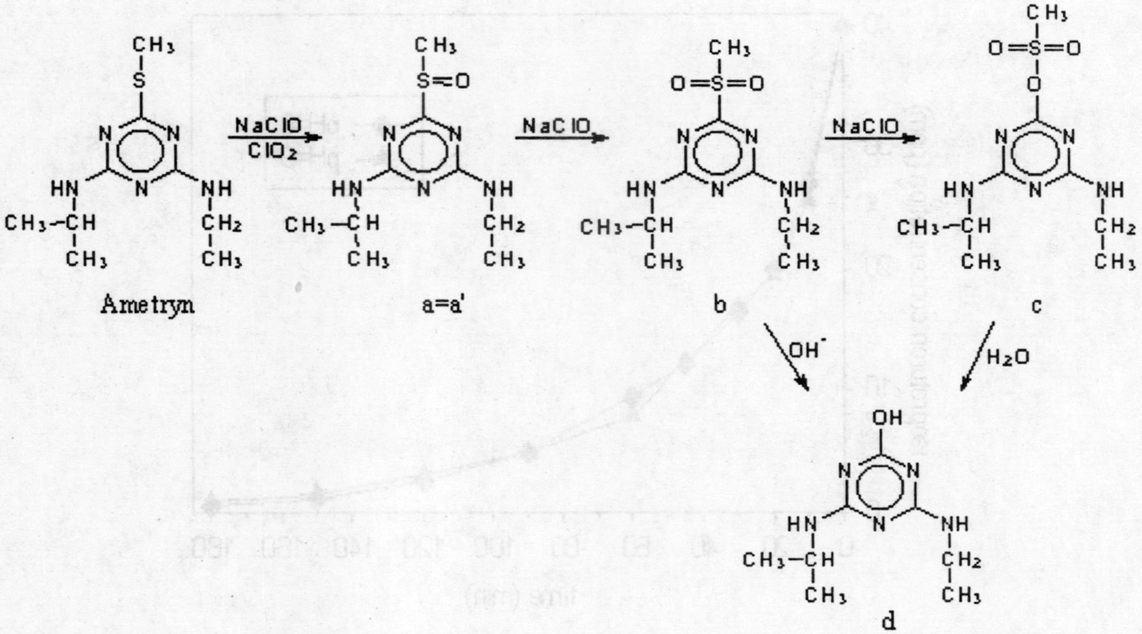

Proposed Pathways for Ametryn Degradation by NaClO and ClO_2
(Details and Experimental in *Lopez et al.,* 1997)

HERBICIDES: aryl,urea-derivatives

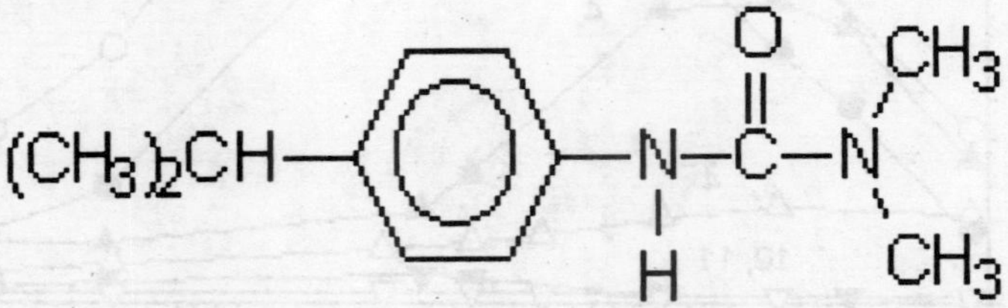

Chemical Structure of the Herbicide Isoptroturon

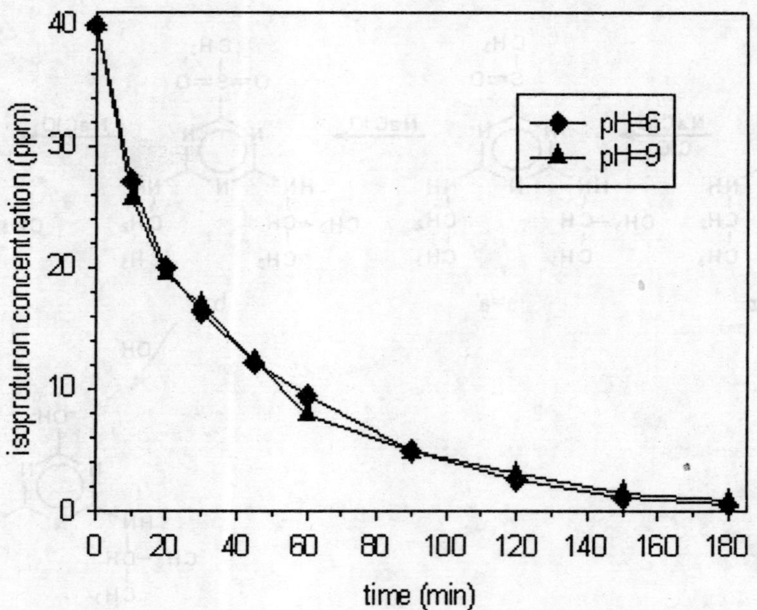

Isoproturon Decays with Hypochlorite $(NaClO)_o$ = 160 mg/l as Cl, T = 20°C
(Details and Experimental in *Mascolo et al.,* 2001a)

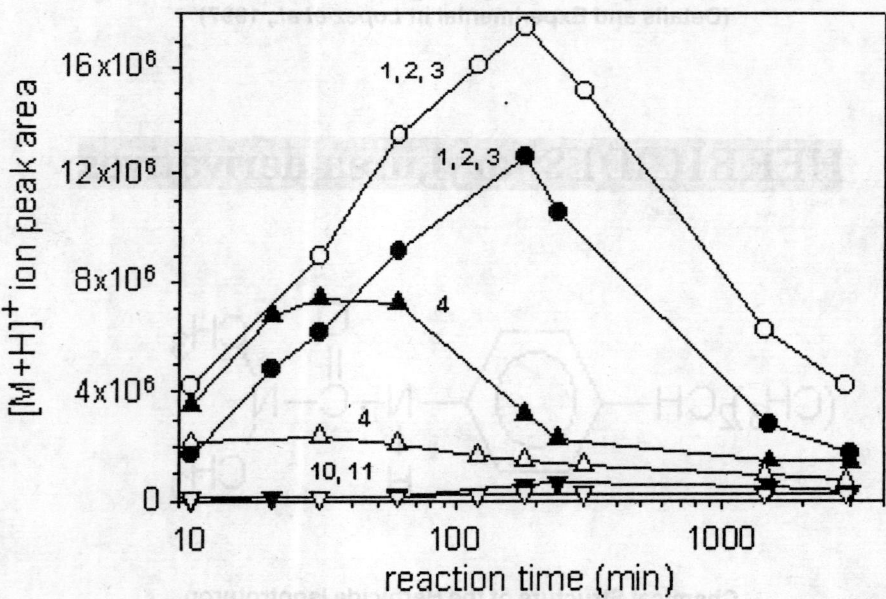

By-products Evolution During the Reaction of Isoproturon with HClO at pH = 6 (full symbols) and at pH = 9 (open symbols).
The curve numbers refer to detected compounds whose identification numbers are those reported for the chemical structures in the following figure.
$(Isoproturon)_o$ = 400 mg/l, $(NaClO)_o$ = 160 mg/l
(Details and Experimental in *Mascolo et al.,* 2001a)

isoproturon

NaClO

1 2, 3 4

5 6, 7, 8 9

ring opening

10, 11

Proposed Pathway for Isoproturon Degradation by Hypochlorite (Details and Experimental in *Mascolo et al.*, 2001a)

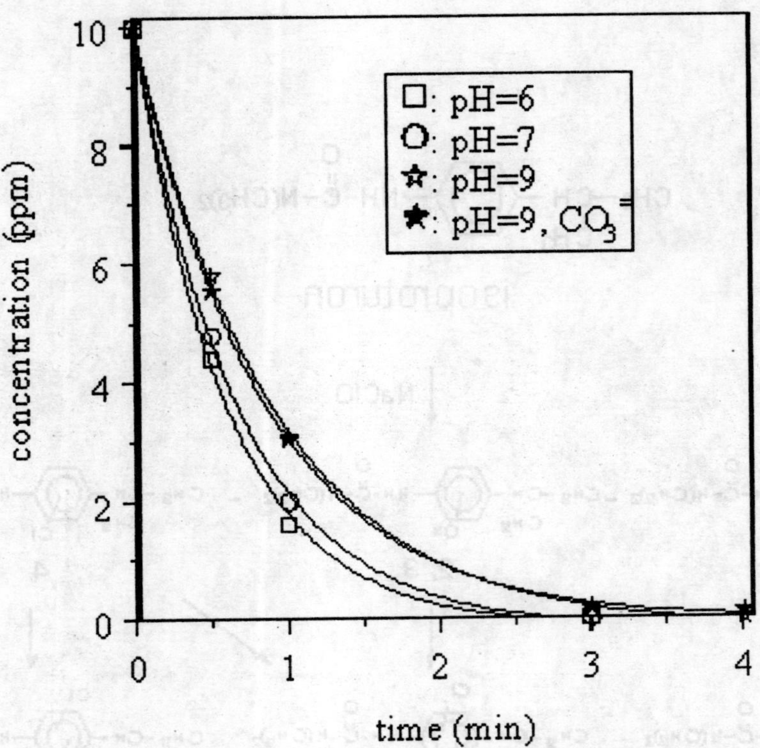

Isoproturon Decays During with Ozonation
T = 20°C, $(\text{Isoproturon})_o$ = 10 mg/l, $(\text{Ozone}_{\text{in air}})_o$ = 90 mg/l at 1 l/min.
(Details and Experimental in *Mascolo et al.*, 2001b)

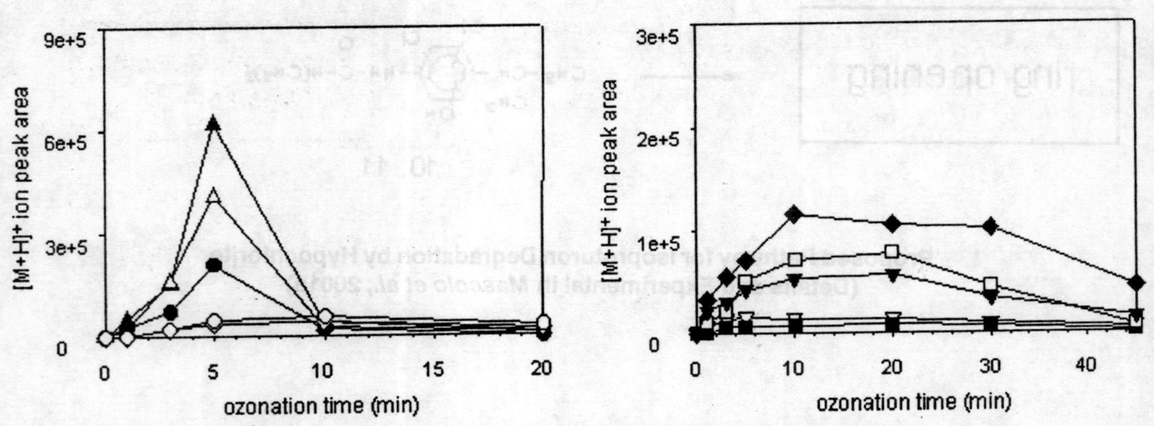

Evolution of Isoproturon By-products During Ozonation at pH 7
$(\text{Isoproturon})_o$ = 0.001 mg/l, $(\text{Ozone}_{\text{in air}})_o$ = 9 mg/l at 0.1 l/min.
(Details and Experimental in *Mascolo et al.*, 2001b)

Proposed Pathway for Isoproturon Degradation by Ozone (Details and Experimental in *Mascolo et al.*, 2001b)

SYNTHETIC DYES

O NH$_2$ SO$_3$Na

O NH— SO$_2$CH=CH$_2$

Uniblu-A

cellulose/OH$^-$

O NH$_2$ SO$_3$Na + O NH$_2$ SO$_3$Na

O NH— SO$_2$CH$_2$CH$_2$OH O NH— SO$_2$CH$_2$CH$_2$O—cellulose

Uniblu-OH Uniblu-cellulose

The above scheme shows the structure of the dye Uniblu-A, that of its hydrolyzed form (Uniblue-OH), and the how this latter binds fibres like cellulose

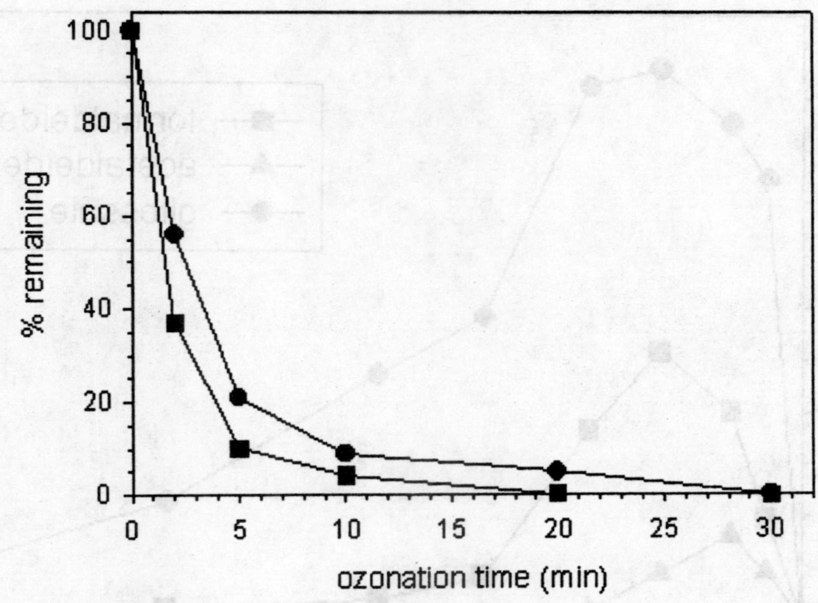

Uniblu-OH Decay During Reaction with Ozone Without ( ) and with ( ) t-butanol (50 nM), Initial Reagents Concentrations: Uniblu-OH = 40 mg/l, Ozone in the Feed Gas (Bubbled at 0.1 l/min) = 16 mg/l (Details and Experimental in *Mascolo et al.*, 2002)

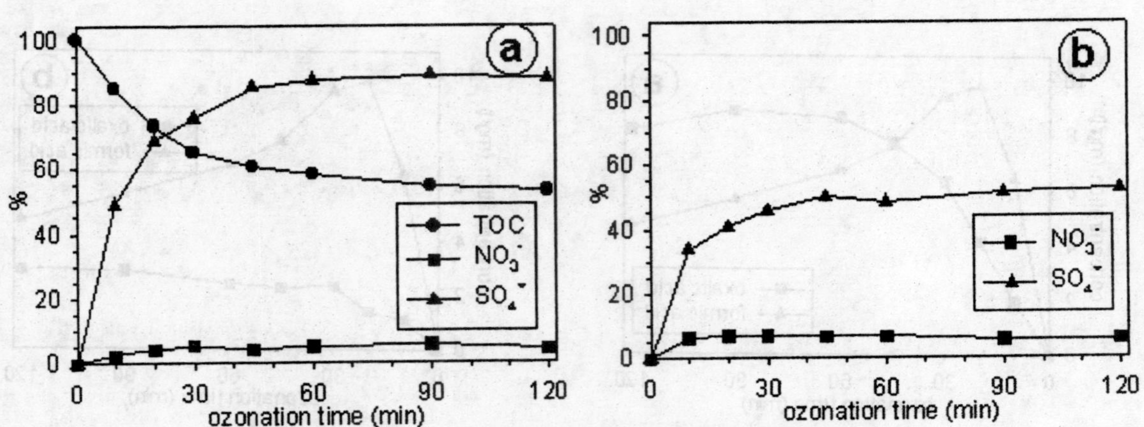

TOC Decrease, SO_4^- and NO_3^- Formation
All expressed as % respect to the theoretical value, during Uniblu-OH degradation by ozone without (a) and with (b) t-butanol (50 mM). Initial reagents concentrations: Uniblu-OH = 40 mg/l, Ozone in the geed gas (bubbled at 0.1 l/min) = 16 mg/l (Details and Experimental in *Mascolo et al.*, 2002)

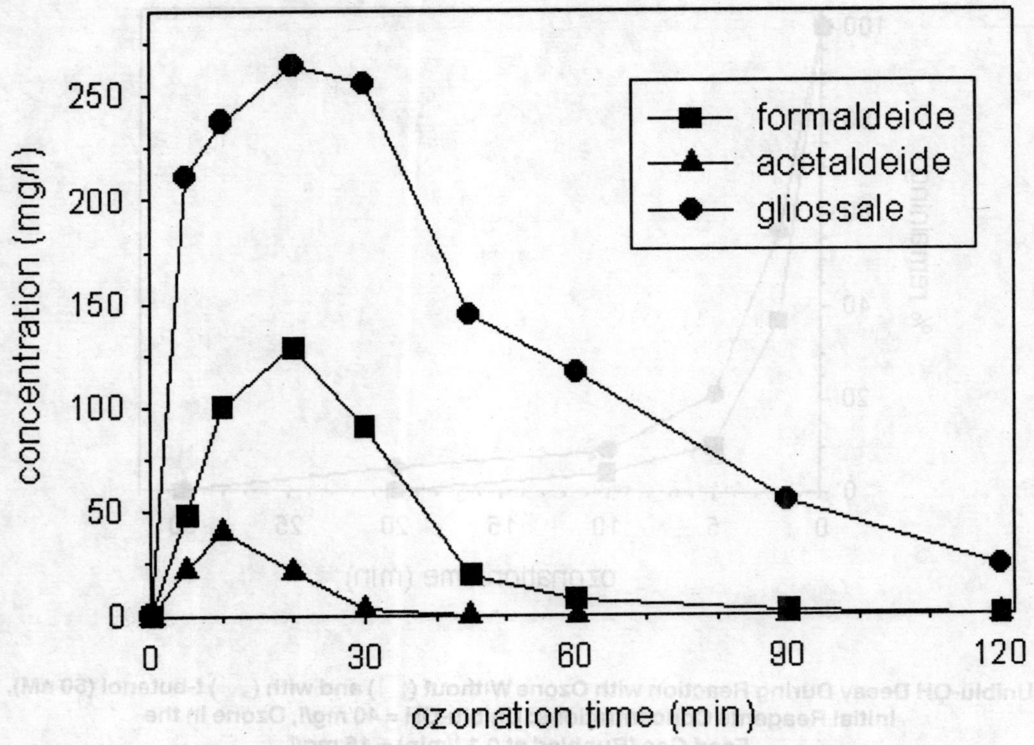

Low Molecular Weight Aldehydes Formation During Uniblu-OH Ozonation Initial reagents concentrations: Uniblu-OH = 40 mg/l, Ozone in the geed gas (bubbled at 0.1 l/min) = 16 mg/l (Details and Experimental in *Mascolo et al.,* 2002)

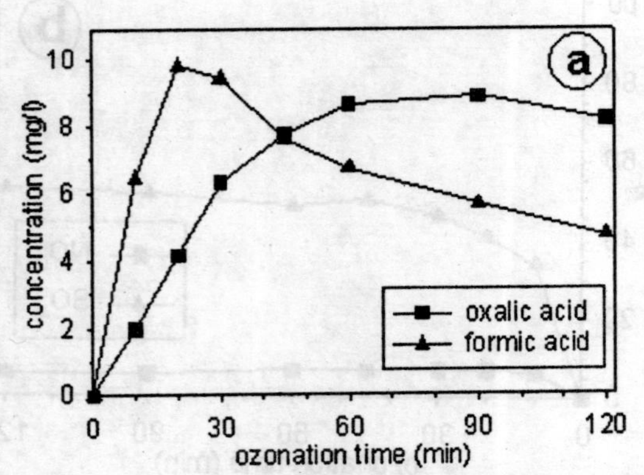

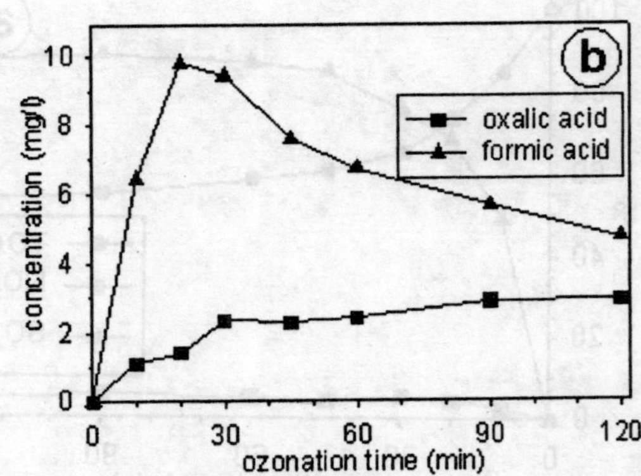

Low Molecular Weight Organic Acids Formation During Uniblu-OH Ozonation without (a) and with (b) t-butanol (50 mM). Initial reagents concentrations: Uniblu-OH = 40 mg/l, Ozone in the geed gas (bubbled at 0.1 l/min) = 16 mg/l (Details and Experimental in *Mascolo et al.,* 2002)

Chemical Structure	*(M-H)*$^-$	*Formula*
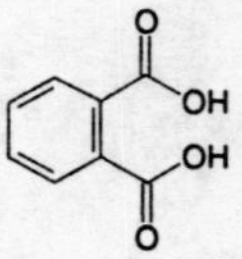	165	$C_8H_6O_4$
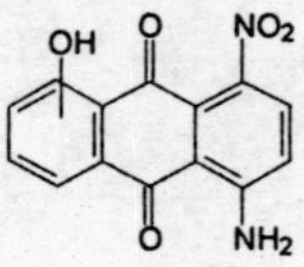	283	$C_{14}H_8N_2O_5$
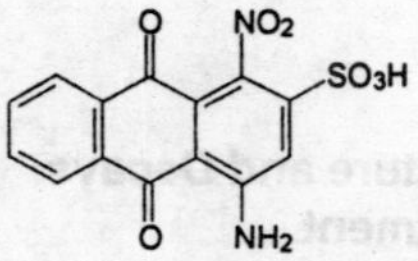	347	$C_{14}H_8N_2O_7S$
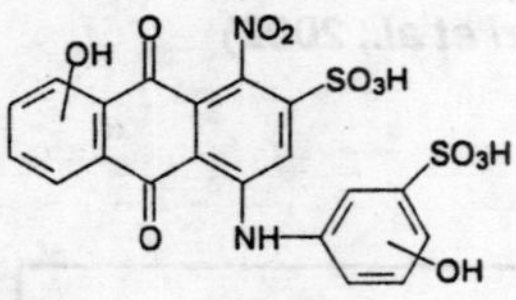	535	$C_{20}H_{12}N_2O_{12}S_2$
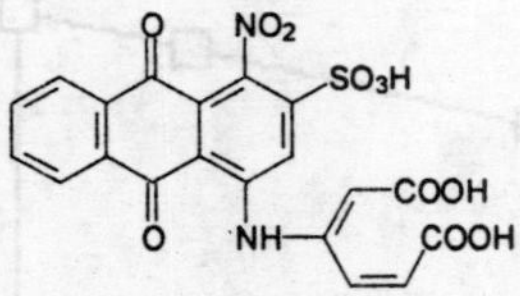	487	$C_{20}H_{12}N_2O_{11}S$
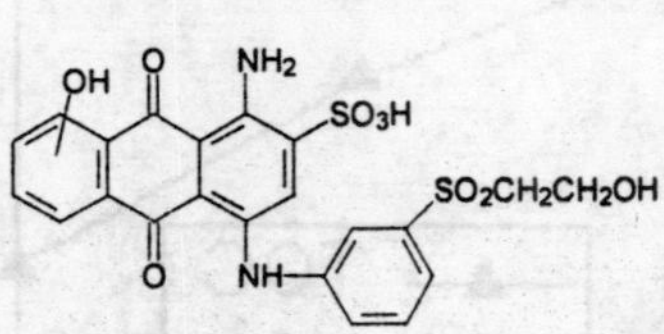	517	$C_{22}H_{18}N_2O_9S_2$
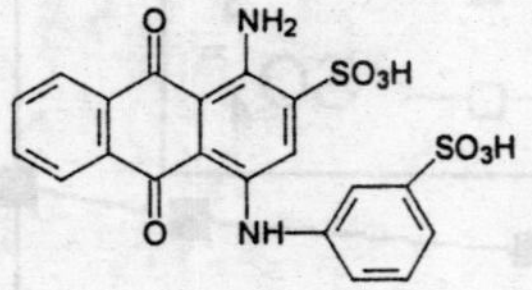	473	$C_{20}H_{14}N_2O_8S_2$
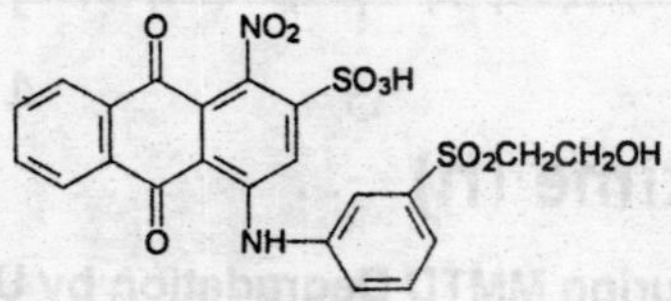	531	$C_{22}H_{16}N_2O_{10}S_2$

Chemical structures, (M-H)$^-$ molecular ions weights and molecular formula of the main by-products identified during the reaction between Uniblu-OH and ozone. (Details and experimental in *Mascolo et al., 2002*)

PHARMACEUTICAL DRUGS

MMTD (5-methyl-1,34-thiadizole-2-thiol) Structure and Decays During H_2O_2, UV and UV/H_2O_2 Treatment
$(H_2O_2)_o$ (when used) = 26 mg/l, UV Hg lamp: 17 W-low pressure
(Details and Experimental in *Bozzi et al.*, 2002)

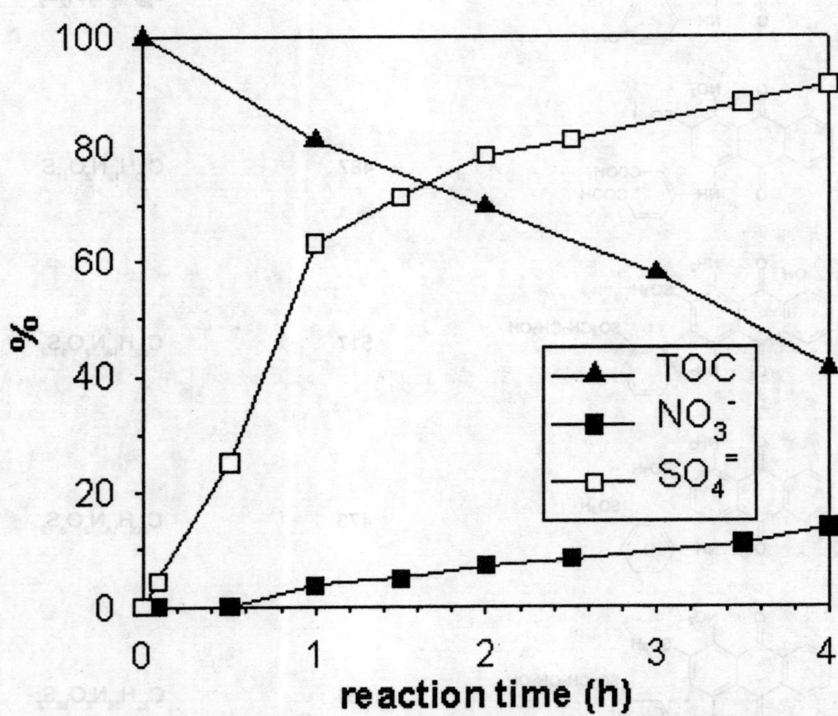

TOC Decrease, $SO_4^=$ and NO_3^- Formation During MMTD Degradation by UV/H_2O_2
All expressed as % with respect to the theoretical value.
Experimental conditions: 17W- UV-low pressure mercury lamp,
$(MMTD)_o$ = 50 mg/l, $(H_2O_2)_o$ = 1.3 g/l
(Details and Experimental in *Bozzi et al.*, 2002)

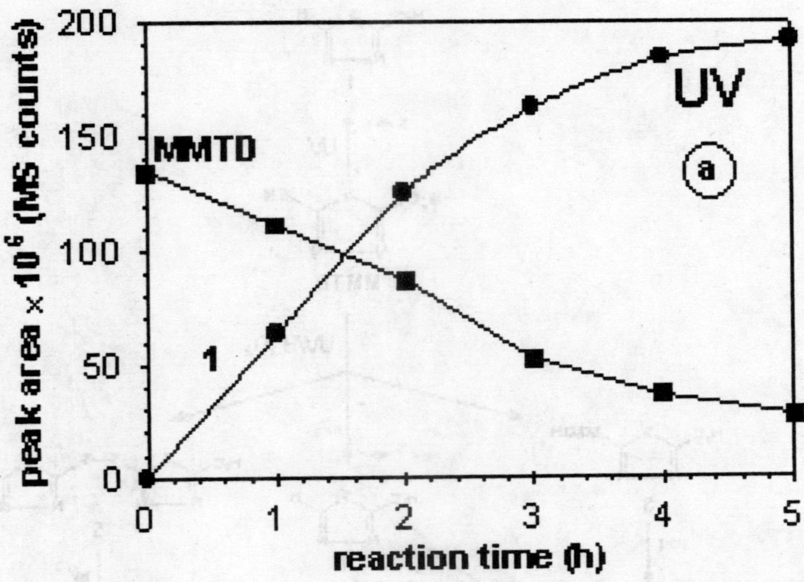

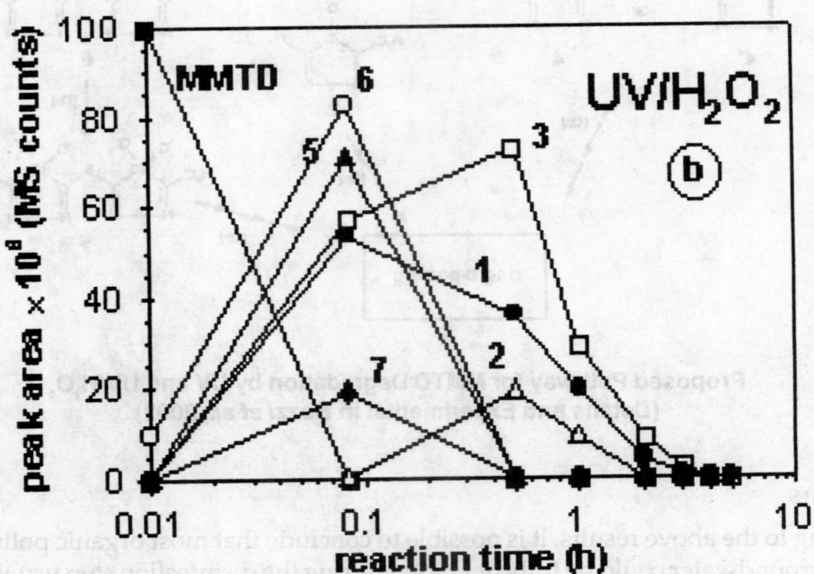

Evaluation of MMTD by-products during the treatment with UV (a) and UV/H_2O_2 (b). The curve numbers refere to the identified compounds numbered in Figure. Experimental conditions: low pressure mercury lamp, $(MMTD)_0$=50 mg/L, (H_2O_2) 0 = 1.3 g/L, (Details and experimental in Bozzi *et al.*).

Proposed Pathway for MMTD Degradation by UV and UV/H_2O_2
(Details and Experimental in *Bozzi et al.*, 2002)

Conclusions

According to the above results, it is possible to conclude that most organic pollutants presently occurring in groundwater could be fully degraded during the disinfection step usually carried out at drinking water treatment plants. This result could have significant implications as for technological issues concerning drinking water treatments. However, the achievement of pollutants full degradation it is not enough. In fact, very often, the complete degradation of target pollutants gives rise to the formation of several by-products.

Whether the toxicity of such by-products is lower or greater than that of the parent compounds is a question of great relevance that can be assessed only after by-products identification has been carried out.

Of course, further studies are necessary to evaluate the health effects of such by-products as well as the of such an approach under experimental conditions actually occurring at real drinking water facilities.

REFERENCES

1. Baier, J. H., Lykins, B. W., Fronk, C. A. and Kramer, S. J., 1987, Using reverse osmosis to remove agricultural chemicals from groundwater. *J. Am. Wat. Wks. Ass.*, 79, 55-60.
2. Bozzi A., Lopez A., Mascolo G., Ciannarella R., Passino R., 2002, "UV and H2O2/UV degradation of a pharmaceutical intermediate in aqueous solution", *Annali di Chimica*, 92, 1-2, 41-51.
3. EEC, 1980, Quality of Water Intended for Human Consumption. *Directive* 80/778.
4. EEC, 2000, EU Water Framework Directive. *Directive* 2000/60/EC.
5. El Dib M.A. and Aly O.A., 1977, Removal of phenylamide pesticides from drinking water I. Effect of chemical coagulation and oxidants. *Wat. Res.*, 11, 611-616.
6. Foster D.M., Rachwal A.J. and White S.L., 1992, Advanced treatment for the removal of atrazine and other pesticides. *Water Supply*, 10, 133-146.
7. Glaze, W. H., Kang, J. W. and Chapin, D. H., 1987, The chemistry of water treatment processes involving ozone, hydrogen peroxide and ultraviolet radiation. *Ozone Sci. Engg.*, 9, 335-352.
8. Hyde, R. A., 1980, Removal of trace organics from water-1. Adsorption of haloforms and chlorinated pesticides by granulated activated carbon. *Water Research Centre Technical Report 136*.
9. Lopez A., Kiwi J., 2001, "Modeling the performance of innovative membrane based reactor. Abatement of azo-dye (Orange II) up to biocompatibility", *Ind. & Engin. Chem. Res.*, 40, 1852-1858.
10. Lopez A., Mascolo G., Tiravanti G., Passino R., 1997, "Degradation of herbicides (ametryn and isoproturon) during water disinfection by means of two oxidants (hypochlorite and chlorine dioxide), *Wat.Sci.Tech*, 35, No.4, 129-136.
11. Lynkins B.W., Koffskey W.E. and Miller R.G., 1986, Chemical products and toxicological effects of disinfection. *J. Am. Wat. Wks. Ass.*, 78 (11), 66-75.
12. Mascolo G., Lopez A., Bozzi A., Sonnante G., Tiravanti G., 2002, "By-products formation during the ozonation of the reactive-dye Uniblue-A", *Ozone Science & Eng.* (in press)
13. Mascolo G., Lopez A., H. James, M. Fielding, 2001b, "By-products formation during degradation of isoproturon in aqueous solution. I: Ozonation", *Wat.Res.*, 35/7, 1695-1704.
14. Mascolo G., Lopez A., H. James, M. Fielding, 2001a, "By-products formation during degradation of isoproturon in aqueous solution.II:Chlorination", *Wat.Res.*, 35/7, 1705-1713.

36

Photocatalytic Removal of Arsenic from Water Environment

A.K. DIKSHIT, M. BANDYOPADHYAY & J.C. SAHA

Department of Civil Engineering, Indian Institute of Technology, Kharagpur - 721302

ABSTRACT

In the present study, the effectiveness of semiconductor photocatalysis, an advanced physico-chemical process, has been evaluated to remove arsenic from water environment. It was found that 80% of As (III) could removed from water using TiO_2 as a catalyst in the presence of sunlight. This paper discusses the various aspects of this emerging technology.

Introduction

In developing countries, arsenic is found to be a potent pollutant entering into the environment through different types of industrial wastes from pesticides and insecticides manufacturing, copper smelting, thermal power plants, weathering of rocks, volcanic activity, smelting operation etc. In many places, groundwater gets contaminated through naturally occurring arsenic ores. But the removal of the same from the water environment is not so easy. Removal of arsenic from aquatic environment got special attention due to its well-known toxic effects on flora and fauna, and the stringent water and wastewater standards. Though, the conventional methods such as physical, chemical and biological treatment are able to degrade most of the organic pollutants, these techniques mostly transfer the pollutants from one phase to another besides consuming enormous resources such as chemicals and non-renewable energy. Hence, any technique based on renewable energy source, like solar energy will be a welcome option for developing countries, where solar energy is available in all parts of the country.

Semiconductor photocatalysis appears to be a promising technology that has a number of applications in environmental system such as air purification, water purification, water disinfection, and hazardous waste remediation. In addition, the basic research that underlies the application of this technology is forging a new understanding of the complex heterogeneous photochemistry of metal oxide system in multiphasic environments.

Solar photosensitizer technology is one such method, where the semiconductor photo-sensitizers like TiO_2 act as catalyst for irreversible oxidation of impurities present in water using sunlight. Catalytic TiO_2 has strong photocatalytic oxidation properties and good stability over repeated runs. It is also inexpensive and non-toxic to the ecological environment (Mills, 1993).

The present study outlines the performance evaluation of semiconductor photocatalysis to remove arsenic from water environment using catalytic TiO_2.

Materials

Three plastic trays of dimension 40 × 30 × 8 cm were used for the experiments. All the working arsenic solutions were prepared in freshly prepared double distilled water. All chemicals were analytical grade and were used without further purification. As (III) stock solution was prepared from diluting $NaAsO_2$ (E-Merck, Germany) from 1000 mg/l in double distilled water to 100 mg/l. Prior to each analysis, intermediate standard As (III) solutions were diluted with distilled water to 10 mg/l from the arsenic stock solutions. Secondary standard solutions of 1 mg/l were also freshly prepared for each experiment from the 10 mg/l intermediate stock solutions.

Analysis of Arsenic

Sample has been prepared using standard method (APHA-AWWA-WPCF, 1989). Arsenic was reduced to arsine by zinc powder in hydrochloric acid and produced arsine reacted with absorber solution containing SDDC reagent (which was prepared by silverdiethyldithiocarbamate and morpholine dissolved in chloroform). A red colour complex was developed by the above reaction, which is suitable for photometric measurement. The intensity of the colour complex was measured at 520 nm with digital Spectrophotometer [UV-160A, Shimadzu, Japan)] equipped with 1 cm quartz cells. This method provides reliable data for total inorganic arsenic in fairly polluted waters.

Methodology

Arsenic compounds are found sensitive to photochemical transformation on semiconductor surfaces. Semiconductors (*e.g.* TiO_2, ZnO, CdS, ZnS, Fe_2O_3) can act as sensitizers for light-reduced redox process due to their electron structures characterized by a filled valance band and an empty conduction band. Experiments have been conducted with As (III) at a concentration of 1 mg/l both with and without the sunlight. Cu^o was prepared from $CuSO_4$ solution through reduction by sodium borohydride. TiO_2 powder particles were spread over the Cu^o particles. This compound was then spread on the arsenic contaminated water and kept for 2-3 hr in the sunlight and similar set of experiments was conducted in dark place.

Results and Discussion

Results show that 80% arsenic as As (III) was removed from the samples which were under sunlight whereas only 20 % As (III) was removed from the samples kept in dark. At the same time, there was no appreciable reduction in arsenic concentration without TiO_2 in the presence of sunlight. Thus further thorough investigations are warranted on this technology using alternative semiconductors to achieve higher removal efficiency.

Conclusions

Semiconductor photocatalysis using TiO_2 was found to remove up to 80% of As (III) from water containing initially 1 mg/l of As (III). This advanced treatment method is also suitable for both wastewater and waste sludges.

REFERENCES

1. Mills, A., 1993, Water purification by semiconductor photocatalysis. *Chem. Soc. Rev.*, 22 (6), pp. 417-425.

2. APHA-AWWA-WPCF, 1989, *Standard methods for the examination of water and wastewater*. American Public Health Association, NY.

37

Applications of Synthetic Polyelectrolytes (Anionic/ Cationic) for Treatment of Herbal Pharmaceutical Wastewater

A.P. VANERKAR, SHANTA SATYANARAYAN, D.M. DHARMADHIKARI & S.N. KAUL

National Environmental Engineering Research Institute, Nehru Marg, Nagpur

ABSTRACT

India possesses the longest and the most glorious traditions in the field of medicine. Indians are accustomed to use the herbal drugs synthesized by the Ayurvedic system of medicine. Ayurveda, the science of long life has a history, which can be traced back to 4500 to 1600 BC. According to World Health Organization (WHO), considerable percentage of the people in developing and developed world are using medicinal plant remedies, and the number is on increase. Ayurvedic system of Medicine uses mostly herbs and its derivatives to cure diseases.

In India, there is a network of small and medium scale industries manufacturing herbal pharmaceutical products based on Ayurvedic system. Around, 3000 industries are distributed in all the segments of the country. An increase in the use of these herbal originated drugs have been reported. Pharmaceutical wastewater has a wide range of pollution characteristics and thus poses great problem in its disposal.

The work presented in this paper describes the application of synthetic polyelectrolytes (anionic/cationic) along with conventional coagulants (alum and lime) for treatment of wastewater from these industries. Present investigation reveals that, the need of pretreatment of the wastewater to reduce high load of

BOD, COD and SS is significantly catered by using synthetic polyelectrolytes (anionic/ cationic) coupled with conventional coagulants. Further, this pretreatment is useful for effective biological treatment. Lime and Alum dosages in the range of 100-900 mg/L were used independently and also in combination of synthetic polyelectrolytes (anionic/cationic).

It has been found that, alum dose of 300 mg/L resulted in the reduction of 26.80% (COD), 35.20% (BOD) and 50.30% (SS). However, the same dose of Alum in combination with synthetic polyelectrolytes (anionic/cationic) (0.3 mg/L) results in the reduction of 58% (COD), 64.39% (BOD) and 71.32% (SS).

Similar degree of reduction in COD, BOD and SS have been found during the use of lime in combination with synthetic polyelectrolyte (anionic/cationic).

Keywords: *Herbal Pharmaceutical, Ayurvedic Medicine, Coagulation/Flocculation, Synthetic Polyelectrolyte (anionic/ cationic)*

Introduction

Of all the sciences, India possesses the longest and most glorious tradition in the field of medicine. Ayurveda, the Science of long life, body and mind, is said to have divine origin. Based upon equilibrium (health) or dis-equilibrium (disease) of humors (Vata, Pitta and Kapha), aiming not only at the cure of disease but also at preserving health of the normal population, the Ayurveda has an integrated view of man, his body and mind, as also of surroundings which include the plants, minerals and animals, uses mostly medicinal plants in diseases to bring back the body to equilibrium.

The sages of India mastered an un-parallel knowledge of medicinal plants. "Rig Veda" was followed by "Atharva Veda" in which various medico-religious uses of plants have been given. Over the years, the traditional practice or system of health care is more commonly known as 'Ayurveda' in India.

In India, around 3000 small and medium scale industries manufacturing herbal pharmaceutical products are distributed in all the segments of the country. According to World Health Organization, in both developed and developing countries a considerable percentage of the people use medicinal plant remedies and the number is on the increase, especially among younger people (1-5).

Herbal pharmaceutical wastewater, being highly polluted in terms of 26, 000 mg/L (COD), 15, 660 mg/L (BOD), 7, 370 mg/L (SS) and 42 mg/L (sulphides), it should not be discharged without treatment. The countrywide scenario of herbal pharmaceutical industries reveals that the products are produced on batch basis. The process involves the cleaning of natural herb parts, cut or ground and are further subjected to extraction, trichuraction, fermentation, distillation, decoction and percolation as per the requirement.

The raw material used can be listed as the part of medicinal plants like herbs, skin, stumps, fruits, flowers, root, seeds, gum, oils etc. The synthetic chemicals added are alcohol, sugar, gelatin, lactose, solvents and special type of minerals and clays. Thus, due to the variation in the types of medicines, manufactured and the raw materials used, the wastewater characteristics, varies as per the process. Also the wastewater flow fluctuates, depending on the process. Because of this, the conventional biological treatment units, which employ anaerobic filter or activated sludge process (ASP) or trickling filters usually malfunction. Since the stringent effluent requirements have to be met during operation of the factory, an alternative treatment process is tried which can be used either independently or in tandem with the conventional treatment units as a pretreatment or a tertiary polishing process. Treatment of herbal pharmaceutical wastewater by biological methods (anaerobic) has been reported in the literature, (6-8).

One of the treatment technologies, which seemed viable for the reduction of pollution problems of herbal pharmaceutical industry is physico-chemical treatment. Literature survey reveals that, studies have been reported using higher dosages of coagulants. However this treatment proposal is not cost effective (9). Some workers reported that herbal pharmaceutical waste, when treated by physicochemical techniques result in moderate removals. According to their studies, alum with polyelectrolyte gave better removals (10).

Materials and Methods

Choice of Industry

Herbal pharmaceutical industry was selected for this study. The raw water consumption of the industry ranged between 17 to 20 m^3/day and about 95% of this quantity is generated as wastewater having high COD, BOD and SS load.

Characterization of Wastewater

The characterization of herbal pharmaceutical wastewater, was carried out on composite samples of raw wastewater from one of the herbal pharmaceutical industries in Maharashtra state. The raw and neutralized wastewater was characterized for detail physico-chemical parameters. The analysis was carried out as per the standard methods for water and wastewater (11). The physico-chemical characteristics are shown in Table 1.

Physico-chemical Treatment

The physico-chemical treatment of the wastewater was carried out using jar test apparatus. Experimental studies were carried out using one litre wastewater sample. While using the jar test apparatus, initially the samples were flash mixed for one minute at 100 rpm and later at 40 rpm for half an hour. After half an hour of mixing, the samples were settled for 30 minutes and the supernatant liquid was drawn and subjected to various parameters and the settled sludge volume was also noted.

The conventional coagulants, lime and alum were used independently and optimum dose essential for coagulation was determined by conducting jar-tests. The treatability studies with these coagulants were also carried in combination with synthetic polyelectrolyte (anionic & cationic) and optimum dose of polyelectrolyte was evaluated by repeating the experiments several times.

Results and Discussion

The herbal pharmaceutical wastewater shows large variation in COD, BOD, totals solids, suspended solids, oil and grease, total nitrogen, total phosphates etc., this is due to the different additives used for various drug preparation and also depending on the nature of the products i.e, liquids, solids and semi-solids. The COD, BOD, and SS values of wastewater varied widely and are usually in the range between 5, 000 to 80, 000 mg/L, 4, 300 to 74, 000 mg/L and 900 to 18, 800 mg/L, respectively, as reported in the literature (10).

Table 1: Characteristics of Raw and Neutralized Wastewater from Herbal Pharmaceutical Industry

Sl. No.	*Parameters**	*Raw Wastewater*	*Neutralized Wastewater*
1.	Colour (Visual)	Dark Yellow	Black
2.	Total Acidity	3000	-
3.	pH	3.9-4.2	7.0
4.	Suspended Solids	7370	5406
5.	Total Dissolved Solid	2285	1157
6.	Total Solids	8380	6040
7.	COD	26,000	20,000
8.	BOD	15,660	11,910
9.	Sulfides as S-2	42	38
10.	Sulphates as SO_4	82	70
11.	Phosphates as PO_4	260	210
12.	Total Nitrogen as N	389	379
13.	Oil and grease	140	85
14.	Sodium as Na	155	130
15.	Potassium as K	140	110
16.	Copper	0.649	0.447
17.	Iron	65.70	42.60
18.	Magnesium	3.22	2.10

* All the values are expressed in mg/L except pH and colour.

Studies on the effectiveness of lime and alum individually and in combination with addition of different synthetic (anionic/cationic) polyelectrolytes were carried out using different dosages. The detailed studies involved two stages treatment *i.e.* First stage neutralization with lime followed by other coagulants including lime.

Lime being the most cost effective chemical, it was studied first. Initially the pH of the wastewater was adjusted to 7, which was in the range of 3.9 to 4.2. Lime dose of 4.2 mg/L was required to bring the pH to 7.0. The wastewater was further subjected to chemical treatment (second stage) using different coagulants.

Studies were carried out using Lime dose in the range of 100-900 mg/L. At 200 mg/L of lime dose COD, BOD and SS reductions of 25.20%, 32.90% and 43.0% respectively, was observed. However when the lime dose was increased to 300 mg/L COD, BOD and suspended solids showed appreciable reduction of 32.40% 40.0% and 53.32% respectively. Beyond 300 mg/L, removals were only marginal and thus the dose of 300 mg/L has been selected as optimum and is indicated in Table 2.

Alum dosages tried were in the range of 100 to 900 mg/L as indicated in Table 3. At a dosage of 300 mg/L reductions of COD, BOD and SS achieved were 26.80%, 35.30% and 50.30%, respectively. Beyond this only marginal improvement was observed.

Recently more stress is being given to the use of anionic/cationic synthetic polyelectrolytes instead of individual use of lime and alum as coagulants to further increase flocculation and settling with a view to reduce the cost and quantity of chemical required. Eight different combinations of polyelectrolytes and conventional coagulants like [Alum + Cationic synthetic polyelectrolyte (A)], [Alum + Cationic synthetic polyelectrolyte (B)], [Lime + Cationic synthetic polyelectrolyte (A)], [Lime + Cationic synthetic polyelectrolyte (B)], [Alum + Anionic synthetic polyelectrolyte (A)], [Alum + Anionic synthetic polyelectrolyte (B)], [Lime + Anionic synthetic polyelectrolyte (A)] & [Lime + Anionic synthetic polyelectrolyte (B)] were tried at the optimised dose of coagulants.

Table 2: Treatability of Combined Wastewater from Pharmaceutical Industry (Herbal)
Coagulant: Lime (mg/L)

Parameters	*Effluent Concentration and Coagulant Dose (mg/L)*								
	100	*200*	*300*	*400*	*500*	*600*	*700*	*800*	*900*
PH	7.61	8.12	8.49	8.83	9.21	9.68	9.98	10.52	11.1
Colour	Black	Black	Black	Black	Black	Black	Black	Black	Black
Sulfide	39.6	37.42	37.42	33.06	30.88	30.88	30.88	28.46	28.46
SS	3399	3081	2524	2384	2254	3130	2027	1930	1892
% Removal	37.12	43.0	53.32	55.9	58.3	60.6	62.5	64.3	65.0
BOD	8885	7992	7146	6601	6360	6041	5800	5550	5312
% Removal	25.4	32.9	40.0	44.58	46.6	49.28	51.3	53.4	55.4
COD	15840	14960	13520	12560	12040	11440	11040	10960	10480
% Removal	20.8	25.20	32.4	37.2	39.8	42.8	44.8	45.2	47.6

All values are expressed in mg/L except pH and colour

Optimum dose of Lime and Alum of 300 mg/L in combination with various cationic and anionic synthetic polyelectrolytes (A) & (B) were tried with various doses of synthetic polyelectrolytes (0.1, 0.2, 03 and 0.6 mg/L). Results are depicted in Figures 1 to 8.

Table 3: Treatability of Combined Wastewater from Pharmaceutical Industry (Herbal)
Coagulant: Alum (mg/L)

Parameters	*Effluent Concentration and Coagulant Dose (mg/L)*								
	100	*200*	*300*	*400*	*500*	*600*	*700*	*800*	*900*
pH	7.16	7.12	7.05	7.00	6.91	6.87	6.84	6.81	6.80
Colour	Brown Black	Brown Black	Brown Black	Black	Black	Black	Black	Black	Black
Sulfide	39.6	39.6	37.42	35.24	35.24	33.06	30.88	30.88	30.88
SS	3616	3281	2687	2519	2339	2189	2081	1989	1930
% Removal	33.11	39.3	50.3	53.4	56.8	59.5	61.5	63.2	64.3
BOD	9361	8718	7706	7277	6896	6693	6479	6241	6122
% Removal	21.4	26.8	35.3	38.9	42.1	43.8	45.6	47.6	48.6
COD	16800	16480	14640	13920	13120	12560	12160	11920	11760
% Removal	16.0	17.6	26.8	30.4	34.4	37.2	39.2	40.4	41.2

All values are expressed in mg/L except pH and colour

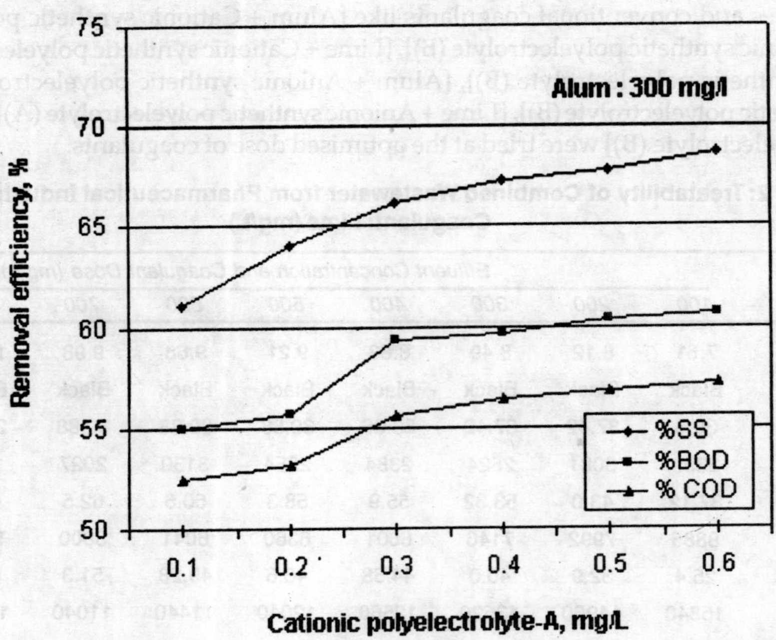

Fig. 1: Effect of Cationic Polyelectrolyte-A and Alum on Organics Removal

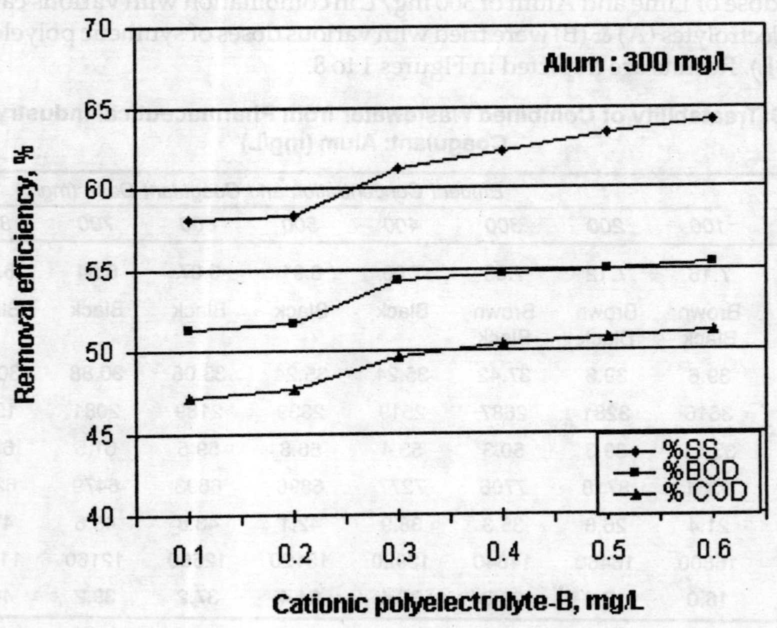

Fig. 2: Effect of Cationic Polyelectrolyte-B and Alum on Organics Removal

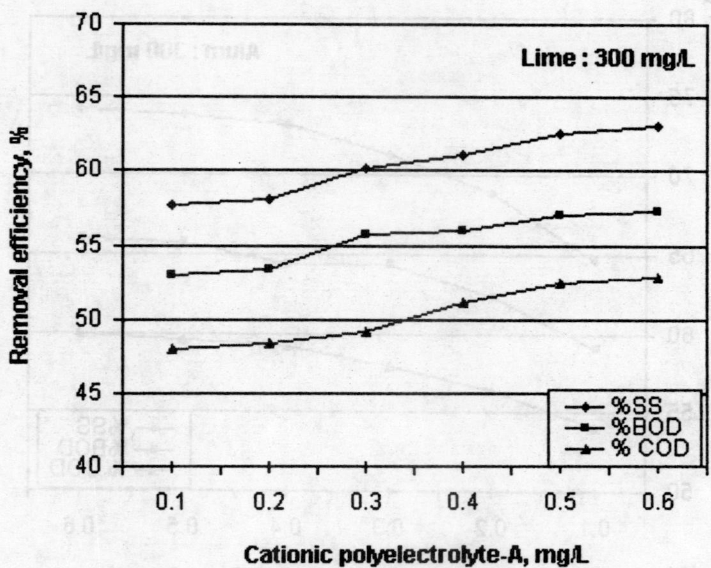

Fig. 2: Effect of Cationic Polyelectrolyte-A and Lime on Organics Removal

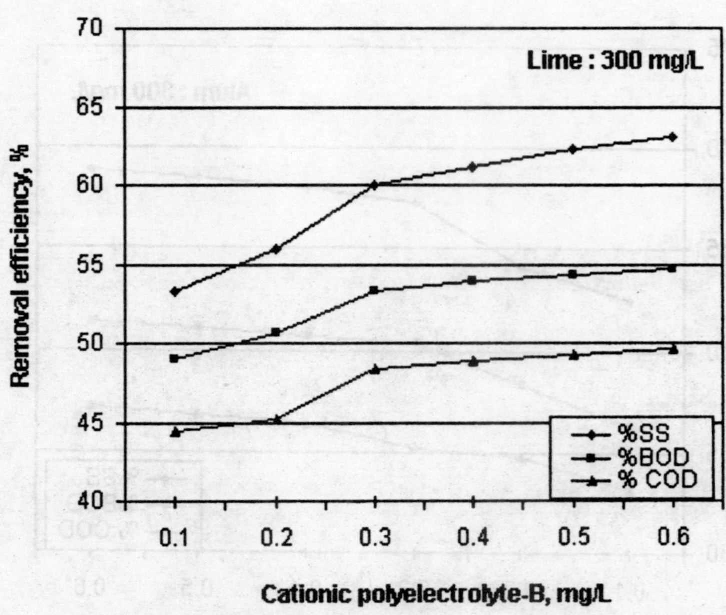

Fig. 4: Effect of Cationic Polyelectrolyte-B and Lime on Organics Removal

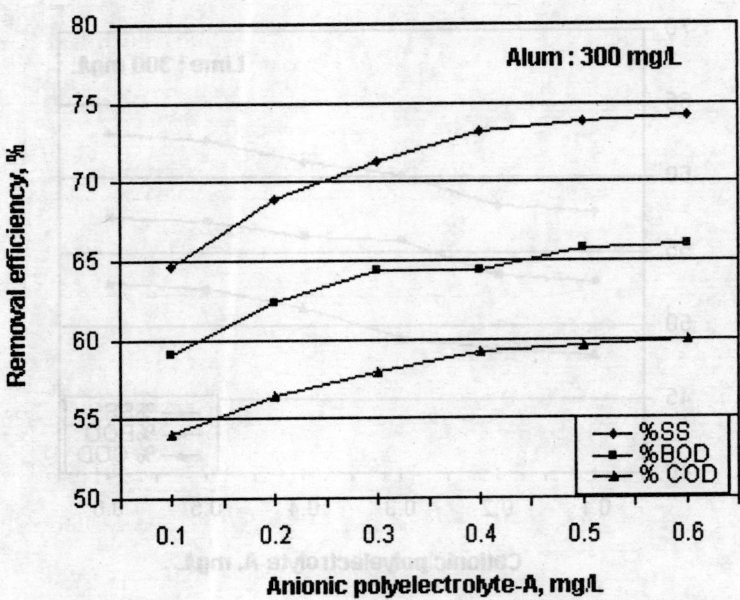

Fig. 5: Effect of Anionic Polyelectrolyte-A and Alum on Organics Removal

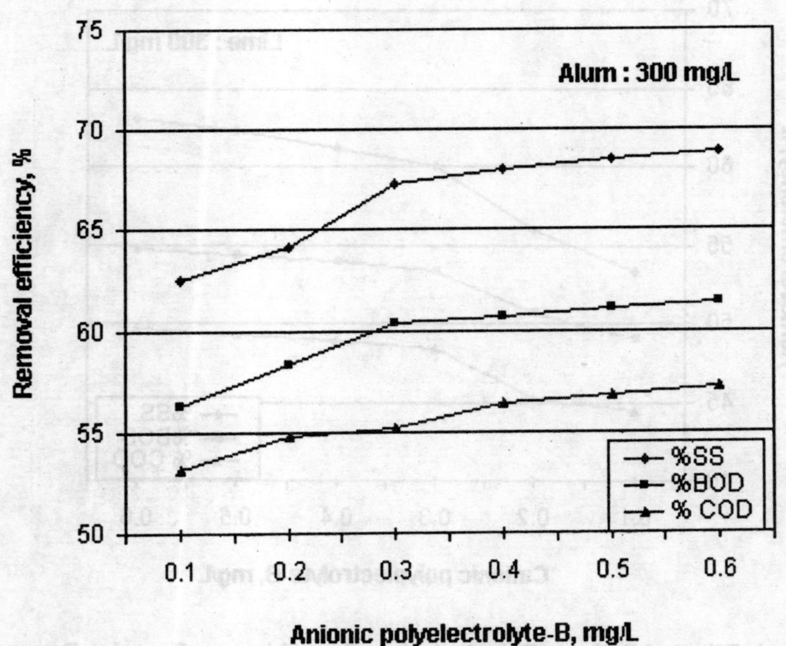

Fig. 6: Effect of Anionic Polyelectrolyte-B and Alum on Organics Removal

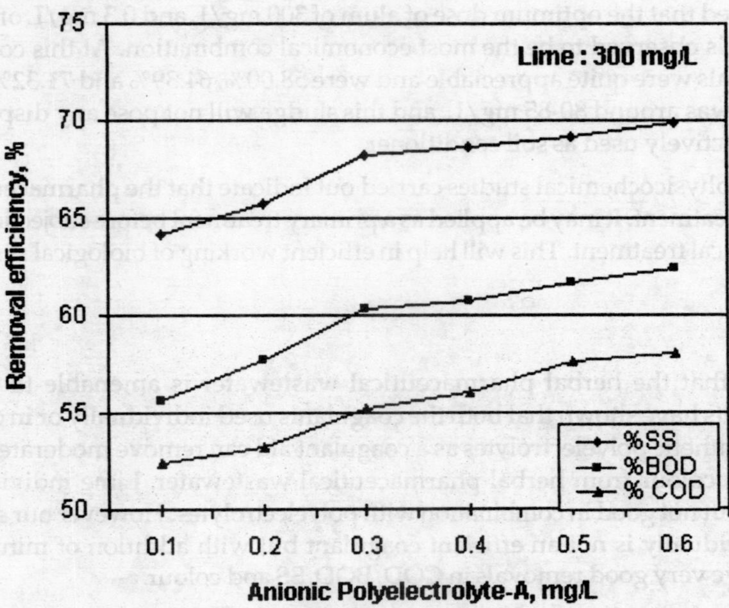

Fig. 7: Effect of Anionic Polyelectrolyte-A and Lime on Organics Removal

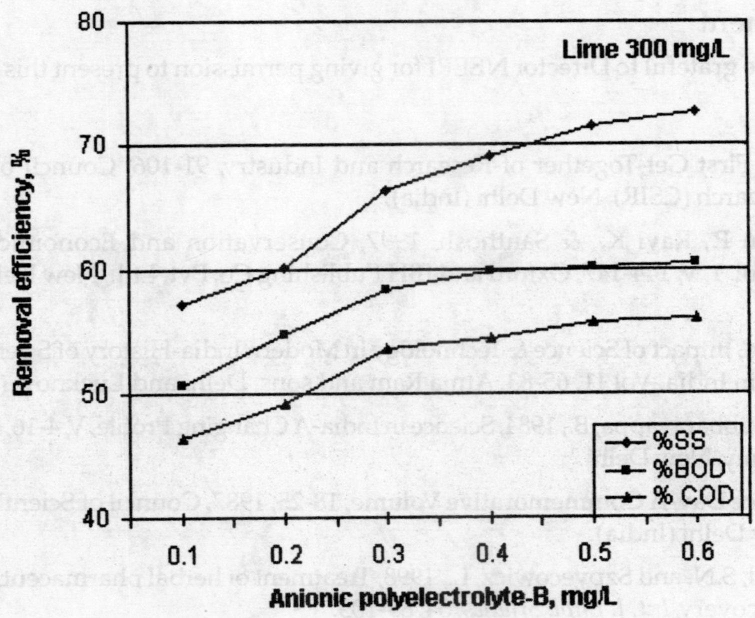

Fig. 8: Effect of Anionic Polyelectrolyte-B and Lime on Organics Removal

Results indicated that the optimum dose of alum of 300 mg/L and 0.3 mg/L of anionic synthetic polyelectrolyte (A) is observed to be the most economical combination. At this combination, COD, BOD and SS removals were quite appreciable and were 58.00%, 64.39% and 71.32% respectively but sludge production was around 80-85 mg/L, and this sludge will not pose any disposal problems, as the same can be effectively used as soil conditioner.

In general, the physicochemical studies carried out indicate that the pharmaceutical wastewater is amenable to this treatment. It may be applied as a primary treatment before subjecting the wastewater to secondary biological treatment. This will help in efficient working of biological system, by reducing the organic load.

Conclusions

Studies show that the herbal pharmaceutical wastewater is amenable to physicochemical treatment. The results have shown that both the coagulants used individually or in combination with anionic/cationic synthetic polyelectrolytes as a coagulant aid can remove moderate to high degree of COD, BOD, SS and colour from herbal pharmaceutical wastewater. Lime individually acts as an efficient coagulant but not good in combination with polyelectrolytes. However our studies reveal that though, alum individually is not an efficient coagulant but with addition of minimum amount of polyelectrolytes gave very good removals in COD, BOD, SS and colour.

The efficiencies of coagulants in their increasing order are: Alum, Lime,

(Lime 300 mg/L + Cationic polyelectrolyte B), (Lime 300 mg/L + Cationic polyelectrolyte A), (Alum 300 mg/L + Cationic polyelectrolyte B), (Lime 300 mg/L + Anionic polyelectrolyte B), (Alum 300 mg/L + Cationic polyelectrolyte A), (Alum 300 mg/L + Anionic polyelectrolyte B), (Lime 300 mg/L + Anionic polyelectrolyte A), (Alum 300 mg/L + Anionic polyelectrolyte A).

Acknowledgement

The authors are grateful to Director NEERI for giving permission to present this work.

REFERENCES

1. April 4, 1966, First Get-Together of-Research and Industry, 91-106, Council of Scientific and Industrial Research (CSIR), New Delhi (India).
2. Pushpangadan P., Ravi K., & Santhosh, 1997, Conservation and Economic Evaluation of Biodiversity, Vol. 1, V, 144-147, Oxford and IBH Publishing Co. Pvt. Ltd., New Delhi and Calcutta (India).
3. Jaggi, O.P., 1984, Impact of Science & Technology in Modern India-History of Science, Technology and Medicine in India, Vol 11, 65-83, Atma Ram and sons, Delhi and Lucknow (India)
4. Mukerji, S.K., Subbarayappa, B., 1984, Science in India-A Changing Profile, V, 4-16, Indian National Science Academy, New Delhi.
5. National Science Day-A Commemorative Volume, 18-25, 1987, Council of Scientific & Industrial Research, New Delhi (India).
6. Nandy, T., Kaul, S.N. and Szpyecowicz, L., 1998, Treatment of herbal pharmaceutical wastewater with energy recovery, *Int. J. Envi. Studies*, 54, 83-105.
7. Debroy Mukti, Nandy T. and Kaul, S. N., 1993, Anaerobic up-flow bioreactor for Pharmaceutical wastewater treatment, *Indian Journal of Envir. Protection* 13 (1), 41-46

8. Naphade, P.G., Bhole, A.G. and Raman, V., 1983, Pharmaceutical wastewater treatment by anaerobic packed bed reactor. *IAWPC* Tech. Annual X, 97-103.

9. Vanerkar, A.P., Satyanarayan Shanta, Dharmadhikari, D.M. and Kaul, S.N., 2001, Herbal pharmaceutical wastewater treatment by coagulation and enhancement of flocculation by synthetic polyelectrolydes. *Proceedings of the National Seminar on Recent Trends in Industrial waste Treatment*, LIT, Nagpur, (India) 119-132.

10. Jain Garima, Satynarayan Shanta, Nawghare, P., Kaul, S.N. and Szpyrocowicz, L., 2001, Treatment of pharmaceutical wastewater (herbal) by coagulation/flocculation process. *Intern. J Environ. Studies*, 58, 313-330.

11. Clesceri, L.S., Greenberg, A.E. and Eaton, A.D., 1998, Standard method for the examination of water and wastewater. 20th Ed. *American Public Health Association. Washington. DC.*

38

Treatment of Coloured Wastewater from Textile Pulp & Paper and Chemical Industries by Chemical Coagulants

R.S. DONGRE, P.P. PATHE & S.N. KAUL

National Environmental Engineering Research Institute, Nehru Marg, Nagpur - 440 020

ABSTRACT

Due to low biological oxygen demand (BOD), and high concentration of chemical oxygen demand (COD), the ratio of BOD to COD became very low-less than 0.25. The presence of recalcitrant organic polymeric materials *viz.* lignin in pulp & paper mill wastewater, dyes in the textile and chemical wastewater makes these wastewaters non-biodegradable. Very low pH value for colored effluent from chemical industry wastewaters *viz.* H-acid (sodium salt of 1-amino-8-hydroxynaphthalene-3, 6-disulphonic acid) does not favour biological treatment, which also involves several operational and maintenance problems.

Physico-chemical treatment methods, *viz.* chemical coagulation, have great potential to remove color, colloidal materials and low-density solids from wastewater and therefore can be applied for the treatment of wastewaters having low BOD/COD ratios. These methods can also be implemented because of simplicity and ease. Four different chemical coagulants were used in the laboratory treatability study *viz.* lime, alum, ferrous sulfate and PAC (Poly-Aluminium Chloride). Out of four coagulants used, only lime and PAC resulted in satisfactory removal of color, SS, COD and BOD from colored textile wastewater. However alum and PAC results in significant removal in color, SS, COD and BOD from pulp & paper wastewater. Chemical wastewater does not respond to any of these chemical coagulants. This paper presents detailed investigations on removal of Color, SS, COD and BOD from colored wastewater using various chemical coagulants.

Keywords: Chemical coagulants, lime, alum, FeSO4, PAC, textile effluent, pulp & paper mill effluent, H-acid effluent, color removal.

Introduction

Pulp & Paper, textile and chemical industries are using high amount of water during manufacturing process and in general, produced highly colored wastewater. Generation of wastewater in textile manufacturing process is mainly from various process operations/units, *viz.*, desizing, scouring mercerizing, bleaching, dyeing and printing (Pathe P P. *et al.* 2001). The chemical industry wastewater *viz.*, H-acid is highly complexed in nature with acidic pH, intense color, very high COD and low BOD (Dongre R.S. *et al.* 2002). This wastewater on being discharged into the surface water bodies without any treatment creating serious water pollution problems like toxicity to the aquatic life, deterioration in water quality. The textile wastewater from bleaching, dyeing, printing units have alkaline pH, dense color, low BOD, high total dissolve solids (TDS) and COD. The pulp and paper mill wastewater has intense brown color, moderate COD and low BOD due to presence of organic polymeric compound – lignin (Subrahmanyam P.V.R 1990). Chemical wastewater has also dark intense color due to presence of dyes and color causing chemicals. The color from colored wastewater causes aesthetic pollution and persists in the water body for long distances (Powels, R. *et al.* 1997). Generally, the wastewater is being treated by biological treatment systems, particularly by activated sludge process and/or by aerated lagoons for removal of biodegradable organics before discharge into receiving water bodies (Nemerow N.L 1978). Very little color and COD is removed by conventional biological processes. The response of biological processes to this colored wastewater is low due to non-biodegradability toxicity and also because of low BOD/COD ratio. The limitations to biological treatment systems are skilled manpower, highly technical operation and maintenance, higher land requirement, process sensitivity, toxicity to microorganisms due to presence of toxic chemicals and heavy metals (Eckenfelder W.W 1965). To treat such non-biodegradable and high colored wastewaters from textile, pulp & paper and chemical industry, a methodology is needs to be explored to improve treatment efficiency. In view of this a technique involving flocculation, coagulation, sedimentation and filtration were considered to remove color, SS, COD and BOD from colored wastewaters before biological treatment. In general, membrane techniques are reported to be used as tertiary treatment to remove high TDS and color from wastewater, generating concentrate, which needs appropriate treatment before being discharged (Chen, G. *et al.* 1997). Currently the treatment employed for chemical wastewater from manufacturing of H-acid is by thermal oxidation/incineration/evaporation which is energy intensive technologies (Kaul, S.N *et al.* 1992). It is reported that advance oxidation processes (AOP) are feasible for the treatment of high strength colored wastewaters containing intense color and high refractory organic pollutants (Lei, L. *et al.* 1996). Both these treatment options are energy intensive and the cost of treatment is exorbitantly high. Therefore, a treatability study for intense colored and high strength wastewater by using conventional chemical coagulants was thought to remove color, SS, COD and BOD.

Materials and Methods

Wastewater from textile, pulp & paper and chemical industry was collected as grab samples and analyzed for various parameters as per standard methods (APHA, AWWA, WPCF 1989). The range of characteristics of wastewater from textile pulp & paper and chemical (H-acid) are presented in Table 1. The experiments with chemical coagulants were conducted using jar test apparatus (Phipps & Bird, USA). About 500 ml of wastewater sample after settling for 60 minutes was used for the experiments using various chemical coagulants *viz.* lime, alum, ferrous sulfate ($FeSO_4$) and poly-aluminium chloride (PAC). The experimental conditions such as reaction pH, reaction time (flash mixing, slow mixing) and sedimentation time were studied.The various doses of chemical coagulants in the range of 200 to 4000 mg/L with an incremental dose of 200 mg/L were added to one liter wastewater. Only poly-aluminium chloride was studied with low dose of coagulant, ranging from 25 to 500 mg/L with an incremental dose of 25 mg/L. To 500 ml pH adjusted wastewater sample from textile, pulp & paper and chemical industries was taken in a separate beaker, and a dose of 200 mg/L lime was added. The admixture was subjected to flash mixing at 100 rpm for 20 seconds followed by slow mixing at 30 rpm for 30 seconds and then the floc was allowed to settle for 60 minutes. After settling of flocs in the form of sludge, the supernatant was carefully decanted and analyzed for various parameters.The experiments were repeated for other doses of lime, and also with other chemical coagulants *viz.* alum, ferrous sulfate at the same experimental conditions. Characteristics of treated wastewater from textile and pulp & paper mill at the optimal dose of lime, alum ferrous sulfate and PAC are presented in Tables 2 and 3 respectively. To 500 ml pH adjusted wastewater sample from textile, pulp & paper and chemical industries 25 mg/L dose of PAC was added and the admixture was treated at the similar experimental conditions. Chemical wastewater without adjusting pH was also subjected to chemical coagulation treatment using lime, alum, ferrous sulfate and PAC and the results are given in Table 4. As the results obtained were not encouraging, it was decided to repeat the experiment with chemical wastewater at various pH. Chemical wastewater gave better response at pH 7.Therefore the experiments were repeated with pH adjusted chemical wastewater at pH 7. Characteristics of treated wastewater from chemical industry at the optimal dose of lime, alum, ferrous sulfate and PAC are given in Table 5.

Table 1: Range of Characteristics of Wastewater from Textile, Pulp & Paper and Chemical Industries

Sl. No.	*Parameters*	*Combined Wastewater from*		
		Textile Industry	*Pulp & Paper Mill*	*Chemical Industry*
1.	Color, Pt-Co unit	2010-2080	500-550	209000-220500
2.	pH	8.0-9.0	7.8-8.2	1.56-1.6
3.	Total Alkalinity	525-550	290-320	-
4.	Suspended Solids	270-300	218-240	8190-8350
5.	Total Dissolved Solids	4180-4320	2235-2310	436980-440780
6.	COD	1920-2000	750-980	139000-140600
7.	BOD_{5d}, 20° C	488-470	210-300	19800-20780

*All values are expressed in mg/L except color & pH

The quantity of the sludge produced at experiments conducted at various doses of different chemical coagulants used to treat textile, pulp & paper and chemical wastewater was measured. Sludge volume at optimal chemical doses of lime, alum, ferrous sulfate and PAC for the treatment of textile, pulp & paper and chemical wastewater is presented in Tables 2 to 5.

Table 2: Characteristics of Treated Effluent from Chemical Coagulation Treating Textile Wastewater at Optimal Dose of Various Coagulants after pH Adjustment

Sl. No.	*Parameters*	*Textile Wastewater Influent #*	*Coagulant dose, mg/L*			
			Lime (1250)	*$FeSO_4$ (1600)*	*Alum (900)*	*PAC (112)*
1.	Color, Pt-Co unit	2060	1026	1655	1569	1009
2.	pH	7.0	9.2	8.3	7.9	7.7
3.	Total Alkalinity	540	1880	490	350	300
4.	Suspended solids	288	89	106	152	59
5.	Total Dissolved Solids	4230	3162	93204	3380	2950
6.	COD	1954	963	1143	1395	697
7.	BOD_{5d}, 20° C	450	158	180	242	113
8.	Sludge volume, ml/L	—	80	90	65	160

* All values are expressed in mg/L except color, pH and sludge volume.
# After adjusting pH of wastewater.

Table 3: Characteristics of Treated Effluent from Chemical Coagulation Treating Pulp & Paper Mill Wastewater at Optimal Dose of Various Coagulants after pH Adjustment

Sl. No.	*Parameters*	*Pulp & Paper Wastewater Influent #*	*Coagulant dose, mg/L*			
			Lime (1100)	*$FeSO_4$ (1150)*	*Alum (800)*	*PAC (100)*
1.	Color, Pt-Co unit	520	417	447	419	362
2.	pH	7.0	10.3	7.6	7.3	6.9
3.	Total Alkalinity	310	2570	280	230	130
4.	Suspended solids	225	139.5	142	157	78
5.	Total Dissolved Solids	2240	1796	1824	1960	1580
6.	COD	996	733	657	797	515
7.	BOD_{5d}, 20° C	280	154	168	190	55
8.	Sludge volume, ml/L	—	40	65	70	305

* All values are expressed in mg/L except color, pH and sludge volume.
# After adjusting pH of wastewater.

Results and Discussion

Effect of Chemical Coagulants on Textile Wastewater

Lime at a dose of 1250 mg/L removed about 50.2% 69.8% 50.7% 64.9 % color, SS, COD & BOD from textile wastewater. About 23.9% color 47.2% SS, 28.6% COD and 41.6%BOD was removed from textile wastewater at an optimum alum dose of 900 mg/L. Ferrous sulfate, at higher dose of1600mg/L removed about 19.7% color, 63 % SS, 50.7%COD & 64.9%BOD from textile wastewater. However at much lower dose of PAC, 112 mg/L was observed to be effective to remove about 51.0% color, 44.5% SS, 64.3% COD & 74.9 % BOD from textile wastewater. Percent removal of color, SS and COD are depicted in Figures 1 to 4 respectively.

Table 4: Characteristics of Treated Effluent from Chemical Coagulation Treating Chemical Wastewater at Optimal Dose of Various Coagulants without pH Adjustment

Sl. No.	*Parameters*	*Chemical Wastewater Influent #*	*Coagulant dose, mg/L*			
			Lime (1600)	*$FeSO_4$ (2200)*	*Alum (2333)*	*PAC (250)*
1.	Color, Pt-Co unit	210000	171360	204540	204750	199920
2.	pH	1.60	1.64	1.61	1.62	1.61
3.	Total Alkalinity	—	—	—	—	—
4.	Suspended Solids	8202	6758	7742	7812	7709
5.	Total Dissolved Solids	437580	424890	419650	435840	427960
6.	COD	140080	130570	127570	138580	133330
7.	BOD_{5d}, 20° C	20000	17820	17500	18960	18060
8.	Sludge volume, ml/L	—	22	16	5	11

*All values are expressed in mg/L except color, pH and sludge volume.
Without pH adjustment of wastewater.

Table 5: Characteristics of Treated Effluent from Chemical Coagulation Treating Chemical Wastewater at Optimal Dose of Various Coagulants after pH Adjustment

Sl. No.	*Parameters*	*Chemical Wastewater Influent #*	*Coagulant Coagulant Dose, mg/L*			
			Lime (1600)	*$FeSO_4$ (2200)*	*Alum (2333)*	*PAC (250)*
1.	Color, Pt-Co unit	211800	170180	202450	202310	195540
2.	pH	7.0	7.2	6.9	6.9	6.7
3.	Total Alkalinity	250	290	220	210	180
4.	Suspended Solids	7800	6304	7262	7310	7084
5.	Total Dissolved Solids	443088	430398	425148	441340	433192
6.	COD	137900	126868	124386	135583	130177
7.	BOD_{5d}, 20° C	19750	17360	17004	18389	17736
8.	Sludge volume, ml/L	-	25	18	7	13

*All values are expressed in mg/L except color, pH and sludge volume.
After adjusting pH of wastewater.

Effect of Chemical Coagulants on Pulp & Paper Wastewater

During the treatment of pulp & paper mill wastewater about 1100 mg/L lime removed 19.8 % color, 38.0 % SS, 26.4 % COD and 45 % BOD. Only at 800 mg/L dose of alum could remove about 19.4 %color, 30.2 % SS, 19.8 % COD and 52 % BOD from pulp & paper wastewater. About 14.1 % color

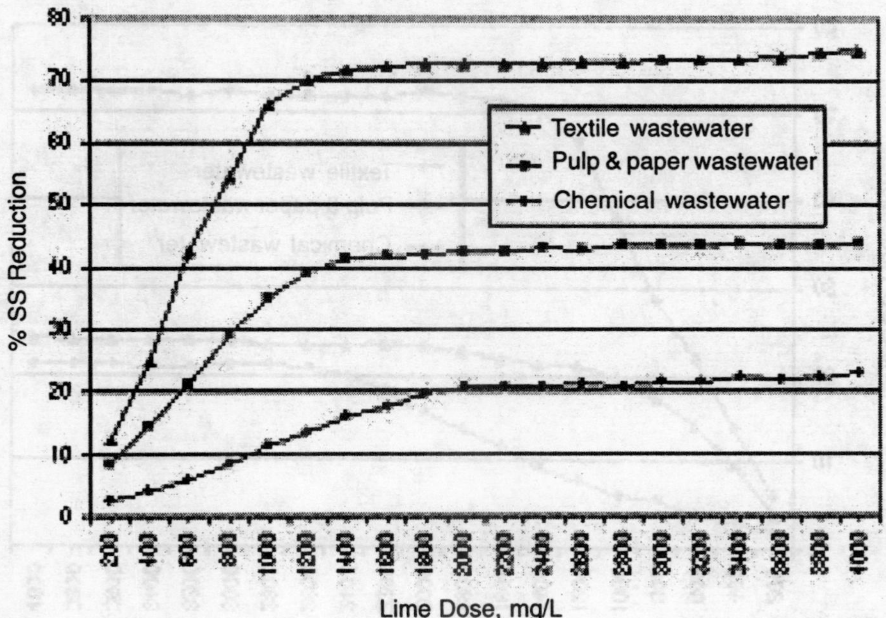

Fig. 1: Removal of Suspended Solids from Various Wastewater at Various Lime Doses

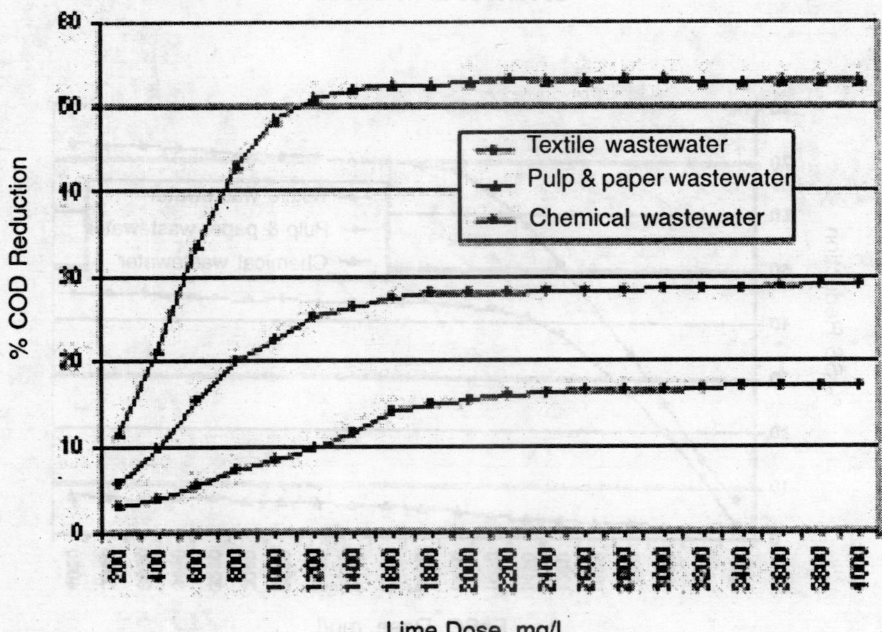

Fig. 2: Removal of COD from Various Wastewater at Various Lime Doses

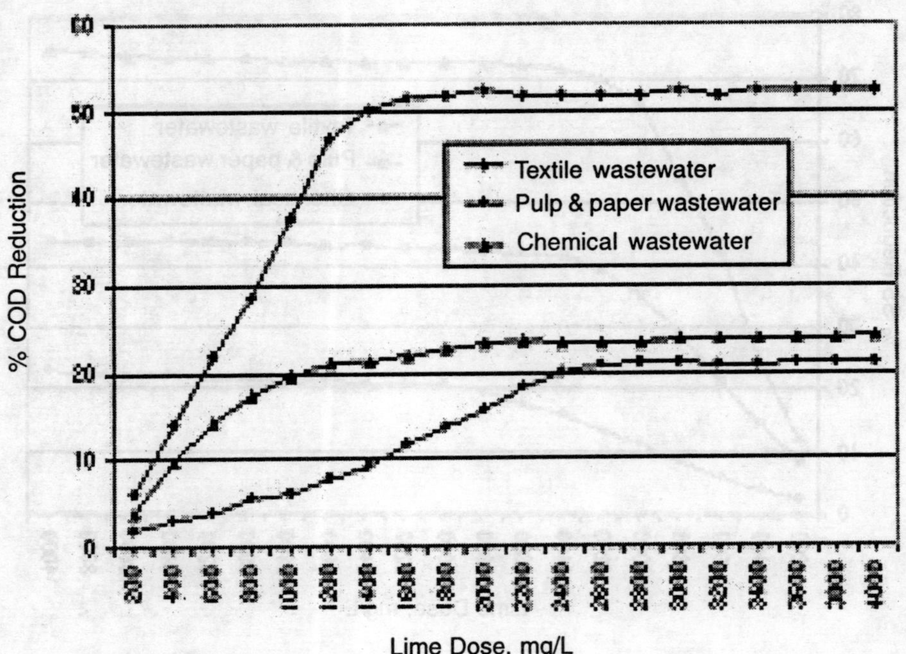

Fig. 3: Removal of Colour from Various Wastewater at Various Lime Doses

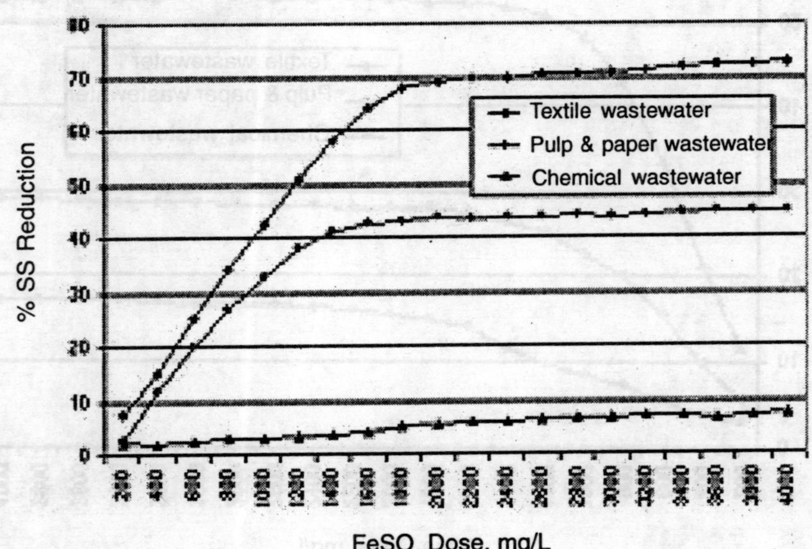

Fig. 4: Removal of Suspended Solids from Various Wastewater at Various Ferrous Sulphate Doses

36.9 % SS, 34.1 % COD and 40.0% BOD was observed to be removed on using 1150 mg/L dose of ferrous sulfate. However, 100 mg/L dose of PAC could remove 30.4 % color, 65.3 % SS, 48.3 % COD and 80.0 % BOD from pulp & paper mill wastewater. Percent removal of color, SS and COD are graphically represented in Figures 5 to 8 respectively.

Effect of Chemical Coagulants on Chemical Wastewater

Lime at a dose of 1600 mg/L remove about 18.4 % color 17.6 % SS, 6.8 % COD and 10.9 % BOD from chemical wastewater without pH adjustment. Alum at much a higher dose of 2300 mg/L removed about 2.5 % color, 4.7 % SS, 1.1 % COD and 5.2 % BOD from chemical wastewater without pH adjustment. However a similar dose, 2200 mg/L ferrous sulfate could remove 2.6 % color, 5.6 % SS, 8.9 % COD and 12.5 % BOD. Poly-aluminium chloride at significantly lower dose of 250 mg/L was observed to be removed about 4.8% color, 6.0% SS, 4.8 % COD and 9.7 %BOD. This indicates that the response of chemical coagulants to treat chemical wastewater was not encouraging. Per cent removal of color, SS and COD are graphically represented in Figures 9 to 12 respectively.

As insignificant response was observed to treat chemical wastewater without pH adjustment, it was decided to adjust pH to neutral 7.0 by addition of requisite dose of sodium hydroxide. Lime at a dose of 1600mg/L removed about 19.7% color 19.2 % SS, 8.0 % COD and 12.1% BOD from pH adjusted chemical wastewater. About 4.4 % color, 6.9 % SS, 8.2 % COD and 9.0% BOD was observed to be removed from pH adjusted chemical wastewater by applying alum dose of 2300 mg/L. Ferrous sulfate at a dose of 2200mg/L removed 4.5%color, 6.3% SS 10.0% COD and 14.5 % BOD from pH adjusted chemical wastewater. Poly-aluminium chloride at a much lower dose of 250 mg/L removed about 7.8 % color, 9.2 % SS, 6.3 % COD and 11.2 % BOD from pH adjusted wastewater. On pH adjustment, the response of chemical coagulants in the reduction of color, SS, COD, and BOD was observed to be significantly improved. But total removal of color, SS, COD and BOD was not encouraging. This

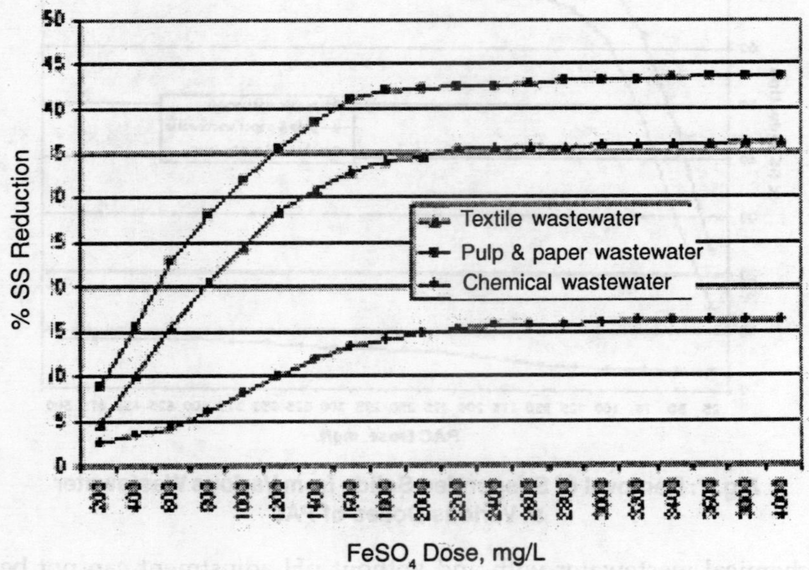

Fig. 5: Removal of COD from Various Wastewater at Various Ferrous Sulphate Doses

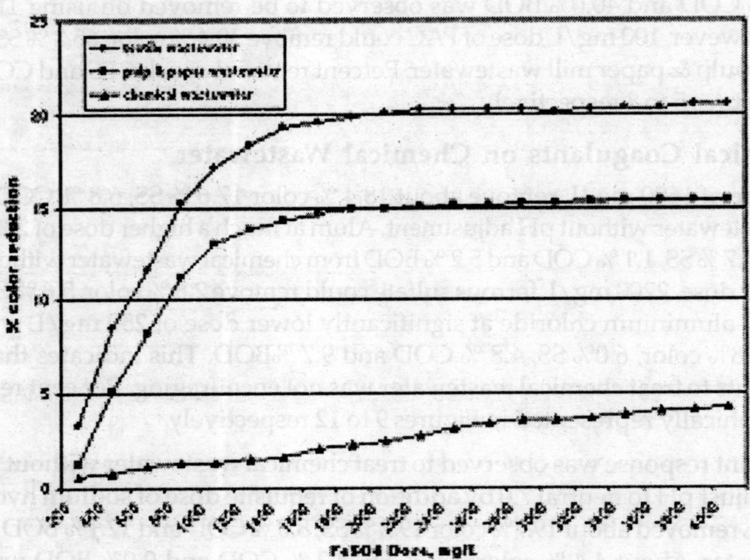

Fig. 6: Removal of Colour from Various Wastewater at Various Doses of Ferrous Sulphate

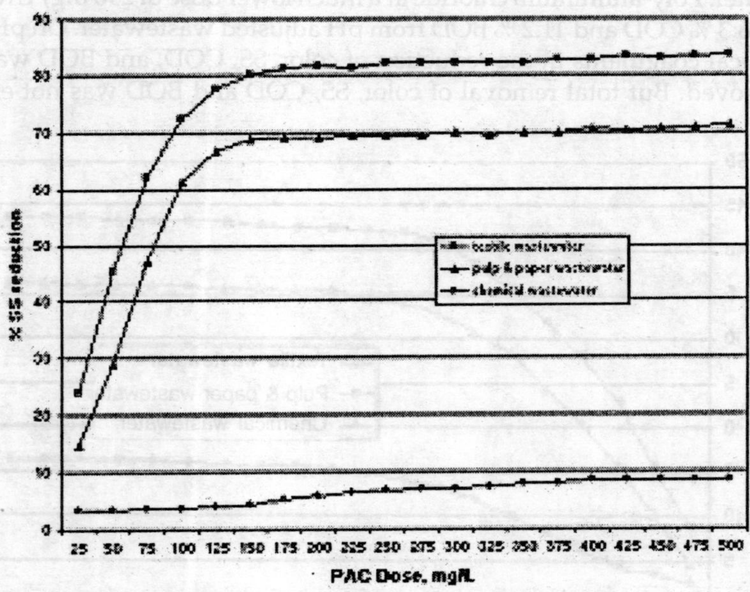

Fig. 7: Removal of Suspended Solids from Various Wastewater at Various Doses of PAC

indicates that chemical wastewater with and without pH adjustment can not be treated using conventional chemical coagulants, *viz.* lime, alum, ferrous sulfate and PAC.

Conclusion

From the chemical coagulation studies conducted for the treatment of colored, non-biodegradable and low BOD/COD ratio wastewaters *viz.* textile, pulp and paper etc., Poly-aluminium chloride was

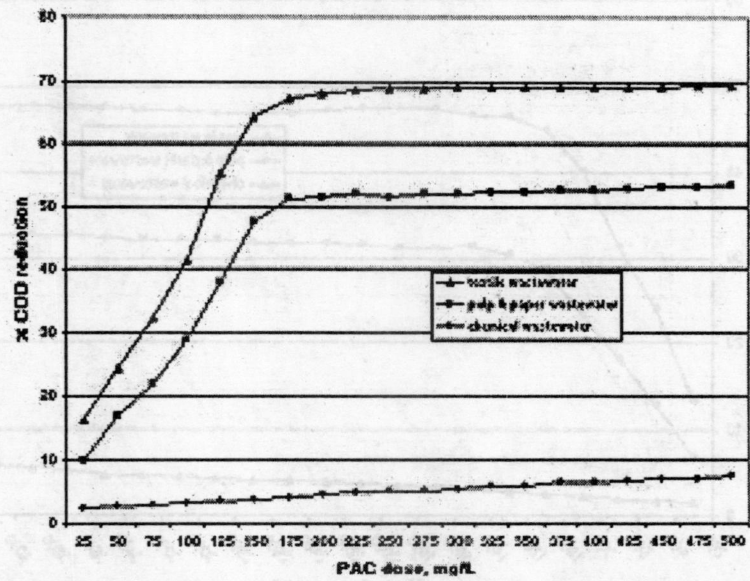

Fig. 8: Removal of COD from Various Wastewater at Various Doses of PAC

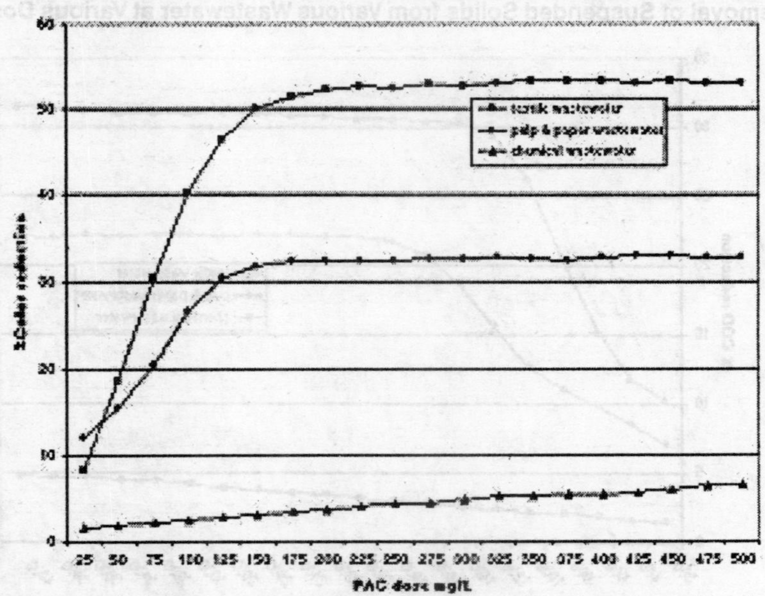

Fig. 9: Removal of Colour from Various Wastewater at Various Doses of PAC

observed to be effective at lower doses ranging from 100 to 112 mg/L resulting in significantly high removal of color, SS, COD and BOD. This also results in generation of low quantity of sludge. This also helps in mitigation of sludge storage and handling problems. Effluents for chemical coagulation treatment require further treatment by using biological or advanced oxidation treatment to meet prescribed discharge standards by regulatory agencies. Due to poor response in removing pollutants,

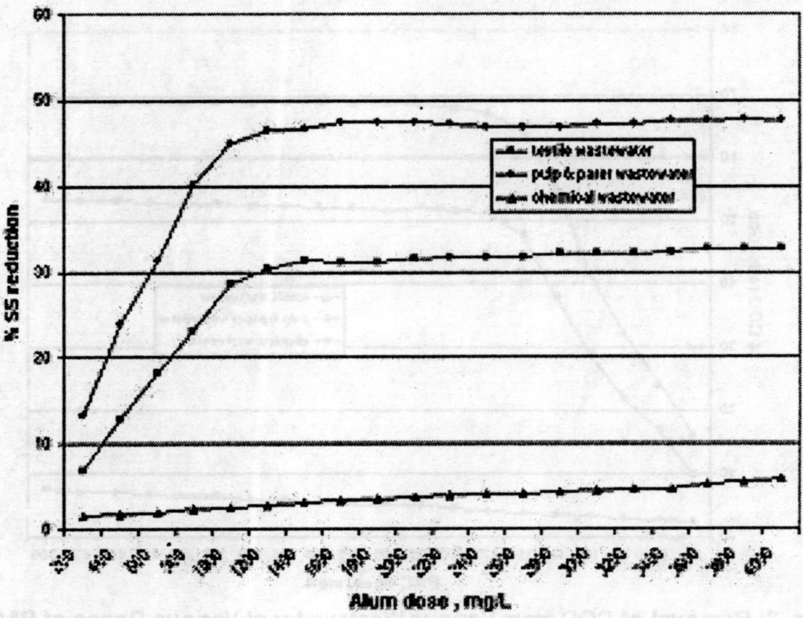

Fig. 10: Removal of Suspended Solids from Various Wastewater at Various Doses of Alum

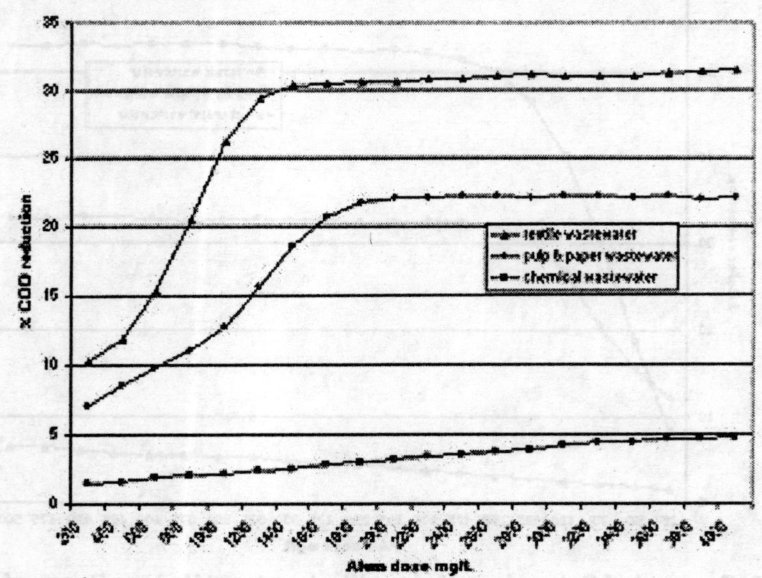

Fig. 11: Removal of COD from Various Wastewater at Various Doses of Alum

conventional chemical coagulation technique can not be used to treat wastewater form chemical (H-Acid) manufacturing industries even at very high doses of coagulants.

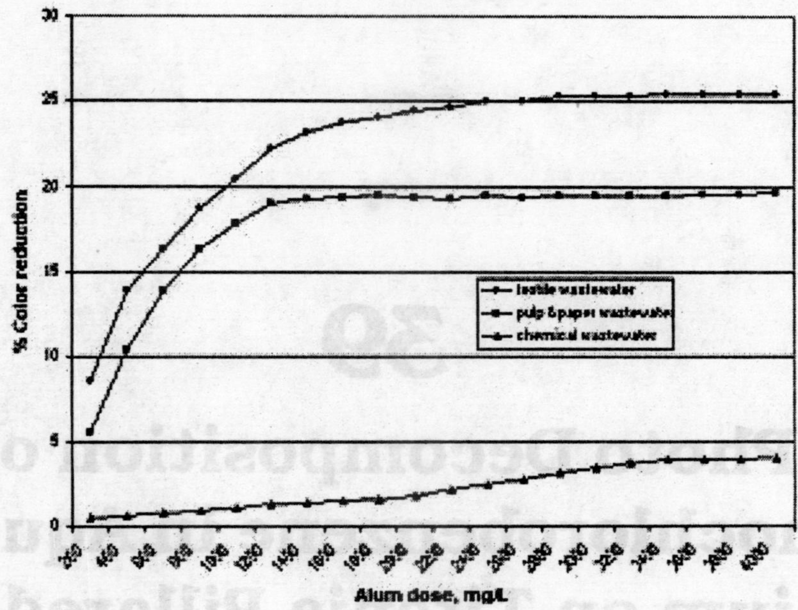

Fig. 12: Removal of Colour from Various Wastewater at Various Doses of Alum

REFERENCES

1. APHA, AWWA, WPCF, 1989, Standard methods for the examination of water and wastewater (17th edn.), Washington DC.
2. Chen G. Chai, × Yue P.l. Mi, Y., 1997, " Treatment of Textile Wastewater by pilot scale Nanofiltration membrane.J.Membrane.Sci., 127 (1), pp.93-99.
3. Dongre R.S., Rao N.N. & Kaul S.N., 2002, "Fenton Oxidation of System for Chemical Treatability of H-Acid Wastewater" Proceedings of National Symposium on Pollution Prevention & Control in India. March 2-3, VRCE, Nagpur, India.
4. Eckenfelder W.W., 1965, "Industrial waste pollution control" International Text Book Co. Washington, D.C. p. 128.
5. Kaul S. N. Nandy. T. Deshpande. C. V. Srivastava. A., Shastry. S., Szpyrkowicz. L., 1999, "Full scale treatment of hazardous wastewater through thermal reduction", Toxicol. & Environ. Chemistry. 68., pp. 247-251.
6. Lei, L., Hu. X., Chen G., Porter J.F. & Yue P.L., 1996, Wet air oxidation of desizing Wastewater from textile industry." Asia. Pac. Conf. Sustain. Energy. Environ. Technol., Singapure., 165.
7. Nemarow N.L., 1978, "Industrial water pollution-origins, characteristics and treatment", Addison Wesley publishing Co., Massachusetts pp. 242.
8. Powels R. & Bhole, A.G., 1997, "Use of Synthetic Poly-electrolytes in treatment of Craft paper mill wastewater" Indian. J. Environ. Health., 39. (3) pp. 177-181.
9. Pathe P.P, Biswas A.K. and Kaul, S.N., 2001, "Physico-chemical treatment for wastewater from Cluster of small scale cotton textile industries " Indian J.Env.Prof.21 (7), 557-587.
10. Subhramanyam, P.V.R., 1990, "Waste management in pulp & paper industry" J.Indian Env. Management, 17 (2), 70-94.

39

Photo Decomposition of Monochlorobenzene in Aqueous Medium on Titania Pillared Clay

SHOBHANA AWATE & VEDA RAMASWAMY

Catalysis Division, National Chemical Laboratory, Pune – 411 008

ABSTRACT

Titania pillared montmorillonite clay was prepared by conventional ion exchange method and post hydrothermal treatment after ion exchange. The calcined clay samples were characterized by different physico-chemical techniques such as XRD, EDX, low temperature (77K) nitrogen adsorption and UV-Visible diffuse reflectance spectroscopy. The content of pillared titania remained unchanged irrespective of the method of preparation. The method of preparation resulted in alteration of the crystallinity and crystallite size of anatase titania. The behavior of titania pillared clays in photo decomposition of monochlorobenzene (MCB) in aqueous medium was studied as a function of their physico-chemical characteristics. The titania pillared clay prepared by post hydrothermal route has shown enhanced photo-catalytic oxidation of MCB as compared to that prepared by conventional route.

***Keywords:** Titania pillared clay, photocatalysis, decomposition, monochlorobenzene, hydrothermal*

Introduction

The titania pillared clays have created tremendous potential for photo-catalytic removal of trace contaminants from air and water which is recognized as a growing research area. Organic compounds such as phenol, trichloroethylene, di-Bu phthalate in water, also toluene, benzaldehyde associated with pharmacaeutical and perfume industries (1-3), discharged effluents from the textile dyeing and finishing industry (4) are important pollutants usually found in air and or/wastewater (5-7). Though photo-oxidation by titania is well documented, the use of titania pillared clay as a photocatalyst for removal of organic pollutants has not been exhaustively reported in view of the method of their preparation and its influence on the efficiency in photocatalysis. In view of this, attempts were made to investigate the performance of titania pillared clay in photo-catalysis of monochlorobenzene as a function of the method of its preparation and their physico-chemical characterization. Monochlorobenzene (MCB) which is used as a solvent in the manufacture of adhesives, paints, polishes, waxes, diisocyanates, pharmaceuticals and natural rubber and for the synthesis of o- and p-chloronitrobenzenes, phenol, DDT and aniline, add to the pollution of atmosphere from fugitive emissions of MCB. Human exposure to this pollutant is from ambient air, especially near point sources. It percolates into ground water if soil is sandy and biodegrades very slowly or not at all in the soil. It has been reported recently by the authors (8) that the photodecomposition of MCB on nanocrystallineTiO_2 increases with decrease in the particle size of the sample. Complete decomposition of MCB was seen within 2 h of exposure to UV irradiation with smaller particle size (~8 nm) anatase titania sample.

We are reporting the performance of titania pillared montmorillonite prepared by conventional ion exchange method, with and without post hydrothermal treatment in photo-catalytic degradation of monochlorobenzene in aqueous medium.

Experimental

Titania pillared clay was synthesized by following the procedure reported by Yamanaka et al. [9] using Kunipia-F montmorillonite clay (Na form, Kunimine Industries) with a cation exchange capacity 1.28 meq. g^{-1} and having structural formula $(Na_{0.35}\,K_{0.01}\,Ca_{0.02})\,(Si_{3.89}\,Al_{0.11})\,(Al_{1.60}\,Fe_{0.08}\,Mg_{0.32})\text{-}O_{10}\,(OH)_2\,.n\,H_2O$. Titanium tetraisopropoxide was added to 1M HCl solution under vigorous stirring at 323 K keeping HCl to alkoxide molar ratio 1. The resulting slurry was peptized till clear TiO_2 sol was obtained. The TiO_2 sol was mixed with aqueous clay slurry (1 %) keeping molar ratio of TiO_2/CEC of the clay 30. The suspension was stirred for 3h at room temperature. For hydrothermal treatment the intercalated pillared clay was dispersed in water resulting into slurry containing 2% solid. Then it was further hydrothermally treated at 523 K for 1h. The hydrothermally treated product was centrifuged and washed till effluent showed no sodium and then dried in air-oven maintained at 373 K for 4 h. The intercalated clay samples synthesized under hydrothermal treatment (HT) was further calcined in an air-oven at 773 K for 3 h. Another sample hydrothermally untreated (HUT) was prepared following the above procedure except giving the hydrothermal treatment at 523K (10).

Chemical Analysis

The content of TiO_2 was determined by EDX (HORIBA MESA-500NA)

Powder X-ray Diffraction

The calcined clay was characterized by powder XRD (Rigaku 3070E using CuKα radiation). The percentage crystallinity due to anatase titania phase in pillared clay sample was evaluated by taking ratio of areas under (101) reflection of the test sample to the reference sample which was prepared by mixing clay homogeneously with pure anatase titania which contributes 57.3 % part (w/w) of the mixture. The crystallite size of the anatase in pillared clay was determined from the powder X-ray patterns using Scherrer equation (11).

Low Temperature Nitrogen Adsorption

From low temperature nitrogen adsorption (BELSORP 28SA automatic Gas adsorption apparatus) isotherms BET surface area, pore diameter; total pore volume and pore size distribution was determined. Prior to the determination of adsorption isotherm, the sample was evacuated at 10^{-2} torr and at 423 K to remove all of the physisorbed species from the surface of the adsorbent.

UV-Vis Spectroscopy

The samples were scanned on UV-Vis diffuse reflectance spectrometer (Shimadzu UV-2500 PC) using barium sulfate as the standard.

Photo-catalytic Oxidation

The experimental photo decomposition cell was a conical flask made of borosilicate glass with a capacity of 50 ml. 10 ml of water and 1 ml of monochlorobenzene were placed in the conical flask containing 0.2 g of titania pillared clay. The slurry was stirred with a magnetic stirrer for 30 minutes in dark. The clear sample was analyzed on gas chromatograph (Shimadzu, GC-14 A with a HP-5 capillary column and FID) with helium as carrier gas. The UV irradiation was sourced from a 400-Watt medium pressure Hg vapour lamp at a wavelength of ≥ 300 nm. The experiment was carried out at 300 K. Aliquots of reaction mixture were taken with a micro syringe after an interval of 30 minutes till 4 h. The concentration of MCB was determined by gas chromatography.

Results and Discussion

Powder X-ray Diffraction

The sample designations, preparation conditions and product characterization are listed in Table 1.

Table 1: Sample Designation and Powder X-ray Diffraction Data

Sample Designation	*Preparation Conditions*			*% Anatase*	*Crystallite*	*% TiO_2*
	Method	*Temp. (K)*	*Time (h)*	*Crystallinity*	*Size [b] (Å)*	*Content*
S-1	HUT	R.T.	-	38	~ 19.5	57.8
S-2	HT	523	1	85	40.0	57.2

HUT, HT-Hydrothermally untreated, Hydrothermally treated respectively

$$\% \text{ Anatase crystallinity} = \frac{\text{Area under the (101) reflection of the test sample}}{\text{Area under the (101) reflection of the referencesample}}$$

Titania pillared clay prepared without / with hydrothermal treatment is labeled as HUT and HT respectively. Fig. 1 shows the XRD profiles of the prepared pillared clays, a) hydrothermally untreated and b) hydrothermally treated in the 2è range from 20 to 60°. Hydrothermally untreated sample (HUT) has shown a broad, less intense (101) reflection due to anatase phase at around ~ 25.3° 2θ. Hydrothermally treated (HT) sample, however, has exhibited a sharp and more intense (101) reflection. It is inferred that the hydrothermal treatment has enhanced the crystallinity of anatase phase. Also the additional peaks at 37, 48 and 55° 2θ of anatase titania were found to appear in the hydrothermally

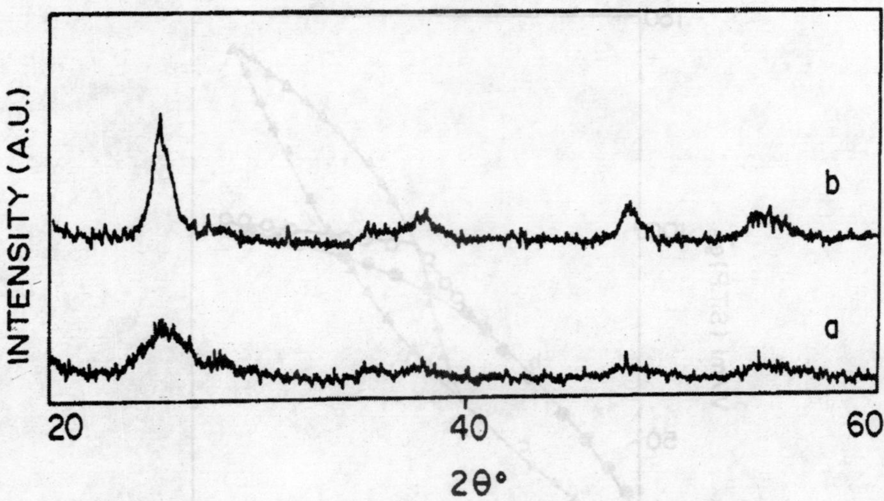

Fig. 1: Powder XRD Profiles of Pillared Clay Samples (a) HUT and (b) HT

treated sample. A prominent peak around 7° (not shown in XRD) in the original clay disappeared in the pillared clays indicating the successful pillaring. Percentage anatase crystallinity increased from 38 to 85% due to hydrothermal treatment. The crystallite size of hydrothermally treated sample is higher (~40 Å) as compared to untreated sample (< 20Å, not possible to determine from XRD). Hydrothermal treatment is effective in increasing both crystallinity of the anatase titania phase as well as the crystallite size.

Nitrogen Adsorption

Fig. 2 depicts the nitrogen adsorption-desorption isotherms at 77 K for HUT and HT samples. The sharp increase in the adsorbed N_2 volume at P/P_s=0.2 and a steadyincrease in the relative pressure region 0.1 to 0.5 in the untreated sample (HUT) as compared to treated sample (HT), indicates a bimodal pore size (both micro and meso) distribution in the former. The influence of hydrothermal treatment in obtaining more mesopores is seen. Table 2 summarizes the textural properties of the HUT and HT samples. HUT showed higher BET surface area than HT sample. Hydrothermal treatment parameters, *viz.* temperature, period etc., may be responsible for lowering the surface area due to disorder in the silicate layers. Even though, the specific BET surface area decreased in hydrothermally treated sample, the total pore volume was found to increase. Micropore volume of HT was less than HUT indicating major contribution of mesopore volume in HT. Fig. 3 shows the pore size distribution curves of HUT and HT samples calculated using Cranston-Inkley method (12). Sample HUT showed the presence of both micropores and mesopores supporting the bimodal shaped N_2 adsorption curve. Contribution due to mesopores however, is lower than micropores. Employing hydrothermal treatment mainly introduced mesopores in the sample.

UV-Vis studies

Band gap energy of HUT and HT samples calculated from UV-Vis diffuse reflectance spectra are 3.34 and 3.41 eV respectively (Table 2). Both the samples were blue shifted as compared to pure anatase. HT sample showed higher band gap energy than HUT, which supports the higher oxidation capability of HT sample as compared to HUT sample.

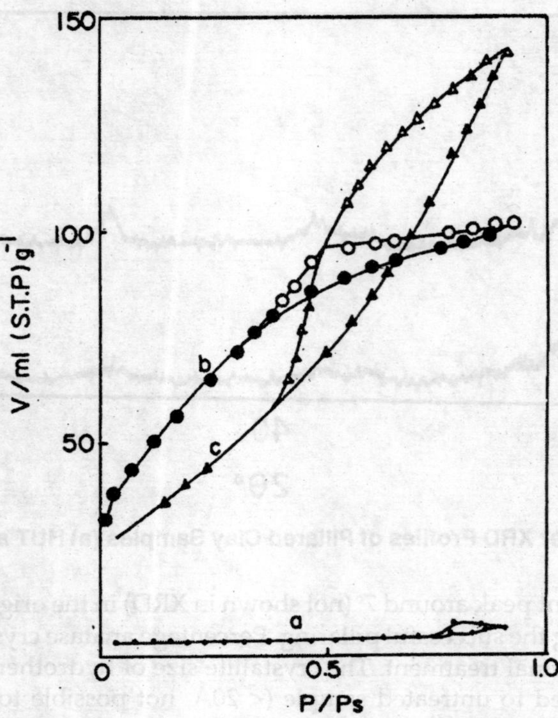

Fig. 2: Nitrogen Adsorption-desorption Isotherms of the Samples (a) clay (adsorption (•), desorption (□)], b) HUT (adsorption (•), desorption (o)], c) HT (adsorption (¯), desorption (¨)]

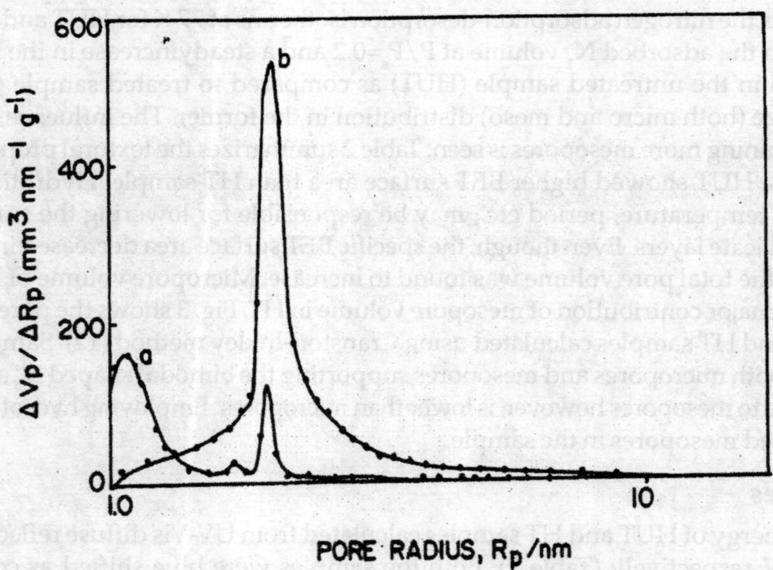

Fig. 3: Pore Size Distribution Curve of Pillared Clay Samples (a) HUT, (b) HT

Table 2: Data Derived from Analysis of Low Temperature Nitrogen Adsorption and UV-Vis Spectroscopy

Sample	*Specific BET Surface Area*[a] *(m^2 g^{-1})*	*Average Pore Diameter*[b] *(Å)*	*Total Pore Volume*[c] *(cm^3 g^{-1})*	*Micropore Volume (cm^3 g^{-1})*	*Band Gap Energy*[e]*, eV*
UT	216	22.0	0.1581	0.1286	3.34
HT	140	38.8	0.2147	0.0187	3.41
Clay	*21*	—	—	—	

[a] Determined from N_2 adsorption.
[b] Determined from pore size distribution curve of N_2 adsorption isotherm.
[c] Calculated from N_2 adsorption isotherm.
[d] Calculated from 't' plot of N_2 adsorption isotherm.
[e] Determined from UV-spectra

Photo-catalytic oxidation

Fig. 4 depicts the results on the photodecomposition of MCB. Period 'A' is the initial time (30 min for both the samples) for which the mixture of sample and MCB was stirred in dark. Period 'B" denotes the results obtained by subjecting the mixture for exposure with UV irradiation after regular interval of 30 min.

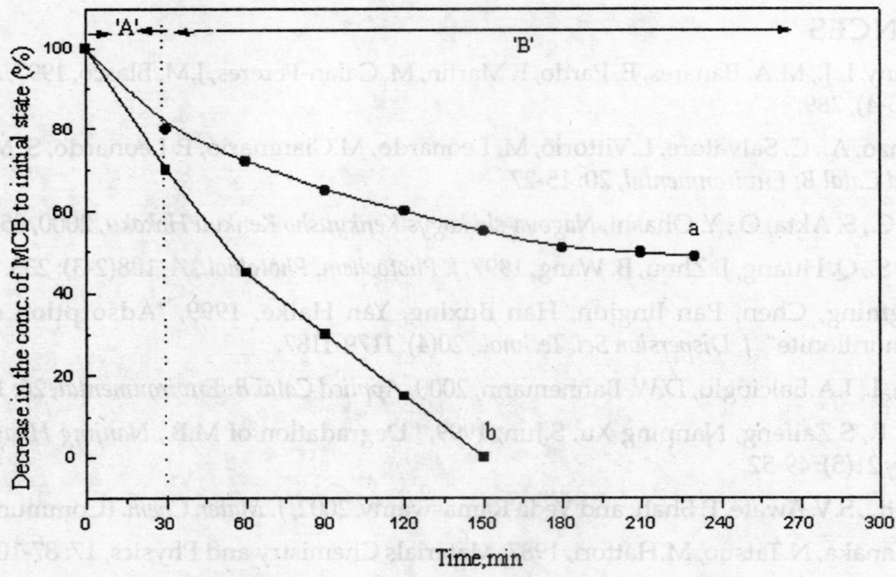

Fig. 4: Relative Concentration of Monochlorobenzene to Initial State (%) by (a) HUT and (b) HT samples with time on stream

The performance of the titania pillared clays prepared by two different routes on photodecomposition of MCB is compared. It is clearly evident from Fig. 4 that HUT and HT samples have shown 80% and 70% decrease in initial concentration of MCB respectively in 30 min stirring in dark. Sample HUT has decreased the concentration linearly after next four 30 minutes interval of UV irradiation to 45%, 30%, 15% and 0% while HUT sample decreased the concentration of MCB to 72%, 65%, 60% and 55 % respectively. For HUT sample even after 4 h UV irradiation, the concentration of MCB decreased only upto 49%. The higher photo-oxidative capability of HT is due to higher anatase crystallinity as compared to HUT. Original clay has not shown any photo-oxidation activity for MCB.

Conclusion

The titania pillared clay was prepared by conventional ion exchange method and also by employing post hydrothermal treatment. The content of pillared TiO_2 was found to be similar irrespective of the method of preparation. Hydrothermal treatment increases the anatase crystallinity and crystallite size determined by powder X-ray diffraction studies. Samples prepared by both methods were blue shifted in UV-Vis DRS spectra with higher bandgap energy for HT sample. From N_2 adsorption studies BET surface area, total pore volume, micropore volume and pore size was determined. Hydrothermally untreated sample showed only 55% decrease in the concentration of MCB to initial state at the end of 4h of UV irradiation, while hydrothermally treated sample oxidized MCB completely in 2h. This observation clearly exhibit the potentiality of hydrothermally treated titania pillared clay as an efficient, economic and environmental friendly photocatalyst for the degradation of organic compounds in aqueous medium.

Acknowledgements

We acknowledge the Department of Science and Technology, New Delhi for financial support.

REFERENCES

1. Alemany, L.J., M.A. Banares, E. Pardo, F. Martin, M. Galan-Fereres, J.M. Blasco, 1997, *Appl. Catal., B*, 13 (3-4), 289.
2. Vincenzo, A., C. Salvatore, L.Vittorio, M. Leonardo, M.Gianmario, P. Leonardo, S. Mario, 1999, *Applied Catal B: Environmental*, 20: 15-27.
3. Ooka, C., S. Akta, O., Y. Ohashi, *Nagoya-shi kogys Kenkyusho Kenkyu Hokoku*, 2000, 85, 13-17.
4. Zheng S., Q. Huang, J. Zhou, B. Wang, 1997, *J. Photochem. Photobiol., A*, 108(2-3): 235.
5. Guangming, Chen, Pan Jingjun, Han Buxing, Yan Haike, 1999, "Adsorption of M.B. on Montmorillonite". *J. Dispersion Sci. Technol.*, 20(4): 1179-1187.
6. Arslan, I., I.A.Balcioglu, D.W. Bahnemann, 2000, *Applied Catal B: Environmental*, 26: 193-206.
7. Yiqun, F., S.Zaifeng, Nanping Xu, S.Jun, 1999, "Degradation of M.B., *Nanjing Huagong Daxue Xuebao*, 21(5): 49-52.
8. P. Awati,., S.V. Awate, P. Shah, and Veda Ramaswamy, 2002, *J. Mater. Chem.* (Communicated).
9. S. Yamanaka, N.Tatsuo, M.Hattori, 1987, Materials Chemistry and Physics, 17: 87-101.
10. S.V. Awate and K. Suzuki, 2001, Adsorption, 7(4): 319-326.
11. B. D. Cullity, 1978, *'Elements of X-ray diffraction'*, Adderson-Wesley Reading, MA.
12. Cranston, R.W. and Inkley, F.A., 1957, *Adv. Catal.*, 9: 143-154.

40

Ionic Membrane Technologies for the Recovery of Valuable Chemicals from Wastewaters

GUIDO SARACCO

Dipartimento di Scienza dei Materiali ed Ingegneria Chimica, Politecnico di Tornio, Corso Duca degil Abruzzi 24, Italy

ABSTRACT

An overview of various processes is provided in which use of ionic membranes and related technologies (dialysis, electrodialysis, electrolysis, etc.) enable the recovery of valuable chemicals. The most relevant literature in the field is surveyed and some elucidating case studies are discussed, also accounting for the results of some research programmes carried out in our laboratories.

Introduction

The treatment of industrial wastewaters has become a diffuse need from an environmental viewpoint. Among the various possible treatment routes, those which allow waste minimisation by recovery of valuable chemicals for recycling in the main production process or elsewhere, are playing an increasingly important role. In this context, electrochemical recovery processes, and particularly ionic membrane technologies, have proved successful in several cases and new applications are being studied by a growing number of researchers. Electrochemical methods have the advantage of easy controllability, efficiency and versatility, to set against the implied cost of electrical power.

This paper focuses on ionic-membrane-based processes. The nature and properties of ionic membranes, as well as the basic principles and taxonomy of electrodialysis, dialysis and electrolysis, are assumed to be well known to readers. Updated information about these topics can be found in (1-3).

Figure 1 shows the three types of ionic membranes available in the market. *Standard membranes* [Fig. 1(a)] are made of an homogeneous matrix (often reinforced with an internal tissue of PVC or similar materials), carrying fixed charges of the opposite sign of those of the permeating ions (counter-ions). Co-ions, which have charges of the same sign of the above fixed charges, are rejected by the membrane owing to the so-called Donnan exclusion mechanism (4). *Monovalent–ion–permselective membranes* [Fig. 1(b)] were developed for edible-salt production from sea water by electrodialysis, so as to limit the presence of divalent ions in the concentrate, thereby avoiding membrane scaling by $CaSO_4$ (5). Permselectivity is obtained by deposition, on both membrane sides, of a thin layer of a resin which is either highly-cross-linked (monovalent-anion permselectivity), or carrying fixed charges with opposite sign compared with the bulk membrane material (monovalent-cation permselectivity). The deposited layers disturb the transport of ions, favouring the passing of monovalent ones. Finally, *bipolar membranes* [Fig. 1(c)] are made of a cationic and an anionic membrane joined together. By imposing a suitable electrical potential gradient, water splitting takes place inside these membranes (6, 7). Protons and hydroxyls are driven to opposite compartments where acids and bases can be recovered from solutions of their salts. The membrane stack is assembled in this case by placing cationic, anionic and bipolar membranes in series.

All these types of membranes have found applications in recovery treatments of industrial wastewaters, via either electrodialysis, electrolysis or diffusion dialysis. Several of these applications are described and discussed hereafter.

Recovery of Metals

The need to treat industrial effluents polluted by metal ions is widely spread. Many manufacturing plants and metal-plating industries produce wastewaters containing metals at a concentration too

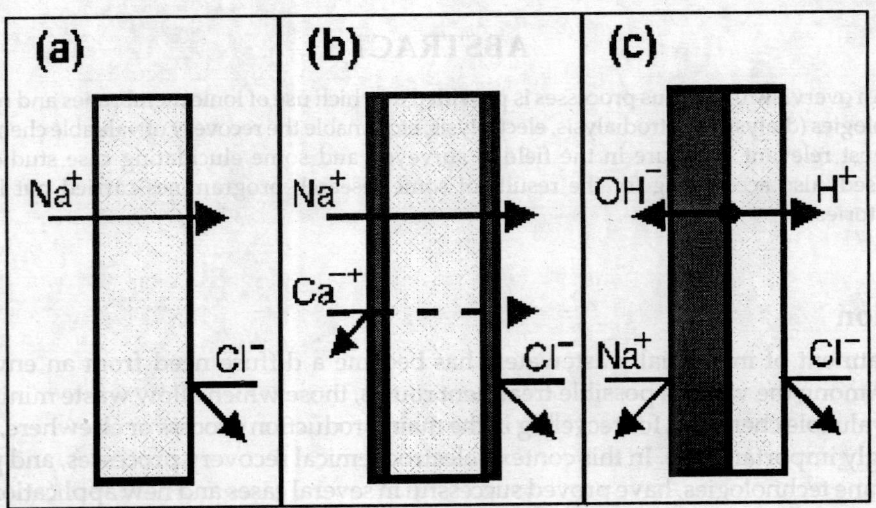

Fig. 1: Classification of Ionic Membranes
(a) Standard type, (b) Monovalent-ion permselective, (c) Biopolar

low for process purposes, but too high for discharge into drainage systems. Table 1 lists the effluent limits concerning metal ions in different countries, together with the residual concentrations of these ions attainable by hydroxide and sulphide precipitation. From these figures it appears clearly that the imposed limits cannot be always obtained by the conventional neutralisation process. Precipitation of sulphide salts may in principle solve the problem, though, apart from more complex plants, sulphide precipitation at such low concentrations would require long contact times since the precipitation agents themselves (*e.g.* H_2S, Na2S) must have quite low end-of-pipe concentrations (< 1 ppm). The use of ion-exchange techniques is also hampered by the lack of regenerable resins for several metals.

Table 1: Effluent Limits and Metal Residual Concentrations After Hydroxide and Sulphide Precipitation

Metal	*Effluent Limits (ppm)*				*After Hydroxide Precipitation (pH = 8)*	*After sulphide Precipitation (pH = 6-7)*
	Italy L. 319/1976 Tabella A	*Germany Abwasser-VWV-1990*	*USA 1983*	*CH 1983*		
Pb	0.2	0.5	0.5	0.5	21	<0.001
Cd	0.02	0.2	0.3	0.1	1500	<0.001
Cu	0.1	0.5	0.5	0.5	1	<0.001
Ni	2	0.5	0.5	2	340	<0.001
Hg	0.005	0.05	—	0.01	—	0.11
g	—	0.1	—	0.1	—	—
Zn	0.5	2	0.5	2	2.6	0.003
Sn	10	2	—	2	—	<0.001

A possible solution of the problem lies in the deposition of such metals on cathodes of electrolytic cells. By these means, pure metals can be recovered instead of salts, which is generally more convenient for recycling. The electricity costs of removing such small quantities are reasonably low, even though low concentrations imply high resistances and energy dissipations. The investment costs are more important being the space-time yield limited by diffusion processes: metal ions have to reach the cathode surface, adsorb there, and migrate into a suitable position on the growing crystal lattice of the metal deposit. The relative rates of these three processes greatly influences the nature of the deposit, which can be either powdery or dendritic or coherent (8).

Depending on inlet and outlet metal concentrations, different cells have been designed. For concentrations higher than 10 $g \cdot l^{-1}$ the classical *tank cell* can be employed: the cathodic and anodic compartments are divided by a cation-exchange membrane, and mechanically stirred; the electrodes are simple plates dipped into the anolyte and the catholyte. For lower concentrations more sophisticated electrodes have been designed so as to reduce mass transfer limitations.

The *fluidised bed electrolyser* [Fig. 2(a)] is suitable for metal ion concentrations up to 5-10 $g \cdot l^{-1}$ (9). Graphite particles are fluidised by the electrolyte and charged by the feeder cathode. Metal ions are deposited on such particles. On increasing weight the heavier particles reach the bottom of the bed where they are removed and replaced by fresh ones. For lower concentrations, up to 1 $g \cdot l^{-1}$, *packed–bed cathodes* are generally preferred, made of graphite, reticulated vitreous carbon (RVC) or metals, in the form of granule beds, meshes, foams, clothes or felts (10).

Recovery of silver from photographic spent baths is a typical application of such electrodes (11). The *enViro cell* [Fig. 2(b)] is particularly attractive for very dilute effluents (9, 12). The cross section of packed-bed of carbon particles increases in the flow direction, thus allowing higher residence times where the metal content has already been depleted to some extent, and mass transfer limitations become more severe. This cell has also been successfully employed for electrocatalytic oxidation of dissolved organics. Applications to the removal/recovery of Hg, Pb and Cd are reported in (13).

A fruitful combination between 3D electrodes for metal deposition and gas diffusion electrodes was found by Njau and coworkers (14). The electrochemical reduction of nickel ions in dilute industrial wastewater from a galvanic nickel plating plant was carried out on a three-dimensional electrode in a gas diffusion electrode packed bed electrode cell (GBC; see Fig. 3) and also on a rotating disc electrode (see later). To explain the experimental results, concentration profiles for selected species were calculated during nickel deposition.

It was found that, depending on the solution composition and electrolysis conditions, metallic nickel, nickel oxide and nickel hydroxide, as well as nickel oxyhydroxide, can be deposited. Metallic nickel dendrite formation and its growth towards and on the Nafion membrane were identified as a major problem in a cell with a small interelectrode gap due to the damage caused on the gas diffusion electrode (GDE) membrane assembly. Moreover, it was found that the nickel selectivity ratio depends on the total current density as well as on the Ni2+ concentration, and that the surface of the rotating disc electrode can be completely blocked by green nickel hydroxide deposit.

A schematic representation of the just mentioned *rotating cylinder electrode* is providede in Fig. 4. Such electrode enables high mass transfer rates by promoting a turbulent flow in the cathodic compartment (12). Such reactors have been operated on various scales (electrode diameter: 50-1000 mm, rotation speed: 100-1500 rpm; current: 1-10000 A).

A typical application of these electrodes is the recovery of photographic silver.

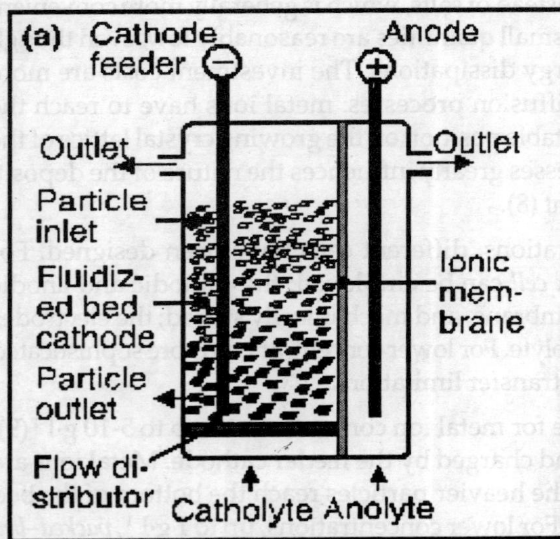

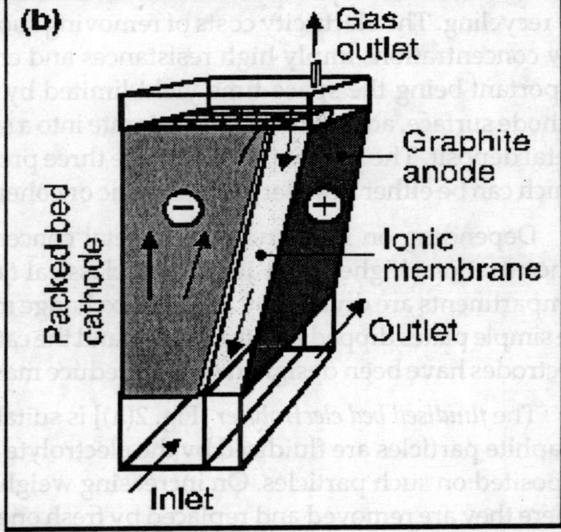

Fig. 2: Electrodeposition of Trace Materials
(a) Fluidised-bed electrolyser, (b) The Enviro cell

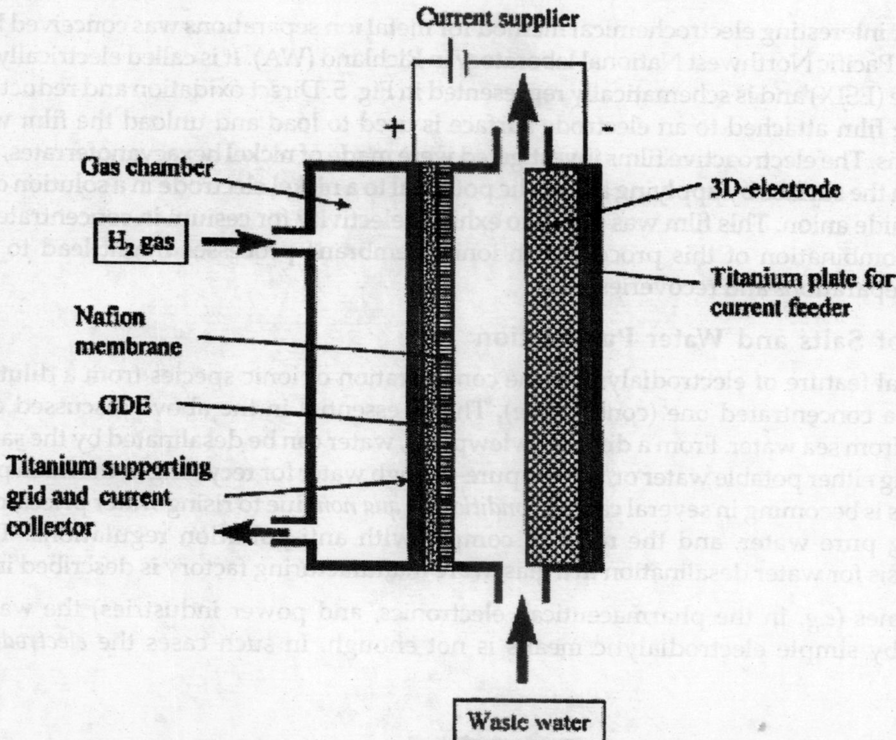

Fig. 3: Schematic Representation of the GBC Cell for Ni Deposition

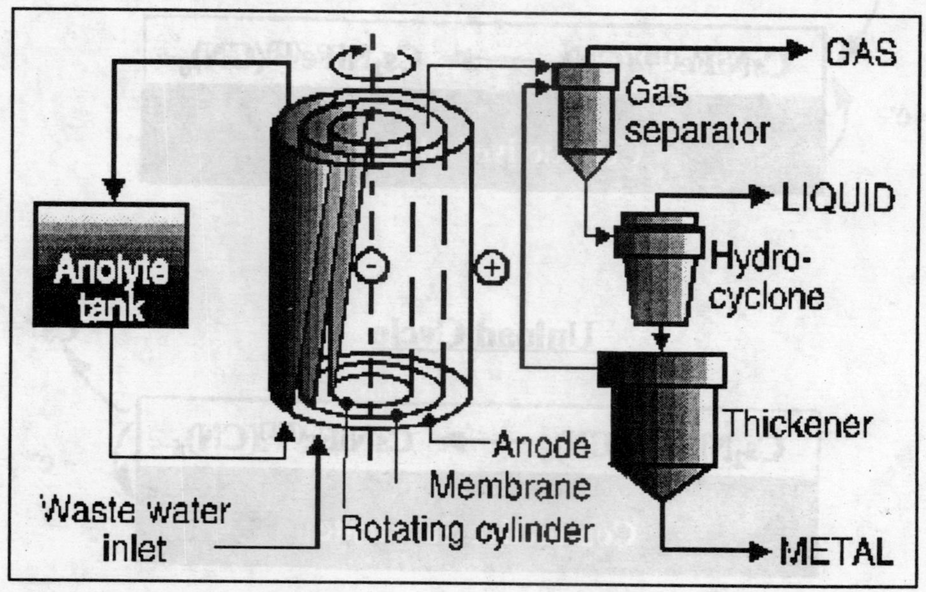

Fig. 4: Recovery of Metal via the Rotating Cylinder Cell

Another interesting electrochemical method for metal ion separations was conceived by Lilga *et al.* (15) at the Pacific Northwest National laboratory in Richland (WA). It is called electrically switched ion exchange (ESIX) and is schematically represented in Fig. 5. Direct oxidation and reduction of and electroactive film attached to an electrode surface is used to load and unload the film with alkali meatal cations. The electroactive films investigated were made of nickel hexacyanoferrates, which are deposited on the surface by applying an anodic potential to a nickel electrode in a solution containing the ferricyanide anion. This film was found to exhibit selectivity for cesium in concentrated sodium solutions. Combination of this process with ionic membrane processes might lead to new (16), interesting separations and recoveries.

Recovery of Salts and Water Purification

A typical feature of electrodialysis is the concentration of ionic species from a dilute solution (diluate) to a concentrated one (concentrate). This is essential in the above discussed edible-salt production from sea water. From a different viewpoint, water can be desalinated by the same means thus attaining either potable water or, at least pure-enough water for recycling in the main production process. This is becoming in several cases a *conditio sine qua non* due to rising water prices, difficulties in obtaining pure water, and the need to comply with anti-pollution regulations. The use of electrodialysis for water desalination in a glassware manufacturing factory is described in (17).

Sometimes (*e.g.* in the pharmaceutical, electronics, and power industries) the water purity achievable by simple electrodialytic means is not enough. In such cases the *electrodeionisation*

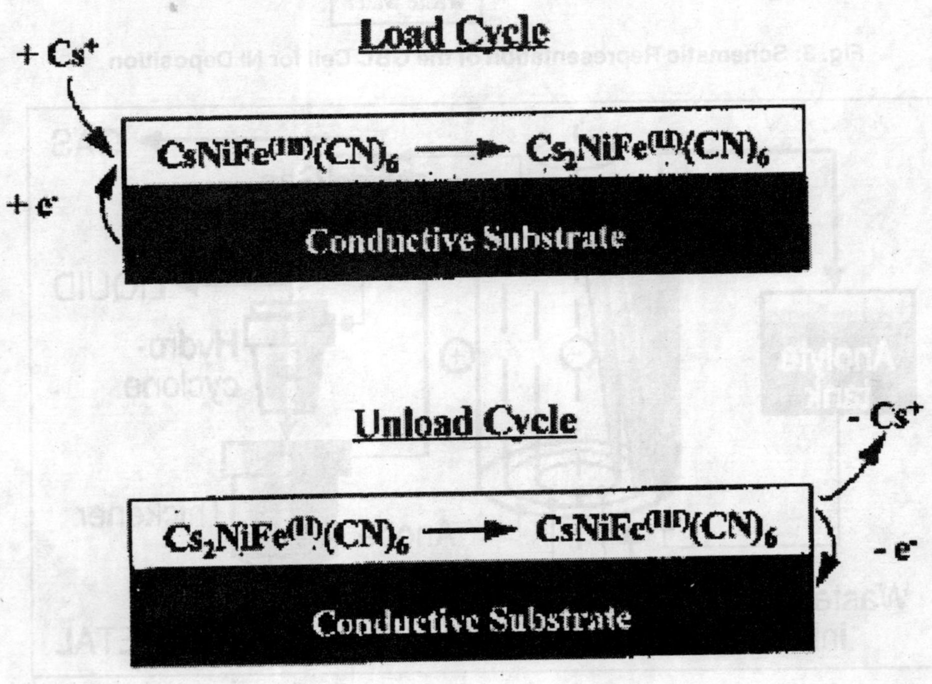

Fig. 5: The ESIX Concept for Metal Ion Leading and Unloading

(commercialised by Millipore as the Ionpure process) may solve the problem (18). This process combines in a single unit electrodialysis and ion-exchange resins (Fig. 4). The most distinctive feature is that the central chamber of each unit cell of the membrane stack is filled with ion-exchange resins. The water to be deionized is fed to this central chamber. The ions, due to the applied electrical potential gradient, migrate through the resin and the membranes, and finally reach the concentrate. Hydrogen and hydroxyl ions, generated by water hydrolysis, give rise to an *in situ* regeneration of the resin. As a consequence, the water purity reaches levels generally associated to mixed-beds of ion-exchangers, with no need of periodic regeneration of the ion-exchangers.

Another interesting application of ionic membranes is the so called *electrochemical–ion exchange*. In this process an ion exchange membrane is bonded directly on the surface of a metal mesh electrode. By applying a suitable current ion-exchange can be enhanced leaving, in some favourable cases, only part per billions of metal ions in the effluent. The regeneration of the resin can be simply attained by reversing the current. Originally developed by AEA Harvell to brackish water desalination, it has afterwards been applied to the removal of ions such as Co^{2+}, Cs^{+}, Li^{+}, Cl^{-}, $SO4^{2-}$ and borates in nuclear waste decontamination, to Ca^{2+} and Mg^{2+} elimination for water softening, to the recovery of precious metals, etc. (12).

Furthermore, Janssen and Koene (19) showed that ion-exchange-assisted electrodialysis could lead to low residual heavy metal concentrations (10^{-2} $mol m^{-3}$), which are though one order of magnitude than those achievable by direct deposition on bi-or tri-dimensional electrodes.

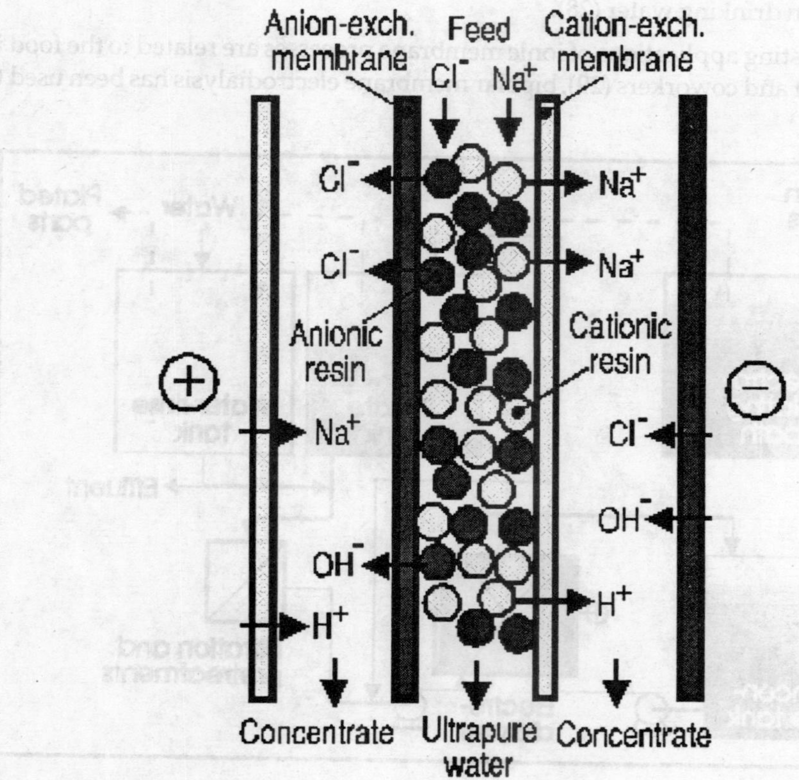

Fig. 6: The Ionpure Process

Coming to valuable salts, a typical application of electrodialysis is the integrated recovery and recycling of nickel from electroplating spent baths (Fig. 7). Similar processes can be applied the plating industry for the recovery of other metals (20), also in combination with acid recovery.

Sometimes separation between different ions is desired. In such cases the complexation of some ions (especially those of higher charge) can be promoted by addition of suitable agents (EDTA, Poly acrylic alcohol, etc.). The obtained complexes possess lower mobilities in the solutions as well as in the membranes, thus enabling separation from those ionic species which do not undergo complexation. Examples of application of these principles are given in (21, 22), concerning *e.g.* electrodialytic recovery of copper from spent solutions of electroless plating onto printed circuit boards, or the separation of metal ions by diffusion dialysis.

Another promising tool to achieve separation between ions of different charge is the use of *monovalent–ion permselective membranes* [Fig. 1(b)]. The author studied the transport properties of these membranes in earlier works (23, 24). Their application to $C_2O4_2^-/Cl^-$ separation was successfully tested in an integrated process for sodium oxalate recovery and recycling from electrolytic titanium leaching solutions (23, 25). Other researchers have used such membranes for Ag+/ (Cu^{2+}, Zn^{2+}) separation (26), achieving a separation factor equal to 4. By Na-EDTA addition and consequent complexation of Cu^{2+} and Zn^{2+} ions a three-fold higher separation factor was attained, though the current efficiency was 50 % lower due to the presence of the sodium deriving from the added EDTA salt. Other practical applications of monovalent-ion-permselective membranes deal with chloride removal from wastewaters derived from the Kraft pulping process of wood (27), or in nitrate and hardness removal from drinking water (28).

Some other interesting applications of ionic membrane processes are related to the food industry. As reported by Bazinet and coworkers (29), bipolar membrane electrodialysis has been used to inhibit

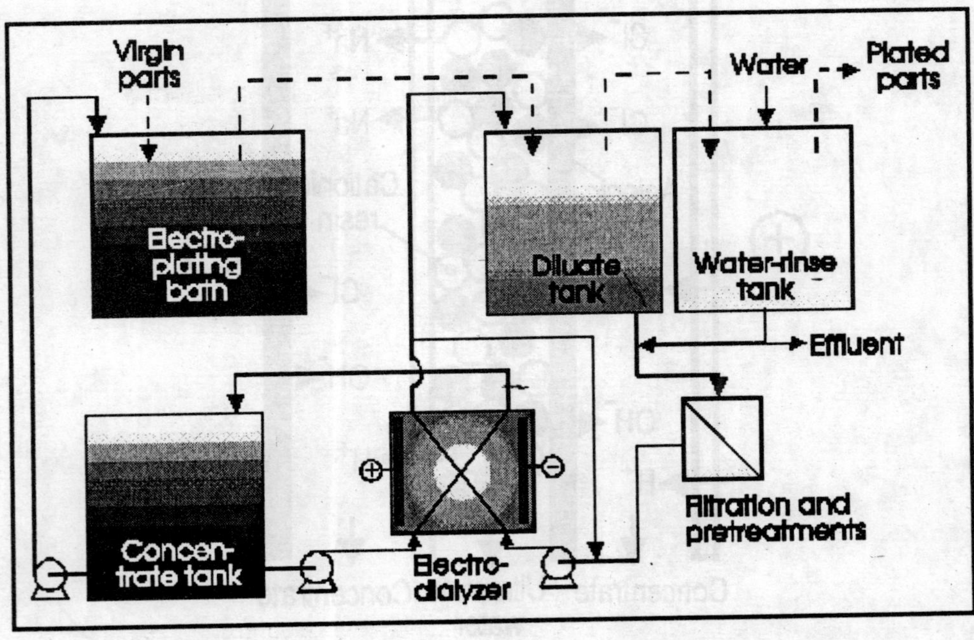

Fig. 7: Recovery of Nickel from Electroplating Spent Solutions

polyphenol oxidase in apple juice, the enzyme responsible for the enzymatic browning of cloudy juice, and to separate soybean proteins from other components without denaturing them, in order to produce protein isolates. Fig. 8 refers to this last case and shows the major advantages of the ionic-membrane-based process compared to the conventional one: no added acids or bases during the process to adjust the pH of the protein solution. In a similar context, bipolar membranes have recently been employed in the production of lactic acid from a fermentation broth (30).

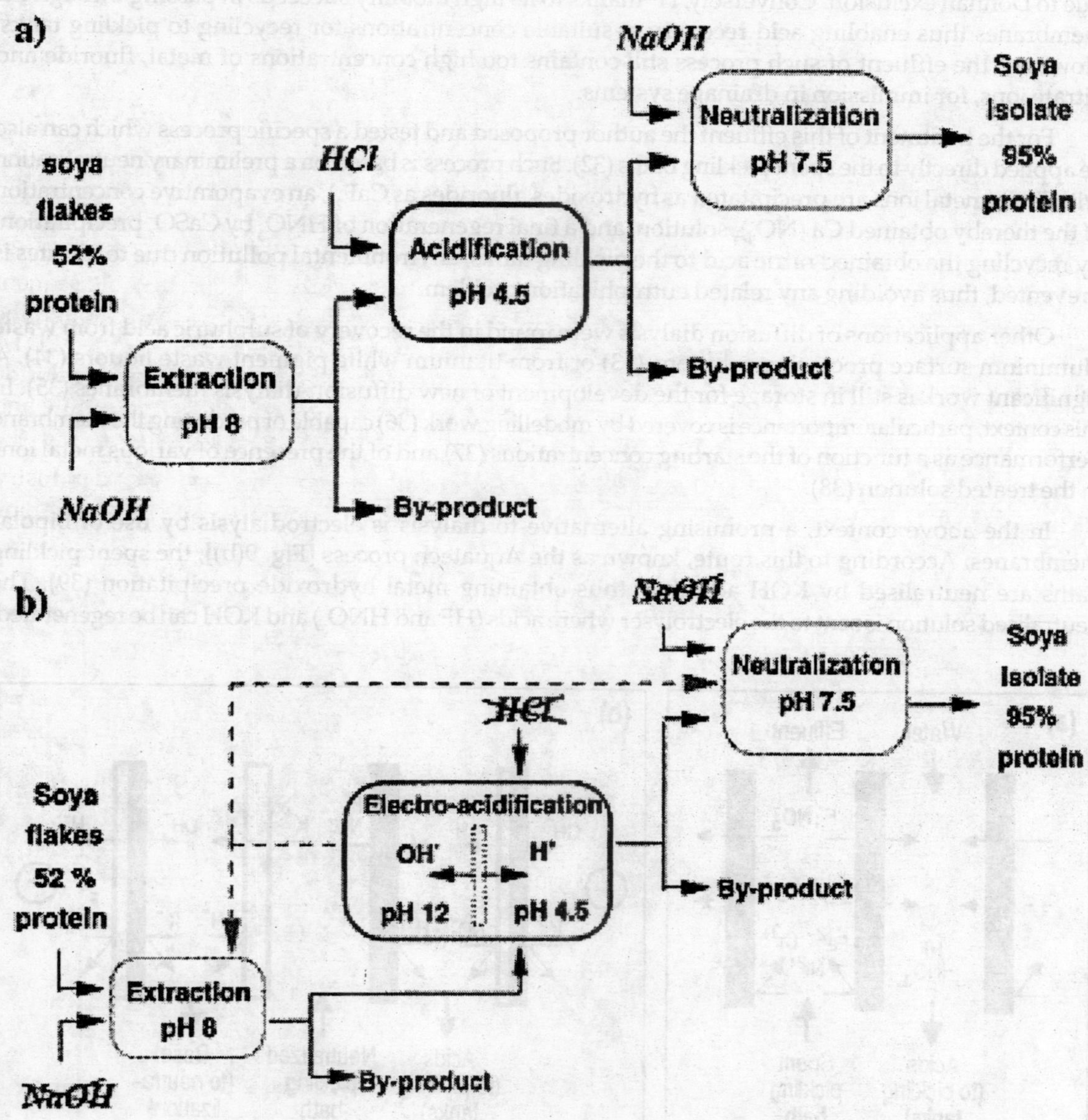

Fig. 8: Comparison Between (a) conventional and (b) bipolar-membrane-based processes for the production of soya protein isolate

Recovery of Acids and Bases

Diffusion dialysis using anion-exchange membranes has found extensive application in the recovery of acids from pickling wastewaters. Figure 9(a) refers to the recovery of HF and HNO_3 from the spent baths of stainless steel pickling (30, 31). The spent solution and pure water are countercurrently fed to neighbouring compartments of the membrane stack. Anions diffuse across the anion-exchange membranes towards the water-side. Metal ions are rejected by the same membranes due to Donnan exclusion. Conversely, H^+ thanks to its high mobility succeeds in passing through the membranes thus enabling acid recovery at suitable concentrations for recycling to pickling tanks. However, the effluent of such process still contains too high concentrations of metal, fluoride and nitrate ions, for immission in drainage systems.

For the treatment of this effluent the author proposed and tested a specific process which can also be applied directly to the spent pickling baths (32). Such process is based on a preliminary neutralisation with lime (metal ions are precipitated as hydroxides, fluorides as CaF_2), an evaporative concentration of the thereby obtained $Ca(NO_3)_2$ solution, and a final regeneration of HNO_3 by $CaSO_4$ precipitation. By recycling the obtained nitric acid to the pickling tanks, environmental pollution due to nitrates is prevented, thus avoiding any related eutrophication problem.

Other applications of diffusion dialysis were found in the recovery of sulphuric acid from waste aluminium surface processing solutions (33) or from titanium white pigment waste liquors (34). A significant work is still in storage for the development of new diffusion-dialysis membranes (35). In this context, particular importance is covered by modelling work (36) capable of predicting the membrane performance as a function of the starting concentrations (37) and of the presence of various metal ions in the treated solution (38).

In the above context, a promising alternative to dialysis is electrodialysis by use of bipolar membranes. According to this route, known as the Aquatech process [Fig. 9(b)], the spent pickling baths are neutralised by KOH addition, thus obtaining metal hydroxide precipitation (39). The neutralised solution is sent to the electrolyser where acids (HF and HNO_3) and KOH can be regenerated,

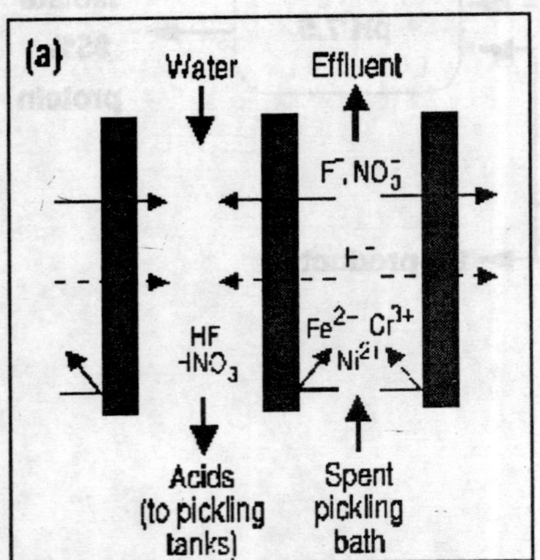

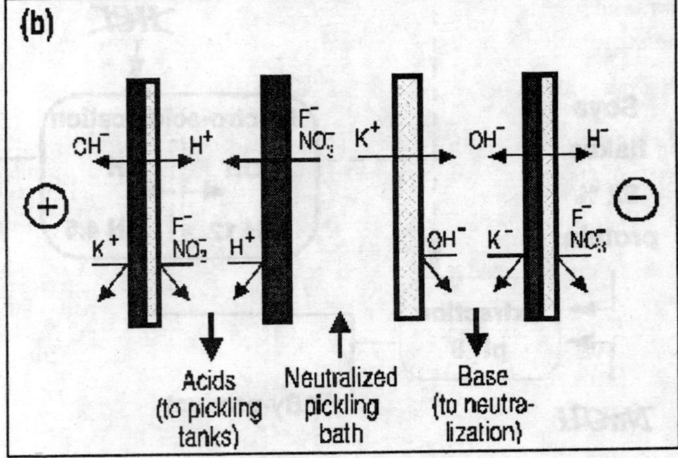

Fig. 9: Recovery from Spent Pickling Baths of Stainless Steel
(a) The diffusion dialysis process, (b) The Aquatech process

in separate streams, for recycling to pickling and neutralisation units respectively. This process allows complete acid and base recycle, though it implies relevant energy requirements for water hydrolysis in the bipolar membranes, whose stability have still to be improved. Other applications of bipolar membranes to the recovery of acids and bases are NaOH synthesis from Na_2CO_3 or $NaHCO_3$, ethylenediamine and HCl production from ethylenediamine dihydrochloride, etc. (21).

Bipolar membrane electrodialysis was also tested to recover acetic acid from dilute wastewaters showing technical feasibility and proving a 70% recovery even if at low electric energy efficiency due to back diffusion of acetica acid through the anion-exchange membranes coupled with the bipolar ones in the stack (40). Another application was in the recovery of high-molecular-weight acids (41). Since an increasing number of bipolar membranes are being commercialised, more applications of such technology are sure to come in the near future.

Some final considerations have to be addressed to chloralkali electrolysis, a well known process for NaOH and Cl_2 production from NaCl. In this context, the recent developments in the synthesis of fluorinated cation-exchange membranes, capable of withstanding NaOH concentrations up to 50%, allowed to outperform the conventional amalgam process, whose environmental impact was relevant due to Hg emissions.

Conclusions

A number of applications of ionic membranes in the recovery of valuable chemicals from process liquid wastes were reviewed. Some of them were rather consolidate, others just promising opportunities. From this survey ionic-membrane technologies clearly appear to be versatile and capable of solving quite different problems.

The author feels that most of the future developments in the area will come from those developers or users that will look at these technologies as tools to cope with their specific treatment requirements. Such tailored systems will probably be the basis on which ionic membranes will gain a deeper penetration in the wastewater-treatment market. In this context, wide margins seem still to be left for innovative processes and applications.

Further research is though needed to improve the corrosion resistance of electrodes materials, to achieve higher membrane stabilities, to enhance selectivities for specific separations, and to develop new module designs allowing higher energy efficiencies.

REFERENCES

1 Rautenbach R. and Albrecht R., 1989, Membrane processes, Wiley, Chichester.

2. Porter M.C., 1990, Handbook of industrial membrane technology, Noyes Publ., Park Ridge, N.J.

3. Mizutani Y., 1990, Structure of ion-exchange membranes, *J. Membr. Sci.*, 49, 121-144.

4. Donnan F. and Guggenheim E.A., 1932, Die Genaue Thermodynamic der Membrangleichgewichte, *Z. Phys. Chem.*, 162, 346-360.

5. Bauer B. and Strathmann H., 1990, Ion-exchange membranes, Proceedings of the 8th ESMST Summer School in Membrane Science, Gargnano, Italy.

6. Trivedi G.S., Shah B.G., Adhikary S.K., Indusekhar V.K., Rangarajan R., 1996, Studies on bipolar membranes, *React. & Funct. Polym.*, 28, 243-251.

7. Pinacci P., 2002, Development of electromembrane processes for waste stream treatment, *Membrane Technol.*, 134, 11-13.

8. Campbell D.A., Dalrymple I.M., Sunderland J.G. and Tilson D., 1994, The electrochemical recovery of metals from effluent and process streams, *Resources, Conservation and Recycling*, 10, 25-33.
9. Kreysa G. and Jüttner K., 1994, Towards a cleaner environment using electrochemical techniques, in Electrochemical engineering and energy, Plenum Press, NY, in press.
10. Bockris J. O'M., Bhardway R.C. and Tennakoon C.L.K., 1994, Electrochemistry of waste removal. A review, *Analyst*, 119, 781-789.
11. Vatistas N., Marconi P.F. and Bartolozzi M., 1993, Removal of silver using graphite-felt electrodes, *J. Appl. Electrochem.*, 23, 390-393.
12. Walsh F. and Mills G., 1994, Electrochemical methods for pollution control, *Chem. Technol.*, 1, 1, 13-18.
13. Grimm J., Bessarabov D., Sanderson R., 1998, Review of electro-assisted methods for water purification, *Desalination*, 115, 285-294.
14. Njau K.N., v.d. Woude M., Visser G.J., Janssen L.J.J., 2000, Electrochemical removal of nickel ions from industrial wastewaters, *Chem. Eng. J.*, 79, 187-195.
15. Lilga M.A., Orth R.J., Sukamoto J.P.H., Haight S.M., Schwartz D.T., 1997, Metal ion separations using electrically switched ion exchange, *Sep. & Purif. Technol.*, 11, 147-158.
16. Ritchie S.M.C., Bhattacharyya D., 2002, Membrane-based hybrid processes for high water recovery and selective inorganic pollutant separation, *J. Haz. Mat.*, 92, 21-32.
17. Asahi Glass Co., 1990, Reuse of wastewater in industrial processes by means of electrodialysis technique, private communication.
18. Muralidhara H.S., 1994, Enhance separations with electricity, *Chemtech*, May 1994, pp. 36-41.
19. Janssen L.J.J., Koene L., 2002, The role of electrochemistry and electrochemical technology in environmental protection, *Chem. Eng. J.*, 85, 137-146.
20. Morgan S.M., Lee C.M., 1997, Metal and acid recovery options for the plating industry, *Res., Cons. & Recycl.*, 19, 55-71.
21. Davis T.A., 1990, Electrodialysis, Chapter 8 in (2), pp. 482-510.
22. Tomida T., Ikawa T. and Masuda S., 1993, Recovery of metal ions from dilute solutions using water-soluble chelating polymer and dialysis membranes, *J. Chem. Eng. Jpn.*, 26, 575-577.
23. Saracco G., Zanetti M.C. and Onofrio M., 1993, Novel application of monovalent-ion-permselective membranes to the recovery treatment of an industrial wastewater by electrodialysis, *I.&E.C. Research.*, 32, 657-662.
24. Saracco G. and Zanetti M.C., 1994, Ion transport through monovalent-anion-permselective membranes, *I.&E.C. Research.*, 33, 96-101.
25. Saracco G. and Onofrio M., 1993, Electrodialytic recovery of NaCl and Na2C_2O_4 from an electrolytic-Ti-leaching waste solution by means of permselective membranes, Proceedings of the ReC'93 Congress, Geneva, Switzerland, Vol. III, pp. 153-165.
26. Cherif A.T., Elmidaoui A. and Gavach C., 1993, Separation of Ag^+, Zn^+ and Cu^{2+} ions by electrodialysis with a monovalent cation specific membrane and EDTA, *J. Membr. Sci.*, 76, 36-49.
27. Rapp H.-J., Pfromm P.H., 1998, Electrodialysis for chloride removal from the chemical recovery cycle of a Kraft pulp mill, *J. Membr. Sci.*, 146, 249-261.

28. Hell F., Lahnsteiner J., Frischherz H., Baumgartner G., 1998, Experience with full-scale electrodialysis for nitrate and hardness removal, *Desalination*, 117, 173-180.

29. Bazinet L., Lamarche F., Ippersiel D., 1998, Bipolar-membrane electrodialysis: applications of electrodialysis in the food industry, *Trends in Food Sci. & Technol.*, 9, 107-113.

30. Franken T., 2001, Bipolar membrane technology and its application, *Membrane Technol.*, 129, 8-11.

31. Kobuchi Y., Motomura H., Noma Y. and Hanada F., 1986, Application of ion-exchange membranes to the recovery of acids by diffusion dialysis, *J. Membr. Sci.*, 27, 173-179.

32. Saracco G. and Onofrio M., 1993, A new process for HNO_3 recovery from spent pickling solutions, *Resources, Conservation and Recycling*, 9, 141-152.

33. Lin S.H., Lo M.C., 1998, Recovery of sulphuric acid from waste aluminium surface processing solution by diffusion dialysis, *J. Haz. Mat.*, 60, 247-257.

34. Tongwen X., Weihua Y., 2001, Sulfuric acid recovery from titanium white (pigment) waste liquor using diffusion dialysis with a new series of anion exchange membranes-static runs, *J. Mem. Sci.*, 183, 193-200.

35. Stachera D.M., Childs R.F., Mika A.M., Dickson J.M., 1998, Acid recovery using diffusion dialysis with poly (4-vinylpyridine) -filled microporous membranes, *J. Mem. Sci.*, 148, 119-127.

36. Kang M.-S., Yoo K.-S., Oh S.-J., Moon S.-H., 2001, A lumped parameter model to predict hydrochloric acid recovery in diffusion dialysis, *J. Mem. Sci.*, 188, 61-70.

37. Oh S.-J., Moon S.-H., Davis T., 2000, Effect of metal ions on diffusion dialysis of inorganic acids, *J. Mem. Sci.*, 169, 95-105.

38. Sato K., 2002, Effect of the feed solution concentration on the separation degree in Donnan dialysis for binary systems of amino acids, *J. Mem. Sci.*, 196, 211-220.

39. Mc Ardle J.C., Piccari J.A. and Thornburg G.G., 1989, Aquatech Systems' pickle liquor recovery process. First year of operation at Washington Steel, 1989 AISE Annual Convention, Pittsburg, PA.

40. Yu L., Guo Q., Hao J., Jiang W., 2000, Recovery of acetic acid from dilute wastewater by means of bipolar membrane electrodialysis, *Desalination*, 129, 283-288.

41. Novalic S., Kongbangkerd T., Kulbe K.D., 2000, Recovery of organic acids with high molecular weight using a combined electrocatalytic process, *J. Membr. Sci.*, 166, 99-104.

41

Removal of Cadmium on Zeolite Developed from Flyash

S. RAYALU*, P. KUMAR**, S.M. DHOPATE**, M.Z. HASAN**, S.N. KAUL** AND A. KUMAR*

***National Environmental Engineering Research Institute, Nagpur - 440020*
**Environmental Biology Research Unit, P.G. Department of Zoology, S.K. University, Dumka-814 101, Jharkhand*

ABSTRACT

Flyash a combustion residue from thermal power industries, has been successfully used as a raw material for synthesis of low cost zeolites. This product exhibits good sorption potential for Cadmium. The equilibrium data fits well in the Langmuir model of sorption showing the formation of monolayer coverage at outer surface of the zeolite material at lower concentration of 500mg/l with maximum adsorption capacity of 270.27mg/g. The rate constants have been derived using Lagergren equation and the values are almost constant at different concentration of Cadmium indicating the reaction to be of first order.The mass transfer analysis carried out using Mckays equation indicate that pore diffusion is the rate limiting step above a concentration of 500 mg/l. The pore diffusion co-efficient at higher concentration of 800 & 1000 mg/l is 2.83 and 4.89×10^{-11} cm^2/sec which is comparable with that reported in literature for other heavy metals. To summarise, it maybe mentioned that the exchange process in zeolite is governed by two mechanisms viz; mass transfer of the sorbate for exchange on surface of zeolite and diffusion within the pores of zeolite for exchange with the mobile ions in the pores.Mass transfer for exchange on the surface of the zeolite is the rate limiting step at lower concentration and at higher concentration the predominant rate limiting is pore diffusion but it is not the only one.

Keywords: Flyash, Zeolite, Cadmium, Sorption, Kinetics, and Mechanism.

Introduction

Cadmium (Cd) is a soft white metal, which has been widely used in special alloys pigments, metal plating, plastics, stabilizers coatings and rechargeable batteries. The effects of cadmium exposure in humans are very serious with cumulative effects similar to those of mercury poisoning. Cadimum causes nausea, vomiting, diarrhea, muscle cramps, salivation, sensory disturbances, liver injury, convulsions, shock, and renal failure from acute exposures; and also has the potential to cause kidney, liver, bone and blood damage from long-term exposure. The maximum permitted concentration of Cd in domestic water should not exceed 0.005 mg/L (1). The discharge limits for wastewater in surface water bodies into land is 2mg/l in India where as the maximum permitted concentration of Cd in domestic water should not exceed 0.01mg/l (as per Indian Standards).

Cd removal from water is mainly achieved by chemical precipitation, electrolysis, reverse osmosis, or ion exchange. However, the cost of such systems may prove to be too high for treatment of effluents to an acceptable level (2). In addition, resulting high pH conditions may also limit their application. Although synthetic ion exchange resins are effective, they are expensive to use on large scale. Several studies have established that the utilization of naturally occurring zeolite may be a relatively inexpensive alternative even at low pH conditions (3).

Sorption on activated carbon is proving to be a very promising technique for removal of heavy metal from solutions. For recovery and remediation of heavy metal from water and wastewater, sorbents *viz.* activated carbon, resin etc. have been tried effectively (4, 5, 6). Some literature is available on use of zeolite for heavy metal removal (7, 8, 9). However, the use of these materials, in water and wastewater treatment is limited due to prohibitive cost and procurement problems. National Environmental Engineering Research Institute (NEERI), Nagpur, India has developed a process for synthesis of low cost zeolites from flyash which can be exploited as sorbent for removal of heavy metals (10, 11, 12). Information on usage of flyash based zeolite-A (FAZ-A) for removal of heavy metal has not been reported to the best of our knowledge. This paper communicates useful data obtained for designing systems for removal of Cadmium using FAZ-A, and also elucidates the mechanism, rate constants and diffusion parameters using Lagergren and Mckay models governing sorption of Cadmium.

Experimental

Flyash based zeolite-A was synthesized in sufficient quantity using procedure patented and published elsewhere (10, 11, 12). The procedure is as follows:

Synthesis of Flyash based Zeolite-A

The FAZ sample was synthesized by fusing flyash with sodium hydroxide. A homogenous fusion mixture was prepared by proper grinding and mixing of flyash and caustic soda in 1:1.2 ratio. This mixture was heated to 550-600°C for 1-1.5 hr. The resultant fused mass was cooled, milled, and mixed thoroughly in distilled water with simultaneous addition of sodium aluminate. The slurry was then subjected to aging for 8-10 hr. This amorphous alumino-silicate gel was then subjected to crystallization between 90-110°C for 2 hrs. The solid crystalline product was recovered by filtration and washed thoroughly till the filtrate pH was 10-11 and dried at a temperature of 50-60°C

Characterization

The calcium binding capacity (CBC) of alumino-silicate was determined following the standard procedure (12). Powder XRD analysis was employed to monitor zeolite formation process using CuKA as a source of X-rays using X-ray diffractometer: Model: Philips PN-1830. d-spacing values reported in literature for Zeolite-A (12.2±0.20, 8.60±0.20, 7.05±0.15, 4.07± 0.8, 3.68± 0.6, 3.26± 0.5, 2.96± 0.5, 2.73± 0.5, 2.60± 0.5) have been used as basis for identification and quantification of crystalline phases. For elemental analyses the FAZ samples were extracted with Lithium metaborate with subsequent dissolution in nitric acid and analyzed by ICP-AES (Model:YJ24) for Al_2O_3 while SiO_2 was estimated using instrumental/conventional methods. Na_2O was estimated using flame photometer (Modiflame 127 with FPM compressor unit 122).

The particle size analysis was conducted by using Fritsch particle size analysette 22. The sorption capacity was determined as loss on ignition (LOI) at 800°C for 2 hrs. The detailed characteristics of the zeolite are provided in Table 1.

Table 1: Physico-chemical Characterization of Zeolites

Reaction parameters	*FAZ-A*
SiO_2/Al_2O_3 ratio	1.4-2.0
Calcium binding capacity (CBC) (meq/100g)	500-520
Surface area (m²/g)	580
Particle size (Mm)	1-3
Crystal structure	Cubic
SiO_2 %	32.10
Al_2O_3 %	28.00
Fe_2O_3 %	1.23
TiO_2 %	0.90
MnO %	0.003
CaO %	-
MgO %	-
Na_2O %	16.20
K_2O %	-
Loss on ignition %	21.00

Simulated Solutions

Stock Cd solution were prepared by dissolving 2g of Cadmium Sulphate salt in distilled water. Different concentrations of Cadmium varying from 500-1000 mg/l were prepared by appropriate dilution with distilled water. The pH of the solution was maintained at 6.0 ± 0.2 with addition of 1N sulphuric acid H_2SO_4. The concentration of Cd in solution was analysed by using atomic absorption spectroscopy (Model SH-8000, TDA)

Batch Experiments for Cd Removal

Pre weighed quantity of FAZ-A and Cadmium solution were mixed in a jar testing unit for varying time (5-120 min), concentration (500, 800 & 1000 mg/l) and doses varying between (1-2 g/l). The stirring speed was maintained at 100 rpm for all the experiments. Also the experiments were carried out at room temperature (30 ± 2°C). The parameters, which warrant investigation, are Cadmium concentration in solution, quantity of FAZ-A mixed, time of contact between adsorbate and sorbent and pH of solution. The schedule of experiments undertaken to evaluate these parameters are:

1. Effect of initial solution concentration and contact time and
2. FAZ-A Cadmium sorption system model

Results and Discussion

Effect of Initial Concentration

Figure 1 illustrates the fraction adsorbed with respect to time. It is apparent from this plot that the fraction of Cd adsorbed from solution of lower initial concentration are higher than from higher initial concentration solution. It indicates that the residual concentration of Cd solution decreases with time upto 45 minute and subsequently remains almost constant for each concentration. This shows that equilibrium is attained in about 45 minutes irrespective of the initial concentration.

Sorption Dynamics

The rate constant for sorption of the Cadmium on the FAZ-A was determined using Lagergren's (13) equation (1898):

$$\text{Log}(q_e\text{-}q) = \log q_e (K_{ad}/2.303).\, t \qquad (1)$$

where,

q_e and q (both in mg/g) are the amounts of Cadmium adsorbed at equilibrium and at time t (min), respectively, and K_{ad} (min^{-1}) is the rate constant for sorption of Cadmium. The values of K_{ad} at different concentration were calculated from the slopes of the respective linear plots of log (q_e – q) V_s t (Figure 2) and presented in Table 2. The applicability of the model to the system indicates that the reaction taking place is of the first order.

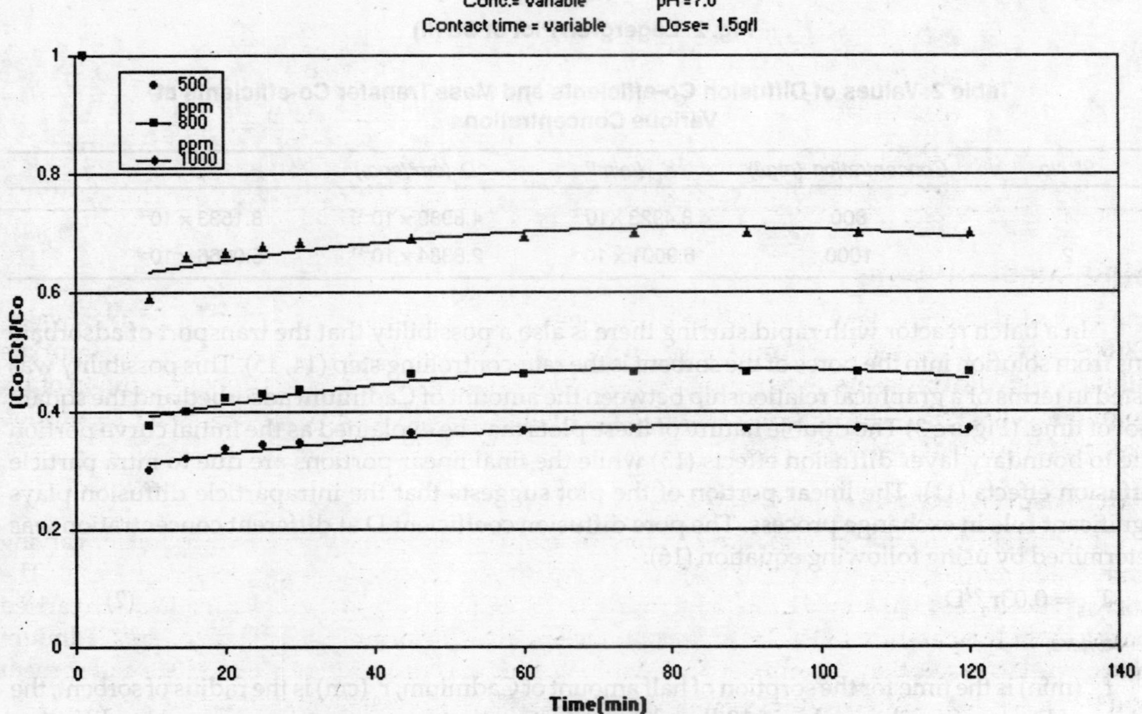

Fgi. 1: Fraction of Cd (II) Removed as a Function of Time

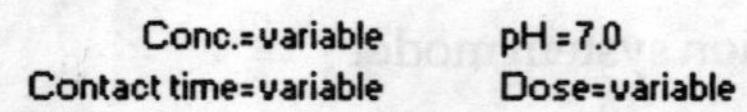

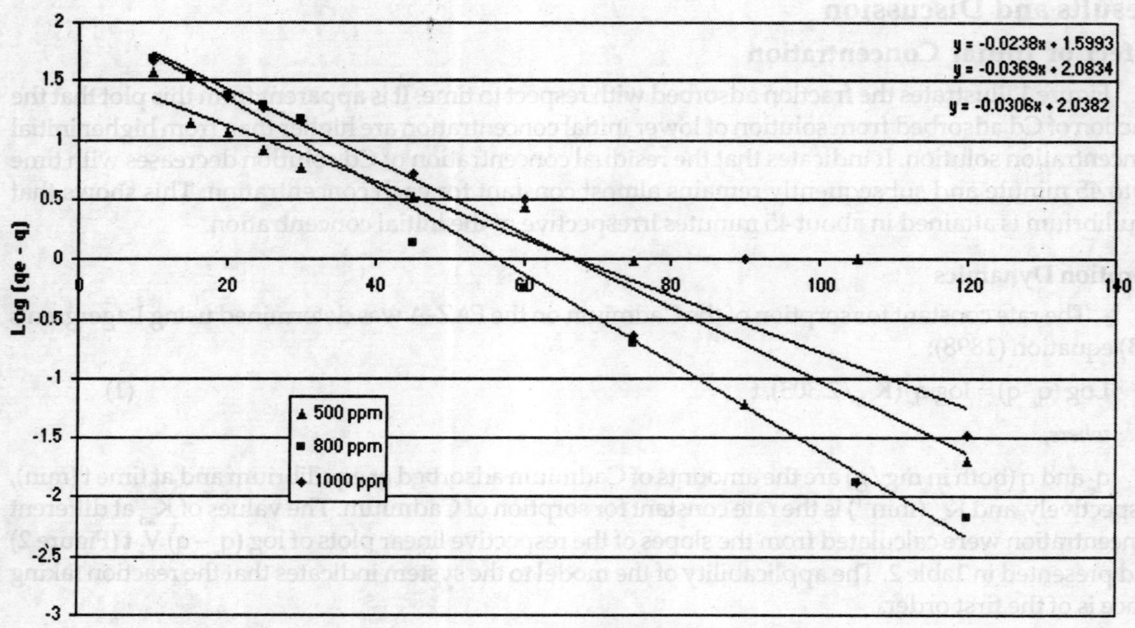

Fig. 2: Legergren Plot of Cd (II)

Table 2: Values of Diffusion Co-efficients and Mass Transfer Co-efficients at Various Concentrations

Sl. No.	*Concentration (mg/l)*	*K_{ad} (min^{-1})*	*D (cm^2/sec.)*	*β*
1	800	8.4923×10^{-2}	4.8989×10^{-11}	8.1533×10^{-2}
2	1000	6.9001×10^{-2}	2.8384×10^{-11}	5.0666×10^{-2}

In a batch reactor with rapid stirring there is also a possibility that the transport of adsorbate ions from solution into the pores of the sorbent is the rate-controlling step (14, 15). This possibility was tested in terms of a graphical relationship between the amount of Cadmium adsorbed and the square root of time. (Figure 3) The double nature of these plots may be explained as the initial curve portion due to boundary layer diffusion effects (13) while the final linear portions are due to intra particle diffusion effects (11). The linear portion of the plot suggests that the intraparticle diffusion plays significant role in exchange process. The pore diffusion coefficient D at different concentration was determined by using following equation (16):

$$T_{0.5} = 0.03 r_0^{2}/D \quad (2)$$

where,

$t_{0.5}$ (min) is the time for the sorption of half amount of Cadmium, r_0 (cm) is the radius of sorbent, the values of D were found in order of 10^{-11} cm^2 S^{-1} indicating that the process is governed by diffusion

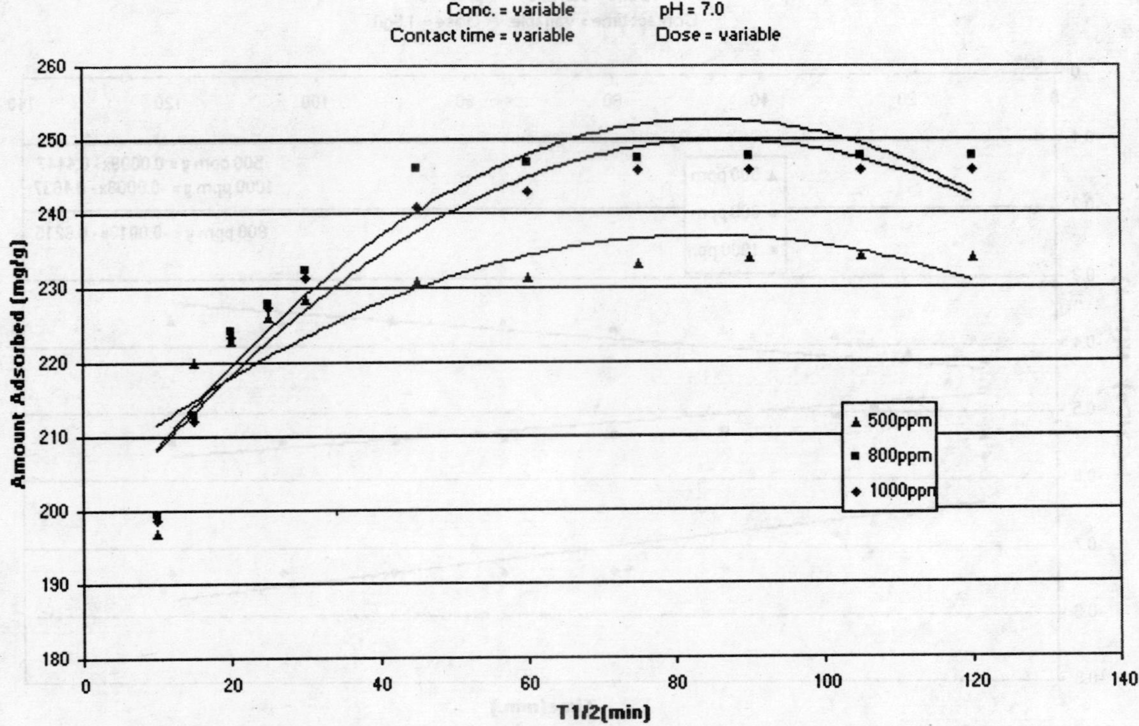

Fig. 3: Weber and Morris Plot for the Rate Constants of Pore Diffusion

(16). According to Michelson et.al (17) for pore diffusion to be rate limiting the value of pore diffusion co-efficient should be in the range of 10^{-11} to 10^{-13} cm^2 S^{-1} for heavy metals.

Mass Transfer Analysis

The mass transfer analysis for the sorption of Cadmium metal ion FAZ-A on sorbent was carried out using Mckay (1981) equation (18)

$$\ln (C_t/C_o-1/1+mk) = \ln (mk/1+mk) - (1+mk) / (mk.\beta_1.S_s.t)$$

Where m (g/l) is the mass of FAZ-A per unit volume of particle free slurry, k ($Q^o b$, l/g) is the Langmuirs constant and β_1 (cm/s) and S_s (cm^{-1}) are the mass transfer coefficient and the outer surface of FAZ-A per unit volume of particle free slurry, respectively. A straight-line plot of ln $((C_t/C_o)$ -{1/ (1+mk) }) Vs t (Figure 4) shows the applicability of the above equation for 800 & 1000 mg/l in the present system. The value of β_1 was calculated from the plot and found to be 12.24×10^{-4} and 7.6×10^{-4} min^{-1} for 800 and 1000 mg/l respectively. The values of B decreases with increasing concentration indicating that the mass transfer of Cd (II) is favoured at low concentration for the adsorbent used.

Adsorption Isotherm

The analysis of equilibrium data for the sorption of the Cadmium on FAZ-A has been done in the light of rearranged Langmuir isotherm model.

$$1/qe = 1/bQ^o.1/C_e + 1/Q^o$$

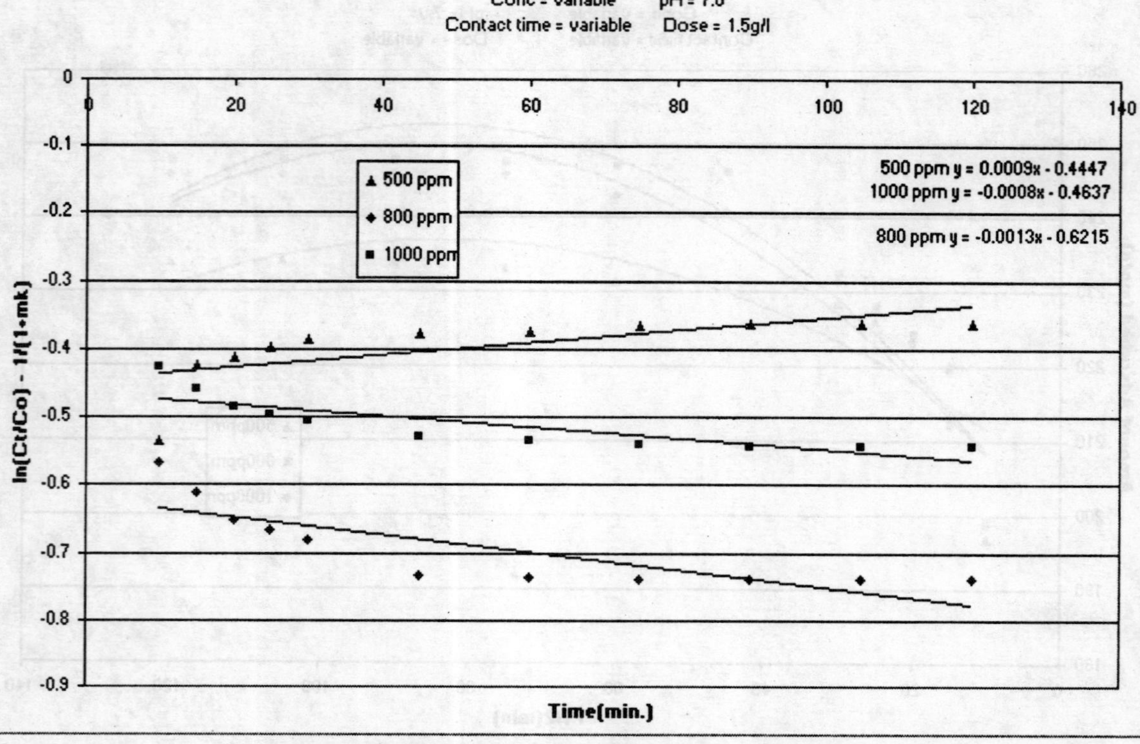

Fig. 4: Mckage Plot for Masss Transfer Co-efficient

where,

C_e (mg/l) is the equilibrium concentration of Cadmium and Q° and b are the Langmuir constants related to the capacity and energy of sorption, respectively. The linear plot of $1/q_e$ V_s $1/C_e$ at a concentration at 500 mg/l (Figure 5) suggests the applicability of the above model for present system, showing formation of monolayer coverage of the adsorbate at the outer surface of the Zeolite. The values of Q° and b at different concentration were determined from the plots and are given in Table 3. The values of Q° are almost constant for the concentration range. These values suggest that upto 500 mgl^{-1} the predominant rate governing step is mass transfer of the sorbate on the surface of the zeolite. Beyond a concentration of 500 mgl^{-1} *i.e.* at 800 mgl^{-1} and 1000 mgl^{-1} pore diffusion is also the rate governing step.

Table 3: Values of Langmuir Constants

Sl. No.	*Concentration (mg/l)*	*Q_o (mg/g)*	*R_L*
1.	500	270.27	4.305×10^{-3}
2.	800	263.15	2.98×10^{-2}
3.	1000	250.00	2.58×10^{-2}

The essential characteristic of the Langmuir isotherm may be expressed in terms of a dimension less equilibrium parameter R_L (19) using the following equation:

$$R_L = (1+bC_o)^{-1}$$

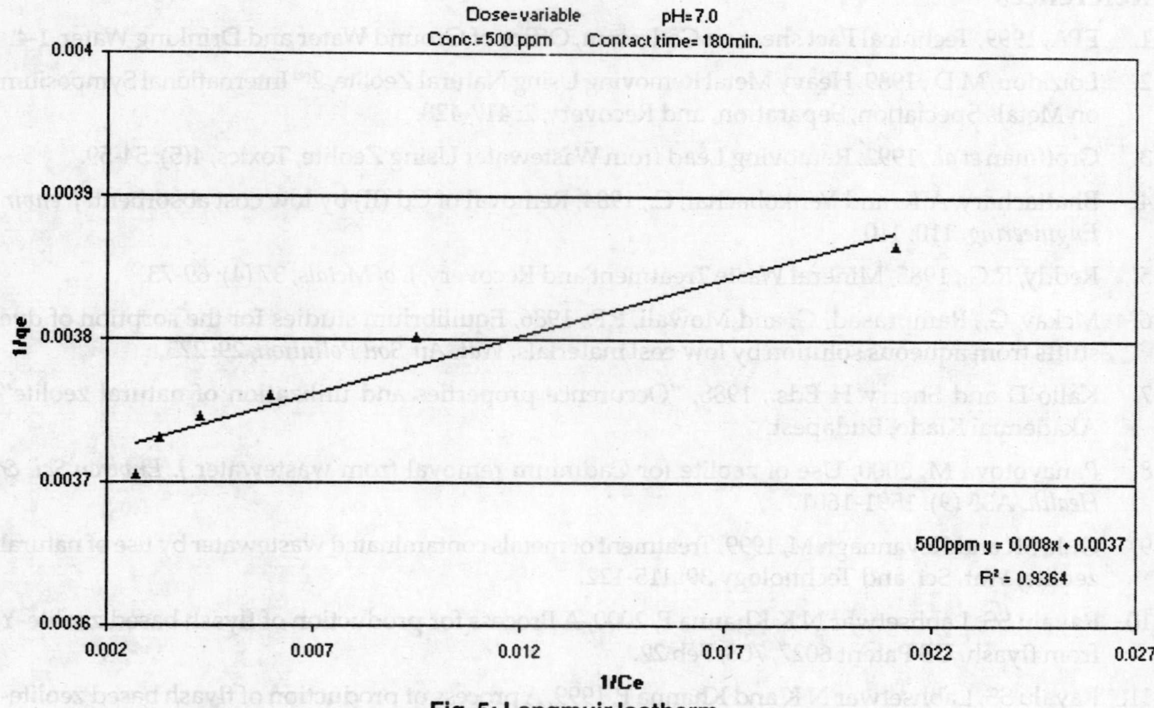

Fig. 5: Langmuir Isotherm

The values of R_L for the studied system at different concentration were found to be in between zero and one (Table 3) showing favorable sorption of Cadmium on FAZ-A.

Conclusion

Zeolites have exchangeable cations distributed on the surface as well as in the pores. At lower concentration the exchange process is limited to exchange on the surface of the zeolite whereas at higher concentraiton diffusion within the pores of the zeolite for exchange with the mobile cations is also the rate limiting step. Though exchange process is dominating in the remvoal of cadmium; surface complexation due to heterogenous nature of zeolites may also be influencing it to certain extent (unpublished work). Cadmium metal ions can be removed from solution by sorption on FAZ-A very effectively at neutral pH. The quantity of metal removed, increases with increase in metal concentration, whereas, the fraction removed increases with decrease in concentration. The sorption kinetics has been elucidated on the basis of adherence of cadmium and intraparticle diffusion. The mechanism involves the first order kinetics and initial rapid rate of removal for the Cadmium ion due to surface exchange followed by infraparticle pore diffusion into the pores of the zeolite, which appears to be the rate-limiting step. The fitness of Langmuir's models in the present system shows the formation of monolayer coverage on the outer surface of the adsorbate at moderate concentration of 500 mgl^{-1} of cadmium ion.

Acknowledgement

The authors are thankful to Dr. N. Labhasetwar and Mr. R. Biniwale, Scientist, NEERI, for their useful suggestions and also acknowledge the help rendered by Dr. D.G Garway, Scientist Sophisticated Instrumentation Center (RSIC), Nagpur in carrying out XRD analysis.

References

1. EPA, 1999, Technical Fact sheet on Cadmium, Office of Ground Water and Drinking Water, 1-4.
2. Loizidou, M.D., 1989, Heavy Metal Removing Using Natural Zeolite, 2nd International Symposium on Metals Speciation, Separation, and Recovery, 2: 417-428.
3. Groffman *et al.*, 1992, Removing Lead from Wastewater Using Zeolite, Toxics, 4(5): 54-59.
4. Bhattachary, A.K. and Venkobachar, C., 1984, Removal of Cd (II) by low cost absorbent. *J. envir. Engineering*. 110: 110.
5. Reddy, R.G., 1985, Mineral Waste Treatment and Recovery, *J. of Metals*, 37 (4): 69-73.
6. Mckay, G., Ramprasad, G. and Mowali, P.P., 1986, Equilibrium studies for the sorption of dye stuffs from aqueous solution by low cost materials. *Wat. Air Soil Pollution*, 29: 273.
7. Kallo D and Sherry H Eds., 1988, "Occurence properties and utilization of natural zeolite" Akademiai Kiado, Budapest.
8. Panayotova M, 2000, Use of zeolite for Cadmium removal from wastewater *J. Environ.Sci. & Health.* A35 (9): 1591-1601.
9. Ouki SK and Kevannagh M, 1999, Treatment of metals contaminated wastewater by use of natural zeolite, Wat. Sci. and Technology 39: 115-122.
10. Rayalu SS, Labhsetwar N K Khanna P, 2000, A Process for production of flyash based zeolite-Y from flyash. US Patent 6027, 708, Feb 22.
11. Rayalu SS, Labhsetwar N K and Khanna P, 1999, A process of production of flyash based zeolite-A. US Patent 5, 965, 105 October 13.
12. Rayalu SS, Meshram S U and Hassan M Z, 2000, "Highly crystalline faujasitic zeolite from flyash". *J. of Hazardous material* B77: 123 –133.
13. Panday K K Prasad G and Singh V N, 1984, Removal of Cr (VI) from aqueous solution by sorption on flyash-Wollastonite *J. Chem.tech.* Biotechnol. 34A: 367.
14. Weber W J Jr. and Morris J C, 1963, Kinetics of sorption on carbon from solution. *J. saint. Engng Div. Am. Soc.civ. Engineers* 89 SA2: 31.
15. Poots V.J.P., Mckay G. and Healy J.J., 1978, Removal of Basic dye from effluent using wood as an adsorbent. Wat. Pollut. Control Fed 50: 926.
16. Bhattacharya A.K and Venkobachar C., 1984, Removal of Cadmium (II) by low cost adsorbants J. Envir. Engng., 410: 110.
17. Michelson L.D., Gideon E. Pace E.G., and Kutal L.H., 1975, Removal of soluble mecury from wastewater by complexing Techniques. US Dept. Induestry. Office of Water Research & Technology. Bull No. 74.
18. Mckay G, Otterburn M. S, and Sweeney A.G., 1981, Surface mass transfer process during colour removal from effluent using Silica. Wat. Res., 14 15.
19. Hall K.R., Eagleton, L.C., Acrivos A. and Vermulen T., 1966, Pore and Solid diffusion kinetics in fixed bed adsorption under constant pattern conditions. Ind. Engg. Chem. Fund., 5: 212.